The International Library of Environmental, Agricultural and Food Ethics

Volume 33

Series Editors

Michiel Korthals, Wageningen University, Wageningen, The Netherlands

Paul B. Thompson, Michigan State University, East Lansing, USA

The ethics of food and agriculture is confronted with enormous challenges. Scientific developments in the food sciences promise to be dramatic; the concept of life sciences, that comprises the integral connection between the biological sciences, the medical sciences and the agricultural sciences, got a broad start with the genetic revolution. In the mean time, society, i.e., consumers, producers, farmers, policymakers, etc, raised lots of intriguing questions about the implications and presuppositions of this revolution, taking into account not only scientific developments, but societal as well. If so many things with respect to food and our food diet will change, will our food still be safe? Will it be produced under animal friendly conditions of husbandry and what will our definition of animal welfare be under these conditions? Will food production be sustainable and environmentally healthy? Will production consider the interest of the worst off and the small farmers? How will globalisation and liberalization of markets influence local and regional food production and consumption patterns? How will all these developments influence the rural areas and what values and policies are ethically sound?

All these questions raise fundamental and broad ethical issues and require enormous ethical theorizing to be approached fruitfully. Ethical reflection on criteria of animal welfare, sustainability, liveability of the rural areas, biotechnology, policies and all the interconnections is inevitable.

Library of Environmental, Agricultural and Food Ethics contributes to a sound, pluralistic and argumentative food and agricultural ethics. It brings together the most important and relevant voices in the field; by providing a platform for theoretical and practical contributors with respect to research and education on all levels.

More information about this series at http://www.springer.com/series/6215

Bernice Bovenkerk · Jozef Keulartz
Editors

Animals in Our Midst: The Challenges of Co-existing with Animals in the Anthropocene

 Springer

Editors
Bernice Bovenkerk
Wageningen University and Research
Wageningen, The Netherlands

Jozef Keulartz
Radboud University
Nijmegen, The Netherlands

This work is part of the research programme 2017-I BOO with project number 023.010.030, which is (partly) financed by the Dutch Research Council (NWO)

ISSN 1570-3010   ISSN 2215-1737   (electronic)
The International Library of Environmental, Agricultural and Food Ethics
ISBN 978-3-030-63525-1   ISBN 978-3-030-63523-7   (eBook)
https://doi.org/10.1007/978-3-030-63523-7

This Springer imprint is published by the registered company Springer Nature Switzerland AG
The registered company address is: Gewerbestrasse 11, 6330 Cham, Switzerland

The original version of the book was revised: The author's name is corrected from "O'Neill, J.S. and M.H. Hastings" to "David A. Leavens" in reference cross citations and list for Chapters 1 and 3. The correction to the book is available at https://doi.org/10.1007/978-3-030-63523-7_32

Acknowledgments

This edited volume has benefitted considerably from the workshop that was held in Wageningen in April 2019, which was made possible by an innovative research grant from the Netherlands Organization for Scientific Research (OZSW), titled "Anthropocene Ethics: Taking Animal Agency Seriously" (projectnr 016.Vidi.185.128) and an Aspasia grant (grant nr 015.014.023).

The editors would like to thank Eva Meijer for her help with editing a number of the chapters. We would like to thank all the authors of this book for their interesting contributions, stimulating discussions, and collaborative spirit. We would also like to thank Inge Ruisch for her support in the organization of the workshop. We are very grateful to the two anonymous reviewers for their helpful comments. We would furthermore like to express our gratitude to the editorial team of Springer, in particular Floor Oosting and Christopher Wilby, for their trust and the pleasant collaboration.

Contents

Editors and Contributors

About the Editors

Bernice Bovenkerk is Associate Professor of philosophy at Wageningen University, The Netherlands. Her research and teaching deal with issues in animal and environmental ethics, the ethics of climate change, and political philosophy. Current topics are animal agency, the moral status of animals, and other natural entities, with a particular focus on fish and insects, the ethics of animal domestication, animal (dis)enhancement, and deliberative democracy. In 2016 she co-edited (together with Jozef Keulartz) *Animal Ethics in the Age of Humans. Blurring Boundaries in Human-Animal Relationships* (Springer) and in 2012 she published *The Biotechnology Debate. Democracy in the Face of Intractable Disagreement* (Springer). She received her Ph.D. title in political science at Melbourne University and her Master's title in environmental philosophy at the University of Amsterdam. Her homepage is bernicebovenkerk.com.

Jozef Keulartz is Emeritus Professor of Environmental Philosophy at the Radboud University Nijmegen, and senior researcher Applied Philosophy at Wageningen University and Research Center. He has published extensively in different areas of science and technology studies, social and political philosophy, bioethics, environmental ethics, and nature policy. His books include Die verkehrte Welt des Jürgen Habermas [The Topsy-Turvy World of Jürgen Habermas, Junius, 1995] and *Struggle for Nature—A Critique of Radical Ecology* (Routledge, 1998). He is co-editor of *Pragmatist Ethics for a Technological Culture* (Kluwer, 2002), *Legitimacy in European Nature Conservation Policy* (Springer, 2008), *New Visions of Nature* (Springer, 2009), *Environmental Aesthetics. Crossing Divides and Breaking Ground* (Fordham University Press, 2014), *Old World and New World Perspectives in Environmental Philosophy* (Springer, 2014), and *Animal Ethics in the Age of Humans* (Springer, 2016).

Contributors

Raymond Anthony Department of Philosophy, University of Alaska Anchorage, Anchorage, AK, USA

John Basl Department of Philosophy and Religion, Northeastern University, Boston, MA, USA

Henk van den Belt University of Western Australia, Crawley, WA, Australia

Charlotte E. Blattner University of Bern, Bern, Switzerland

Hidde Boersma Amsterdam, The Netherlands

Bernice Bovenkerk Wageningen University and Research, Wageningen, The Netherlands

Sabrina Brando AnimalConcepts, Teulada, Spain;
University of Stirling, Stirling, Scotland, UK

Leonie Cornips NL-Lab, Humanities Cluster (KNAW), Amsterdam, The Netherlands;
Faculty of Arts and Social Sciences, Maastricht University, Maastricht, The Netherlands

Martin Drenthen Institute for Science in Society, Radboud University, Nijmegen, The Netherlands

Clemens Driessen Wageningen University & Research, Wageningen, The Netherlands

Erno Eskens Noordboek, Gorredijk, The Netherlands

Charles Foster Green Templeton College, Oxford, UK;
University of Oxford, Oxford, UK

Louis van den Hengel Faculty of Arts and Social Sciences, Maastricht University, Maastricht, The Netherlands

Elizabeth S. Herrelko National Zoological Park, Smithsonian Institution, Washington, DC, USA;
University of Stirling, Stirling, Scotland, UK

Ned Hettinger College of Charleston, Charleston, SC, USA

Hugh A. H. Jansman Wageningen Environmental Research, Wageningen, The Netherlands

Jozef Keulartz Wageningen University and Research, Wageningen, The Netherlands;
Radboud University, Nijmegen, The Netherlands

Yulia Kisora Wageningen University & Research, Wageningen, The Netherlands

Michiel Korthals University of Gastronomic Sciences, Pollenza/Bra, Italy

Nathan Kowalsky St. Joseph's College, University of Alberta, Edmonton, Canada

Joost Leuven The Hague, The Netherlands

Franck L. B. Meijboom Ethics Institute, Utrecht University, Utrecht, The Netherlands

Eva Meijer Wageningen University and Research, Wageningen, The Netherlands

Joachim Nieuwland Faculty of Veterinary Medicine, Utrecht University, Utrecht, The Netherlands

Susan Ophorst University of Applied Sciences Van Hall Larenstein, Leeuwarden, The Netherlands;
Radboud University, Institute for Science in Society, Nijmegen, The Netherlands

Clare Palmer Texas A&M University, College Station, TX, USA

Andreia De Paula Vieira Animal Welfare Scientist and One Health Researcher, Curitiba, Brazil

Christopher J. Preston University of Montana, Missoula, MT, USA

Ronald Sandler Department of Philosophy and Religion, Northeastern University, Boston, MA, USA

Adam Shriver Uehiro Centre for Practical Ethics, Oxford, UK

J. A. A. Swart University of Groningen, Groningen, The Netherlands

Paul B. Thompson Michigan State University, East Lansing, MI, USA

Mateusz Tokarski Independent researcher, Nijmegen, The Netherlands

Lauren E. Van Patter Queen's University, Kingston, ON, Canada

Cor van der Weele Wageningen University, Wageningen, The Netherlands

Jennifer Welchman University of Alberta, Edmonton, AB, Canada

Chapter 1
Animals in Our Midst: An Introduction

Jozef Keulartz and Bernice Bovenkerk

Abstract In this introduction we describe how the world has changed for animals in the Anthropocene—the current age, in which human activities have influenced the planet on a scale never seen before. In this era, we find many different types of animals in our midst: some—in particular livestock—are both victims of and unwittingly complicit in causing the Anthropocene. Others are forced to respond to new environmental conditions. Think of animals that due to climate change can no longer survive in their native habitats or wild animals that in response to habitat loss and fragmentation are forced to live in urban areas. Some animals are being domesticated or in contrast de-domesticated, and yet others are going extinct or in contrast are being resurrected. These changing conditions have led to new tensions between humans and other animals. How can we shape our relationships with all these different animals in a rapidly changing world in such a way that both animal welfare and species diversity are not further affected? We describe how animal ethics is changing in these trying times and illustrate the impacts of Anthropocene conditions on animals by zooming in on one country where many problems, such as biodiversity loss and landscape degradation, converge, the Netherlands. We conclude by giving an overview of the different chapters in this volume, which are organised into five parts: animal agents, domesticated animals, urban animals, wild animals and animal artefacts.

Other contributions in this volume in which the Netherlands will function as a textbook example are the chapters of Hidde Boersma, Susan Ophorst and Bernice Bovenkerk, Eva Meijer and Martin Drenthen.

The original version of this chapter was revised: The author's name is corrected from "O'Neill, J.S. and M.H. Hastings" to "David A. Leavens" in reference cross citation and list. The correction to this chapter is available at https://doi.org/10.1007/978-3-030-63523-7_32

J. Keulartz (✉) · B. Bovenkerk
Wageningen University and Research, Wageningen, The Netherlands
e-mail: Jozef.keulartz@wur.nl

B. Bovenkerk
e-mail: bernice.bovenkerk@wur.nl

© The Author(s) 2021, corrected publication 2021
B. Bovenkerk and J. Keulartz (eds.), *Animals in Our Midst: The Challenges of Co-existing with Animals in the Anthropocene*, The International Library of Environmental, Agricultural and Food Ethics 33,
https://doi.org/10.1007/978-3-030-63523-7_1

1

1.1 Introduction

As we are preparing this book for the publication, the world is in the grips of Covid-19, the corona virus. Scientists have traced the virus back to bats and some argue that the bats become particularly infectious when they are stressed. This stress is caused amongst other things by waking them up prematurely from their hibernation or by keeping them in captivity.[1] This virus does not only have implications for humans, but also for other animals. Some have also been infected, most famously a tiger in the Bronx zoo in New York.[2] Others are being used as animal models in medical experiments to find a vaccine for Covid-19. In a number of European countries, infections of mink have been met by a massive culling of all animals on mink farms in order to prevent them becoming a reservoir for COVID-19. This was done after a comparison of viral DNA that suggested that mink had infected a small number of employees (Oreshkova et al. 2020). In the Netherlands, the corona crisis has led to a debate, spurred on by the Party for the Animals, about the intensive way in which production animals are kept, with virologists stating that the Netherlands is 'full of potential hosts that can transmit a virus'.[3] Globalization has certainly been a driver of the massive outbreak of this disease, and as we will argue in the introduction has had major implications for animals and for the human-animal relationship.

In the (Australian) summer of 2019/2020, large parts of Australia are on fire. So far, the devastating bushfires killed 26 people and over a billion animals.[4] In particular one of Australia's main symbolic animal species, the koala, is hit hard by the fires, as its strategy is to move further up the tree in case of danger. With the fires still blazing, the Australian government is denying its own contribution to climate change and downplaying the link between climate change and the bush fires.[5] Australia ranked last in the Climate Change Performance Index out of the 57 countries that are responsible for the lion's share of greenhouse gas emissions.[6]

In 2015, a dentist from Minnesota killed famous lion Cecil just outside of Hwange National Park in Zimbabwe. The hunt was legal, but nevertheless caused a public outcry against trophy hunting.[7] This case raises difficult ethical questions, such as 'should we allow trophy hunting and even support game farms in order to raise revenue for wildlife conservation?' It also sheds interesting light on the changing

---

[1] https://www.nrc.nl/nieuws/2020/02/07/het-spoor-van-corona-leidt-naar-een-gestreste-vleermuis-a3989708.

[2] https://www.theguardian.com/world/2020/apr/06/bronx-zoo-tiger-tests-positive-for-coronavirus.

[3] https://www.ad.nl/binnenland/viroloog-nederland-is-vol-met-gastheren-die-een-virus-over-kunnen-dragen-br~afbc59f5/.

[4] https://aeon.co/essays/we-cant-stand-by-as-animals-suffer-and-die-in-their-billions.

[5] https://www.vox.com/world/2020/1/8/21051756/australia-fires-climate-change-coal-politics.

[6] https://newclimate.org/wp-content/uploads/2019/12/CCPI-2020-Results_Web_Version.pdf.

[7] https://wildlife.org/survey-shows-u-s-citizens-increasingly-humanize-animals/?fbclid=IwAR30Dxvy7q-B5Q_cxGrh9j-1eINgQWML4hLO9DjYKl7e3YwYui6H40qP69E.

attitude of humans towards other animals. A recent study suggests that people increasingly attribute human-like characteristics to wild animals, and that this anthropomorphism results in changing strategies for wildlife management. The researchers of this study notice 'a shift in values from domination, in which wildlife are for human uses, to mutualism in which wildlife are seen as part of one's social community' (Manfredo et al. 2019, 1).

In Cincinnati, two police officers sitting in a car, were approached by a goose, who started tapping on the police car's door. The goose did not respond to the food offered her, but kept on tapping, walking off, and tapping on the door again, as if asking for help. When the police officers followed her, they found her baby goose tangled up in string, and rescued the baby goose, under the watchful eye of his mother.[8] This case raises interesting questions about animals' cognitive capacities and capacity for self-willed action, or agency. The goose clearly went up to the human beings planning to ask for help, expecting them to be able to afford help. Does this mean the goose is capable of intentional action? And does intentional action presuppose second-order thought? Could she even be said to possess Theory of Mind?

These and many other recent cases exemplify the changing and ambivalent relationships we have with the diverse animals that live in our midst in the Anthropocene. The term Anthropocene, which was introduced in 2000 by chemist and Nobel Prize laureate Paul Crutzen and biologist Eugene Stroemer, refers to the current age, in which human activities are so omnipresent that humanity itself has developed into a global geophysical force, at least as influential as natural forces. Even though the term would suggest otherwise, nonhuman animals in fact play an important role in the Anthropocene, not only as victims of our treatment, but also as actors in their own right. Some—in particular farm animals—unwittingly are driving forces of the Anthropocene, others are forced to respond to new environmental conditions. Think of animals who due to climate change can no longer survive in their native habitats or wild animals that in response to habitat loss and fragmentation are forced to live in urban areas. Other types of animals in our midst are (extreme breeds of) companion and sports animals, previously domesticated animals that are becoming de-domesticated, exotic and invasive species, animal species faced with extinction, and vice versa, extinct animals faced with resurrection. We witness changing relationships between these different groups of non-human animals and between human and non-human animals.[9] Our integration with many of these animals has become stronger, leading to problems such as zoonotic diseases, invasions of exotic species, and human-wildlife conflicts. At the same time, our knowledge about animals and their (mental) capacities is increasing and this raises new questions about our treatment of them and about the possibilities to create shared life worlds with animals.

---

[8] https://www.homesluxury.net/goose-kept-pecking-cop-until-he-decided-to-follow-her-she-lead-him-to-her-trapped-baby/?fbclid=IwAR0ofz3B7OINuAezBAhzQKIe3sRXx3Zjran513N08rPI57Pvydxx4g31f-0.

[9] For simplicity's sake we will mostly refer to non-human animals as animals and human animals as humans in this chapter.

1.2   Animal Ethics in the Anthropocene

In the Anthropocene we face a tension in our dealings with nonhuman animals. On the one hand current research shows that animals are capable of 'self-willed action'—in other words that they possess agency (Irvine 2004). Characteristics that used to be seen as human-specific have been discovered in certain other animal species, including language use, morality, a sense of justice, altruism, complicated hierarchies, and cognition (Meijer 2019; Bekoff and Pierce 2009; Brosnan and de Waal 2012; Leavens 2007; Wasserman and Zentall 2012). This calls into question the sharp division between human and animal minds on which human exceptionalism rests—the view that humans are essentially (rather than just in degree) different from and superior to other animals (De Waal 2016; Lurz 2009; Gruen 2011). On the other hand, despite this heightened awareness of animals and their capacities, animals are increasingly limited in their agency. So even though the Anthropocene has shaped the knowledge and technology for us to realize that animals have more agency than has been assumed, ironically it is also an epoch where animal agency is increasingly curtailed. In the Anthropocene, most (vertebrate) animals live in captivity as livestock or companion animals, where humans control their movements and genetic make-up. Domesticated animals such as cows, pigs, and chickens often live in 'simple, predictable and monotonous environments', where they are hardly challenged and their agency is not stimulated (Špinka and Wemelsfelder 2011, 27). In the wild, through habitat loss and fragmentation, urban sprawl, and climate change, animal agency is also curtailed, as animals have less room to maneuver in environments suitable to their species. Due to the detrimental impact of livestock production on the environment, humans are forced to seek alternative means of producing proteins, such as cultured meat. In the Anthropocene, technology increasingly mediates our interactions with animals—from milking robots to gene editing and cloning—and this also has implications for the human-animal relationship and for animals' capability to exert their agency.

Whereas human–nature relationships figure prominently in discussions about the Anthropocene (Rolston 2012), human–animal relationships remain underdeveloped (except for Bovenkerk and Keulartz 2016; Tønnessen et al. 2016). Animal ethicists have criticized human exceptionalism (Gruen 2011), but have not yet formulated a coherent response to the specific and urgent challenges of this new epoch. The animal ethics field arose as a response to ill treatment of farm and laboratory animals, explaining its focus on individual domesticated animals, overlooking dilemmas occurring on the level of species (Bovenkerk 2016). Traditional animal ethicists (Singer 1975; Regan 1983) perceived animals as passive victims of our treatment, rather than active agents with whom we need to negotiate our common lifeworld. These theories have insufficiently thought through what it means for human–animal relationships when we take animal agency more seriously. Moreover, they could not have anticipated the variety of human-animal relationships and corresponding challenges we would be facing in this new epoch.

In this edited volume, we have brought together authors from a range of disciplines—philosophy and ethics, ecology, communication science, linguistics, cultural geography, animal welfare science, history and law—to shed light on the changing human-animal relationships in the Anthropocene. The reader will encounter starving polar bears, greeting cows, stray cats, pedigree dogs, backyard rats, laughing chimps and roaming wolves. The question throughout is how we can give shape to new relationships with these animals. Can we find consolation in philosophy when confronted with discomforting wildlife? How do we know what is the right dog to take into our homes? Should we feed polar bears who are victim to changing climatic conditions? How can we conserve the biodiversity of animal species? Should we resurrect extinct species? These and many other questions need to be raised in our efforts to create new meaningful relationships with the animals in our midst.

This volume is divided into five parts: animal agents, domesticated animals, urban animals, wild animals, and animal artefacts. Each part is followed by a commentary. With this division we base ourselves in part on the categories of animals proposed by influential animal philosophers Sue Donaldson and Will Kymlicka. In Zoopolis (2011) they plead for a political turn in animal ethics and suggest that we grant different types of political rights to three groups of animals: domesticated, liminal, and wild animals. All animals have universal rights, such as the right not to be killed or unnecessarily harmed, but different groups of animals have differentiated additional rights. Domesticated animals are full members of our political communities and should be treated as such. We have more specific duties towards them than towards for example wild animals, because they are already part of our community; we have made decisions about the way they live and even about whether or not they live and about what their genetic composition is. They work alongside humans and fulfil important functions in our societies. According to Donaldson and Kymlicka, justice requires that we grant them citizenship rights. In their view, animals that have not been domesticated, but that live among humans, such as urban wildlife, should be termed 'liminal animals'. Think of the squirrels in our parks and the rats in our backyards. We are witnessing an increasing 'liminalisation' of wild animals, raising the question of whether this might be a precursor to their elimination (Donaldson and Kymlicka 2016, 226). In order to prevent their elimination, we will need to develop a way of cohabitation with them that protects their basic rights but that also gives us opportunities to combat the nuisance they are sometimes causing. We owe these animals similar duties as we owe to tourists who are visiting our country; they have residency without citizenship rights and are so-called denizens. Finally, wild animals have an interest in being able to live their lives as unimpeded by human interference as possible. Justice between human communities and wild animal communities could be compared to justice between countries. Wild animals should be granted sovereignty rights; the right to autonomy over their own territories. Many of the chapters in this book relate to Donaldson and Kymlicka's views in one way or another. For example, in his search for meaningful co-existence with wolves in the Netherlands, Martin Drenthen draws on their model and argues that we should see wolves as members of a sovereign community. Eva Meijer uses the concept of liminal animals to analyse the question of how we should deal with stray cats in the city. Susan Ophorst and

Bernice Bovenkerk discuss what the notion of citizenship rights for dogs tells us about what the right choice is when obtaining a dog.

In the remainder of this introduction, we will sketch the terrain in more detail. What impacts do Anthropocene conditions such as climate change and biodiversity loss have on animals? What different groups of animals are impacted and in what ways? We will illustrate these developments by zooming in on one country where many problems, such as biodiversity loss and landscape degradation, converge, the Netherlands.

1.3   The Netherlands as Mirror of Biodiversity Problems

This is by no means an arbitrary choice. In the Netherlands only about 15% of the original biodiversity remains. This means that the loss of biodiversity is considerably greater than elsewhere in Europe and the world.

Almost two-thirds of the Dutch territory is used for agriculture, and two-thirds of that is used for cattle breeding. The Dutch landscape is therefore increasingly dominated by monocultures of drained rye-grass, a uniform green billiard cloth that has taken the place of flowery meadows full of birds. Landscape degradation is not only caused by intensive livestock farming, but also by the enormous number of distribution centers. The Netherlands is a distribution country par excellence. At the beginning of 2017, the country had 1760 distribution centers with a total surface area of 28 million square meters. This proliferation of distribution centers is accompanied by a substantial expansion of the transport infrastructure, making the Netherlands the most fragmented region in the whole of Europe, with disastrous consequences for the diversity of species in the Netherlands.[10]

Finally, there is a danger that the remaining natural landscape will be completely transformed into a recreational landscape. As a result of the increase in prosperity and mobility and the decline in leisure time, we are living in what sociologists have termed an 'experience society'. The more nature becomes an attraction, complete with mountain bike routes, pancake restaurants, souvenir shops, campsites and holiday homes, the more biodiversity will be lost.

The Netherlands, then, scores very poorly on all kinds of nature rankings. By taking a closer look at this country, one gets a sharper picture of the biodiversity problems that people and animals elsewhere also have to contend with. This is done in this introduction, in which a group portrait is sketched of the many species of animals currently found in the Netherlands, whereby we sometimes make trips to other parts of the world.

---

[10] A study from 2016 calculated that half of the European territory is situated within the critical distance of 1.5 km from (rail) roads. Within this area, the number of birds decreases by a quarter and the number of mammals even halves. In the Netherlands, where the distance to transport infrastructure is much smaller than the European average, we have to expect a much greater loss of biodiversity (Torres et al. 2016).

1.3.1   The Recovery of Wildlife

We start this group portrait with a phenomenon that gives rise to some optimism. There is currently a spectacular comeback of wildlife, not only in Europe, but also in North America. The conditions for the recovery of wild populations were created by the large-scale reforestation and revegetation that took place more or less simultaneously on both continents as a result of rural depopulation. The European Common Agricultural Policy (CAP), established in 1962, has led to a far-reaching intensification of agriculture and the associated depopulation of low-productive farmland, particularly in mountain areas. Thus, while intensive agriculture leads to an impoverishment of the landscape and species diversity, it has ironically led to the return of wild animals at the same time. Another important driving force behind the return of wild populations was the emergence of the environmental movement in the 1970s, which created support for European legislation for the protection of species and habitats.[11]

As a result of these changes in agricultural and nature policy, the number of large grazers increased sharply. This was a prerequisite for the recovery of the populations of large predators, which were almost completely extinct in the course of the eighteenth and nineteenth centuries. A recent study, based on data from all European countries (with the exception of Russia and Belarus) for the period 2012–2016, comes to the following estimate: wolves are the most abundant with 17,000 individuals; second is the brown bear with between 15,000 and 16,000 individuals; next is the lynx with between 8000 and 9000 individuals; last is the wolverine with between 1000 and 1250 individuals (Linnell and Cretois 2018).

Also in the Netherlands many species have returned or have recovered considerably. Some species have returned spontaneously, such as the eagle owl, the crane, the white-tailed eagle and the wild cat, and also the wolf has settled here by now (see also the chapters by Jansman and by Drenthen in this Volume). Other species have been helped by reintroduction, such as the raven, the stork, the beaver, the otter, the badger, and quite recently, the bison.

1.3.2   Exotic Species and Climate Refugees

There is another category of wild animals that is increasingly making its presence felt in the Netherlands. These are not native species that return here after a long absence, but exotic species, that do not originally belong here, such as the muskrat, the collar parakeet and Japanese knotweed. In fact, migrating plants and animals are nothing new, but since globalization has taken off, more plants and animals have turned into 'globetrotters' than ever before. As traffic and transport, trade and tourism grow, the significance of boundaries is diminishing, and at the same time vulnerability to

---

[11] Such as the 1982 Berne Convention of the Council of Europe and the 1979 and 1992 Birds and Habitats Directives of the European Union respectively.

the massive arrival of exotic species is increasing enormously everywhere. In this context, there is even talk of a 'mass migration'.

The number of registered exotic species in the Netherlands is almost 2400, an extremely conservative estimate because we do not have sufficient information for a number of species groups. With this high figure, the Netherlands is European champion of exotic species, and has the dubious honour of even being among the world's top. This is the ecological downside of international trade and the transit of goods and raw materials to the European hinterland (Leuven 2017).

Apart from globalisation, there is another important cause for the influx of exotic species, namely climate change. While cold-loving species migrate northwards, heat-loving species come our way from the south. Among these newcomers are a number of pest species that can cause a lot of nuisance and also serious public health problems. One of these species that made the headlines in the spring of 2019 is the oak processionary caterpillar. This moth caterpillar, whose numbers has tripled in the Netherlands since 2018, causes complaints such as itching, rash, and irritation to the eyes or respiratory tract. The pine processionary caterpillar, which has even more burning hairs than the oak processionary caterpillar, has already moved into the Belgian Ardennes. Other species that threaten to settle here are the Asian Tiger Mosquito, which can transmit tropical diseases such as chikunguna, dengue and zika, and the Hyalomatick, a giant tick that can transmit the dangerous Crimean Congovirus.

While thermophilic species are advancing, cold-loving species are increasingly getting into trouble. This is especially true for plants that can only move very slowly to colder areas because they can only move by spreading their seeds. It also applies to cold-blooded animals, such as insects, fish, reptiles and amphibians, whose body temperature depends on the ambient temperature and who are therefore very sensitive to climate change. These animals all too often encounter obstacles that are difficult to overcome in the form of natural barriers, such as mountains and rivers, or in the form of infrastructural works, such as motorways, railways and viaducts. In order to save these animals from extinction, wildlife managers sometimes resort to what is known as assisted migration or assisted colonisation: the deliberate relocation of these 'climate refugees' to new habitats that they cannot reach on their own (see Larson and Barr 2016).

1.3.3   The Sixth Mass Extinction

All in all, a lot of wild animals, spontaneously or otherwise, have returned to the Netherlands or emigrated here. But let's not get too excited yet: in the Anthropocene, the age of humans, we are confronted with a worldwide extinction wave. We must therefore assess the recent increase in wild populations in the light of dramatic historical declines: most of the populations of species in the process of comeback are still far from reaching a genetically and demographically sustainable size, while many species are still declining in size. In fact, we are in the midst of a new mass extinction. Mass extinction occurs when the earth loses more than three-quarters of its species. This has only happened five times in the course of Earth's history, the last time 65 million years ago when the dinosaurs disappeared from the face of the earth. Experts estimate that the current rate of species extinction is 100–1000 times higher than the so-called background extinction rate.

In order to be able to assess the situation properly, we need to pay attention not only to the extinction of species but also to the decline in the number and size of populations. WWF's latest *Living Planet Report* (2018) shows that the size of vertebrate populations fell by no less than 60% between 1970 and 2014.[12] When we realise that population extinction should be seen as a prelude to species extinction, it becomes clear that "the window for effective action is very short, probably two or three decades at most" (Ceballos et al. 2017).

When we now turn our gaze to Dutch nature, we see a rather dramatic picture. This is evident from the Living Planet Report: *Nature in the Netherlands*, a report published by WWF at the end of 2015, in collaboration with a large number of nature organisations. According to the report, animal populations in nature reserves have declined by an average of 30% since 1990. The most recent *State of Nature in the EU*, a report published by the European Environment Agency in 2015, shows that the Netherlands, with no less than 96% of nature areas in an unfavourable condition, is at the very bottom of the list of all 26 EU member states.

1.3.4 Rewilding and De-extinction

It is a widespread misunderstanding that the extinction crisis only began during the era of the great discoveries, from the end of the fifteenth century until the eighteenth century. This crisis began as early as the transition from Pleistocene to Holocene, around the end of the last Ice Age, some 11,000 years ago. The main victims were the megafauna species, a group of mammals considerably larger than the current mammals, such as mammoths, mastodons, sabre-toothed tigers, ground sloths, cave bears, giant wolves and giant deer.[13]

To explain this tremendous loss of mega-fauna species, four hypotheses have been put forward, known tongue-in-cheek as 'overkill', 'overchill', 'overill' and 'overgrill'. The 'overkill' hypothesis blames the extinction of megafauna species on the spread of modern humans (Homo sapiens). The 'overchill' hypothesis puts the blame on climate change at the end of the Pleistocene, the 'overill' hypothesis on some 'hyper disease', and the 'overgrill' on a comet impact or shock wave above North America (Wolverton 2010). Because there is hardly any support for the last two hypotheses ('overill' and 'overgrill'), the research has mainly focused on the role of climate change and the role of the hominids. The outcome of this research is that the

---

[12] Invertebrate populations are also in a precarious situation; their size has decreased by 45% over the last 4 decades (Dirzo et al. 2014).

[13] The term megafauna refers to large mammals weighing more than 45 kg.

extinction of megafauna species is strongly related to the prehistoric geographical distribution of hominins and only weakly related to interglacial climate changes.

There is a significant difference in the magnitude of megafauna extinctions between sub-Saharan Africa, where hominins and megafauna have long coexisted, and Australia and America, where Homo sapiens were the first hominin species to arrive. While megafauna extinctions were universally low in sub-Saharan Africa, they were exceptionally high in Australia and the Americas. Eurasia, where megafauna came into contact with hominins long before the arrival of Homo sapiens, falls between these extremes. It has been suggested that these differences stem from the 'naivety' of prey animals that had not (yet) learned how to defend themselves against a new predator with advanced hunting techniques (Sandom et al. 2014).

The great loss of megafauna species is ecologically disastrous. Large carnivores, who are at the top of the food pyramid, exert a strong influence on the animal and plant populations that are at a lower level of the food pyramid and therefore play a key role in the regulation of ecosystems. They regulate the size and behaviour of prey populations, the small and medium-sized herbivores and the so-called 'meso-predators' that hunt smaller animals. Large herbivores, which are largely resistant to predation, also have an important influence on ecosystems, especially on the structure and composition of their vegetation (Terborgh et al. 2010; Svenning et al. 2016).

To compensate for the dramatic loss of megafauna, rewilders use a multitude of methods and techniques. In the case of globally extinct species one can try to make use of ecological substitutes for these species. Under the title of 'Pleistocene rewilding', American restoration biologists are considering using the megafauna still present in Africa and Asia, such as cheetahs, lions, camels and elephants, as replacements for the extinct American species.

More recently, a number of new technologies have been developed that make it possible to revitalise extinct species, a practice known as de-extinction. Two methods from synthetic biology are currently in vogue to bring back extinct species: Cloning via somatic cell core transplantation (SCNT) and genetic modification using CRISPR/Cas technology. The most spectacular de-extinction project using the latter technology aims to bring the woolly mammoth back to life.[14]

In addition to cloning and genetic modification, there is a third, less controversial method of de-extinction, namely back breeding: the crossing and selection of domesticated breeds with the aim of bringing back the traits of the wild extinct ancestors of these breeds. Such semi-wild breeds played an important role in Dutch nature policy. In 1983, 32 Heck cattle were introduced to the 'new nature area' Oostvaardersplassen. The cattle were named after the brothers Heinz and Lutz Heck, who in the 1920s and 1930s tried to breed back the aurochs (Bos primigenius), that went extinct in 1627, by means of crossbreeding with primitive breeds such as the Camargue cattle, the Hungarian Steppenrund and the Scottish Highlander. One year later, 20 konik horses followed; they are the result of attempts by the Polish agronomist and biologist Tadeasz Vetulai of the University of Poznań to breed back the wild tarpan horse (Equus ferus ferus), which became extinct in 1887.

---

[14] See https://reviverestore.org/projects/woolly-mammoth/.

1.3.5   Intensive Livestock Farming[15]

The extent to which wild animals have become oppressed worldwide is abundantly clear from the fact that the biomass of all people on earth is ten times greater than that of all wild land mammals combined, while the biomass of domestic animals—farm animals and companion animals—is as much as 35 times greater than that of all wild land mammals combined. Among the vertebrates, Homo sapiens and Bos Taurus have become by far the most dominant species on earth.

Biomass of all the land mammals on planet Earth:

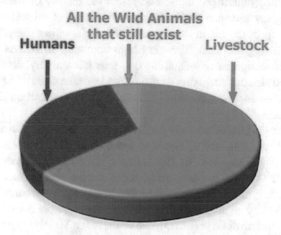

Based on Vaclav Smil (2011)

In the Netherlands those proportions are even a lot more unbalanced. With 502 inhabitants per km^2 our country ranks fourth among the most densely populated countries in the world (after Bangladesh, Taiwan and South Korea) but at the same time it is also the second largest exporter of agricultural products in the world (after the US).[16] In terms of livestock density, the Netherlands is even at the top of the world's rankings. The number of farm animals in 2016 was more than 126 million. The largest group is made up of 105 million chickens, followed by 12 million pigs and 4 million cattle. However, the share of livestock farming in the Gross Domestic Product (GDP) is only a mere 0.6%.[17]

The large amount of land taken up by livestock farming is causing serious biodiversity problems. If we do not include the production forests, only 13% is currently left for nature, half of which consists of large inland waters such as the Wadden

---

[15] We would like to thank Joost van Herten for his critical reading of and comments to this section.

[16] This ranking does not take into account small city-states and small islands or island groups with very high population densities.

[17] This is despite strong productivity growth as a result of the processes of intensification and economies of scale, the contribution of agriculture to GDP fell from 15 to 1.5% in the period 1950–2015, less than half of which is accounted for by livestock farming. http://www.clo.nl/indicatoren/nl2125-productiewaarde-landbouw.

Sea and lake IJsselmeer. The size of our livestock population causes biodiversity problems outside of the Netherlands as well. Our livestock is so large that it can only be partially fed on fodder grown here. The Netherlands is the world's second largest importer of soy after China. Although two-thirds of the soy is exported to the European hinterland, with what remains behind, the Netherlands is still the fifth largest soy user in Europe. To grow the soy that we import annually for our livestock, we need an area as large as three-quarters of the Netherlands. A lot of tropical rainforest is deforested, with disastrous consequences for the richness of local species.

The high density of livestock not only compromises the space for nature, but also leads to major environmental problems that affect the quality of nature. The Netherlands has the highest nitrogen and phosphate surpluses of all EU member states.[18] Nitrogen ends up in the air in the form of ammonia, either directly from the stables or after fertilising the land. The very high ammonia emissions lead to acidification and eutrophication of nature areas and thus have a very detrimental effect on biodiversity and also on the quality of ground and surface water.[19] Livestock farming is also responsible for the emission of large quantities of carbon dioxide, methane and nitrous oxide.

The high livestock density causes not only biodiversity problems but also all kinds of animal welfare problems. The lack of room to move of the tightly housed farm animals leads to aggressive behaviour of the animals, which in turn necessitates controversial interventions such as the dehorning of cows and calves, the docking of pig tails to prevent tail biting, and the cutting of chicken beaks to prevent cannibalism. In addition, the high density of livestock regularly leads to outbreaks of animal diseases such as foot and mouth disease (FMD), and swine fever. Mainly for economic reasons, animals are usually 'culled', i.e. slaughtered on a large scale in order to make the herd disease-free again. And then there are the many barn fires in which sometimes more than two hundred thousand animals are burnt alive in one year.

Finally, the high density of livestock also causes public health problems. Some animal diseases, so-called 'zoonoses', such as Q-fever and bird flu, are transmissible to humans. The excessive use of antibiotics in animal husbandry also has harmful consequences for public health.[20] It can lead to bacterial resistance and thus to a decrease in the efficacy of antibiotics that are also used in human health care. Finally, livestock farming, through the emission of ammonia which is converted into particulate matter in the air, increases the risk of respiratory problems for people living near

---

[18] Our country does not even manage to meet the conditions of the so-called 'derogation', which allows Dutch livestock farmers to use 250 instead of 170 kilos of nitrogen from animal manure per hectare of land, in derogation of the European Nitrates Directive.

[19] In recent years fertilisation has led to environmental standards being exceeded in 86 of the 220 drinking water wells in the Netherlands.

[20] Fortunately, due to stricter regulations, the use of antibiotics in animal husbandry has decreased by 63.8% since 2009 in the Netherlands. See https://cdn.i-pulse.nl/autoriteitdiergeneesmiddelen/userfiles/sda%20jaarrapporten%20ab-gebruik/AB-rapport%202018/sda-rapportage-2018-def-err.pdf. In the whole of Europe it has decreased by 32% between 2011 and 2017. See https://www.ema.europa.eu/en/veterinary-regulatory/overview/antimicrobial-resistance/european-surveillance-veterinary-antimicrobial-consumption-esvac.

animal farms. The conclusion seems inescapable: what is needed first and foremost to create space for wild animals is a substantial shrinkage of the herd.

1.3.6   The Ecological Impact of Large-Scale Hunting

However, more space for nature is not only a matter of the number of hectares, but above all concerns the management of nature areas. This management must provide space for wild animals to fulfil their ecological role, thereby giving natural processes that are essential for the much-needed restoration of biodiversity a chance again. In current wildlife management, however, that space is severely restricted because of the central position of large-scale hunting. Take the fox, which is currently hunted in order to protect meadow birds. Bird numbers have fallen sharply since the varied, flowery grasslands, on which the meadow birds depend for food, mating and nesting, gave way to monocultures of drained English rye-grass lands, from which all life has disappeared. It seems unfair and disproportional to punish foxes for the birds' demise.

Besides the fox, many other wild animals are hunted. For decades, the Netherlands has had a 'zero tolerance' policy for large game, such as fallow deer, red deer and wild boar. The animals are only tolerated in a very limited number of nature reserves. If they leave these areas, they will be shot. But even within these areas, large numbers of deer and boars are shot. By far the most important justification for large-scale hunting is the prevention of damage to agricultural crops and of injuries as a result of traffic accidents. As damage increases with the number of animals present in an area, the density of game populations is hunted down to a level acceptable to farmers, foresters, private landowners and other interest groups. This is also referred to as 'societal carrying capacity'.

The socially determined target densities are extremely low: for red deer and wild boar, target densities of 2–3 animals per 100 ha are by no means unusual. In order to achieve these low targets, huge numbers of large ungulates have to be shot every year. Shooting percentages of 60–70% for red deer and 80–90% for wild boar occur regularly. As a result, the animals can hardly have any influence on their habitat and are in fact ecologically eliminated. Some animal species that were locally extinct, such as the beaver, have been reintroduced with a lot of publicity (and subsidy) as an asset to Dutch nature, but then declared outlawed when the success of this action started causing us discomfort.

1.3.7   Companion Animals

Besides farm animals, people also keep large numbers of pets in the Netherlands. North America is the world's leading country in pet-keeping; the country has over 300 million pets, which is four times the number of children. In the Netherlands no less than 60% of households have a pet. In total, about 35 million animals are involved. The list is led by fish, carrier pigeons and singing and ornamental birds, followed by cats and dogs. In addition, rabbits, other rodents and reptiles are also kept in large numbers.

Many of our pets suffer from serious health and welfare problems. This is especially true for pedigree dogs, which make up about a quarter of the two million dogs in the Netherlands. They are condemned to an inadequate and unhealthy existence by breeding for extreme external characteristics and breeding with a gene pool that is too small. Just think of the English Bulldog, who suffers from shortness of breath because of his characteristic large head with a flat snout; the German Shepherd Dog, who has a very high risk of hip dysplasia and wear and tear of the dorsal vertebrae because of his sharply sloping hip; the Shar-Pei, who suffers from all kinds of allergies and skin infections due to his exaggerated folds; the Shih Tzu, who suffers from respiratory problems due to his flat snout and also has a risk of corneal inflammation due to his protruding eyes, and the list goes on.

The discussion on the issue of rearing pedigree dogs should not only address the complaints and ailments which these animals face, but also the underlying question of whether we should be allowed to control the genetic composition of animals to such an extent that we actually create human artefacts (Bovenkerk and Nijland 2017).

But pets do not only have problems, they also cause problems themselves. This applies to dogs, who can be involved in biting incidents and the transfer of zoonoses (see also Ophorst and Bovenkerk in this volume) but also for example to cats, who pose a serious threat to biodiversity in our country. In their book *Cat wars: The devastating consequences of a cuddly killer*, published in 2016, Peter Marra, director of the Smithsonian Migratory Bird Center, and journalist Chris Santella defend the proposition that cats cause serious ecological damage worldwide. These cuddly animals are in fact super-predators that threaten many endemic species with extinction, a problem that is particularly acute on isolated islands. In the United States, cats are said to devour 1.3–4 billion birds and 6.3–22.3 billion mammals annually, with stray cats accounting for 70% and domestic cats for the rest. In the United States, according to Marra and Santella, "more birds and mammals die in cats' mouths than from pesticides and toxins, from collisions with wind turbines, cars and skyscrapers, and from all other anthropogenic causes combined" (Marra and Santella 2016, 69). Three to four million domestic cats live in the Netherlands and the number of stray cats is estimated at 135,000–1.2 million (for a discussion on how to co-exist with stray cats, see Eva Meijer in this volume). According to a rather conservative estimate, we must assume that domestic and stray cats kill 100 million and 50 million prey annually respectively, mainly birds, but also mammals, reptiles and amphibians, including endangered species (Knol 2015).[21]

---

[21] Given the ecological impact of cats, Marra and Santella argue for house cats to be kept indoors and obliged to be chipped, registered and sterilised, and for stray cats to be offered for adoption. This appears to be a declaration of war to cat lovers for whom these animals are sacred. Marra now receives death threats and is portrayed as a Josef Mengele propagating the mass extermination of cats. The Huffington Post even suggested that he would call for cats to be clubbed to death and shot. In the Netherlands, recently two legal scholars have called for a duty to keep cats on a leash when

1.3.8 The 'Liminalisation' of Wildlife

One of the most important consequences of the scaling up and intensification of agriculture is the migration of many wild animals to villages and towns. One example is the oystercatcher, which had already gone through a transition from coastal bird to meadow bird and is now seeking refuge in the city, on the roof of hardware stores and schools. Another example of the migration to the built and inhabited world is the stone marten. The disappearance of small-scale agriculture is to blame for its relocation. Particularly because modern stables are no longer suitable as a place to rest and sleep, the stone marten has moved to villages and cities where, in addition to sufficient food, it can also find a warm shelter.

There are many other examples of wildlife advancing to urban areas. With many thousands of boars, Berlin has now been proclaimed the 'capital of the wild boar'. When it comes to foxes, London is undoubtedly the first city to qualify for such a title. In the Netherlands there is similarly a steady increase in city foxes and boar, and also city deer, just think of the fallow deer that roam in a number of coastal cities, such as Zandvoort.

As a result of the forced migration to urban areas due to a growing lack of space, more and more wild animals acquire the status of liminal animals (Donaldson and Kymlicka 2011). Although liminal animals are our co-inhabitants, according to Donaldson and Kymlicka we should not treat them as fellow citizens, as we should do with domesticated animals. Since deportation (to the wilderness) or domestication (to farm or pet) are not real options for the majority of liminal animals, we must accept their presence; we must give liminal animals a certain right of residence and not treat them as pariahs and make their lives unnecessarily difficult or completely impossible. On the other hand, they are not entitled to full citizenship rights. We must exercise extreme restraint in our contacts with liminal animals; we must not feed them or make friendly relations with them. This only leads to conflicts of which the animals themselves are the victims. We must develop forms of co-existence in which the right of residence of liminal animals is protected but in which we can also combat the nuisance these animals may cause, for example by reducing populations through contraceptive methods, or by restricting access to buildings by means of fences and grids (see also Drenthen in this Volume).

1.3.9 The Struggle for Nature Between People

As we hope to have made clear, the growing presence of animals in our midst raises many questions to which we often do not have the answer. How to deal with the

---

outside (Trouwborst and Somsen 2019). This was also met with fierce criticism. As Eva Meijer convincingly argues in Chapter 16, stray cats have agency which should be taken into account. However, there is a clear tension with the agency of their prey animals and this tension at least should be the topic of public debate in our view.

environmental and welfare problems in the increasingly intensive livestock farming? What about the many millions of companion animals that inhabit our country? How to live together with wild animals such as geese and wild boar? Should we help the climate refugees among the animals through assisted migration? How do we treat foxes, peregrine falcons and other wild animals that are currently moving into our cities? How to deal with the many exotic species that come to us in large numbers through world trade and mass tourism? And what about the animals that are in danger of disappearing from our midst due to extinction? In short: how can we shape our relationships with all these different animals in a rapidly changing world in such a way that both animal welfare and species diversity are not further affected?

The answers to these questions vary strongly. The fight for the animals is mainly a battle between people, also between people who personally care about the animals and/or are professionally involved in the policy and management of animals. All too often animal protectionists are at odds with nature conservationists. But even between conservationists it is not always easy, as there are often fundamental differences.

Conflicts between animal protectionists and conservationists are mainly about the contrast between individualism and holism. Animal protectionists generally give priority to the welfare and rights of individual animals and tend to downplay the importance of species conservation and the prevention of biodiversity loss. Nature conservationists oppose this individualistic approach; they embrace a more holistic view in which individual organisms are seen as part of a larger whole, such as communities, species or ecosystems, and tend to subordinate animal welfare and rights to the importance of species conservation.

Conflicts between conservationists are mainly about the contrast between the separation and the interweaving of nature and culture. This contrast is central to the discord between the primitive and the pastoral representation of the ideal—Arcadian—landscape. While primitive Arcadia is inhabited by people who behave like wildebeests, all dangerous creatures (such as the lion and the snake) have been banished from pastoral Arcadia and the ideal animals (such as the cow and the bee) behave like dutiful and industrious citizens. Because both imaginations of the ideal landscape are at odds, there is a constant struggle between them, which, according to the British historian Schama (1995), even extends into debates within nature conservation and the environmental movement, 'between the brighter and paler shades of green'.

This thesis also holds true for Dutch nature policy, which for decades has been torn between the pastoral and the primitive representation of the ideal landscape. The Dutch model of the pastoral imagination par excellence is formed by the pre-industrial landscape of yesteryear, which can only be maintained by old agricultural techniques. The primitive imagination is not about maintaining patterns that have arisen in the course of the history of human habitation and exploitation, but about striving to keep natural processes as undisturbed as possible. To achieve this, human intervention must be kept to a minimum: 'hands off' is the motto. While the pastorals advocate an integration of culture and nature, the primitives are in favor of a strict separation between the two. In other words, the pastorals have a human-inclusive vision of nature and cherish an interventionist management of nature, while the primitives have a human-exclusive vision and are explicitly non-interventionist.

The contrast between separation and integration also plays a role in the debate between the established, traditional conservationists and the so-called ecomodernists. The ecomodernists distance themselves from the doom and gloom of the old nature movement, which the Anthropocene sees exclusively as a potential ecological disaster. Instead, they welcome this era as a new step in the progress of mankind. In their view, man is not a pest species, as some traditional conservationists seem to think, but a 'God species'. The techno-phobia of the traditional nature movement is giving way to a pronounced techno-triumphalism among the ecomodernists. And ecomodernists see industry not so much as a culprit but rather as an ally (see Keulartz and Bovenkerk 2016).

Whereas the traditional nature movement generally advocates nature-inclusive forms of agriculture, such as organic or ecological agriculture, ecomodernists want to 'decouple' man and nature and save space for nature by further intensifying agriculture. Here the contrast between separation and integration returns, in the form of the duo land sparing and land sharing.

1.4 Overview of the Volume

Besides the land sparing versus land sharing tension, a number of other tensions are thematised—implicitly or explicitly—in this book. Firstly, we find the aforementioned tension between our awareness of animal agency and the efforts of humans to curtail this agency. How can we do research into animals' agency if these animals are raised in conditions that impoverish their capacities for expressing their agency? How can we take animal agency seriously, while at the same time dealing with potential harmful expressions of this agency (such as in the case of stray cats)? Should we emphasize the differences or rather the similarities between humans and other animals?

Secondly, there is the aforementioned tension between those focussing on individual animals who tend to view all animals as domestic animals that we need to take care of and those focussing on wild animals, who tend to have a laissez-faire attitude. Do we allow wild zoo animals to exhibit behaviours that are not considered wild, but that do appear conducive to their welfare? How to deal with wild animals that due to climate change can only survive if they are fed by humans, such as polar bears, if this means they lose part of their wild status?

Thirdly, we find a tension between different responses to the Anthropocene, including the role of technology. Traditional conservationists call for a modest attitude towards other animals and nature. They emphasize the negative consequences of human actions and look for ways of righting the wrongs we have committed in the past. Ecomodernists on the other hand, welcome the Anthropocene as an opportunity to increase human welfare worldwide through careful governing and management of nature: because of modern technology, we no longer need to worry about planetary boundaries, but can simply redesign Earth for our own benefit. With their exclusive focus on humans, ecomodernists may ignore animal agency. Some plead for intensification of agriculture, including livestock farming. However, because livestock farming is driven by efficiency, it increasingly upscales to ever larger facilities and becomes more technology dependent. Such circumstances increasingly limit animal agency. Other ecomodernists, however, such as Hidde Boersma in this volume, argue for a decrease in meat consumption and the development of cultured meat. Technology does not necessarily curtail animal agency, but could potentially be used to augment it as well. Is a middle road possible between technology-shy conservationists and technology optimistic ecomodernists, like Cor van der Weele argues?

This volume is divided into five parts, on the basis of different types of animal: animal agents, domesticated animals, urban animals, wild animals, and finally, animal artefacts. After this introduction we have included a second introduction, written by Hugh Jansman from a wildlife-ecological perspective. The purpose of this chapter is to give the reader some theoretical and empirical background to the conservation issues that form the backdrop of many of the chapters in this volume. Jansman argues, again taking the Netherlands as an example, that our conservation efforts should focus more on the level of viable ecosystems. He proposes a strategic plan to do so, which is called Cores, Corridors and Carnivores. This involves setting aside more space for nature and natural processes, including top-down forcing by apex consumers (in particular large predators) and finding a way of coexistence with our fellow creatures. He discussed the cases of red deer and wolves in the Netherlands by way of illustration and concludes with a vision of what the Netherlands could look like a hundred years from now.

1.4.1   Part 1: Animal Agents

The first part of this volume deals with the capacities of animals and what they mean for our treatment of them. Many discussions about animal welfare betray a limited understanding of animals. For example, only the way animals' lives are ended are regarded as problematic. This seems based on a narrow understanding of animal welfare as one of being free from pain, hunger and thirst. Meijer and Bovenkerk present a broader vision of animals as agents. In their chapter, they defend a relational approach to animal ethics, viewing other animals as subjects capable of co-shaping relations. Charlotte Blattner examines how animal agency is dealt with in the law and encounters deep-seated anthropocentric biases. The authors of both chapters call for new forms of research that take animal agency seriously. The following three chapters each deal with foundational issues in animal ethics that have an influence on how we regard animal agency. Nathan Kowalsky argues that the common strategy in animal ethics to base moral consideration on the similarities humans share with non-human animals is really a disguised form of anthropocentrism. He proposes an ethic of animal difference that does justice to animals' otherness without being imperialist. Jozef Keulartz analyses a problem that has been at the heart of animal ethics ever since animal rights were proposed: the predation problem. If we take animal agency seriously, we should grant predators the opportunity to manage their own affairs and refrain from protecting prey animals against them. Perhaps counter-intuitively, he argues firstly, that if we want to take prey animals' agency seriously we should also allow predation and secondly, that this is only possible in the Anthropocene if we manage wild populations. The latter raises a tension for animal agency: conservation may necessitate keeping wild animals in zoos, but this could interfere with their wild animal agency. John Basl and Ronald Sandler revisit the central discussion in animal ethics about what entities should be attributed moral status. They defend the view that species partiality in consideration and treatment of animals is possible without the need to adopt a species-membership or human-privilege view on moral status. Finally, Michiel Korthals places the discussion about animal agency in a broader context. Focussing on the complexity of relationships between different living organisms he highlights the ways in which organisms select and value specific items in their network of living and non-living entities. In his comment to this part, Joost Leuven calls attention to the societal and moral urgency of more research into animal agency, in a human-dominated Anthropocene.

1.4.2   Part 2: Domesticated Animals

Part two focuses on domesticated animals. Intensive livestock farming in particular is greatly implicated in creating the Anthropocene, and current farming systems are organised in such a way that animals' agency is hardly taken into account, influencing the relationship between humans and farmed animals. Plant-based meat substitutes and in vitro meat and milk are increasingly pursued, but technological solutions to improve intensive farming systems are also sought. How can we reshape our relationship with the animals we have domesticated, be it for companion or for their products? Hidde Boersma sets the scene of this part by zooming in on the production and consumption of animal protein and its detrimental impact on the planet. He gives the reader insight into the ecomodernist mindset and proposes a diverse strategy: on the consumption side he argues for a move from beef to chicken and pork, a reduced consumption of meat and the consumption of lab grown meat. On the production side he argues for intensification of the production process through the closing of global yield gaps between the best and the worst performers and through the development of novel integrated indoor systems like agroparks. Drawing on a combination of ethnographic fieldwork and critical theoretical inquiry, Leonie Cornips and Louis van den Hengel examine how dairy cows, whose freedom is profoundly restricted by bars and fences, nevertheless enact social and linguistic agency. The chapter thereby reflects on the role of language in changing human-animal relationships. The third chapter in this section that focuses on the conditions of animal protein production is written by renowned agricultural and food ethics professor Paul Thompson. He calls on philosophers to not simply reject intensive livestock farming, but to engage with producers and scientists in order to improve animal welfare standards. Andrea

De Paula Vieira and Raymond Anthony focus on a neglected topic in animal ethics: how to prepare for emergencies and disasters that not only affect humans, but also non-human animals? What happens to animals when disaster strikes? The authors present six ethical stewardship caretaking aims for emergency preparedness and response and a number of recommendations for reasonable decision-making the face of emergencies. The final chapter of this section, by Susan Ophorst and Bernice Bovenkerk deals with the quintessential companion animal: the dog. They point out several problems related to dog-keeping, such as aggression, zoonosis, health-and welfare problems due to exaggerated breeding standards. As dogs were domesticated so long ago these problems might be a precursor to problems other human-animal relationships may face in the Anthropocene. In their view, many of these problems can be traced back to one moment: the decision-moment of wannabe dog keepers. In his comment, Erno Eskens focuses on the opposition between animal rights defenders on the one hand and 'non-ideal animal ethicists' on the other. The latter are of the opinion that we can improve animal welfare gradually by appealing to standards of human decency. Eskens is rather sceptical of this position, because it might in fact move us in the opposite direction of an ideal situation. Yet, he acknowledges no-one has a clear vision of what such an ideal situation actually is.

1.4.3 Part 3: Urban Animals

In part three urban animals are central. One of the most important consequences of upscaling and intensification of agriculture, in combination with management of wild animals, and habitat fragmentation is the colonization of urban areas by wild animals. So, although wildlife becomes extinct in many places, it magically appears in the midst of urban life—in zoos, nature parks, and popular culture. How can we defend ourselves against the nuisance these animals may cause while taking into account their interests and agency as well? Possible examples could be reducing out of control populations with birth control or limiting access to buildings with fences and screens. But how far are we allowed to go with these measures? What meaningful ways of cohabitation can we develop? Besides liminal animals, we find another category of urban animals: wild animals living in captivity, most notably in zoos. The justification of keeping animals in zoos is increasingly up for discussion and at the same time zoos are keeping up with the times by changing their raison d'être, their designs and management practices, focussing more on conservation and more closely mimicking natural habitats. Is it justified to take away animals' liberty? Are we allowed to 'sacrifice' the individual animal for the sake of its species? How can zoos take into account animals' agency better? In her chapter, Eva Meijer narrates the story the Amsterdam Stray Cat Foundation, who in their practices and views they challenge common assumptions about cat subjectivity and agency, the cats' right to a habitat and social relations, as well as the idea that there is a strict difference between cats and humans. Joachim Nieuwland and Franck Meijboom deal with liminal animals that are even less wanted in cities: rats. They consider the question in what way the aversion people feel towards rats affects moral deliberation about pest management and about animal political theory in general. Next, we turn to zoos. Yulia Kisora and Clemens Driessen examine narratives about two orangutans in zoos: Jinga, who has become famous on Youtube for laughing at a magic trick, and Jacky, whose video where he plays 'catch the banana' with a zoo visitor has gone viral on Youtube as well. Their interpretation of the comments underneath the videos shows that such virtual interactions between humans and zoo animals can serve to cast doubt on the division between the human and the animal. It also shows the ethical potential of these interactions to either reinforce or question common practices of dealing with wild animals. Animal welfare specialists Sabrina Brando and Elizabeth Herrelko in their chapter give the reader a look behind the scenes of zoos and ask the question what role zoos can play in connecting people with nature. They discuss a number of dilemmas that modern zoos face. For example: how to give zoo animals sufficient choice and control over their environment and activities? How to deal with animals who choose to engage in behaviour that does contribute to their welfare, but that appears to undermine their wild status? And how can we make sure that the conservation message still comes across even if zoo animals are perceived as less wild? In her comment to this section, Lauren van Patter ponders the challenges different types of animals face due to urbanisation. Asking how animals can make a living within the city is both a spatial question, involving human judgments about who belongs where, who is wanted and who is a pest, and an ethical question about what rights animals have to the city.

1.4.4   Part 4: Wild Animals

Part four deals with our relationship to wild animals. Habitat fragmentation, urban sprawl and species invasions resulting from globalisation have curtailed animals' freedom of movement, but at the same time our interactions with 'wild' animals have become more numerous. Many people rejoice about the return of wild animals such as the wolf and sea eagle, while having doubts about how to co-exist with these animals. Modern (bio) technologies are employed to deal with species loss and to enable co-existence with animals, raising moral questions. We can discern two opposing trends: On the one hand, nature conservation efforts, intensification of agriculture—leading to an exodus out of rural areas—and global traffic, trade and tourism—bringing in exotic species—have led to a comeback and proliferation of wild animals in Europe and North-America. On the other hand, we are witnessing the sixth mass extinction of species. How can we help wild animals to cope with challenges posed by the Anthropocene: climate change, habitat loss, and biodiversity loss? How can we adapt animals and their lifeworld in a responsible manner so that they can survive in a world dominated by humans? Should we engage in assisted migration or intervening in wild populations by supplementary feeding or giving veterinary care? The latter question is addressed by well-known animal ethicist Clare Palmer, who looks into the plight of polar bears. How should we respond to the suffering of individual polar bears due to the effects of climate change? Would aid to the bears result in park-like management of some bear populations and the associated loss of wildness? Ned Hettinger in his chapter defends the preference for native species against criticism of being, amongst others, incoherent and xenophobic. What are the implications of such a preference for non-native, sentient animals? Both Martin Drenthen and Mateusz Tokarski in their chapters discuss the recolonisation and repopulation of wild animals, in particular wolves, into humanized cultural landscapes. What does such recolonisation mean and how can we learn to co-exist with these animals? Drenthen, drawing on Donaldson and Kymlicka, argues that we should regard wolves as sovereign beings belonging to a sovereign community. In his view, fences should be regarded as communicative devices that help parallel sovereignties arrive at a common understanding. Tokarski suggests that environmental philosophy can offer consolation for the discomfort that we will undoubtedly experience as a result of co-existence with wild and sometimes dangerous animals. Finally, veterinarian, legal scholar, ethicist and former hunter Charles Foster examines intuitions about the human enjoyment of killing and eating animals, from a philosophical but primarily from a personal perspective. This chapter has a different character than the other chapters, as it builds not on academic arguments, but rather on anecdotal knowledge. Hereby it aims to give the reader an experiential account of what goes on in the mind of a (former) hunter. Commentator Sjaak Swart recounts the broad range of implications of the Anthropocene for wild animals. As the cases in this section show, for some animals Anthropocene conditions are threatening, while for others they may be favourable. Can the traditional view in animal and environmental ethics that we should leave wild animals alone as much as possible be upheld when wild animals are increasingly affected by our actions? Moreover, can the categories of wild versus domesticated animals be maintained with the increasing presence of wild animals in 'human' cultural landscapes?

1.4.5 Part 5: Animal Artefacts

The final part focuses on 'animal artefacts': animals whose genetic make-up is changed by humans, raising the question of whether they are at least in part human artefact. Different responses are possible to the mass extinction of species we are facing in the Anthropocene. While conservationists may focus on saving species from extinction, ecomodernists want to call in the aid of technology, for example by engaging in resurrection ecology. Bringing back extinct species with the aid of technologies, such as gene drives, raises a host of moral and philosophical questions about for example dealing with risks, about human control, and about our view of nature. What would be the moral status of such 'animal artefacts'? Of course, biotech-nologies are employed in other areas as well, such as reproductive technologies in livestock, and cloning in equestrian sports. How does the employment of such tech-nologies impact human-animal relationships? Are biotechnological interventions in animal genomes instances of animal enhancement or rather animal disenhancement?

Christopher Preston, author of *The Synthetic Age*, takes another look at the concept of agency, albeit on the level of the genome. Focussing on the technologies of gene sequencing, gene synthesis, and genome editing, that make possible practices such as gene drives and de-extinction, Preston critically examines the promises that such technologies hold in store for us. He argues that 'speculative ethics' around these technologies overlook their problems, caused by reductive and non-relational thinking and by a neglect of non-human agency. Jennifer Welchmann takes a closer look at the practice of de-extinction, illustrated by the case of the Heath Hen. Can it be coherently argued that we owe it to species we have driven to extinction to bring them back? Adam Shriver revisits a number of famous thought experiments around genetic modification of animals, such as the football bird, blind chickens, and painless pigs, in order to critique the way in which the concepts of enhancement and disenhancement have been demarcated. He defends a welfarist definition of disenhancement, which implies that some cases that are often regarded as disenhancement, such as painless livestock, should instead be regarded as enhancement. In her chapter, Cor van der Weele comes back to the opposition between techno optimists, such as ecomodernists, and techno sceptics, or in the words of an influential book by Charles Mann: the wizards and the prophets. Illustrating her argument with the case of cultured meat, she aims to break through this opposition and to find constructive ways of dealing with dualisms and ambivalence. Ecomodernism is also the red thread of Henk van den Belt's comment. Each of the technological practices discussed in this section forms an instance of rewriting the biosphere and engages one way or another with the techno-optimism of ecomodernists.

References

Bekoff, M., and J. Pierce. 2009. *Wild justice: The moral lives of animals*. Chicago: University of Chicago Press.

Bovenkerk, B. 2016. Animal captivity: Justifications for animal captivity in the context of domestication. In *Animal ethics in the age of humans. Blurring boundaries in human-animal relationships*, ed. B. Bovenkerk and J. Keulartz, 151–171. Dordrecht: Springer.

Bovenkerk, B., and J. Keulartz (eds.). 2016. *Animal ethics in the age of humans: Blurring boundaries in human-animal relationships*. Dordrecht: Springer.

Bovenkerk, B., and H. Nijland. 2017. The pedigree dog breeding debate in ethics and practice: Beyond welfare arguments. *Journal of Agricultural and Environmental Ethics* 30 (3): 387–412.

Brosnan, S.F., and F. de Waal. 2012. Fairness in animals: Where to from here? *Social Justice Research* 25 (3): 336–351.

Ceballos, G., P. Ehrlich, and R. Dirzo. 2017. Biological annihilation via the ongoing sixth mass extinction signaled by vertebrate population losses and declines. Proceedings of the National Academy of Sciences (Early Edition).

De Waal, F. 2016. *Are we smart enough to know how smart animals are?*. New York: W.W. Norton & Company.

Dirzo, R., H. Young, M. Galetti, G. Ceballos, N. Isaac, and B. Collen. 2014. Defaunation in the Anthropocene. *Science* 345 (6195), 401–406.

Donaldson, S., and W. Kymlicka. 2011. Zoopolis: A political theory of animal rights. Oxford: Oxford University Press.

Donaldson, S. and W. Kymlicka. 2016. Comment: Between wild and domesticated: Rethinking categories and boundaries in response to animal agency. In *Animal Ethics in the age of humans: Blurring boundaries in human-animal relationships*, eds. B. Bovenkerk and J. Keulartz, 225–239. Dordrecht: Springer.

Gruen, L. 2011. *Ethics and animals: An introduction*. Cambridge: Cambridge University Press.

Irvine, L. 2004. *If you tame me: Understanding our connection with animals*. Philadelphia: Temple University Press.

Keulartz, J., and B. Bovenkerk. 2016. Changing relationships with non-human animals in the Anthropocene—An introduction. In *Animal Ethics in the Age of Humans*, ed. B. Bovenkerk and J. Keulartz, 151–171. Dordrecht: Springer.

Knol, W. 2015. Verwilderde huiskatten: effecten op de natuur in Nederland. Koninklijke Nederlandse Jagersvereniging, Amersfoort. Intern rapport nummer 15-01.

Larson, B.M.H., and S. Barr 2016. The flights of the monarch butterfly: Between. In *Situ and Ex Situ Conservation. In Animal Ethics in the Age of Humans*, ed. B. Bovenkerk and J. Keulartz, 355–368. Dordrecht: Springer.

Leavens, D.A. 2007. Animal cognition: Multimodal tactics of Orangutan communication. *Current Biology* 17 (17): 762–764.

Leuven, R. 2017. *Over grenzen van soorten. Inaugurale rede*. Nijmegen: Radboud Universiteit.

Linnell, J.D.C., and B. Cretois. 2018. Research for AGRI committee—The revival of wolves and other large predators and its impact on farmers and their livelihood in rural regions of Europe. European Parliament, Policy Department for Structural and Cohesion Policies, Brussels.

Lurz, R.W. (ed.). 2009. *The philosophy of animal minds*. Cambridge: Cambridge University Press.

Manfredo, M.J., E.G. Urquiza-Haas, A.W. Don Carlos, J.T. Bruskotter, and A.M. Dietsch. 2019. How anthropomorphism is changing the social context of modern wildlife Conservation. *Biological Conservation*. https://doi.org/10.1016/j.biocon.2019.108297.

Marra, P., and C. Santella. 2016. *Cat wars: The devastating consequences of a cuddly killer*. Princeton: Princeton University Press.

Meijer, E. 2019. *When animals speak: Toward an interspecies democracy*. New York: New York University Press.

Oreshkova, N., R.J. Molenaar, S. Vreman, F. Harders, B.B. Oude Munnink, R.W. Hakze-van der Honing, N. Gerhards et al. 2020. SARS-CoV-2 infection in farmed minks, the Netherlands, April and May 2020. *Eurosurveillance* 25 (23).

Regan, T. 1983. *The case for animal rights*. Berkeley: University of California Press.

Rolston III, H. 2012. *A new environmental ethics: The next millennium for life on earth*. New York: Routledge.

Sandom, C., S. Faurby, B. Sandel, and J.-C. Svenning. 2014. Global late quaternary megafauna extinctions linked to humans, not climate change. *Proceedings of the Royal Society B* 281: 20133254.

Schama, S. 1995. *Landscape and memory*. London: HarperCollins Publishers.

Singer, P. 1975. *Animal liberation: A new ethics for our treatment of animals*. New York: New York Review.

Špinka, M., and F. Wemelsfelder. 2011. Environmental challenge and animal agency. In *Animal Welfare*, ed. M.C. Appleby, I.A.S. Olsson, and F. Galindo, 27–44. Wallingford, UK: CAB International.

Svenning, J.-C., P. Pedersen, C.J. Donlan, R. Ejrnaes, S. Faurby, M. Galetti, et al. 2016. Science for a wilder Anthropocene: Synthesis and future direction for rewilding research. *Proceedings of the National Academy of Sciences* 113 (4): 898–906.

Terborgh, J., R.D. Holt, and J.A. Estes. 2010. Trophic cascades: What they are, how they work, and why they matter. In *Trophic cascades: Predators, prey, and the changing dynamics of nature*, ed. J. Terborgh and J.A. Estes, 1–18. Washington, DC: Island Press.

Tønnessen, M., K. Armstrong Oma, and S. Rattasepp (eds.). 2016. *Thinking about animals in the age of the Anthropocene*. Lanham: Lexington Books.

Torres, A., J.A.G. Jaeger, and J.C. Alonso. 2016. Assessing largescale wildlife responses to human infrastructure development. *Proceedings of the National Academy of Sciences* 113 (30): 8472–8477.

Trouwborst, A., and H. Somsen. 2019. Domestic cats (Felis catus) and European nature conservation law—Applying the EU Birds and Habitats Directives to a significant but neglected threat to wildlife. *Journal of Environmental Law*, 1–25.

Wasserman, E.A., and T.R. Zentall. 2012. *Comparative cognition: Experimental explorations of animal intelligence*. Oxford: Oxford University Press.

Wolverton, S. 2010. The North American Pleistocene overkill hypothesis and the re-wilding debate. *Diversity and Distribution* 16: 874–876.

Websites *(All Retrieved April 8, 2020)*

https://www.nrc.nl/nieuws/2020/02/07/het-spoor-van-corona-leidt-naar-een-gestreste-vleermuis-a3989708.

https://www.theguardian.com/world/2020/apr/06/bronx-zoo-tiger-tests-positive-for-coronavirus.

https://www.ad.nl/binnenland/viroloog-nederland-is-vol-met-gastheren-die-een-virus-over-kun nen-dragen-br~afbc59f5/.

https://aeon.co/essays/we-cant-stand-by-as-animals-suffer-and-die-in-their-billions.

https://www.vox.com/world/2020/1/8/21051756/australia-fires-climate-change-coal-politics.

https://newclimate.org/wp-content/uploads/2019/12/CCPI-2020-Results_Web_Version.pdf.

https://wildlife.org/survey-shows-u-s-citizens-increasingly-humanize-animals/?fbclid=IwAR30 Dxvy7q-B5Q_cxGrh9j-1eINgQWML4hLO9DjYKl7e3YwYui6H40qP69E.

https://www.homesluxury.net/goose-kept-pecking-cop-until-he-decided-to-follow-her-she-lead-him-to-her-trapped-baby/?fbclid=IwAR0ofz3B7OINuAezBAhzQKIe3sRXx3Zjran513N08rP I57Pvydxx4g31f-0.

https://reviverestore.org/projects/woolly-mammoth/.

https://cdn.i-pulse.nl/autoriteitdiergeneesmiddelen/userfiles/sda%20jaarrapporten%20ab-gebruik/ AB-rapport%202018/sda-rapportage-2018-def-err.pdf.

https://www.ema.europa.eu/en/veterinary-regulatory/overview/antimicrobial-resistance/european-surveillance-veterinary-antimicrobial-consumption-esvac.

Jozef Keulartz is Emeritus Professor of Environmental Philosophy at the Radboud University Nijmegen, and senior researcher Applied Philosophy at Wageningen University and Research Centre. He has published extensively in different areas of science and technology studies, social and political philosophy, bioethics, environmental ethics and nature policy. His books include Die verkehrte Welt des Jürgen Habermas [The Topsy-Turvy World of Jürgen Habermas, Junius, 1995] and Struggle for Nature—A Critique of Radical Ecology (Routledge, 1998). He is coeditor of Pragmatist Ethics for a Technological Culture (Kluwer, 2002), Legitimacy in European Nature Conservation Policy (Springer, 2008), New Visions of Nature (Springer, 2009), Environmental Aesthetics. Crossing Divides and Breaking Ground (Fordham University Press, 2014), Old World and New World Perspectives in Environmental Philosophy (Springer, 2014), and Animal Ethics in the Age of Humans (Springer, 2016).

Bernice Bovenkerk is associate professor of philosophy at Wageningen University, the Netherlands. Her research and teaching deals with issues in animal and environmental ethics, the ethics of climate change, and political philosophy. Current topics are animal agency, the moral status of animals and other natural entities, with a particular focus on fish and insects, the ethics of animal domestication, animal (dis)enhancement, and deliberative democracy. In 2016 she co-edited (together with Jozef Keulartz) Animal Ethics in the Age of Humans. Blurring boundaries in human-animal relationships (Springer) and in 2012 she published The Biotechnology Debate. Democracy in the face of intractable disagreement (Springer). She received her PhD title in political science at Melbourne University and her Master's title in environmental philosophy at the University of Amsterdam. Her homepage is https://bernicebovenkerk.com.

Chapter 2
Animal Conservation in the Twenty-First Century

Hugh A. H. Jansman

Abstract Biodiversity on Earth is rapidly decreasing and the situation in the Netherlands is in that perspective a textbook example. The main causes for species extinction are habitat loss, landscape degradation and overuse. Conservation efforts should focus more on the level of viable ecosystems. A strategic plan to do so is called Cores, Corridors and Carnivores (rewilding's three C's). This requires strong Cores of nature, mutually connected via robust Corridors. Based on island biogeography theory it can be calculated that if we want to conserve roughly 85% of the current biodiversity, 50% of the Earth's surface needs to be protected, 'Nature needs half'. For healthy ecosystems we need to get top-down forcing by apex consumers back in ecosystems. These apex consumers are mainly large Carnivores, and bringing them back asks for coexistence. If we want to keep our living conditions on planet Earth healthy we have to change our unsustainable way of living and change our way of thinking with respect to nature, natural processes and our relation with other species. The loss of biodiversity can only be halted or reversed if we save more space for nature and natural processes including top-down forcing and last but not least, find a way of coexistence with our fellow creatures.

2.1 Introduction

Conditions for life as we know it are exceptionally favourable on Earth compared to other known planets in the universe. In billions of years, evolution has created a very rich biodiversity. Biodiversity includes biological variation, whether it is at a genetic, species, population or community level, or even at their ecosystem-level interactions (Wilson 1992). Yet, the biodiversity that happens to coexist with us humans being, the dominant life form in the so-called Anthropocene, faces the 6th mass extinction (see Bovenkerk and Keulartz in this Volume). World Wildlife Fund (WWF) reported

H. A. H. Jansman (✉)
Wageningen Environmental Research, Wageningen, The Netherlands
e-mail: Hugh.jansman@wur.nl

© The Author(s) 2021
B. Bovenkerk and J. Keulartz (eds.), *Animals in Our Midst: The Challenges of Co-existing with Animals in the Anthropocene*, The International Library of Environmental, Agricultural and Food Ethics 33,
https://doi.org/10.1007/978-3-030-63523-7_2

that population sizes of wild animals on average have been reduced with 60% since 1970 (WWF living planet report 2018). Main cause is the rapid growth of the human population in the last centuries, in combination with an unsustainable way of living by humans, especially in 'Western' societies. Since human population growth is still continuing and developing countries rapidly adopt Western consumption patterns, the living conditions for many species on planet Earth are gradually decreasing (Intergovernmental Science-Policy Platform on Biodiversity and Ecosystem Services; IPBES report 2019). Humans dominate the global ecosystem in three ways: by land use, the nitrogen cycle and the atmospheric carbon cycle (Primack and Sher 2016). Firstly, human land use, mainly for agriculture, and our need for resources, especially forest products, have transformed as much as half of the Earth's ice-free land surface from natural to cultural lands. Regionally this can be more than 90%. Secondly, each year human activities release more nitrogen into terrestrial systems than natural biological and physical processes, for instance by cultivating nitrogen-fixing crops, using nitrogen fertilizers and burning fossil fuels. And thirdly, human use of fossil fuels and the unsustainable cutting down of forests will result in a significant increase of the concentration of carbon dioxide in the Earth's atmosphere. Scientists have determined ten planetary boundaries that should not to be exceeded if we want to keep the living conditions on earth favourable for us and many other species. Three of those boundaries are already exceeded: biodiversity loss, climate change and the nitrogen cycle (Fig. 2.1). It is no surprise that within ecosystems those boundaries are all

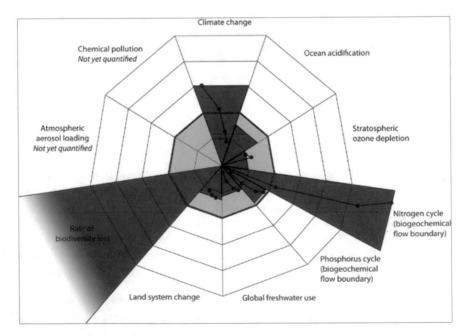

Fig. 2.1 Estimate of quantitative evolution of control variables for seven planetary boundaries from pre-industrial levels to the present (from Rockström et al. 2009)

interconnected. They are different faces of the same central challenge: the increasingly dangerous impact of our choices on the health of our natural environment.

2.2 Viable Populations

The main causes for species extinction are habitat loss (destruction), habitat degradation (e.g. by pollution, fragmentation or invasive species) and overuse (unsustainable hunting, fishing, logging etc.). Due to the destruction of large parts of their habitat, many populations of wildlife have decreased in size. Besides the demographic risk of being more prone to extinction by occasional drops in numbers due to dramatic events (e.g. disease or wildfire), such small populations will also gradually loose genetic variation. Reduced mating choice, and therefore a higher risk of inbreeding, will further reduce diversity and potentially result in reduced viability and/or reproductive capacity (i.e. inbreeding depression). Furthermore, the loss of genetic variation limits a population's adaptability to change, while gradually moving along with changing climate zones is for many species impossible due to barriers in the landscape. Not to mention that climate change is currently going much faster than the speed in which most species can change their distribution area. Finally, barriers for dispersal between fragmented habitat patches also limit the natural restoration of local diversity by (re)immigration (Frankham et al. 2010).

At the end of the 1970s, the rapid increase of species extinctions gave rise to a new field of science: conservation biology. This young discipline deals with the management of nature and of earth's biodiversity with the aim of protecting species, their habitats and ecosystems, from excessive rates of extinction and the erosion of biotic interactions. It is an interdisciplinary research area drawing not only on natural but on social sciences as well, and also on the practice of natural resource management.

Conservation biologists conduct monitoring programmes to evaluate the status of populations and ecosystems. They label species that have significantly been reduced in number and/or distribution area as threatened. Depending on specific criteria, these species are listed on a conservation priority list, the Red List. This list of threatened species was established by the International Union for Conservation of Nature (IUCN 1964) and has evolved to become the world's most comprehensive information source on the global conservation status of species. What followed were international agreements for biodiversity conservation, such as the Convention on International Trade in Endangered Species (CITES 1973), the Bern Convention (1982), and the Convention on Biological Diversity (Earth summit; Rio de Janeiro, 1992).

These agreements strive to protect the most endangered species. The populations of endangered species are frequently divided in small subpopulations due to habitat fragmentation. As a result conservation is in most cases focussed on important subpopulations and not the whole population. If conservation of a threatened species is intensified, a four step approach of restoration is launched in an attempt to get a red listed species viable again. Step 1 is to secure the area in which a threatened (sub)population is living; its distribution area. Step 2 is to find out which specific factors are negatively influencing the population, and mitigate them. Step 3 is to enlarge the current distribution area with additional suitable habitat for the species. Step 4 is the connection of the isolated subpopulation with a corridor to another subpopulation, allowing for natural dispersal and gene flow. To some extend this 4-step approach is adopted in (inter)national nature policies. Europe's Natura 2000 directive, as implemented in its member states, focuses on the protection of remaining habitat and strives to reconnect them via e.g. fauna passages and corridors.

2.3 Sufficiently Large Numbers and the Amount of Area They Require

One of the biggest questions in conservation biology is which qualities populations must have in order to be able to survive in the long term. How many individuals are needed for a population to reduce the risk of extinction to a bare minimum? And furthermore, what size of habitat is required to sustain such a population?

The term *Minimum Viable Population* Size (MVP) was first introduced by Shaffer (1987) and defined by him as "the smallest isolated population having a 99% chance of surviving for 1000 years despite the foreseeable effects of demographic, environmental, and genetic stochasticity, and natural catastrophes". Variation in the size of a population depends on those factors. They all may have a temporary or permanent negative impact on the population size. Chance events play a strong role here (Shaffer 1987). A natural catastrophe may lead to abnormally high mortality rates, climate conditions may fluctuate and genetic variation may be lost as a result of chance effects in the presence of particular gene variants (genetic drift; Nei 2005). In addition, negative demographic, environmental and genetic influences may produce a synergistic effect which in extreme cases may result in ever increasing contraction: the extinction spiral (Blomqvist et al. 2010). A viable population must therefore be sufficiently large to avoid finding itself in such a spiral as a result of chance events. Unsustainable use, invasive species, pollution or bad luck are otherwise easy executioners.

Ideally, genetic-, demographic- and environmental factors will be taken into account in an estimate of the MVP, through what is known as a *Population Viability Analysis* (PVA). As part of the analysis, the likelihood of a population becoming extinct within a certain number of years is calculated on the basis of context-specific assumptions, like the mating system of the species, sex-ratio in the local population and population dynamics. Since a greater number of potential risks are taken into consideration, MVP estimates based all factors usually result in higher numbers than estimates based on genetic risks alone (Ottburg and van Swaay 2014). Traill et al. (2007) compared as many published MVP estimates as possible from the previous 30 years, based both on PVA analyses and on population-genetic models, and found major differences between species and also between populations of the same species. They therefore concluded that context is of overriding importance in practice. Nevertheless they provided average values for each species group. For mammals the safe threshold for a minimum population size was set at ~2.900 individuals; for birds, reptiles and amphibians, and fish the threshold was set at respectively ~3.300, ~4.000 and ~500.000 individuals. With the estimated safe threshold of ~2.900 individuals of a mammal species one can imagine that huge areas are needed to provide sufficient habitat for these species.

Animals need sufficient food and shelter in their habitat, so densities of species depend on the quality of an area. Habitat with poor soil conditions, harsh climate conditions and little cover carry lower densities of species then rich habitats. The threshold numbers mentioned above are relatively easily met for small rodents that need small areas, but for populations of for example deer there are not many areas in Europe large enough to sustain such a large population, not to mention viable populations of carnivores like bears and wolves. This explains why so many species have difficulty surviving, specifically the ones requiring large areas.

So what is the relation between the size of a nature reserve and biodiversity? MacArthur and Wilson (1967) studied the distribution of biodiversity on islands. What they found is that the larger an island, the richer the biodiversity. They developed a formula for the species-area relationship, the so-called island biogeography model. It predicts that islands of 10, 100, 1000 and 10,000 km^2 in size would have 2, 3, 6, and 10 species respectively. Each tenfold increase of the size of an area increases the number of species by a factor of approximately 2. But the opposite is true as well. Reduce the area of an ecosystem to one tenth and you lose roughly half of your biodiversity. Since humans transformed huge areas in a way that ecosystems are highly degraded and fragmented, one can speak of islands of nature in a sea of human dominated landscapes. Therefore the island biogeography model can to a large extent be applied to nature areas on the mainland. However, the extent to which the suboptimal landscape surrounding a patch of key habitat is in fact still used by a species is not always exactly known, and may be underestimated.

What is also important is the level of population fragmentation. If the distribution area of a population is fragmented, we talk about multiple subpopulations. A subpopulation can be isolated, meaning there is no dispersal to surrounding subpopulations, or it can be connected via corridors allowing for exchange of individuals. This exchange is important for survival since it counters stochastic effects in subpopulations and prevents genetic degradation. A cluster of subpopulations with mutual exchange we call a 'metapopulation'. It is clear that many of today's fragmented nature conservation areas provide inadequate resources for self-sustaining populations of thousands of individuals. The solution lies in preventing or ending isolation: creating corridors between nature areas. The above mentioned definitions of MVP's are all based on a self-sustaining, isolated population (Shaffer 1987; Franklin 1980). However, when several populations are combined to form a larger population or metapopulation in which regular dispersal takes place, the variation lost in a subpopulation may be restored by immigration from another subpopulation (Frankham et al. 2010). A criterion of one migrant per generation is often applied to avert the negative consequences of inbreeding and genetic drift (Mills and Allendorf 1996). In short, where there is a regional metapopulation and each subpopulation receives a migrant which contributes to reproduction at least once a generation, the aforementioned genetic guidelines for an MVP will apply to that regional metapopulation as a whole (Mergeay 2012).

The way we manage wildlife can have its effect on the viability of a population and the integrity of a species as well. For instance management of ungulates is mostly done by a random cull of a large proportion of the population each year. In the Netherlands roughly 50% of the red deer and 75% of the wild boar population is randomly shot each year in order to reduce the conflict with human interests like traffic mortality and crop damage (Faunabeheereenheid Gelderland 2019). We can only guess at the consequences of reducing such large numbers for population structure, vitality, genetic variability, adaptation and behaviour. In those heavily managed populations almost all females participate in reproduction. Whereas in an unmanaged population, only the best animals reproduce due to mutual competition for resources. The mechanism of evolution is based on the principle that within a population, individuals have different characteristics. Some of those characteristics are inheritable and some of those characteristics might result in better survival and/or reproductivity. This results in selection and adaptation, survival of the fittest. Recently this process was illustrated in a wild red deer population on the isle of Rhum, Scotland. In this wild population the average parturition date has advanced by nearly 2 weeks in 4 decades in a response to climate change (Bonnet et al. 2019). Is this driving force of life still possible in populations that are predominantly managed by us?

And what about management of populations by 'removing' the individuals that cause trouble? For instance, a bear that learns to associate humans with food might start to eat from trash cans, or feral horses that are being fed by tourist might become pushy. These individuals are often removed from the area, because they showcase behaviour that is unwanted by the public or managers. By doing so we probably select for characteristics that we humans prefer, resulting in a kind of taming or domestication of wild species and therefore interfere with the process of natural selection (Donaldson and Kymlicka 2016).

In my opinion this is where we stand: scientists have a fairly accurate estimation of how many individuals a viable populations should contain and we can estimate how large suitable areas should be to hold those populations. But for more and more species that is hard to achieve, if human demands for land and resources are not reduced. This results not only in dwindling species, but in an increase of conflict potential between nature and humans, since we penetrate more and more into the last remaining nature areas (see the chapters of Drenthen and Tokarski in this Volume).

2.4 Challenges

The focus of conservation is relatively more on individual threatened species rather than on healthy ecosystems, partially due to international agreements. Measures taken for one species can be detrimental for another. As a result species conservation becomes kind of similar to gardening. Per nature reserve we pick a few target species to conserve or, in fact, manage. While even if those target species are carefully selected to represent key functions or habitat needs, this undervalues a system's complexity. Some species, like ungulates and so called pest animals, are managed by culling in order to control their numbers and therefore avoid conflicts with human interests. Disease transmission, naturally occurring in wildlife and potentially spilling over to humans and our livestock (specifically zoonoses like Covid-19) is a topic that gets more and more attention. These management decisions are predominately taken from the perspective of human interest and less so in the interest of nature.

Altogether, while awareness of the need for biodiversity conservation is on the rise, realizing it in practice is very difficult (IPBES 2019), even more so since the pressure of humans and human activities on planet Earth is still increasing. There seems to be a constant and growing conflict between humans and wildlife combined with less than optimal species conservation since the needs of viable biodiversity are not met. As a result extinctions are ongoing. A similar example are efforts to mitigate climate change: while this is a topic that most people are nowadays well aware of, the political and societal will to take preventive measures is meeting resistance, as such measures may directly impact our current life style.

2.5   Trophic Downgrading: "When the Cat Is Away, the Mice Will Play"

Up to now I've mainly discussed the conservation of species. But more and more scientists are becoming aware how important interactions are between organisms in an ecosystem. Erosion of ecosystems rapidly continues to this date, especially due to nature policy often ignoring the fact that ecosystems consist of complex interactions between species. When a species becomes extinct a much more insidious kind of extinction occurs as well: the extinction of ecological interactions (Estes et al. 2011). If the link between all species in the system is weakened, or even gone if species became extinct, resilience of the entire system is affected. This might for instance lead to an overabundance of deer if predators like wolves are absent, or exotic species easily becoming invasive in eroded ecosystems. On a broader scale the reduction in megafauna on earth has severely constrained the flow of nutrients across continents and between the oceans, freshwaters and land (Jepson and Blythe 2020)

Estes et al. (2011) states that one of humankind's most pervasive influences on nature is probably the eradication of species at the top of the food chain. These so-called apex consumers were ubiquitous across the globe for millions of years. Apex consumers are mainly large carnivores, but can be megaherbivores as well, like elephants and rhinos whose adults are largely immune to predation. Recently scientists have become aware how extensive the cascading effects of their disappearance are in marine, terrestrial and freshwater ecosystems worldwide. Miller et al. (2001) explain the importance of large carnivores for healthy ecosystems. The absence of top-down forcing in ecosystems by apex consumers is called trophic downgrading. Ecosystems may be shaped by apex consumers, their impacts spreading downwards through the food webs (Estes et al. 2011; Keulartz 2018). An example is the influence of apex consumers like wolves in supressing herbivory. Regarding biodiversity Estes et al. (2011) mention the fact that most protected nature areas don't function as intended due to the absence of large apex consumers. This may result in species from lower trophic levels spinning out of control, although our current understanding is too limited to predict such effects in detail. As a result, our society may be confronted with ecological surprises, such as pandemics, population collapses of valued species, population eruption of species we dislike, shifts in ecosystem state and loss of ecosystem services. According to Estes et al. (2011) top-down forcing must be included in conceptual overviews if there is to be any real hope of understanding and managing the workings of nature.

2.6 Conservation in Twenty-First Century: 'Cores, Corridors and Carnivores' Meets 'Nature Needs Half'

If we want to conserve our biodiversity we should focus on robust and complete ecosystems, including the presence of large apex consumers. We should change the conservation focus from mainly species oriented management to self-supporting sustainable ecosystems. A strategic plan to do so is called Cores, Corridors and Carnivores (rewilding's three C's; Soulé and Noss 1998). For sustainable conservation, ecosystems require large units of nature (Cores), mutually well connected (Corridors) and the presence of Carnivores for their top-down forcing as apex consumer. For the Netherlands a similar concept was already invented as the three E's of nature development: Ecological core areas, Ecological corridors, and Ecological networks (Baerselman and Vera 1989). For many nature reserves this means that their size should increase, robust corridors should be created allowing for sufficient dispersal potential and gene flow, and apex consumers are returned. This approach is named restoration ecology or rewilding, which overlap. Restoration ecology is the practise of restoring the species, landscapes and ecosystems that occupied a site at some point in the past, but were damaged or destroyed. It normally follows the four step approach mentioned earlier, but frequently with the addition of reintroducing original species as well (www.ser.org). Rewilding, or trophic rewilding, aims at maintaining or even increasing biodiversity through the restoration of ecological and evolutionary processes using extant keystone species or ecological replacements of extinct keystone species that drive these processes (Svenning 2016; Keulartz 2018). Whereas restoration has typically focused on the recovery of plant communities, rewilding often involves animals, particularly large carnivores and large herbivores. Whereas restoration aims to return an ecosystem back to some historical condition, rewilding is forward-looking rather than backward-looking: it examines the past not so much to recreate it, but to learn from the past how to activate and maintain the natural processes that are crucial for biodiversity conservation (Keulartz 2018; Jepson and Blythe 2020). Restoration ecology and rewilding both use reintroductions in their conservation approach (Box 2.1).

> **Box 2.1 Examples of Reintroductions**
>
> (1) *Reinforcements*, involving the release of an organism into an existing population of conspecifics to enhance population viability.
> (2) *Reintroductions*, where the intent is to re-establish a population in an area after local extinction, or, more from the rewilding perspective, has the intent to restore ecological and evolutionary processes.
> (3) *Assisted colonization*, the intentional movement of an organism outside its indigenous range to avoid extinction of populations due to current or future threats.
> (4) *Inter situ*-conservation, the so called One Plan approach which was launched in 2012 by the IUCN. This approach stimulates the interactive exchange of animals between in situ populations (in nature) and ex situ populations (in captivity) to increase the viability of the species.
> (5) *Ecological replacement*, (more from the rewilding perspective) the release of an appropriate substitute species to re-establish an ecological function lost through extinction. Examples are back breeding, taxon substitution and de-extinction, all subject to scientific controversy.

It is clear that the realisation of sustainable ecosystems requires huge areas. Co-inventor of the island biogeography model (before mentioned) and one of the founding fathers of nature conservation E.O. Wilson started the half-earth project. His goal: "With science at its core and our transcendent moral obligation to the rest of life at its heart, the Half-Earth Project is working to conserve half the land and sea to safeguard the bulk of biodiversity, including ourselves" (www.half-earthproject.org). According to Wilson (2016) and based on IUCN data, there are now roughly 160.000 nature reserves on land and 65.000 in sea areas, covering 15% of the continents and 2.8% of the oceans. The island biogeography model is still relevant since more and more nature reserves now function as islands due to isolation. Wilson states that a 90% reduction of the size of nature areas is currently the case in many locations all over the world. This size reduction results in only 50% of the current existing species being able to maintain viable in the long term, a reduction of biodiversity in time with another 50% on top of what's already lost. The opposite can be done as well. Wilson calculated that in order to conserve roughly 85% of the current biodiversity, 50% of the earth surface needs to be protected: therefore the program's name: 'nature needs half'. So the aim is to reserve roughly 50% of the earth's surface for nature in order to prevent further loss of biodiversity and sustainable living conditions for biodiversity and us humans as well. The focus is on biodiversity hotspots around the equator, but all ecosystems should be conserved. According to these figures, nature reserves on earth have to be enlarged with roughly 35% and in seas with 47.2%.

2.7 Viable Ecosystems with Red Deer and Wolf in the Netherlands

The Netherlands is a relative small country with a high human density and an even higher livestock density. It is considered to be the second largest exporting country in the world regarding agricultural products (see Bovenkerk and Keulartz in this Volume). As a consequence there are many environmental problems like nitrogen deposition and pesticides. Still there is wildlife left in the Netherlands, although management is quite intensive and many populations suffer from habitat destruction, fragmentation and high traffic mortality. I will discuss two species in detail, red deer (*Cervus elaphus*) and wolf (*Canis lupus*), as examples of what a future desirable arrangement of the Netherlands would have to look like to hold viable populations of wildlife with self-serving ecosystems including top-down forcing by large ungulates (megafauna) and carnivores, and less conflict potential with human interests.

2.7.1 Current Population of Red Deer in the Netherlands

There are two Dutch nature reserves where large populations of red deer are allowed; the Veluwe (circa 1000 km^2; of which 912 km^2 is a Natura 2000-area) and the Oostvaardersplassen (a Natura 2000-area of circa 56 km^2, of which 20 km^2 is used by herbivores for grazing); see Fig. 2.2. Both areas are more or less fenced in, so they are closed populations.

The Veluwe is a relatively poor soil forest–heather ecosystem. Although it appears from a birds perspective to be one large area, it is fragmented due to many fences. Ecoducts have been built to allow dispersal and to stimulate the mixing of the subpopulations of red deer. However, genetic research shows that these ecoducts do not function fully yet. Genetic research shows that the populations do not mix optimally, probably as a result of these (partial) migration barriers (De Groot et al. 2016). The population of red deer is about 2500 individuals (Groot Bruinderink 2016). Management cull is about 50% of the annual population size in order to prevent crop damage and traffic collisions which leads to conflict with human interest. Forestry and forest rejuvenation is another reason to keep the herbivore density low (Den Ouden et al. 2020). Therefore the population density is much lower than the carrying capacity of the area and as a result mutual competition amongst the deer is low, resulting in all hinds having a calf each year.

The Oostvaardersplassen is very rich in minerals with abundant growth of vegetation. The population of red deer was not managed since there was no conflict with

Fig. 2.2 The map of the Netherlands in 2020 (insert) and the vision for the future in the Netherlands in 2120 (Baptist et al. 2019), with some additional corridors added by H. Jansman, illustrating better connection between cores of nature areas: Oostvaardersplassen (OVP), Veluwe (VL) and Utrechtse Heuvelrug (UH)

human interest due to the absence of public infrastructure and agricultural lands in the area. Natural processes were the main driver of the ecosystem. Therefor the Oostvaardersplassen is, together with Yellowstone National Park, seen as one of the most illustrative examples of rewilding (Jepson and Blythe 2020; Flannery 2018). If I refer in this chapter to the Oostvaardersplassen I refer to the period before 2018 in which natural processes were dominating the development in the ecosystem. Since 2018 management has changed from reactive management (only shooting animals that are in a very poor condition and no longer capable of surviving the week) to proactive management (culling of deer in prime condition), similar to the management at the Veluwe. The change in management was based on a management advice by Van Geel et al. (2018). The commission concluded there was a lack of public support due to the high number of starving animals in winter, and conflict with Natura 2000 goals due to overgrazing by the large number of ungulates. As a result of the changed management, that winter more than 1700 red deer were shot to reduce the population size. In November 2019 a court decision stated that the shooting of red deer had to stop. The court ruled that the management was not sufficiently motivated and that the management advice report by Van Geel et al. (2018) was ecologically inadequately substantiated (Schreuder and Bontjes 2019). Until 2018 numbers fluctuated around the carrying capacity, which was roughly between 2.500 and 4.000 individuals and mortality mainly due to starvation was on average ca. 25% per year. This winter mortality depended on competition with other grazing species in the reserve and climate conditions. Reproduction was affected by this competition as well, resulting for instance in not all hinds having a calf each year.

2.7.2 Current Population of Wolf in the Netherlands

In January 2019 the first wolf settled in the Netherlands after an absence of about 150 years. Conflict with humans and human interest had led to its eradication. Due to better protection within the European Union, conservation programs as Natura 2000 and abandonment of rural areas, wildlife, including wolves, are recolonizing former habitat. In 2000 the first pack of wolves was a fact in Germany, close to the Polish border. In 2018 there were approximately 100 packs and pairs of wolves in Germany and the distribution area was nearing the Dutch border (www.nabu.de). Since 2015 already more than 23 wolves have been visiting the Netherlands (www.wageninge nur.nl/wolven), mainly from the Central European population, but 1 from the Alpine population as well. Some of them settled at the Veluwe and in 2019 and 2020 pups were born, forming the first Dutch pack.

Depending on habitat quality and prey density, wolf packs need about 150–400 km^2 for a territory. Currently most wolves in Central Europe find their territories in robust nature areas and less in human dominated agricultural areas. The reason for that is probably the potential conflict between wolves and humans and livestock. Due to long term persecution wolves probably have learned to keep a safe distance to humans. Although wolves are strictly protected within the European Union, illegal poaching is still a common cause of death for wolves (Liberg et al. 2012). If a wolf forms a serious threat to humans or specializes on livestock and frequently kills well protected livestock, dispensation might be given to remove that wolf by killing it (IPO 2019).

2.7.3 Predator-Prey Relation Between Wolf and Red Deer

Large ungulates like deer are the most prominent food item for wolves. Wolves and deer have evolved together which resulted in behavioural and morphological modifications. Although predation of deer by wolves seems at first glance the most dominant impact of wolves on deer, this is not the case. The presence of wolves results in a change in behaviour by deer. Deer can change the group size and avoid certain areas to reduce the risk of predation. This is called the landscape of fear (Van Ginkel et al. 2019; Jepson and Blythe 2020). As a result there is more structure in grazing density which is good for diversity and vegetation growth. Wolves can also influence the number of mesopredators like coyotes or jackals, which might be beneficial for species that are eaten by coyotes or jackals. Altogether carnivores like wolves have a dominant top down regulation impact, which results in more stable and healthy ecosystems (Atkins et al. 2019). This has been well studied in Yellowstone national park, were wolves were introduced since 1995. Before the return of the wolf, deer numbers had increased enormously, resulting in overgrazing of the landscape. After the return of wolves, the deer population was predated on by wolves and as a result deer avoided dangerous areas. This led to a lean and mean deer population. Although ecosystem processes are very complex and many aspects have to be taken into account, like climate change, forest fires, increase of bears and decrease of coyotes, the positive effect of the return of wolves to this ecosystem and its biodiversity seems impressive (Smith et al. 2016). Since wolves are fiercely territorial and claim large areas, overhunting of their prey populations in natural conditions never takes place.

For a single large nature area the MVP for red deer was calculated to be around 4.000 individuals. For subpopulations with sufficient mutual dispersal and gene flow this was 400 individuals (Van der Grift et al. 2018). Thus it can be concluded that even the largest nature areas in the Netherlands doesn't hold a population large enough for long term survival. The genetic diversity was studied as well and found the population in the Oostvaardersplassen to be more diverse than the Veluwe population (De Groot et al. 2016). In deer from the Veluwe, parts of the genome showed hardly any variation, which is a sign of genetic drift or inbreeding (de Jong 2018). This could be the result of both historic management choices like introductions and restocking, but it could also be caused by current management strategies (proactive versus reactive), since these strategies differ largely. At the Oostvaardersplassen random mating is much easier, due to the absence of barriers in the reserve. With not all females having a calf each year, it is likely that only the most fit animals participate in reproduction which is a strong evolutionary driver for selection and adaptation. At the Veluwe there is still some level of habitat fragmentation. Also, by randomly culling approximately 50% of the population each year, it is questionable if random mating is still possible. Fact is that the population is kept much lower than the carrying capacity, so there is hardly any mutual competition for resources. As a result, all hinds participate in reproduction so there is no clear selection on fitness from that perspective. Therefore it is questionable if adaptation to for instance climate change, as recently shown in the red deer population on Rhum island, is possible in intensely managed populations like the one at the Veluwe.

Stokland (2016) mentions that a MVP for wolves should be 800 individuals in a closed population or 200 in a subpopulation with mutual exchange of individuals. As a small country, it's not likely that the Netherlands will have the capacity to hold 800 wolves. Even 200 wolves is a challenge. Wolf populations are a good example of a species that needs large areas and therefore are expected to cross borders. The Dutch wolves will always be part of the Central European population and they rely on dispersal for the long term viability. Compared to deer, wolves are more agile and a simple fence does not easily stop their migration. Wolves might include human cultivated areas in their territories. So it is less easy to avoid human-wolf conflict than it is to do so for human-deer conflict which can be averted with fences.

2.8   The Netherlands in 2120

The solution for viable ecosystems in the Netherlands and vital populations of red deer and wolf is the Core, Corridor and Carnivore approach in combination with more room for nature. Currently about 13% of the Dutch territory is protected as nature, more than half of which consists of large waterbodies like IJsselmeer and Markermeer. If the Netherlands wants to meet the Aichi biodiversity targets (2010) then it should protect 17% of its land area and 10% of its water area as nature reserves before 2020. Technically this means a doubling of the current size of terrestrial nature areas. If the Dutch landscape is rearranged in a smart way, then it is possible to enlarge the current nature reserves, forming more robust cores of nature. Next, those cores need to be connected via corridors, not only nationally but internationally as well, allowing for transboundary migration of species. If around these cores and corridors buffer zones are created which are extensively managed, for instance nature-inclusive agriculture, or forest for the use of CO_2 buffering or wood production, then conflicts between nature and human interest are less prone in comparison to intensive agriculture situated next to nature reserves. Certain species, such as meadow birds, might even benefit from an extensive level of management like nature-inclusive agriculture. If recreational activities and game management are concentrated in these buffer zones rather than in nature reserves, animals will be much more disturbed by humans in the buffer zone and therefore perceive this zone as scary and probably avoid it more. By doing so, recreation and hunting mimics predator behaviour, resulting in a landscape of fear. The areas with industry and intensive agricultural management as factory farming should best be positioned in areas with less biodiversity value and not neighbouring nature reserves. Finally there should be a good system to provide preventive measures to avoid conflict with wildlife. If that is not sufficient, there should be funding for unforeseen damage by wildlife. This is of importance if we want to coexist with (large) animals that due to their long distance travel potential might show up in areas with intensive human use (Bekoff 2014). In the cores natural processes will be the dominant driver. On the edges with intensive human use, mitigation and management focussed on conflict avoidance will be more dominant.

A vision of how the Netherlands could look like in the future if we allow more room for nature and natural processes was recently created by Baptist et al. (2019; Fig. 2.2). According to the authors, this map illustrates a version of the Netherlands in 2120. The vision is based on a number of criteria: for example, it had to deliver an optimal outcome for the biodiversity, because only then can the country fundamentally thrive. And they had to work as much as possible with solutions in which there is a big role for natural processes. The result is a map of what is possible, i.e. feasible and realistic when future choices on the use and lay-out of the Netherlands are based on understanding natural systems and processes. In order to better connect three major nature areas in the centre of The Netherland, I added two corridors to this map. A corridor connecting the Oostvaardersplassen with the Veluwe and one connecting the Utrechtse Heuvelrug with the Veluwe.

This approach allows for healthy populations of red deer, due to more space and better (seasonal-) migration between cores. Furthermore, their numbers do not need to be managed dominantly by management culling, allowing for more natural processes in the population. Wolves will be able to easily move nationally and internationally via the corridors, allowing for sufficient dispersal and gene flow in their population. The top-down forcing effect of wolves in the nature reserves allows for more stable ecosystems.

2.9 Change

In order to achieve this vision, we really need to change. Change our unsustainable way of living and change our way of thinking with respect to nature, natural processes and our position in relation to other species. In my opinion we humans are not superior, just different from other species. Human-wildlife conflict is in fact most of the time a conflict between opposing human values: what do you consider nature? What is the position of humans in relation to nature? Etc. We need a value-reorientation. Western societies have alienated from what nature is, natural processes (for instance seasons of food scarcity, mortality, only the fittest individuals participating in reproduction), and the feeling that we are part of nature and therefore depend on a stable ecosystem on planet earth. Rewilding not only nature, but our minds as well, is in my opinion a necessity. Albert Einstein already said: "We cannot solve our problems with the same thinking we used when we created them". In my view, our problems, addressed in the first paragraph, are great and therefore there is an urgency for sustainable leadership. We know what is good for us and for biodiversity. But the human mind never needed to evolve in dealing with these challenges, since during most of the history of our species, we were only a minor player in the ecosystem. Nowadays, however, the human population growth curve of the last centuries shows an exponential growth and therefore corresponds very well with that of a plague species. Furthermore, our footprint is still increasing. Currently there is a large imbalance between how fast we consume resources and generate waste, and how fast nature can absorb our waste and generate new resources. The food system is also a major problem. The cost of ecological degradation is not considered in the price we pay for food, yet we are still subsidizing unsustainable fisheries and agriculture. From an ecological point of view, the key solution is managing our human population number and our livestock numbers. But that is quite a taboo topic and difficult to achieve in the short term. We at least need to adapt to a sustainable way of living and co-exist. How can we live in harmony with our fellow species on planet earth?

2.10   Further Reading

In this chapter I have presented many topics and addressed them briefly. In this book, some of these topics are discussed in more detail. Firstly, regarding saving more space for nature one can think of many options like land sparing (for instance by factory farming) versus land sharing (for instance by nature-inclusive agriculture); see the chapter by Hidde Boersma. Another interesting take on this issue regards the switchover from large scale livestock farming and meat consumption to cultured meat as described in the chapter by Cor van der Weele. Secondly, with regard to bringing back top-down forcing in ecosystems, one of the more controversial options is ecological replacement like back breeding, taxon substitution and de-extinction. Christopher Preston in his chapter discusses the speculative ethics' that has arisen around these technologies as gene reading, gene synthesis, and gene editing. Further it is often argued that we "owe it" to species driven to extinction "to bring them back." Jennifer Welchmann discusses whether justice can really require us to make restitution for anthropogenic extinctions. Thirdly, coexistence, in particularly with large carnivores like wolves can be a challenge. The chapter by Martin Drenthen discusses the dualistic idea that culture and nature are two strictly separated realms of reality, and how to learn and negotiate that the landscape as a space that is interpreted and inhabited by many different beings with whom we are always already communicating, even if we are not always aware of it. Mateusz Tokarski explains that environmental philosophy can provide conceptual tools easing the difficulties of cohabitation. He presents practical remarks regarding how environmental consolation could be practiced today in the context of difficult cohabitation with wildlife.

References

Atkins, J.L., R.A. Long, J. Pansu, J.H. Daskin, A.B. Potter, M.E. Stalmans, C.E. Tarnita, and R.M. Pringle. 2019. Cascading impacts of large-carnivore extirpation in an African ecosystem. *Science* 364: 173–177.

Baerselman, F., and F.W.M. Vera. 1989. *Natuurontwikkeling. Een verkennende studie* (Achtergrondreeks Natuurbeleidsplan; No. 6). 's-Gravenhage: S.D.U. Uitgeverij.

Baptist, M., T. van Hattum, S. Reinhard, M. van Buuren, B. de Rooij, X. Hu, S. van Rooij, N. Polman, S. van den Burg, G. Piet, T. Ysebaert, B. Walles, J. Veraart, W. Wamelink, B. Bregman, B. Bos, and T. Selnes. 2019. A nature-based future for the Netherlands in 2120. WUR report. https://doi.org/10.18174/512277. Accessed 23 March 2020.

Bekoff, M. 2014. *Rewilding our hearts: Building pathways of compassion and coexistence.* Novato: New World Library.

Blomqvist, D., A. Pauliny, M. Larsson, and L.A. Flodin. 2010. Trapped in the extinction vortex? Strong genetic effects in a declining vertebrate population. *BMC Evolutionary Biology* 10: 33.

Bonnet, T., M.B. Morrissey, A. Morris, S. Morris, T.H. Clutton-Brock, J.M. Pemberton, and L.E.B. Kruuk. 2019. The role of selection and evolution in changing parturition date in a red deer population. *PLOS Biology.* https://doi.org/10.1371/journal.pbio.3000493. Accessed 23 March 2020.

De Groot, G.A., G.J. Spek, J. Bovenschen, I. Laros, T. van Meel and H.A.H. Jansman. 2016. Herkomst en migratie van Nederlandse edelherten en wilde zwijnen: een basiskaart van de genetische patronen in Nederland en omgeving. Wageningen: Alterra, Wageningen-UR (Alterra-rapport 2724).

De Jong, J.F. 2018. *Genetic variation of wildlife in a human-dominated landscape: Genome-wide SNP analysis of wild boar (Sus scrofa) en red deer (Cervus elaphus) from the European continent.* Dissertation, Wageningen University, Wageningen.

Den Ouden, J., D.R. Lammertsma, and H.A.H. Jansman. 2020. *Effecten van hoefdieren op Natura 2000-boshabitattypen op de Veluwe.* Rapport 3031 Wageningen University and Wageningen Environmental Research.

Donaldson, S., and W. Kymlicka. 2016. Comment: Between wild and domesticated: Rethinking categories and boundaries in response to animal agency. In *Animal ethics in the age of humans,* eds. B. Bovenkerk and J. Keulartz. Wageningen: Springer.

Estes, J., J. Terborgh, J. Brashares, M. Power, and J. Berger. 2011. Trophic downgrading of planet Earth. *Science* 333 (6040): 301–306.

Faunabeheereenheid Gelderland (FBE). 2019. Faunabeheerplan Grote Hoefdieren FBE Gelderland 2019–2025. FBE Gelderland. https://www.faunabeheereenheid.nl/gelderland/PUBLICATIES.

Flannery, T. 2018. *Europe—The first 100 million years.* UK: Penguin Books.

Frankham, R., J.D. Ballou, and D.A. Briscoe. 2010. *Introduction to conservation genetics.* Cambridge, UK: Cambridge University Press.

Franklin, I.R. 1980. Evolutionary change in small populations. In *Conservation biology: An evolutionary ecological perspective*, ed. M.E. Souleand and B.A. Wilcox, 135–119. Sunderland, MA: Sinauer Associates.

Groot Bruinderink, G.W.T.A. 2016. Het Edelhert. In *Atlas van de Nederlandse zoogdieren* (deel 12), eds. S. Broekhuizen, K. Spoelstra, J.B.M. Thissen, K.J. Canters, and J.C. Buys. Leiden: Naturalis Biodiversity Centre and EIS Kenniscentrum.

IPBES. 2019: Link to the global assessment report on biodiversity and ecosystem services. https://ipbes.net/global-assessment-report-biodiversity-ecosystem-services. Accessed 23 March 2020.

IPO. 2019: Interprovinciaal wolvenplan Nederland. https://www.bij12.nl/nieuws/wolvenplan-goedgekeurd-door-provincies/. Accessed 23 March 2020.

IUCN Red list. https://www.iucnredlist.org/. Accessed 23 March 2020.

Jepson, P., and C. Blythe. 2020. *Rewilding—The radical new science of ecological recovery.* London: Icon Books Ltd., Omnibus Business Centre.

Keulartz, J. 2018. Rewilding. *Oxford Research Encyclopedia of Environmental Science.* January 2018.

Liberg, O., G. Chapron, P. Wabakken, H. Pedersen, N. Hobbs, and H. Sand. 2012. Shoot, shovel and shut up: Cryptic poaching slows restoration of a large carnivore in Europe. *Proceedings of the Royal Society B: Biological sciences* 279: 910–915. https://doi.org/10.1098/rspb.2011.1275.

MacArthur, R.H., and E.O. Wilson. 1967. *The theory of Island biogeography.* Princeton, NJ: Princeton University Press.

Mergeay, J. 2012. *Afwegingskader voor de versterking van populaties van Europees beschermde soorten.* Adviezen van het Instituut voor Natuur- en Bosonderzoek, nr. INBO.A.2832.

Miller, B., B. Dugelby, D. Foreman, C. del Rio Marinez, R. Noss, M. Philips, and L. Willcox. 2001. The importance of large carnivores to healthy ecosystems. *Endangered Species Update* 18 (5): 202–210.

Mills, L.S., and F.W. Allendorf. 1996. The one-migrant-per-generation rule in conservation and management. *Conservation Biology* 10: 1509–1518.

Nei, M. 2005. Bottlenecks, genetic polymorphism and speciation. *Genetics* 170: 1–4.

Ottburg, F.G.W.A., and C.A.M. van Swaay. 2014. *Gunstige referentiewaarden voor populatieomvang en verspreidingsgebied van soorten van bijlage II, IV en V van de Habitatrichtlijn* (WOt-rapport; No. 124). Wageningen: Wettelijke Onderzoekstaken Natuur & Milieu, Wageningen UR.

Primack, R.B., and A.A. Sher. 2016. *An introduction to conservation biology.* Sunderland, MA: Sinauer Associates.

Rockstrom, J., W. Steffen, K. Noone, A. Persson, F.S. Chapin, III, E. Lambin, T.M. Lenton, et al. 2009. Planetary boundaries: Exploring the safe operating space for humanity. *Ecology and Society* 14 (2): 32. http://www.ecologyandsociety.org/vol14/iss2/art32/. Accessed 23 March 2020.

Schreuder, A., and A. Bontjes. 2019. Rechter verbiedt afschieten edelherten Oostvaardersplassen. *NRC*, 12 December.

Shaffer, M. 1987. Minimum viable populations: Coping with uncertainty. In *Viable populations for conservation*, ed. M.E. Soule, 69–86. Cambridge, UK: Cambridge University Press.

Smith, D.W., R.O. Peterson, D.R. MacNultry, and M. Kohl. 2016. The big scientific debate: Trophic cascades. *Yellowstone Science* 24 (1): 70–71. https://www.nps.gov/yell/learn/yellowstone-sci ence-24-1-celebrating-20-years-of-wolves.htm Accessed 23 March 2020.

Soulé, M.E., and R.F. Noss. 1998. Rewilding and biodiversity: Complementary goals for continental conservation. *Wild Earth* 8: 19–28.

Stokland, H. 2016. How many wolves does it take to protect the population? Minimum viable population size as a technology of government in endangered species management (Norway, 1970s–2000s). *Environment and History* 22: 191–227.

Svenning, J.C., B.M. Pil, C. Pedersen, J. Donlan, R. Ejrnæs, S. Faurby, M. Galetti, D. M. Hansen, B. Sandel, C. J. Sandom, J.W. Terborgh, and F.W.M. Vera. 2016. Science for a wilder Anthropocene: Synthesis and future directions for trophic rewilding research. *PNAS* 113 (4): 898–906; first published October 26, 2015 https://doi.org/10.1073/pnas.1502556112.

Traill, L.W., C.J.A. Bradshaw, and B.W. Brook. 2007. Minimum viable population size: A meta-analysis of 30 years of published estimates. *Biological Conservation* 139 (1–2): 159–166, September 2007. https://doi.org/10.1016/j.biocon.2007.06.011. Accessed 23 March 2020.

Van der Grift, E., A. Schotman, H. Jansman, and G.A. de Groot. 2018. Uitplaatsing van grote grazers uit de Oostvaardersplassen: Een quickscan van potentiële uitzetgebieden. *Wageningen Environmental Research* (rapport nr. 2903).

Van Geel, P.L.B.A., P.J.M. Poelmann, and H.J. van der Vlist. 2018. Advies Beheer Oostvaardersplassen - Kaders voor provinciaal beleid - provincie Flevoland. Externe Begeleidingscommissie beheer Oostvaardersplassen.

Van Ginkel, H.A.L., D.P.J. Kuijper, J. Schotanus, and C. Smit. 2019. Wolves and tree logs: Landscape-scale and fine-scale risk factors interactively influence tree regeneration. *Ecosystems* 22: 202. https://doi.org/10.1007/s10021-018-0263-z. Accessed 23 March 2020.

Wilson, E.O. 1992. *The diversity of life.* Cambridge, MA: Harvard University Press.

Wilson, E.O. 2016. *Half-earth: Our planet's fight for life.* New York, NY: W.W. Norton and Company Ltd.

WWF. 2018. *Living planet report—2018: Aiming higher*, ed. M. Grooten and R.E.A. Almond. WWF, Gland, Switzerland. https://wwf.panda.org/knowledge_hub/all_publications/living_pla net_report_2018/. Accessed 23 March 2020.

Hugh A.H. Jansman is a wildlife ecologist at Wageningen Environmental Research, the Netherlands. His research is focused on sustainable ecosystems and vital populations of wildlife. Topics are monitoring, population viability analyses, reintroductions, conservation genetics, telemetry, post-mortem examinations, coexistence and management. Current project include monitoring of wolves, conflict management of red deer, wild boar & geese and monitoring the status of the reintroduced population of otters. According to Hugh the formula "Wildlife management = 90% management of people & 10% biology" refers best to his field of work. He received his Master's title at the Medical faculty, University of Leiden, but decided to shift to ecology. More information can be found at https://www.wur.nl/en/Persons/Hugh-drs.-HAH-Hugh-Jansman.htm.

Part I
Animal Agents

Chapter 3
Taking Animal Perspectives into Account in Animal Ethics

Eva Meijer and Bernice Bovenkerk

Abstract Recent years have seen an explosion of interest in nonhuman animal agency in different fields. In biology and ethology, new studies about animal languages, cultures, cognition and emotion are published weekly. In the broad field of animal studies, the symbolic and ontological human-animal distinction is challenged and other animals are presented as actors. These studies challenge existing approaches to animal ethics. Animals are no longer creatures to simply think about: they have their own perspectives on life, and humans can in some instances communicate with them about that. Animal ethics long determined individual moral rights and duties on the basis of nonhuman animal capacities, but this often measures them to human standards and does not take into account that nonhuman animals are a heterogeneous group in terms of capabilities as well as social relations to humans. The questions of whether animals have agency, and how we should morally evaluate their agency, are especially urgent because we live in an age in which humans dominate the lives of large numbers of other animals. The Anthropocene has shaped the knowledge and technology for humans to realize that animals have more agency than has been assumed, but ironically it is also an epoch where animal agency is increasingly curtailed. This leads to new conflicts and problems of justice. How should animal ethics deal with the new knowledge and challenges generated in the Anthropocene? In this chapter we defend a relational approach to animal ethics, viewing other animals as subjects capable of co-shaping relations.

The original version of this chapter was revised: The author's name is corrected from "O'Neill, J.S. and M.H. Hastings" to "David A. Leavens" in reference cross citation and list. The correction to this chapter is available at https://doi.org/10.1007/978-3-030-63523-7_32

E. Meijer (✉) · B. Bovenkerk
Wageningen University and Research, Wageningen, The Netherlands
e-mail: eva1.meijer@wur.nl

B. Bovenkerk
e-mail: bernice.bovenkerk@wur.nl

© The Author(s) 2021, corrected publication 2021
B. Bovenkerk and J. Keulartz (eds.), *Animals in Our Midst: The Challenges of Co-existing with Animals in the Anthropocene*, The International Library of Environmental, Agricultural and Food Ethics 33,
https://doi.org/10.1007/978-3-030-63523-7_3

3.1   Introduction

Prairie dogs, a species of ground squirrel who live in tunnels under the ground, have developed a complex communication system. When an intruder enters their territory, they do not only tell each other whether it's a human, a dog or someone from another species, they also describe this intruder in detail. In the case of a human, they for example mention their height, the colour of their hair and T-shirt, and whether or not they carry an object, such as for example an umbrella or a gun (Slobodchikoff et al. 2009). Prairie dogs are not the only nonhuman animals who have more elaborate systems of communication than humans have always thought. What is exceptional is that their language has been studied in this much detail.

For a long time, nonhuman animal capacities were mostly studied to better understand humans (Meijer 2019). This is changing. Recent years have seen a turn towards studying nonhuman animals' languages, cultures, emotional lives, cognitive capacities and even politics (Meijer 2019; Bekoff and Pierce 2009; Brosnan and De Waal 2012; Leavens 2007; Wasserman and Zentall 2012). These studies show their inner lives are more complex than previously assumed, and that interspecies communication about many issues is possible, raising questions about the relationships between humans and other animals. Humans can no longer assume they know best—other animals have perspectives on their lives, and on their relations with humans, too—or treat nonhuman animals only as objects of study. In order to formulate what is just, ethically or politically, we therefore need to engage differently with them. The first step in this process is to recognize that their agency matters. Similar to humans, other animals have an interest in shaping their own lives. The second step is to get a better understanding of how different animals express themselves, their desires and views, in order to be able to build new, and better, relations with them.

Taking into account animal agency in ethics and politics is however not easy. Current views about nonhuman animals and their agency are shaped by stereotypical or anthropocentric ideas about their capacities, that were formed within a behaviouristic framework. If we believe that we cannot get to know anything about what goes on in nonhuman animals' minds from watching their behaviour or engaging with them, it will be difficult to recognize animal agency in the first place. Furthermore, human treatment of other animals is often focused on curtailing their agency. In farming practices this not only means using fences and other material devices to limit their freedom of movement, but also modifications of their bodies (such as the cutting of beaks and tails) and even genetic interventions. Our stereotypical views on animals and the curtailing of their agency limit our understanding of animals' capacity for agency. Both these aspects imply that humans currently should practice epistemic humility—there is much we do not know about other animals, and much of what we think we do know is formed by stereotypical views. Much scientific knowledge about nonhuman animal capacities reflects this. Our questions determine the answers other animals can give, and we long asked the wrong questions (Despret 2016). For example, if we want to know whether animals can make conscious choices, this will be difficult to find out when the animals are kept in a setting that is devoid of possibilities for making choices. Getting to know more about animals thus requires a critical investigation of existing concepts and knowledge, aimed at decentring the human, and new forms of animal research.

However, we also cannot refrain from considering the question of animal agency when thinking about ethics. We currently live in an epoch in which human actions curtail the agency of animals, and the ways they can shape their own lives, more than ever before. We cannot simply let them be, human and animal lives are thoroughly intertwined. Due to human impact on ecosystems, nonhuman animals and the planet, our current age has been named the Anthropocene: the age of the human. The Anthropocene has shaped the knowledge and technology for humans to realize that animals have more agency than has been assumed, but ironically it is also an epoch where animal agency is increasingly curtailed. Captive animals, for example, are curtailed in their agency when their enclosure is too small or they cannot go out to look for new experiences. This means they cannot express their 'inquisitive exploration' and when they are solitarily housed they cannot 'engage in social play' (Špinka and Wemelsfelder 2011, 36). They will not acquire certain competences and their level of interaction and self-expression will be limited. In contrast, in the wild they are confronted more often with novel challenges for which they have to seek a solution and this stimulates their agency, as well as opportunities for building social relations and moving around. Animals in the wild, however, are increasingly curtailed in their agency as well, due to habitat loss and fragmentation and changing climatic conditions. These conditions lead to new conflicts between humans and other animals, and problems of justice.

In this chapter we explore the questions of animal ethics and animal agency in tandem, because they are interconnected: taking seriously animal agency has ethical consequences and animal ethics should take into account animal agency. Furthermore, as the concept agency currently is used in many different ways in different fields, we also aim to shed light on the concept itself, focusing on its meaning in philosophy. Our focus of inquiry is not animals' moral agency, which we see as part of the larger spectrum of agency; some animals may well be moral agents and others perhaps not. We instead focus on the ethical consequences of the fact that other animals are actors with their own perspective on life, and on relations. Moreover, we do not take animal agency as an 'entry ticket' for the moral community, or as synonymous with moral status. We assume that all animals that have subjective experiences should be attributed moral status and our point is that it is in the interest of all beings with moral status to have their agency taken seriously, to the extent that they have agency.

We begin by reviewing existing approaches to animal agency, and discuss their shortcomings and strengths, formulating a working definition of agency. We then argue for a relational model of ethics that takes animal agency seriously at the micro- and macro level. We end by discussing initial steps towards formulating new relations with other animals. As we do not know the precise scope of animal agency yet, given that humans have so long ignored it, and even oppressed it in so many ways, we do not aim to write a final statement, but we rather see this as an investigation into developing a new, relational, animal ethics for the Anthropocene.

3.2 Conceptualizing Animal Agency: Two Models

Animal agency is currently used as an umbrella term for ways in which animals act and influence the world around them in various directions of study, and the meaning varies between different fields. In this chapter we focus on its philosophical meaning. More specifically, in this section we review two models of conceptualizing agency in relation to animal ethics. The models can be seen as two opposites on a spectrum of approaches. The first we will call propositional agency, following Sebo (2017), in which a specific form of human rationality takes centre stage, leading to anthropocentrism. The second model argues agency is found in everything that has the capacity to move something else, leading to problems for ethical and political theory (Bennett 2010). These two extreme models obviously do not provide a comprehensive overview of all theories on (animal) agency. However, they do bring to light some of the key questions in thinking about nonhuman and human agency, and they show how views of morality have been linked to views about agency in the philosophical tradition.

3.2.1 Propositional Agency

The concept agency has in philosophy traditionally been reserved for intentional human action and is linked to the capacity for propositional thought (Sebo 2017, see also Bermúdez 2007). In an insightful article about agency and moral status, Sebo (2017) calls this conception of agency 'propositional agency'. Propositional agency starts from the common sense idea that there is a difference between action and mere behaviour. This difference is in the philosophical tradition often interpreted narrowly: as the difference between intentional action and mere behaviour. Intentional action is specified as acting on 'judgments about what we have reason to believe, desire, and/or do' (Sebo 2017, 14). Being capable of intentional action thus formulated presupposes cognitive capacities that other animals were long thought not to possess, such as for example second order thoughts.

Many animal philosophers today challenge this interpretation of agency, together with the underlying view of animal subjectivity (Sebo 2017). They argue that at least some nonhuman animal species possess (some of) these capacities and that differences between humans and other animals in this regard are a matter of degree, and not kind (see for example Gennaro 2009). Furthermore, humans often act habitually (Sebo 2017, Donaldson and Kymlicka 2011) and an image of the human as primarily a rational being relies on an idealized view of humans. Introducing agency as intentional agency in the narrow description above does not do justice to other animals and exaggerates the gap between humans and other animals.

In addition to these problems, propositional agency is also based on specific forms of human reasoning, and values these more than other forms of reasoning. While recent research finds that certain other animals are perhaps capable of these forms of reasoning, this does not do justice to the fact that different species have different forms of agency. Dog agency should be understood as dog agency, not as lesser-than-human agency. Current anthropocentric conceptions of agency can function as a starting point for evaluating others and interpreting relations, but they cannot be an end point. For assessing their epistemic value in multispecies contexts, we need to investigate the power relations that led to current formulations, take into account new empirical research about other animals (including insights from narrative ethology and case studies), and engage with them differently in order to foster their agency instead of constraining it (see Blattner et al. (2020) for a longer discussion, see also Calarco 2018). More generally, new multispecies definitions of concepts—such as agency—should not be based on how much the other animals resemble humans, but include respect for their forms of expression and knowledge formation.

3.2.2 Materialist Agency

On the complete other side of the spectrum of propositional agency, we find object-oriented theories. One of their proponents, Jane Bennett, formulates a political ecology, in which nonhumans exercise agency on a spectrum with humans. She argues for a 'vital materialism' (2010, 23) in which objects possess power, and agency is located in 'assemblages': ad hoc groupings of diverse elements that can consist of human and non-human bodies. Bodies are always part of larger networks, which Bennett envisions as webs, or 'knotted worlds' of vibrant matter (2010, 13). Objects, or non-humans, the terms are used interchangeably, are interconnected with human bodies, which are themselves made of matter and influenced by pressure from the outside world. To conceptualize the pressure different bodies exercise—or, in other words: their agency—Bennett uses Spinoza's term 'conatus', which means a trending tendency to persist. According to Bennett and Spinoza, non-human bodies share this conative nature with human bodies. Bennett sees these bodies as associative, or even social (2010, 21), in the sense that each body by nature continuously affects and is affected by other bodies.

While Bennett rightly recognizes that agency can be exercised by different beings in different ways, and that specific forms of agency come into being in and through relations, this approach makes normative judgments difficult. Subjective agency dissolves when we are all just bodies moving, and intentions do not seem to matter anymore, which is counterintuitive. Furthermore, the category 'nonhuman' runs the risk of reinforcing stereotypical ideas about nonhuman animals, who are grouped with things, and as ever contrasted with human (as in the word 'nonhuman'). Agency and subjectivity are highly problematic and exclusionary concepts in the philosophical tradition, but reformulating them in this way does not get to the root of the problem, and creates new problems for thinking about nonhuman animal agency. These problems also matter with regard to human responsibility. In an age in which humans dominate the lives of animals of so many other species, taking responsibility is an ethical priority. While Bennett argues that her work is meant to promote this type of responsibility (2010, Introduction), from within the theory it seems hard to realize.

3.2.3 A Working Definition of Agency

Both the propositional and the materialist approach fail to adequately take into account nonhuman animals' perspectives. The first because human forms of rationality and agency are taken as the standard, which excludes many other animals (and some humans) beforehand, and the second because it fails to offer a framework in which we can make normative or political judgments about others, and does not explicate new forms of engagement with nonhuman animals, instead grouping them with things. How then, should we understand (animal) agency? We do not want to give a fully developed definition from the outset, as we think an understanding of animal agency should come about in the diverse practices of human-nonhuman animal relationships, and more research needs to be done into such relationships. However, we need a loose working definition of agency in order to give our search some direction.

Irvine (2004) defines agency as 'the capacity for self-willed action'. Entities capable of self-willed action should be described as subjects, in the sense that they experience their own world subjectively. We agree with the materialist agency approach that an important aspect of agency is that by the action the subject exerts an influence on the world around her. However, for an action to be more than mere behaviour, it seems that a desire or will should be behind the action. Note that an action that exerts influence and that expresses a desire or will does not presuppose the presence of intentionality, at least not intentionality in the strict second-order thought sense of the word. Consider the following example: a (human or non-human) animal can see a piece of food, say an apple, desire to eat that apple and then move to grab it. This involves the intention to eat the apple, but does not necessarily entail that the animal reflects on his or her own desire for the apple before acting to grab it. An agent can express her will or desires and influence the world around her without necessarily having the capacity to think about how the action will impact on others. Our working definition, which is inspired by Blattner, Donaldson and Wilcox (2020) and Sebo's idea of perceptual agency starts from the idea that agency is the capability of a subject to influence the world in a way that expresses her desires and will. This capability springs from the phenomenology of the individual in question—her genetic make up, capacities, physicality and so on. In addition, and here we add a different layer to Sebo's view of perceptual agency, it is important to realise that desires and wills do not come about in a vacuum. In the constitution of desires and wills, and thus also in agency, relationships—on the individual level, but also social, political and cultural structures—form an important role. In other words, there is always an interaction between agency and an animal's environment. Social, political and cultural structures can limit animals' capability of agency. While they might have the capacity for having agency, animals might be limited in their expression of this agency. Think of a zoo animal that is limited in her ability to roam freely. The animal, say a deer, may still have the capacity for running long distances, but in the zoo environment cannot exercise that capacity. On the other hand, zoo animals may develop new forms of agency in this limited environment, that they did not have before. The capacity for agency, can therefore be developed or hindered in interaction with the animal's environment.[1] In this sense, Anthropocene conditions will also influence animal agency. In our view, in order to do justice to animal agency in our moral deliberations, we need a relational model that takes animals' perspectives into account, as well as the social-historical context, and that does not measure other animals to a human standard.

3.3 Taking into Account Relational Agency in Animal Ethics on the Micro- and Macro Level

Acknowledging that animal agency matters and that we need to make space for their perspectives should lead to a relational, situated approach to ethics, in which not the human subject is the standard, but that focuses on the other. Drawing on insights developed by ecofeminists such as Carol Adams and Lori Gruen, we aim to move the question of how we should take agency into account ethically, past assessing the content of animals' minds and building a judgment on that, to assessing the social context in which agency is curtailed or fostered. While we need more empirical research into animals' minds (cognition and emotion) and cultures, we also need to focus on the social conditions that foster or constrain agency. This latter aspect is often underestimated in philosophy and animal ethics. This is problematic because the conditions of the Anthropocene target not only individual animals but also social groups and even species, so we need to acknowledge human responsibility not just in individual relations, but also on the macrolevel.

3.3.1 Relational Agency and Animal Ethics

(Eco)feminist approaches to ethics take (unequal) relations as the starting point for ethical considerations. Acting morally does not simply involve following rules, as in deontological approaches such as the animal rights theories mentioned above, maximizing happiness, as utilitarian approaches demand, or perfecting one's character as virtue ethics requires, because the focus of our acts should not be on the self but on

---

[1] In this context, we use the term 'capacity' to denote the physical and mental Characteristics necessary to be able to exercise agency and we use the term 'capability' for the actual possibility a being has to exercise agency. The latter is dependent not only on the capacity, but also on the situation the being finds herself in, which is influenced by environmental, social, cultural and political structures. With this distinction we build on insights from the capability theory (as put forward by for example Nussbaum, Sen, and Robeyns) without necessarily fully embracing capability theory.

the other (Held 1990). This other is always a real, situated other, not a universal or idealized human. Ethical judgments are for this reason not universal but always tied to a context. Feminist ethicists emphasize that all of us are born into webs of relations and are dependent on others at several points in our lives, to different degrees. This dependency is not something to shy away from: relations with others are an ontological given, they constitute who we are, and can be a source of strength. As ecofeminists (Adams 2010; Donovan 2006; Gruen 2015) argue, the individuals we stand in relation to are not just human: we are also always entangled in relations with animals of other species, and they with us.

This relational and situated approach to animal ethics adds a different dimension to how we understand agency, compared to the ones sketched above. Agents are always tied to specific circumstances, capacities, and contexts, which influence their options for acting. Humans for example are born as a certain gender, in a specific culture, as a body that has certain cultural advantages or not, with a specific skin colour, in a specific class. They can choose professions, religions, partners, and so on; life will bestow hardships and joy onto them. The same applies to other animals. Social relations, physical dispositions, work and luck can all play a role in one's options for exercising agency. Species characteristics matter, but are never the whole picture. We are furthermore all entangled in different relations with individual others that can create different forms of interdependence, influencing our agency and autonomy.

This way of conceptualizing agency in relation to ethics recognizes that there are different degrees of agency and intentionality, and that there is no one strict line between species when it comes to exercising these. It also shows that in formulating ethics it is not enough to just focus on biological capacities of certain species. Focusing on the social dimensions of relations between humans and other animals can for example help us see the role unequal power relations play and have played in their options for exercising agency. This is perhaps most clear in the case of domesticated animals, but the lives of humans and non-domesticated animals are often also intertwined. Think for example about the animals that reside in our gardens; our actions and theirs exert influence on each other. Understanding that human and nonhuman animal lives are entangled has a normative dimension. Lori Gruen (2015) argues that in order to do justice to others, including other animals, we should develop a caring perception that she calls entangled empathy. This form of empathy is focused on the wellbeing of others, and developing this entangled empathy is a process in which emotion and reason play a part.

Gruen's entangled empathy is helpful in thinking about ethical relations with the individuals we encounter. However, most animals are also entangled in relations with others on social, cultural or political levels, and these entanglements often also strongly influence our scope for decision-making. This matters for ethical judgments. Individual nonhuman animals all belong to certain social groups, similar to humans, and this influences their scope for action. Legislation based on categorization of species as wild or domesticated determines space of movement or protection for individuals. Cultural constructions determine whether some city animals are seen as pests, such as rats (see chapter by Nieuwland and Meijboom in this volume), and others as belonging, such as songbirds, which has a strong impact on their options for self-realization. Cats who are born as companions have a very different set of options for agency than feral cats (see chapter by Meijer in this volume). Above we discussed the different specific harms that concern the agency of farmed animals. In order to adequately conceptualize agency, and to formulate new ethical guidelines, we also need to take these social and cultural aspects into account; in other words, we need to take the macro-level into account as well.

3.3.2 Taking into Account Macro-Relations in Thinking About Agency and Ethics

Recognizing the importance of social and political relations with other animals, Donaldson and Kymlicka (2011) developed a theory of political animal rights. They propose to view different groups of nonhuman animals as social groups based on their relation to human political communities. Specifically, they argue that wild nonhuman animal groups should be seen as sovereign nations, liminal nonhuman animals—those who live amongst humans in cities or rural areas but who do not desire close relations with them, such as mice, crows or feral rabbits—as denizens and domesticated animals as citizens. In all of these groups they emphasize the animals' agency, arguing that they and not humans usually know what is best for them. For wild nonhuman animals, the good life usually means a life without human interference; for liminal animals this can involve contact with humans under certain conditions. For domesticated animals it often involves more contact with humans, because many of them need or desire some human assistance to flourish. This does not mean that they have no interest in shaping key aspects of their lives themselves, nor does it imply that the current power relation, in which humans are hierarchically above them, is just or unavoidable. Donaldson and Kymlicka argue that for these reasons domesticated animals, co-citizens in shared communities with humans, should have the right to be represented in political decision-making and should have the right to be included in the people in whose name the state governs: they should also have democratic agency, meaning the right to co-shape common decisions.

With their theory of animal rights they draw attention to nonhuman political agency, which is often overlooked and erased (Meijer 2019), using examples that range from resistance to co-creating common interspecies communities. They also draw explicit attention to the distinction between micro-agency and macro-agency, arguing that macro-agency is also relevant in the nonhuman context (see also Donaldson and Kymlicka 2013). Micro-agency refers to the scope for making personal decisions, which can for domesticated animals include deciding on what to eat, where to sleep, and who to play with in the park or garden. Only focusing on this type of agency, however, obscures how larger scale power relations, reified in laws, institutions, and political and social processes, form the scope for this micro-agency. Dogs can in many places in the world usually for example not decide to leave the house they live in when they desire, even if their human would allow them to, because cities are not safe for them and there are leash laws and areas where they are not wanted. Horses cannot choose to leave their meadow or barn and are usually not consulted before riding, even when some riders take their preferences into account in deciding whether to engage in dressage, jumping, or other sports (see Meijer 2019 for a longer discussion of this problem). Wild nonhuman animals have no say in the preservation of their habitat. Donaldson and Kymlicka show that taking animal agency seriously implies more than simply reformulating our individual relations with them: humans should also consider their political institutions and processes, laws, and rights, to incorporate nonhuman animal perspectives.

Especially in the Anthropocene taking macro-agency into account is of the utmost importance, because the context shapes how and sometimes even if animals can exercise agency. As we mentioned above, in the Anthropocene animals' agency risks being increasingly curtailed due to upscaling of livestock production, habitat loss and fragmentation, and climate change. In our globalized world, the conditions for our relations with other animals are institutionalized in many ways. Economic, political, legal and cultural structures determine their and our space for movement and for creating new relations. Developing an animal ethics for the Anthropocene thus asks for more than considering our individual duties: it also implies carefully rethinking political and social institutions and practices. In the context of interacting with other animals this means taking into account these macro-factors, in epistemic and ethical judgments. But it also has ethical implications beyond that, such as that we should aim to change these large-scale oppressions.

3.4 Risks for Relational Approaches to Ethics

While a relational approach opens up new ways of engaging with other animals and co-creating communities, there are also risks. One familiar argument against relational approaches is that they are too dependent on context, thereby making it impossible to formulate clear ethical guidelines. Following this, ethical judgment is not universal, but relative to communities. In the case of animal ethics this could lead to unequal treatment, but also even mistreatment of animals, in particular in contexts where the human-animal relationship is an instrumental one, such as in industrialized farming or animal experimentation. We recognize this risk. We understand agency to be important, because it enables subjects to shape their own lives, in line with their subjectivity. This presupposes a certain view of the animal subject, and is part of respectful engagement with them, which is not compatible with killing or abusing them for human benefit. We believe that this has been adequately argued for by animal philosophers before and therefore focus on the next step of the argument (see also Donovan's 2006 reply to such criticism). Arguing that context forms precise secondary rights and duties also does not have to be random: we are in favour of developing clear outlines for this project. As it involves dealing differently with other animals, however, these cannot be provided beforehand; after all, the outcomes are dependent on the input from other animals.

A second risk involves anthropomorphism, falsely attributing human characteristics to other animals. While all theories of animal ethics begin with a human framework to assess animals' standing, relational theories seem to require more interpretation, because they involve communication, a focus on a changing context, and a place for animal perspectives in social and political decisions. Furthermore, even seeing animals as agents in these relations is according to some (compare Sebo 2017) a matter of anthropomorphism. To begin with the second point: not attributing any emotions or cognitive content to other animals is not a neutral stance, but rather the outcome of (Western) power relations. De Waal (1999) calls this 'anthropodenial'. It is a self-serving ideology, because it closes off animal participation beforehand. While we are not sure how relations with other animals can evolve, we do note that if we presuppose they have no agency and are not capable of new forms of interaction, these new forms will never happen. This relates back to the first point. Relational approaches do not require more interpretation. They require a different form of interpretation. Instead of taking a view from nowhere, and from there once and for all defining what animal expressions mean, our aim is to learn more about animal agency from the ground up, in human/non-human animal interactions, step by step. In this process philosophy, ethology, and other fields of study have a role to play, as do actual relations with nonhuman animals. While risks of interpretation and context are something to be aware of in relational approaches, they are inherent in doing animal ethics more generally. As such, they should be given attention, but they should not be overstated.

A final issue for consideration is the relevance of species membership. Focusing on relations might give the impression that biology and ethology no longer have a role to play in animal ethics. It would simply be a matter of social or political philosophy, and experimenting with new forms of co-habitation. However, 'animals' are a heterogeneous group. While we know quite a lot about certain animal species—such as dogs or dolphins—about most species we do not know so much. We therefore need more empirical research, as we will explicate in the next section, on species who are very different from humans, and on those who live close to humans but who are usually not seen as important research subjects for their own sake, such as farmed animals. In order to be able to take specific animals' perspectives more seriously, we need more, and different kinds of, knowledge about the species the animals belong to.

3.5 Further Directions

A relational ethics for the Anthropocene should take into account nonhuman animal voices and perspectives. According to feminist standpoint ethics, the voices of socially suppressed groups need to be heard, as 'their views are found inevitably to be subversive of the ideological system that would render them silent – sexism in the case of women and girls and speciesism in the case of animals' (Donovan 2017, 210). This implies taking responsibility for human actions while at the same time acknowledging the influence and acts of nonhuman animals. This asks for an attitude of empathy, listening and curiosity, both with regard to attending to actual nonhuman animals, and with regard to locating injustice and domination in existing larger scale relations. This might sound utopian, but we already find examples of new relations between humans and other animals. As a conclusion we will highlight these in three different fields: animal research, cultures, and work.

3.5.1   Research

Models of animal research are traditionally human-centred. There is a hierarchy between the human researcher and the animals, both materially—animals are often kept in cages—and epistemologically—animals are not seen as interlocutors but as objects of study. Ethologists such as Smuts (2001) and Bekoff (2007) challenge this model, studying nonhuman animals as subjects, and not as objects. This leads to new methodologies—such as following them in their habitats, and letting them co-shape the conditions of the studies. Smuts for example describes how she had to learn to 'speak baboon' in order to be able to study a group of baboons. Scientists usually try to ignore primates, so as not to let their presence near them influence their interaction (Smuts 2001). Smuts found out that ignoring the baboons was not a neutral act, because baboons are social animals: she had to learn to interact with them on their terms. Furthermore, by interacting with them, she experienced critical aspects of their society, such as hierarchy, personal space and communication, directly. Because the questions asked determine the answers nonhuman animals can give, adopting this kind of attitude towards animals matters greatly to learning about their inner lives and cultures, leading to knowledge that can and should inform ethical theory. Taking animal agency seriously most likely also implies not using them in forms of experimentation that infringe on their liberty or harm their welfare.

3.5.2   Animal Cultures

Another example concerns the increasing recognition of the importance of animal cultures. In recent years animal culture has become a topic of study in many species. Elephants are perhaps the most famous example. They usually travel along the same routes every year (Barua 2014). Knowledge about these routes and specific locations, for example where water can be found in dry seasons is transmitted culturally. Matriarchs teach the younger elephants the ropes of survival. Human activity increasingly disrupts these patterns of knowledge exchange and other cultural processes, in several ways. Humans may designate a certain area as a nature reserve, and use fences to close it off—sometimes with the best of intentions, aiming to protect the nonhuman animals living there—which makes it impossible for elephants to find their way. Poachers also often kill matriarchs and other older elephants in a group, which traumatizes younger elephants, and makes it very difficult for them to rebuild their lives afterwards. Gay Bradshaw researches PTSD in elephants and shows that while intensive human care may help younger elephants to recover from the trauma's they witnessed, it is very difficult for humans to take on the cultural role (Bradshaw 2009). When the matriarchs die, often that specific cultural knowledge dies with them.

Other examples of cultural knowledge being transmitted concern the migration of bighorn sheep, birdsong, and chimpanzee fashion. Conservationists take this cultural dimension increasingly seriously (Laland and Janik 2006), but respect for and knowledge about nonhuman animal cultures is not only relevant for conservationists. It is important for politicians who design new legislation for liminal animals, city planners, animal rescue organizations, and others dealing with groups of nonhuman animals, their travelling routes and habitats.

3.5.3   Animal Workers

New relations are also found in the context of work. While most nonhuman animal workers, for example in factory farms, are exploited, there are also forms of work that can benefit both human and nonhuman animals. An example concerns crow workers. Dutch start-up, Crowded Cities, plans to train city crows to pick up cigarette butts. Using crows as cleaners raises many questions about their working conditions, but when the work is safe and their freedom is not compromised, this working arrangement could be beneficial for crows, humans, and the environment. At this stage, more research should be done on the benefits and burdens of this type of work for them, including monitored pilots of the project, in order to make sure they are not exploited.

Another example of animal work concerns domesticated support animals, such as for example rescued dogs who go to hospitals to distract young patients. The benefits of therapy animals for humans have been proven scientifically (Glenk 2017). The health of these animals, and the possible benefits for them, have however not been studied in detail (ibid.). Many companion animals suffer from boredom and many like to work, so for them working could contribute to better health and happiness. In order to establish which types of engagement are possibly beneficial for humans and other animals, we need more research into the benefits and burdens of care work for nonhuman animals. This could for example lead to establishing labor rights, including the right to play, time off to do stuff with friends, rest, and a pension when they are old (Cochrane 2016).

3.5.4 Further Directions

These three examples show that establishing an ethics for the Anthropocene is not a matter of all or nothing: ethical relations exist and existing relations can be improved. Furthermore, they show that developing an ethics for the Anthropocene should be an interspecies project. In order to give shape to an animal ethics for the Anthropocene we first need to further develop our understanding of animal agency 'from the ground up', through studying animal behaviour and interspecies relationships.[2] We also need to further reflect on the question of how phenomena like agency, intentionality, autonomy, and self-realisation relate to each other, and how they will change once we take seriously the different ways in which other animals relate to these concepts. Animal ethics is not something to be thought out solely by humans behind computers, it is something for which humans also need to engage differently with other animals. This is important for several reasons. We need more, and different forms of, empirical research, to find out the scope of their agency, their view on relations, and to find out how we can build better relations, that are beneficial to all those involved. It is also important to decentre the human and find out how we can theorize together with other animals. Perhaps it even requires a move from animal ethics to an interspecies ethics, at least with regard to the (domesticated and liminal) animals with whom we share our lives, households and cities.

References

Adams, C. 2010 [1990]. *The sexual politics of meat: A feminist-vegetarian critical theory*. London: Continuum.

Barua, M. 2014. Bio-geo-graphy: Landscape, dwelling, and the political ecology of human–elephant relations. *Environment and Planning D: Society and Space* 32: 915–934.

Bekoff, M. 2007. *The emotional lives of animals: A leading scientist explores animal joy, sorrow, and empathy and why they matter*. Novato: New World Library.

Bekoff, M., and J. Pierce. 2009. *Wild justice: The moral lives of animals*. Chicago, IL: University of Chicago Press.

Bennett, J. 2010. *Vibrant matter: A political ecology of things*. Durham: Duke University Press.

Bermúdez, J.L. 2007. *Thinking without words*. Oxford, UK: Oxford University Press.

Blattner, C.E., S. Donaldson, and R. Wilcox. 2020. Animal agency in community: A political multispecies ethnography of VINE sanctuary. *Politics and Animals* 5: 33–54.

Bradshaw, G. 2009. *Elephants on the edge: What animals teach us about humanity*. New Haven, CT: Yale University Press.

Calarco, M. 2018. The three ethologies. In *Exploring animal encounters: Philosophical, cultural, and historical perspectives*, ed. D. Ohrem and M. Calarco, 45–62. New York, NY: Springer.

Cochrane, A. 2016. Labour rights for animals. In *The political turn in animal ethics*, ed. R. Garner and S. Sullivan, 15–31. Lanham, MD: Rowman & Littlefield.

Brosnan, S.F., and F. De Waal. 2012. Fairness in animals: Where to from here? *Social Justice Research* 25 (3): 336–351.

---

[2] This type of empirical research is planned for the follow-up stages of our research project, by way of interpretive research and multi-species ethnography.

Despret, V. 2016. *What would animals say if we asked the right questions?* Minneapolis: University of Minnesota Press.

De Waal, F. 1999. Anthropomorphism and anthropodenial: Consistency in our thinking about humans and other animals. *Philosophical Topics* 27 (1): 255–280.

Donaldson, S., and W. Kymlicka. 2011. *Zoopolis: A political theory of animal rights.* Oxford, UK: Oxford University Press.

Donaldson, S., and W. Kymlicka. 2013. A defense of animal citizenship. Part 1: Citizen canine: Agency for domesticated animals. Unpublished manuscript.

Donovan, J. 2006. Feminism and the treatment of animals: From care to dialogue. *Signs: Journal of Women in Culture and Society* 31 (2): 305–329.

Donovan, J. 2017. Interspecies dialogue and animal ethics: The feminist care perspective. In *The Oxford handbook of animal studies*, ed. L Kalof, 208–226. Oxford, UK: Oxford University Press.

Gennaro, R.J. 2009. Self-awareness in animals. In *The philosophy of animal minds*, ed. R.W. Lurz, 184–200. Cambridge, UK: Cambridge University Press.

Glenk, L.M. 2017. Current perspectives on therapy dog welfare in animal-assisted interventions. *Animals* 7 (2): 7.

Gruen, L. 2015. *Entangled empathy: An alternative ethic for our relationships with animals.* New York, NY: Lantern Books.

Held, V. 1990. Feminist transformations of moral theory. *Philosophy and Phenomenological Research* 50: 321–344.

Irvine, L. 2004. *If you tame me: Understanding our connection with animals.* Philadelphia, PA: Temple University Press.

Laland, K.N., and V.M. Janik. 2006. The animal cultures debate. *Trends in Ecology & Evolution* 21 (10): 542–547.

Leavens, D.A. 2007. Animal cognition: Multimodal tactics of orangutan communication. *Current Biology* 17 (17): 762–764.

Meijer, E. 2019. *When animals speak.* New York, NY: New York University Press.

Sebo, J. 2017. Agency and moral status. *Journal of Moral Philosophy* 14 (1): 1–22.

Slobodchikoff, C.N., B.S. Perla, and J.L. Verdolin. 2009. *Prairie dogs: Communication and community in an animal society.* Cambridge, Mass.: Harvard University Press.

Smuts, B. 2001. Encounters with animal minds. *Journal of Consciousness Studies* 8 (5): 293–309.

Špinka, M., and F. Wemelsfelder. 2011. Environmental challenge and animal agency. In *Animal welfare*, 2nd ed, ed. M.C. Appleby et al., 27–44. Wallingford, UK: CAB International.

Wasserman, E.A., and T.R. Zentall. 2012. *Comparative cognition: Experimental explorations of animal intelligence.* Oxford, UK: Oxford University Press.

Eva Meijer works as a postdoctoral researcher at Wageningen University (NL) in the project Anthropocene Ethics: Taking Animal Agency Seriously. She taught (animal) philosophy at the University of Amsterdam and is the chair of the Dutch study group for Animal Ethics, as well as a founding member of Minding Animals The Netherlands. Recent publications include Animal Languages (John Murray 2019) and When animals speak. Toward an Interspecies Democracy(New York University Press 2019). Meijer wrote nine books, fiction and non-fiction, that have been translated into eighteen languages.

Bernice Bovenkerk is associate professor of philosophy at Wageningen University, the Netherlands. Her research and teaching deals with issues in animal and environmental ethics, the ethics of climate change, and political philosophy. Current topics are animal agency, the moral status of animals and other natural entities, with a particular focus on fish and insects, the ethics of animal domestication, animal (dis)enhancement, and deliberative democracy. In 2016 she co-edited (together with Jozef Keulartz) Animal Ethics in the Age of Humans. Blurring boundaries in human-animal relationships (Springer) and in 2012 she published The Biotechnology Debate.

Democracy in the face of intractable disagreement (Springer). She received her PhD title in political science at Melbourne University and her Master's title in environmental philosophy at the University of Amsterdam. Her homepage is https://bernicebovenkerk.com.

Chapter 4
Turning to Animal Agency in the Anthropocene

Charlotte E. Blattner

Abstract Agency is central to humans' individual rights and their organization as a community. Human agency is recognized in the Universal Declaration of Human Rights through guaranteed rights, such as the right to life, basic education, freedom of expression, and the freedom to form personal relationships, which all protect humans from tyranny and oppression. Though studies of animal agency consistently suggest that we grossly underestimate the capacity of animals to make decisions, determine and take action, and to organize themselves individually and as groups, few have concerned themselves with whether and how animal agency is relevant for the law and vice versa. Currently, most laws offer no guarantee that animals' agency will be respected, and fail to respond when animals resist the human systems that govern them. This failure emerges from profound prejudices and deep-seated anthropocentric biases that shape the law, including law-making processes. Law and law-making operating exclusively as self-judging systems is widely decried and denounced—except in animal law. This chapter identifies standpoint acknowledgement as a means to dismantle these tendencies, and provides instructions on how to ask the right questions. It concludes by calling for an "animal agency turn" across disciplines, to challenge our assumptions about how we ought to organize human-animal relationships politically and personally, and to increase our civic competence and courage, empathy, participation, common engagement, and respect for animal alterity.

---

[1] Inquiries into agency are still largely descriptive—focusing on whether and to what degree someone exhibits agency—and do not ask if their actions are good or bad. Questions like "should person A act/have acted that way" fall under the purview of moral agency. Moral agency can be a dimension or a manifestation of agency *tout court*, and there are some that have asked if animals are moral agents (e.g., Rowlands 2012), but I will not address this topic here.

C. E. Blattner (✉)
University of Bern, Bern, Switzerland
e-mail: charlotte.blattner@oefre.unibe.ch

© The Author(s) 2021   65
B. Bovenkerk and J. Keulartz (eds.), *Animals in Our Midst: The Challenges
of Co-existing with Animals in the Anthropocene*, The International Library
of Environmental, Agricultural and Food Ethics 33,
https://doi.org/10.1007/978-3-030-63523-7_4

4.1 The Centrality of Agency

Agency, the capacity for self-willed action, is central to laws that govern the individual rights of people and their freedom to organize collectively.[1] The 1948 Universal Declaration of Human Rights (UDHR) was a milestone achievement for human rights founded on freedom and justice. In its preamble, the UDHR proclaims that "human beings shall enjoy freedom of speech and belief and freedom from fear and want," to promote "life in larger freedom" (UDHR 1948, preamble). It is on the grounds of our agency that we commit to securing universal rights and foundational freedoms for all humans, including the right to life, basic education, freedom of expression, and the freedom to form personal relationships. These rights, in turn, are an acknowledgment of the need and desire to protect normative dimensions of our agency (Griffin 2004, 2008, 149).[2] Their realization represents "the highest aspiration of the common people" (UDHR 1948, preamble) and they must be protected as a matter of the rule of law, for, without them, humans would be "compelled to have recourse, as a last resort, to rebellion against tyranny and oppression" (UDHR 1948, preamble).

Prior to the UDHR and the emergence of a shared commitment to secure human agency, rebellion was the only form of protest available to humans whose agency was ignored, restricted, or simply not guaranteed by positive action. In theory and under perfect conditions, the rights of the UDHR eliminate the need to rebel because they secure human agency and expand opportunities to exercise it. As we acknowledge the central role played by agency in the organization of human life, we have—so far—failed to extend this concept to nonhuman animals, although there is overwhelming evidence that they resist and rebel against (human) tyranny. Elephants break free from their chains and seek revenge against the people who maltreated them with bull-hooks, tigers leap out of their enclosures and track down visitors who tormented them, whales target trainers who confined them and separated them from their offspring (Hribal 2010). Sheep escape from the slaughterhouse, pigs jump off transports, and cows prefer to swim into the open sea rather than enduring heart-wrenching conditions aboard ship. Animals resist by screaming, running, and defending themselves with horns, teeth, and claws; they express disapproval through eye contact, stiffness, repetitive behavior, depressive ear drooping and reticence, or simply by retreat (Philo

---

[2]There are arguments that agency alone (ought to) ground human rights. Griffin (2008) argues that "human rights should be seen as protections of our normative agency,"; this "is not a derivation of human rights from normative agency; it is a proposal" (p. 1). Liao (2009), however, argues for a wider account of human rights that draws on the notion of agency and other elements of a good life. Griffin considers agency the sole ground of human rights, based on a classic "rationalistic" understanding, and argues that we must autonomously conceive of a worthwhile life (autonomy), be at liberty to pursue this conception (liberty), and have some minimum material provision and education (Griffin 2002, 311). His conception of agency cannot be upheld because it excludes many people (i.e., it is ableist), discriminates against people on the basis of wealth, income, and education (which is untenable, among others, because it directly contradicts article 2 UDHR 1948), and is manifestly anthropocentric (by precluding recognition and consideration of all forms of animal agency).

1998; Wadiwel 2018). In their given environment, animals express many "forms of resistance against human ordering" (Wilbert 2000, 250), and, as such, materialize their "capacity for self-willed action" (see chapter by Meijer and Bovenkerk in this volume).

In a world dominated by humans and governed by laws that further human interests, resistance to curtailments of their agency is still animals' only recourse. Agency, so central to us human animals and the laws governing our relationships, is neither recognized nor secured by the laws governing nonhuman animals and our relations with them. Here, I explore the consideration of *animal agency as a matter of law*, not whether animals have legal capacity[3] and as such, are *agents of the law*.

The law on the books suggests animal agency is not a matter of or for the law. For example, the Dutch Animal Law recognizes the intrinsic value of animals (2011, art. 3 para. I), but posits in its preamble that the law serves to secure animals' welfare and to market animal products (2011, preamble). Worldwide, "animal welfare acts" or "animal protection acts" claim to be primarily preoccupied with securing the welfare of animals or protecting them (Blattner 2019).[4] But do concepts of "welfare" and "protection" include agential action? Generally, an animal's state of welfare is considered good if, as the World Organization for Animal Health (OIE) provides, they are "healthy, comfortable, well nourished, safe, able to express innate behaviour, and [...] not suffering from unpleasant states such as pain, fear, and distress" (OIE 2019, art. 7.1.1). Crucially, however, "animal welfare laws" often still legitimate using and killing animals by laying down how and when they can be bred, taken from the wild, separated from their families, confined, used, maimed, slaughtered, skinned, and turned into convenience products. Since most animal laws do not interfere with these and other majority group practices (Deckha 2012), nonhuman animals are, all things considered, deprived of legal protection (i.e., animal law in a substantive sense) and recourse (i.e., animal law in a procedural sense) (Kymlicka 2017). The almost exclusive focus of the law on the needs of humans thwarts its efforts to be just, equitable, and fair (including fair to all humans, since animal law can be a tool to oppress certain human groups). These crucial dimensions still require translation into mainstream debates about animal law, however, I am here not primarily interested in whether or not the law can deliver on animal welfare grounds. My main criticism is that the law, even in the best case, namely when it is truly designed to protect animals and perfectly enforced, maximally sees animals as *welfare-recipients*—beings who are acted upon, "victims" in need of rescue, "voiceless beings" that require a human voice (Corman 2016)—rather than as actors with their own will and deserving of individual or communal rights that secure their agency.

---

[3]By legal capacity, I mean the capacity of individuals to make binding amendments to their rights, duties, and obligations, e.g., getting married or merging, entering into contracts, making gifts, or writing a valid will.

[4]Blattner (2019, 71–80) has looked at the laws of over 60 states to establish this. Note that especially constitutional laws exhibit a broader variety of rationales or approaches to protecting animals. In India, for example, people are obliged to have compassion toward animals, and in Switzerland, the dignity of animals must be protected (Blattner 2019, 321–334).

The law does not consider animals' desires and preferences for, e.g., where they want to live or with whom, whether they wish to bear and keep their young, or have their organs removed; neither does it require those applying the law to do so. Instead, laws detail the "proper" way to dehorn or debeak, cut off snouts or tails, and remove toes and other body parts that animals need and use to express themselves, navigate their relationships with others, and flourish (see esp. on how birds, specifically chickens, are forced to endure such practices, Davis 2011). In doing so, animal law is not only complicit in disregarding animal agency and failing to respond to its many manifestations, but, above all, operates as a central legitimizing scheme to ignore and silence animals and inhibit their agency. Some might argue that these practices were written into law starting in the 1960s up until, roughly, the 1990s, before there was any scientific evidence of animal agency.[5] Animals' agential capacities could, to some extent, be argued to be recognized by those states that have recently begun to frame animals as *quasi* subjects of the law by recognizing them as "living and sentient beings", notably in their civil codes.[6] Though this is certainly an improvement over labeling animals as "objects," these statements notwithstanding, most states openly declare that they will continue to treat animals as objects of the law (Blattner 2019, 243–244). But what can reasonably be the transformative potential of laws that reject the notion that animals are objects but lay down that, for reasons of convenience, animals are still treated as if they were property? Since there is no functional difference between being *treated like property* under the law and *being property*, animals have not yet benefited from the nominal recognition that they are "living and sentient beings." And likely, they never will.

Animals' agency can play a critical role in facilitating the law's recognition of their subjectivity, by making plain that each and every animal is an agent with robust interests in self-determination. Most people living with companion animals take pleasure in describing the animals' sassiness or pointing out that their companions ask for things that are important to them (e.g., particular foods, being taken out for a walk, or their preferences for and dislikes of particular people). However, these individual insights rarely shape people's views about animals at large, who are often presumed to lack agency. Overall, animals are still seen as reacting in unthinking and deterministic fashion to natural forces guided by scripts predefined by their genes or species membership (e.g., Nussbaum 2006; Rollin 1995). Many believe this "genetic imprint" prevents animals from determining or changing the course of their lives in a meaningful sense; they operate under the assumption that animals' actions and desires are predictable and that they do not have the "necessary free will" to act as agents. This old-fashioned view is based on arguments that have traditionally been

---

[5]The scope and breadth of animal protection laws can be determined, very roughly, on the basis of three different "generations of animal law." The first-generation animal laws only protect the monetary interests of owners. Second-generation animal laws penalize cruelty and abuse of animals, even if committed by an animal's owner. And third-generation animal laws additionally lay down binding rules on the proper care and treatment of animals. See Blattner (2019, 281).

[6]See, for an overview of these recent developments in Austria, Brussels, California, Colombia, France, Germany, New Zealand, Portugal, Spain, Switzerland, and other countries, Blattner (2019, 243–244).

used to deprive others of their rights (e.g., women), and is heavily influenced by confirmation bias as it ignores clear evidence to the contrary.

Our denial of animals' agency often starts with how we talk about them. Animals are farm*ed*, they are domesticat*ed*, and us*ed* for food production, research, or any other purpose. Animals are primarily defined by how we seek to use them (Eisen 2010) and by framing them as mere passives upon whom we do things, we strip them of agency. Our everyday language neither recognizes existing forms of animal agency nor does it, as it is used today, seem to leave room for its recognition in the future. In addition, we typically see and encounter animals only in highly restrictive environments, and this, in turn, influences our judgment of their agential capacities. Everywhere we turn, we see instances of humans suppressing animal agency, insisting on and enforcing the roles that we ascribe to them. We cram them into small quarters to fatten them for food, impregnate them for milk production, train and discipline them to docility; we pen them in restrictive environments that prevent them from exercising agency; and reduce their lives to "simple, predictable and monotonous" actions (Špinka and Wemelsfelder 2011, 27). On the socio-political level, considering animals as belonging into these environments, even just seeing animals in these environments, reinforces the dominant view that they lack agency. On the research level, studying animals in these environments means we ask limited questions and that the answers to those questions are bound to be tainted, biased, and only marginally useful (Blattner et al. 2020). It is these ideological blinders and our pervasive anthropocentric bias that create a vicious circle and reinforce existing power hierarchies, unchecked biases about others, and the continued oppression of animals. If humans—like animals—were forced to live penned up on one square meter, denied the ability to interact with others, tied up by ropes to be forcefully impregnated, or forced into slaughterhouses,[7] we would challenge the claims of those who justify these practices. We would argue that in these instances, any person under such restrictive and oppressive conditions is denied agency, so it seems reasonable to turn this argument around and to point out that, if animals do have agency, it will least evident under restrictive and oppressive socio-political conditions. To advance useful proposals for improving the lot of animals, we need to learn to see the many aspects of our socio-political and interpersonal relationships with animals that are limited by our ignorance and bias. Only then can we formulate ethical and legal arguments that can address the issue of animal agency.

4.2 On Animal Agency and Self-Judging Obligations

Animal studies is an emerging field that builds on scholarship in the humanities, social sciences, and sciences to investigate past and present relations between human and

---

[7]Gillespie (2016), for example, witnessed animals being "beaten, yelled at, kicked, shocked, and crushed against the wall or floor for trying to escape or fight back against humans who were herding them through space" (p. 126).

non-human animals, the representation of those relations, their ethical implications, and their social, political, and ecological effects in and on the world (Wesleyan 2019; Kalof 2017). Researchers in the field seem the most likely candidates for removing the blinders that rationalize and protect human activities of casually confining and eliminating nonhuman animals, but the field is still trapped in the tarpits of anthropocentrism. In animal studies, "studying" is still understood as a unidirectional process: humans study animals, and not the reverse. Humans decide which questions are asked, choose modes of encounter with animal participants, interpret the results, and then represent animals in the products of research. Even when animal studies have positive effects on animals, and even if researchers are well-intentioned and attempt to center the interests of animals in their studies, nonhuman animals are still fully dependent on the goodwill of researchers to ask the right questions, correctly interpret the answers, and communicate them adequately to the public. So far, we have not been able to shift away from this human center of animal studies.

In an era of the Anthropocene, the lives of animals are massively and irreversibly shaped by human action, to the extent that animal losses regularly manifest as human gains. Cows, fish, and chickens die so humans can be happy and well-fed. Beagles, monkeys, frogs, and others are confined and harmed to improve or save human lives. Dogs and cats are disciplined, patronized, and controlled by using force to ensure human society is orderly. In these socio-cultural contexts, raising the argument that animals resist often meets hostility from researchers, who may benefit from misinterpreting, misrepresenting, and systematically neglecting the interests of animals. Even the most well-intentioned researchers, who strive for impartiality and acknowledge the perspective of animals, may hesitate to challenge the larger power structures that dictate research funding, job availability, professional reputation, and outreach (Reichlin et al. 2016). When the whole power structure is arrayed against animal agency, it is difficult to begin and persevere in research projects that look for, or even better, presume this agency.

Standard research structures, and the results they produce, are especially problematic as they shape our understanding of animals (personally and politically). Informed by these views, laws are then set up by humans vis-à-vis animals, so any obligations that flow from them are, without exception, "self-judging." In international law, self-judging obligations are widely decried, as the Separate Opinion in the *Norwegian Loans* case by Judge Lauterpacht, writing in his capacity as a judge for the International Court of Justice (ICJ), shows: "An instrument in which a party is entitled to determine the existence of its obligation is not a valid and enforceable legal instrument of which a court of law can take cognizance. It is not a legal instrument. It is a declaration of a political principle and purpose" (ICJ Norwegian Loans 1957, 43). The structural shortcomings of self-judging obligations that Judge Lauterpacht analyzes, are, *mutatis mutandis*, inherent in any legal system that is organized exclusively by humans and which unilaterally lays down our obligations vis-à-vis animals. Animal studies and the broader scientific inquiries that have an effect on animal agency—be it animal research, food ethics, political theory, environmental ethics, constitutional and human rights theory, or any other field or discipline—exemplify this sort of unchecked power: One group investigates another in a wholly unchecked manner, and determines the rules of interaction, too, unchecked by principles of objectivity. Because we humans are beneficiaries of animal use—be it directly or indirectly—, we are at a perpetual risk of lacking the necessary objectivity to evaluate these competing interests.

This imbalance is not limited to the animal realm. When researchers study children, adults dominate research design, process, and outcome. The difference is that ethical and legal principles govern these interactions. Research Ethics Boards (REBs) ensure that researchers adhere to pre-agreed principles, sanction researchers who violate them, and guarantee that research with human participants truly meets ethical standards. But REBs do not review research conducted on or with animals. Instead, Animal Care Committees and their Animal Use Protocols govern these relations, taking an instrumental, anthropocentric view wherein animals are treated as research objects (Cojocaru and von Gall 2019). These protocols center on welfare and humane use, framed mainly by the 3R principle, i.e. the duty to replace, reduce, and refine the use of animals in research (Herrmann and Wayne 2019). The 3R principle typically requires or, in effect, leads to a cost-benefit analysis (Peters 2012, 31–41), where harms to animal subjects are weighed against benefits to science, humans, other animals, or broader ecological groups/systems. Existing guidelines are rarely concerned with ethical assessments of whether the knowledge we gain merits the use of animals, and even more rarely ask if animals should be used at all (Orlans 2008). As Gillespie and Collard (2015, 205) note, "[a]nimals are considered outside the purview of 'human' ethics, and animal ethics revolves, in most cases, around a presumed 'disposable' animal life".

While we can and should challenge this approach in all animal research, invasive and non-invasive, we still lack principles that can guide us in the pursuit of respectful research with (rather than on) animals in order to, for example, study animal agency. Without such principles, instances of animal agency are unlikely to be seen or looked for, and, as a consequence, we cannot begin to define, let alone move ahead with, more respectful relations with animals. We thus need to design, advance, contest, and discuss principles that guide (human) researchers in their interactions with and representations of animals. In a recent article in *Animals & Society*, Van Patter and Blattner (2020) took a first step, and proposed a set of guiding principles to fill this gap. They suggest principles for designing an ethics protocol for non-invasive research with animal participants based on welfare- and agency-based considerations, which departs from current speciesist institutional animal care conventions. The protocol is guided by respect, justice, and reflexivity and defines three core principles: non-maleficence (including duties of vulnerability and confidentiality), beneficence (including duties of reciprocity and representation), and voluntary participation (mediated informed consent and ongoing embodied assent). Weaved into these three principles are duties to represent animals as subjects with their own agencies, communities, and personalities; to center their stories, thoughts, feelings, and uniqueness; and to study animals' material lifeworlds, use of space, and social interactions with a motivation to acknowledge their agency and subjectivities. The protocol is designed to spark broader scholarly engagement with the topic, which can and should, ideally, permeate into the law. As long as the status quo in research institutions is to resort to a welfarist 3R framework, scholars and practitioners who want to engage in respectful, non-invasive research with animal participants can adopt such protocols on a voluntary basis. As more researchers do so, institutional review boards may gradually incorporate ethical considerations for non-invasive research with animals into their protocols. For, after all, institutionalizing respectful research principles with animals should be the end-goal not only of people dedicated to advancing the fate of animals, but of all people dedicated to solid, impartial research.

4.3 Standpoint Acknowledgement and How to Ask the Right Questions

While we are, as a society and individually, working to ensure that ethical considerations become embedded in research with animals and taken up by law and policy, we must remain attentive to power relations and positionality (Van Patter and Blattner 2020). Researchers must guard against interacting with and representing animals in ways that perpetuate relations of domination and marginalization. We must, for example, stop bending research on animal agency to human supremacy. Current practice is to ask questions that presume and look for differences in human and animal agency. Worse even, we tend to avoid the word "agency" altogether when we talk about animals and when we assess the rules of interaction between us. Or, we admit to the existence of animal agency to the extent that this does not throw out of order the dominant ways in which we use and abuse animals. For example, we usually recognize and respond to dogs' and cats' food preferences since this does not question our right to use them, but we ignore chickens' preference to remain alive as they are swallowed up alive by "chicken harvesting machines" (Wadiwel 2018). This is even though in the human case, we consider our interests in life fundamental, whereas our interests in food types are usually less protected (at least legally speaking). This suggests that also in the case of animals, when they exercise agency in defense of fundamental values, like life and bodily integrity, this should be taken much more seriously (compared to food types etc.).

Again, this tendency to omit looking for agency in animals who find themselves in heavily restrictive environments is explained by the focus on our use of animals rather than on the animals themselves. To dismantle these self-reinforcing practices, we must acknowledge the unequal power relations and our own positionality as humans socialized within one-sided systems of thought. We should also engage in reflexive practices and consider how this inequality and our biases may influence our research design and conduct, asking open-ended questions about animals (rather than questions that serve pre-determined human interests), including:

- Do animals have capacity and interest in self-willed action?
- How important is it for animals to exhibit agency? In what form? What factors facilitate animals' use of agency? And which factors thwart it?

- What are the best social, environmental, political, economic, and other circumstances for studying animal agency?
- If animals do have and value their agency, how must this shift our current ethical and political understanding of human-animal relationships?
- To what extent and how must the law adapt to ensure animals can realize their agency?
- What would laws that respect animal agency look like?

So far, the focus of most scientific inquiries into animal agency was on animal resistance, offering us a richer picture of animals' desires, and throwing into doubt the presumption that animals can be freely used, handled, farmed, or done to whatever humans like to do to them. If resisting animals were taken seriously instead of silenced, it is easy to imagine how the world could become a more just place for them: We would respond to instances in which they do not feel comfortable, adapting our behavior accordingly. However, resistance is not "the only measure for the wellbeing and welfare of animals living, laboring, and dying in service to capital accumulation" (Gillespie 2016, 129). Focusing on resistance alone means that the only agential option for animals is to opt out. Building a political system on this premise is risky as it disregards the structural, institutional, and interpersonal biases against animals that render their environment largely unresponsive to their concerns and reduce their ability to meaningfully resist (Meijer 2016, 66). Animals whose resistance goes unheard will, as a consequence, often develop learned helplessness, which renders them "inarticulate" (Despret 2004, 124). Focusing on resistance as a model for animal agency alone also risks positing animals as reactants, as passive beings to whom things happen. As such, it does not account for the manifold ways in which animals shape and change the world around them and initiate and foster relationships. An exclusive resistance model limits our ability to recognize that animals have much more agential capacity and a much more profound interest in exercising and realizing it than we typically assume (Blattner 2020).

Studies set up to reduce researcher bias against animals have consistently shown that animals have impressive capacities and strong, indeed, intrinsic interests in decision-making, self-willed action, and relational agency, which we tend to heavily underestimate (Blattner 2020). Animals have their own individual preferences for, e.g., specific foods, locations, social partners, activities, and objects (Slocombe and Zuberbühler 2006), and they invest considerably into getting what they like (Hopper et al. 2015). Having choices has a strong positive effect on animals. Giant pandas (Owen et al. 2005), polar bears (Ross 2006), goats and sheep (Anderson et al. 2002), and many other animals are less stressed and show positive behavioral changes when provided with, e.g., more space, access to different rooms, or choice about where to spend time. Rhesus monkeys prefer completing a series of cognitive tasks in a self-chosen order rather than an assigned order (Perdue et al. 2014). The research with giant pandas and polar bears (Owen et al. 2005; Ross 2006) shows that animals prefer choices even when they do not take advantage of them. Chimpanzees and gorillas respond positively to having the choice to go outside (demonstrating positive social behavior like grooming, lower cortisol levels, a steep drop in signs of anxiety and restlessness), even if they chose to stay inside (Kurtycz et al. 2014). By and large, whenever humans have gone the extra mile to inquire about animal agency, they have consistently found that animals have strong instrumental interests in agency and, indeed, intrinsic interests in agency (Blattner 2020).

Because many investigations of animal agency are still carried out in controlled environments with confined individuals or groups, they give us only an incomplete and limited picture of animal agency. Possibilities of agential action are limited, especially for decision-making that could change the macro-dimensions of animals' lives, concerning, e.g., whether they want to live (which, unsurprisingly, most animals do), where and with whom they want to live (humans? nonhuman animals? a multi-species society?), their communities and social structures (with common decision-making structures, hierarchical, or equality-based), and what their daily routines should look like (including daily activities, foods, places and routines of food, sleep, play, greeting, etc.). As we seek to reveal certain glimpses of animal agency without sensitivity to these bigger questions, we run the risk of re-inscribing larger power hierarchies. Demonstrating that dogs who have control over ending electric shocks recover more quickly (Seligman and Maier 1967), for example, does not justify inflicting pain on dogs, exposing them to stressors, or confining them. Rather, proof of agency in these controlled environments shows that animals value self-determined action. Accordingly, we must take full account of animal agency, or, at the very least, aspire to do so.

We need researchers who are committed to critically evaluating existing accounts of agency and to developing a more accurate picture of animal agency, its extent and relevance, especially in environments that provide them with the broadest possibilities for agential action. Innovative research in this area has explored, for example, individual and collective dimensions of animals' agency in sanctuary settings, by studying their use of space and place, their practices and routines, and their social roles and norms, in order to learn whether and how animals might want to live with us, and how we can recognize and support their agency through our relationships (Blattner et al. 2020). Exploring animal languages, too, is a fruitful inquiry that has the potential to reveal previously unknown manifestations of or desires in exercising agency (Meijer 2019).

The emerging research area of animal agency is marked by three distinct tendencies. First, this research begins with the animals' perspective instead of comparing and contrasting the capacities of animals and humans (Meijer and Bovenkerk, in this volume). Second, it explores animals' agency in its positive dimensions—looking for decision-making, intentional action, pro-active behavior, self-willed action, and relational agency on top of instances of resistance—instead of taking the welfarist track and considering them only as pain-avoiders. After all, animals have myriad interests in deciding for themselves what to eat (cabbage? carrots? chickpeas?), whom to live with (humans? animals? no one?), what to do throughout the day (wander around? say hi to different people? forage in the woods? go for a swim at the beach?), whether or not to have relationships with other animals and humans, where and how to sleep, and what ground and property they want to traverse. Third, research into animal agency has the potential to influence and, ideally, change the larger political realities.

If, however, agency is only superficially explored, within the confines of human oppression, our understanding of animal agency becomes a watered-down version of what it truly is. Rather than empowering animals and working toward a multispecies polity, limited accounts of animal agency operate as means for humans not to question the larger actions by which they disenfranchise and oppress animals. Honest, unbiased, and open-ended inquiries into animal agency, on the other hand, can challenge existing power hierarchies and make clear that current injustices are not irreversible, given, or nonnegotiable. One new and particularly promising strand of research where animal agency is studied to ask and answer political questions is "political multispecies ethnography"—an ethnographic participant methodology suited to the study of human and animal interactions, and committed to supporting their agency and advancing interspecies justice (Kymlicka and Donaldson 2014; Blattner et al. 2019). This is a relational methodology that is dedicated to the study of human-animal relations (be they close or distant) to understand power-laden entanglements among species and alter interspecies status and hierarchies (Gillespie 2019). Results produced by political multispecies ethnography can challenge deep-seated biases in the larger socio-political structures—such as the ones I identified in this article—, to make visible animals' views and help us understand how animals' agential actions themselves are challenging broader phenomena, for example, climate change or the expansion of human population into nature and animals' territories.

4.4   Calling for an "Animal Agency Turn"

In the Frankfurt Germany's Fechenheim district, a 22-year old Arabian mare named Jenny roams the neighborhood on her own. Every morning, she takes a leisurely stroll through the streets. Dozens of worried pedestrians have called the authorities, afraid that Jenny has been neglected or poses a danger to herself and others during her morning walks. These worries were dismissed by veterinarians, who testified that Jenny knows very well what she's doing and seems to be satisfied with her activities. Jenny now wears a letter attached to her harness, informing concerned people in town that she knows her way around and is doing her own thing: "I'm called Jenny, not a runaway, just taking a walk. Thanks." Locals got used to seeing her walk around on her own; some even say that more animals should be able to walk freely (DW Newsletter 2019). Jenny's story shows that there is much to be gained from the study of animal agency as it challenges our views about what agency is, who can exercise it, and how it manifests. Such common knowledge, as well as new scientific findings about animals' agency should be integrated into neighboring disciplines, including politics, law, geography, design, and economics. For example, if the law recognized animals as agents, this would crucially change the way we organize human-animal relationships personally and politically: Animals' voice would need to be considered in deciding who deserves legal protection and, relatedly, who gets legal recourse. Building on this, this contribution calls for an "animal agency turn" that we must take, in concert with animals, by educating fellow researchers and exposing friends, family, and the public to instances of animal agency. For animals' acts of agency to be heard, seen, and recognized, we need nothing short of civic competence and courage, empathy, participation, common engagement, and respect for animal alterity.

References

Anderson, U., M. Benne, M. Bloomsmith, and T. Maple. 2002. Retreat space and human visitor density moderate undesirable behavior in petting zoo animals. *Journal of Applied Animal Welfare Science* 5 (2): 125–137.

Blattner, C.E. 2019. *Protecting animals within and across borders: Extraterritorial Jurisdiction and the challenges of globalization.* New York, NY: Oxford University Press..

Blattner, C.E. 2020. Animal labour: Toward a prohibition of forced labour and a right to Freely choose one's work. In *Animal labour: A new frontier of interspecies justice?*, ed. C.E. Blattner, K. Coulter, and W. Kymlicka, 91–115. Oxford: Oxford University Press.

Blattner, C.E., S. Donaldson, and R. Wilcox. 2020. Animal agency in community: A political multispecies ethnography of VINE sanctuary. *Politics and Animals* 5: 33–54.

Cojocaru, M.-D., and P. von Gall. 2019. Beyond plausibility checks: A case for moral doubt In review processes of animal experimentation. In *Animal experimentation: Working towards a paradigm change*, ed. K. Herrmann and K. Wayne, 289–304. Leiden: Brill.

Corman, L. 2016. The ventriloquist's burden: Animal advocacy and the problem of speaking for others. In *Animal subjects 2.0*, ed. J. Castricano and L. Corman, 473–512. Waterloo: Wilfrid Laurier University Press.

Davis, K. 2011. Procrustean solutions to animal identity and welfare problems. In *Critical theory and animal liberation*, ed. J. Sanbonmatsu, 35–53. Plymouth: Rowman & Littlefield Publishers.

Deckha, M. 2012. Toward a postcolonial, posthumanist feminist theory: Centralizing race and culture in feminist work on nonhuman animals. *Hypatia* 27 (3): 527–545.

Despret, V. 2004. The body we care for: Figures of anthropo-zoo-genesis. *Body and Society* 10 (2/3): 111–134.

Dutch Animal Law. 2011. Wet van 19 mei 2011, houdende een integraal kader voor regels over gehouden dieren en daaraan gerelateerde onderwerpen (Wet dieren) [Act of May 19, 2011, containing an integrated framework for rules on animals kept and related subjects (Animal Law)], May 19, 2011 (Neth.).

DW Newsletter. 2019. Horse takes daily stroll through Frankfurt—Without owner.

Eisen, J. 2010. Liberating animal law: Breaking free from human-use typologies. *Animal Law* 17: 59–76.

Gillespie, K. 2016. Nonhuman animal resistance. In *Animals, biopolitics, law: Lively legalities*, ed. I. Braverman, 117–132. Oxon and New York, NY: Routledge.

Gillespie, K., and R.C. Collard. 2015. *Critical animal geographies: Politics, intersections and hierarchies in a multispecies world.* London, UK: Routledge.

Gillespie, K.A. 2019. For a politicized multispecies ethnography: Reflections on a feminist geographic pedagogical experiment. *Politics and Animals* 5: 17–32..

Griffin, J. 2002. First steps in an account of human rights. *European Journal of Philosophy* 9 (3): 306–327.

Griffin, J. 2004. Discrepancies between the best philosophical account of human rights and the international law of human rights. In *Proceedings of the Aristotelian Society*, 1–28.

Griffin, J. 2008. *On human rights.* Oxford, UK: Oxford University Press.

Herrmann, K., and K. Wayne. 2019. *Animal experimentation: Working towards a paradigm change.* Leiden: Brill.

Hopper, L., L. Kurtycz, S. Ross, and K. Bonnie. 2015. Captive chimpanzee foraging in social setting: A test of problem solving, flexibility, and spatial discounting. *PeerJ* 3: e833.

Hribal, J. 2010. *Fear of the animal planet: The hidden story of animal resistance.* Petrolia: CounterPunch.

ICJ Norwegian Loans. 1957. Case of Certain Norwegian Loans. Separate Opinion Sir Hersch Lauterpacht, 1957 ICJ Rep. 34.

Kalof, L. 2017. *The Oxford handbook of animal studies.* Oxford, UK: Oxford University Press.

Kurtycz, L., K. Wagner, and S. Ross. 2014. The choice to access outdoor areas affects the behavior of great apes. *Journal of Applied Animal Welfare Science* 17 (3): 185–197.

Kymlicka, W., and S. Donaldson. 2014. Animals and the frontiers of citizenship. *Oxford Journal of Legal Studies* 34 (2): 201–219.

Kymlicka, W. 2017. Social membership: Animal law beyond the property/personhood impasse. *Dalhousie Law Journal* 40 (1): 123–155.

Liao, S.M. 2009. Agency and human rights. *Journal of Applied Philosophy* 27 (1): 15–25.

Meijer, E. 2016. Interspecies democracies. In *Animal ethics in the age of humans,* ed. B. Bovenkerk and J. Keulartz, 53–72. Cham: Springer.

Meijer, E. 2019. *When animals speak: Toward an interspecies democracy.* New York, NY: Press.

Nussbaum, M. 2006. *Frontiers of justice: Disability, nationality, species membership.* Cambridge, MA: The Belknap Press..

OIE. 2019. Terrestrial animal health code. Paris: World Organization for Animal Health (OIE).

Orlans, F.B. 2008. Ethical themes of national regulations governing animal experiments: An international perspective. In *The animal ethics reader,* ed. S.J. Armstrong and R.G. Botzler, 365–372. New York, NY: Routledge.

Owen, M., R. Swaisgood, N. Czekala, and D. Lindburg. 2005. Enclosure choice and well being in giant pandas: Is it all about control? *Zoo Biology* 24 (5): 475–481.

Perdue, B., T. Evans, D. Washburn, D. Bumbaugh, and M. Beran. 2014. Do monkeys choose To choose? *Learning & Behavior* 42 (2): 164–175.

Peters, Anne. 2012. Rechtsgutachten zu verschiedenen Fragen im Zusammenhang mit der EU-Tierversuchsrichtlinie insb. zur Unionsrechts- und Verfassungskonformität des Entwurfs eines Dritten Gesetzes zur Änderung des Tierschutzgesetzes sowie des Entwurfs einer Verordnung zur Umsetzung der Richtlinie 2010/63/EU. https://www.aerzte-gegen-tierversuche.de/images/pdf/recht/gutachten_eu_richtlinie.pdf. Accessed 25 August 2020.

Philo, C. 1998. Animals, geography, and the city: Notes on inclusion and exclusions. In *Animal geographies: Place, politics, and identity in the nature-culture borderlands,* ed. J. Wolch and J. Emel, 51–71. London, UK: Verso.

Reichlin, T.S., L. Vogt, and H. Würbel. 2016. The researchers' view: Survey on the design, conduct, and reporting of in vivo research. *PLoS ONE* 11 (12): e0165999.

Rollin, B.E. 1995. *Farm animal welfare: Social bioethical and research issues.* Ames, Iowa: Iowa State University Press.

Ross, S. 2006. Issues of choice and control in the behaviour of a pair of polar Bears. *Behavioural Processes* 73 (1): 117–120.

Rowlands, M. 2012. *Can animals be moral?.* Oxford, UK: Oxford University Press.

Seligman, M., and S. Maier. 1967. Failure to escape traumatic shock. *Journal of Experimental Psychology* 74 (1): 1–9.

Slocombe, K., and K. Zuberbühler. 2006. Food-associated calls in chimpanzees. *Animal Behaviour* 72 (5): 989–999..

Špinka, M., and F. Wemelsfelder. 2011. Environmental challenge and animal agency. In *Animal welfare,* ed. M.C. Appleby, I.A.S. Olsson, and F. Galindo, 27–44. Wallingford, UK: CAB International.

Universal Declaration of Human Rights. 1948. G.A. Res. 217 A (III), U.N. GAOR 3rd Sess., Universal Declaration of Human Rights, U.N. Doc. A/RES/810 (December 10, 1948).

Van Patter, L., and C.E. Blattner. 2020. Advancing ethical principles for non-invasive, respectful research with animal participants. *Society & Animals,* 1–20.

Wadiwel, D. 2018. Chicken harvesting machine: Animal labour, resistance and the time of production. *South Atlantic Quarterly* 117 (3): 525–548.

Wesleyan. 2019. *Animal studies.* https://www.wesleyan.edu/animalstudies/. Accessed 25 August 2020.

Wilbert, C. 2000. Anti-this—anti-that: Resistance along a human-nonhuman axis. In *Entanglements of power: Geographies of domination/resistance*, ed. J. Sharp, P. Routledge, C. Philo, and R. Paddison, 238–255. London: Routledge.

Charlotte E. Blattner is a Senior Researcher and Lecturer at the Institute for Public Law, University of Bern, Switzerland. She earned her PhD in international law and animal law from the University of Basel, Switzerland, as part of the doctoral program Law and Animals. From 2017-2018, she completed a postdoctoral fellowship at the Department of Philosophy at Queen's University, Canada, working on animal labour as part of Animals in Philosophy, Politics, Law, and Ethics (APPLE). From 2018–2020, Blattner was a postdoctoral fellow at Harvard Law School's Animal Law & Policy Program, funded by the Swiss National Science Foundation, to explore critical intersections of animal and environmental law. She is the author of *Protecting Animals Within and Across Borders* (2019) and *Animal Labour: A New Frontier of Interspecies Justice?* (2020, coedited with Will Kymlicka and Kendra Coulter), both published by Oxford University Press. Her current research focuses on climate law and studies, in an attempt to challenge conservative estimates, purely market-based mechanisms, and ingrained anthropocentrism.

Chapter 5
Animal Difference in the Age of the Selfsame

Nathan Kowalsky

Abstract In this chapter, I argue that mainstream animal-centered (i.e., "humane") ethics and critical animal studies attempt to account for nonhuman moral considerability in terms of those animals' similarities with human animals. I argue that this emphasis on similarity is a reason why these two fields are generally anti-naturalistic and ultimately (though ironically) anthropocentric. Moreover, on the assumption of a general Levinasian ethic of alterity, this anti-naturalism and anthropocentrism is violently immoral. I propose, therefore, an ethic of animal difference based on an ethically naturalistic reading of intra- and inter-specific behavior sets. However, such naturalism is problematic if the Anthropocene is understood to be a naturalized fact which undermines all (metaphysical or normative) claims to naturalness or wildness. In response, I argue that the Anthropocene is not a naturalized fact but a socially-contingent and constructed fact, and as such is open to moral evaluation. My proposed ethic of animal difference offers one such critique, and one more effective than those found in mainstream humane ethics or critical animal studies.

5.1 Progressivist Anti-naturalism

Peter Singer (1981), the founding father of animal liberation ethics, sees the so-called "circle of ethics" as expanding over the course of history, moving outwards from the individual human self as normative center. Relying on the nineteenth-century historian William Lecky, Singer sees ethics in general as a growth of concern from one's own well-being towards one's family, and eventually out towards all people and even animals. As the story goes, human beings are, by nature, egotistic and "inherently partial" to themselves (Kemmerer 2011, 73), as is the rest of the animal kingdom. Eventually, however, it dawned on our species (at least) that our own self-interests were better served by mutual co-operation, in spite of our innate distaste for getting

N. Kowalsky (✉)
St. Joseph's College, University of Alberta, Edmonton, Canada
e-mail: nek@ualberta.ca

© The Author(s) 2021
B. Bovenkerk and J. Keulartz (eds.), *Animals in Our Midst: The Challenges of Co-existing with Animals in the Anthropocene*, The International Library of Environmental, Agricultural and Food Ethics 33,
https://doi.org/10.1007/978-3-030-63523-7_5

along with each other. Morality was thus invented to keep beneficial social groups operating smoothly, even though it was initially limited to small groups. Over time, however, those groups got bigger and moral inclusivity increased; new human subsets were included in the definition of the self-interested ego. Therefore, recent and more enlightened generations have seen movements advocating the moral and legal equality of women, African-Americans, LGBTTQQPIANU+ persons,[1] and other oppressed minorities. Today, enlightened or progressive persons find themselves at the point where all of humanity is within the community of moral concern, and they face the question of expanding morality further to include nonhuman animals.

The narrative of moral progress is not merely a description of how humans have, in fact, morally developed. It is a normative claim about how morality should have developed (and thankfully, is developing). Ethics is and ought to be *self-interest increasingly generalized over time.* The ego is necessarily the only intrinsically valuable thing, and "higher ethical consciousness" simply expands the boundary of the ego to include other selves within its own self-definition (Singer 1997). Progressive ethics which include at least some nonhuman animals (hereafter called "humane ethics")[2] criticize classical Enlightenment moralities for not being progressive enough—the latter ethics are "anthropocentric," an egoism of humanity. While other chauvinisms recognize no values outside a narrowly defined self, anthropocentrism broadens that self until it is continuous with a conception of the entire human species that recognizes no inherently valuable things outside itself (Midgley 1994). For humane ethics, the solution to anthropocentric chauvinism is to expand the definition of the ego yet further, beyond the boundary of the human species. Thus does John Clark, a critic, identify this move as *moral extensionism*, "the project of applying ethical theories based on anthropocentric (and usually ethical individualist) presuppositions to greater-than-human and larger-than-individual moral realities such as species, ecosystems, and the biosphere" (Clark 2014, 171, n. 46).

There is tension, however, between moral extensionism and nonhuman animals. Up to the species barrier it was comparatively easy for the circle of ethics to expand, because the differences between one's own self and other human beings could be clearly shown to be surmountable. But crossing the species barrier presents progressive morality with an unprecedented obstacle: generally speaking, animals are not capable of behaving in accordance with the dictates of generalized egoism. In Singer's terms, they do not and indeed cannot act in accordance with the principle of utility. Of

---

[1] This abbreviation stands for Lesbian, Gay, Bisexual, Transgender, Two-Spirit, Queer, Questioning, Pansexual, Intersex, Asexual, Non-Binary, Unlabelled, and more. Source: https://www.su.ualberta.ca/services/thelanding/.

[2] A convenient shorthand to denote "animal-centered ethics" has been hard to come by. Neither "animal welfare" nor "animal rights" will suffice, because these terms denote exclusively utilitarian or deontological frameworks. "Animal activists" and "animal advocacy movement" have been proposed, but neither term gives an indication of the sort of ethics operative therein. Thence my proposed plural "humane ethics," as it captures (as I shall argue below) the anthropocentrism implicit in moral extensionism ("the word 'humane' is just a dressed-up version of the word we use for ourselves" [Seitz 2010, 75]) while being colloquially associated with nonhuman animals (e.g., the Humane Society), albeit with unnecessary utilitarian connotations.

course, humane ethics do not claim that animals should voluntarily follow Enlightenment norms—at this point Singer (1975, 237), Tom Regan (2004, xxxvi–xxxviii), and Lori Gruen (2011, 182–183) grant animals the autonomy to be what they are and behave in their own ways—but Singer, at least, cannot but "regret that this is the way the world is." In other words, it's a lamentable shame that nonhuman animals do not fit well within the sphere of morality that expands outwards to include them (Raterman 2008).

This regret is the crux of the well-known clash between humane and environmental ethicists (Hargrove 1992). While many environmental philosophers have used the language of expansion when encouraging the broadening of human moral horizons to include ecology (Leopold 1949; Naess 1989; Rolston 2012), in holistic environmental ethics the individualism assumed by Enlightenment ethics was seen (albeit controversially) to be relativized by encompassing natural systems (Rodman 1977; Goodpaster 1979; Callicott 1980). By contrast, humane ethics understood their expansive transcendent self to be a *de jure* indivisible thing, a norm of moral inviolability, an 'individual.' But the naturalistic holism of land ethics illuminated a recalcitrant reality: "Nature…is not fair; it does not respect the rights of individuals" (Callicott 1989, 51). If it did, every food chain that exists would shut down: "The most fundamental fact of life in the biotic community is eating…*and being eaten*" (Callicott 1989, 57). There is no right to life evident in nature, nor any tendency to alleviate suffering. Nature (or at least the processes of wild or undomesticated ecologies) does not line up very well with an ethic of generalized self-interest where the primary duty is not harming whatever counts as an ego.

Environmental ethics thus diverged from humane ethics for the same reason that nature has fared poorly in Western ethics generally: naturalness, like tradition, functions as a limit to (putative) reason and progress. To be associated with nature or the body is, as ecofeminists have pointed out, to be considered 'irrational' or, in the socio-political sense, 'backwards.' A case in point is critical animal theorist James Stanescu's (2012a) advocacy for "the Gothic's resistance to the natural order" because "a dark animal studies needs to dissociate itself from the tyranny of the natural order" (p. 44)…"We are now about as far away from [Michael] Pollan's notion of having 'a respect for what is' as we can be" (p. 46). This anti-naturalism is even more boldly articulated by vegan food writer Stefany Anne Golberg (2011): "Nature is an asshole. We know this and other animals don't." Antipathy towards nature is presupposed at the outset of morally progressive narratives, scuttling attempts at resolving the impasse between humane and environmental ethics.

5.2 Sameness and Anthropocentrism

Enlightenment progressivism sees itself as *discontinuous* with what it conceives of as nature, be it vicious wild animals or humanity's own primitive animality. On the other hand, Enlightenment progressivism expands by uncovering *continuities* between itself and entities not yet included within its boundaries. Between humans, particular differences (such as age, gender, class, or creed) are conceptually discarded as accidental, while universals like 'humanity' are held to be the basis of our unalienable rights as individuals. The task of humane ethics is to show that this core notion of self—an ideal derived from the Enlightened human exemplar—shares relevant commonalities with some nonhuman animals. We have already seen this to be the case with Singer's expanding circle, but it is also the case for a wide and representative swath of non-welfarist humane ethics.

The morally relevant commonality for Regan's deontological ethic is being a "subject-of-a-life," which is supposed to engender moral duties in human animals to respect the desires of nonhuman animals to not be used, harmed, or killed—duties which are analogous to how human animals are obliged to treat each other. Ecofeminist Carol Adams uses clearly progressive language: "Color will lose its character as a barrier, just as 'animals' will lose their otherness, and join human animals as a 'we' rather than a 'they' or a collective of 'its'" (Adams 1994, 78). Gruen also argues on the basis of similarity with humans: "Other animals matter because, *like us*, their lives can go better or worse for them. They are sentient beings who have interests and well-beings. They can be harmed when their interests are thwarted and their wellness undermined" (Gruen 2011, 33, emphasis mine). Sue Donaldson and Will Kymlicka (2011) offer a theory of citizenship and universal rights that applies, via analogy, from human to nonhuman animals: just as certain human beings are granted different rights depending on the sort of citizenship they have in a political community, nonhuman animals are granted different rights depending on whether they are analogous to human *citizens* of one's own country, human citizens of a *foreign* nation, or human *denizens* of one's own country.

Similarity and sameness between human animals (at the presumptive moral core) and nonhuman animals (recently relocated from the outside to the inside of the moral sphere) is at the root of the humane ethics mentioned above, and can be also found within a wide and representative swath of critical animals studies. Even though Nik Taylor (2011) criticizes moral expansionism for "simply maintain[ing] dualist conceptions while moving the boundary slightly…and as such, ultimately reinforc[ing] traditional anthropocentrism" (pp. 206–207), she goes on to advocate for "the removal of animal oppression and the serious inclusion of animals themselves into our intellectual sphere" (p. 219) by "waging war on essential differences" (p. 210) and allowing "the cognitive capacities of humans to migrate to objects" (p. 211), as if she forgot her point about reinforcing traditional anthropocentrism. Richard Twine, meanwhile, simply assumes that "what we share with other animals both socially and corporeally ought to be enough to transgress the human/animal dualism of moral considerability" (Twine 2014, 199). Critical animal theorists generally assume that human exceptionalism is the only philosophical obstacle they face, and that its solution is an inclusive appeal to cross-species commonalities. Doing so, however, operates on the assumption that sameness with humans is good while difference is not: "What is powerful is not what makes us unique, but what makes us in-common. What is exhilarating is not what individuates us, but rather what brings us together" (Stanescu 2012b, 576–577); "In other words, we invest a vast amount of intellectual work in trying to figure out what separates and individuates the human species, rather than in what makes us a part of a commonality with other lives" (Stanescu 2012b, 569). Therefore, as Lisa Kemmerer (2011) adds, "working to define human beings as distinct from other animals [has] the hidden agenda of justifying human supremacy, dominion, and exploitation" (p. 70).

I contend that the progressive search for commonalities in humane ethics (including critical animal studies generally) follows the logic of Hegelian dialectic: in the beginning there is the self, the subject. But subjective consciousness does not know very well what it is (or what its ethic should be), and so it puts forth a proposal of selfhood, externalizing itself, articulating its loosely formed idea in the objective realm. But objective consciousness is not subjective consciousness and so there is always an incongruity between the two; the objective other-than-self is different than the selfsame, and the self is rocked back upon itself, disgusted by its poorly realized (not inclusive enough) ethic, and so is forced to revise its understandings and to try again anew. This often violent relation between subjective thesis and objective antithesis (Hegel [1956, 21] calls it a victimizing "slaughter-bench") is the engine of progress, driving the self forward dialectically as it encounters recalcitrant objectivity, appropriating it, and creating new syntheses therefrom. However, the final goal—the Absolute—is when the negativity inherent in objectivity is overcome by the self's *discovery of itself* in the other.

On my reading of Hegel, other-modification takes priority over self-modification. As much as Hegelians might want to say that the subject discovers alterity within itself, this is not what provides the subject relief. The horrors of the objective stage are resolved by the balm of the selfsame, not alterity. Even if, for Hegel, the Absolute stage were to achieve perfectly reciprocal representation of both difference and sameness whereby both subject and object are mutually modified by each other, the normative standards of expansionist moral progressivism are *not* altered by the encounter with suitably similar nonhuman animals. In Kemmerer's unequivocal words:

> We ought not to theorize about "others."… If we can look into the bright eyes of a calf and see into a mirror – if we can see in this individual a person – complete with interests, hopes, and fears – not unlike ourselves, then our theorizing is likely to have a greater degree of validity. If we theorize about self whenever we theorize about fish or a dice snake, crab-eating mongoose, or killfish, our theories are more likely to be grounded in reality – the reality that there is no "other," the reality that we are all animals, and therefore are fundamentally alike, particularly in morally relevant ways, such as our ability to suffer and our innate desire to live without suffering…. Those who look at another human being, or another animal, and see "other" must not theorize about those "others."… If we are to theorize about oxen and sheep, then we must theorize about *self*…. *Please, do not theorize about "other" animals.* (Kemmerer 2011, 79–82 original emphasis)

Humane ethics thus reach the satisfaction of the Absolute when they find *sameness* at the heart of the other animal. Difference qua difference is simply opposition, negativity, even evil. The good is that which the self can find in the other to be in line with itself.

5.3   Violence Against Otherness

For the purposes of this chapter, I will take for granted a broadly Levinasian ethic of alterity, whereby moral wrongdoing is paradigmatically defined by violence, which is in turn defined as the reduction of the Other to the Same. On this account, my Hegelian reading of humane ethics implicates them in a violently immoral opposition to animal otherness. This starts with the creeping significance of species difference into human ethics. While differences between human persons are supposed to be morally irrelevant in progressive ethics generally, the expansion of the human(e) ego is not as seamless when encountering animal difference. While moral consideration can be extended to socially marginalized humans without modification, it cannot be extended without modification to even our closest "evolutionary comrades" (Vera 2008). Crossing the species barrier *is* morally relevant, even for moral progressives whose rhetoric suggests otherwise. Even though Singer (1974, 104) does not want to admit that species difference is morally relevant, he is clear that animals should not vote. So *prior* to the particulars of Singer's argument (and indeed regardless of whether this sensitivity to difference is consistent with the expanding circle), we can already see that as the ethic of moral sameness extends outward from the core of the human individual, it must be adjusted if it is to apply to nonhuman animals. The kinds of moral standing we recognize for nonhumans will *depend also on the differences between humans and nonhumans*. Natural difference means that human moral sameness cannot be the absolute moral standard after all.

Moreover, the ethic of sameness can only be extended so far before it exhausts itself. There are minimum requirements of similarity that must be met before moral recognition will be extended; failing those, the circle of ethics stops expanding. Humane ethics set minimum standards for moral considerability (for Singer, the line is somewhere in-between shrimp and clams [Singer and Mason 2006, 133–134, 275–276], while Regan is largely concerned with adult higher mammals), but at some point the differences between humans and certain animals—to say nothing of plants or nonliving ecosystemic components[3]—are just too great for humane ethics to include. *The more different a being is in comparison to the human, the less it will count within the scheme of expansionistic moral progress.*

What this means, then, is that some animals simply do not benefit from the expansion of human egoism. Their difference is such that insufficient commonalities are recognized between them and the transcendent Self. In addition to being excluded from moral considerability, some animals actually stand in clear opposition to the egoism being extended by moral progress, particularly predators.[4] Some humane ethicists (e.g., Singer, Regan and Gruen) fall back on the lack of moral agency

---

[3]Gruen (2011) draws the line between animals and plants, for while plants "can have their interests negatively affected," unlike us and (some?) other animals "they will never be interested in that impact" (p. 29). Regardless of whether clams or mosquitoes should be counted among plants, subjective rather than objective interests are Gruen's touchstone.

[4]The issue of predators is explored explicitly and at length in the chapter by Jozef Keulartz's "Should the Lion Eat Straw Like the Ox? Animal Ethics and the Predation Problem" in this volume.

in nonhuman animals—that is, their difference—to avoid advocating the policing of wild animal behaviors, but Martha Nussbaum's ethic is more progressive than that. Her capabilities approach "calls for the gradual formation of an interdependent world in which all species will enjoy cooperative and mutually supportive relations. Nature is not that way and never has been. So it calls, in a very general way, for the gradual supplanting of the natural with the just" (Nussbaum 2006, 399). She therefore requires that nonhuman predators, for example, be treated in ways analogous to how human sexual predators are to be treated (i.e., incarceration and behavioral modification [Nussbaum and Faralli 2007, 157]). After all, a violent animal's lack of moral culpability doesn't mean that it shouldn't be stopped. Therefore, when animals stand in opposition to the extension of human moral standards, those animals need to be corrected, i.e. forced into alignment with the progressive moral order (Wissenburg 2011).

So while many if not most people have intuitions that there are right and wrong ways to treat nonhuman animals, and while there are meaningful similarities between ourselves and many if not most nonhuman animals, difference nevertheless raises its ugly head and the progressive humane ethic has to backpedal. While there is no need to modify rights when they are extended from some (adult) humans to other (adult) humans, there is a need to modify rights when they cross the species barrier. Because species difference is ethically relevant even to anti-speciesists, it undermines the expansionistic model of moral progress. Progress is not supposed to have limits, and yet species difference does constitute a limit. The circle of ethics reaches a point where it can expand no further: because clams or trees (and whatever lies below them on the *scala natura*) *do not possess anything like the most basic element of what counts morally* for the humane ethicist, they cannot be directly morally considerable. When there is no self to be found in the other, difference outweighs sameness and inclusion stops. Beyond the boundary, the radically nonhuman can be instrumentally valued, benignly neglected, morally lamented, or coercively policed. The dark side of the ethic of sameness is its anti-naturalism: the more something can be included within the expanded human self, the better, whereas the less amenable something is to inclusion within that sphere, the more naturally problematic it is.

Progressivist anti-naturalism thus ends in the oppression of 'insufficiently human' nonhuman animals, just as (on my reading) Hegel's encounter with alterity violently reduces the Other to the Same. The Enlightenment ego values things (other persons, animals) insofar as they *cease* to be considered different from it and rather come to be seen (at least in the morally relevant aspects) as the same as it. Moral progressivism assumes egoism as the ethical starting point, basic to human nature and unavoidably rampant in the State of Nature, and it sees the solution to egoism as a more inclusive and broad egoism. Moral progress is the aggregation of egos whereby more and more things which had previously been excluded from the realm of moral sameness are included. Things that were on the outside are now on the inside; things that were other are now incorporated into the self. This logic views difference as a threat; Hans Jonas calls it "the negative experience of otherness" (Jonas 2001, 332). It tries to affirm variety and diversity by making it all morally homogenous. Anything outside Enlightened, democratic, liberal tolerance that resists assimilation is vilified as uncivilized, barbaric, or even savage. The form of subjectivity advanced by humane ethics replicates the imperialist logic of colonialism.

Granted, moral progressivism starts from the reasonable supposition that we value our selves for their own sakes, and some things (e.g., oxygen) are clearly better for ourselves than other things (e.g., hydrogen sulfide). It is trivial to say that things may be either good or bad in relation to the self (instrumentally valuable), but if we go on to claim that goodness itself is *completely defined* in terms of what is good for the self, then we claim the self to be absolute—the only being that matters—rather than one limited being among many. If the self is (*de jure*) an absolute being and an absolute unity, then anything different from the self or anything which threatens its absolute one-ness, is absolutely bad—the very definition of evil itself. Things that do not fit with the sameness of the self are cosmically out of order, because the self is the standard around which the cosmos should be ordered. Anthropocentrism thus reveals itself to be more than human exceptionalism or the simple denial of direct moral considerability for nonhumans; it is rather the species-level absolutization of otherwise reasonable self-preservation. Because they see alterity antagonistically, aspirationally nonanthropocentric humane ethics self-defeatingly replicate anthropocentrism by imposing human-modelled sameness onto nonhuman animals. *Animals only count in so far as they approximate human beings.* Human beings remain at the center of the circle, the absolute moral standard for all things, only to find this ideal increasingly frustrated the further it moves outward into nonhuman territory. Against this self-defeating ethic of expansionistic sameness, I propose a direction for ethics where animal differences are viewed positively rather than as obstacles to be overcome, where animals possess *independent* standards of value for themselves rather than being beholden to standards centering on us.

5.4   A Proposal for an Ethic of Animal Difference

Homes Rolston, III is a naturalistic environmental ethicist who offers an ethic which insists on the axiological relevance of "discontinuity" between animal species as the touchstone for our evaluation of animals (Rolston 1989). Such discontinuity vexes humane ethics, because it entails natural animal behaviors that do not appear to conform to the models derived from human civil society. Predation, parasitism, cannibalism, coprophagy, and cuckoldry are but a few examples of animal alterity that cannot be made to fit into the progressive moral order. Coprophagy—the eating of feces or dung—might just strike us as disgusting (although lagomorphs and juvenile iguanas apparently both enjoy and benefit from it), but carnivory, parasitism, cuckoldry and cannibalism all turn out rather badly for the particular individuals at the receiving end: prey (or cannibalized cubs) are painfully killed and eaten; hosts to parasites can suffer greatly before eventually dying; cuckolded parents struggle to feed their inadvertently adopted offspring, while their own offspring are often fatally outcompeted. Any ethic of generalized egoism cannot look kindly on such *de facto* violations of *de jure* inviolate individuals, and thus falls into anti-naturalism. My proposed solution, then, is to encounter animal otherness without trying to force it into alignment with individualistic subjectivity.

For Rolston, the key to an ethic of animal difference is recognizing the *wildness* of animals as a legitimate form of alterity. If humans should not reduce the value of animals to what they (or some of them) have found valuable about themselves, then they should espouse a value pluralism—or species relativism—in nature: "There are myriad sorts of things and they are differently made" (Rolston 1992, 253). Indeed, the etymology of the word 'species' is indicative of this plurality: each species is specific and special, and there are millions of species. Each one is different from the other in certain important aspects. There are many degrees of similarity between species too, of course, but what constitutes them *as species* is their specificity or specialty, their unique differences from other species.[5] Earlier, I argued that anthropocentrism should be understood as the imposition of human-modelled sameness onto the other-than-human. Here, this means that anthropocentrism should be understood as a denial of legitimate species-specificity: progressive moral expansionism sees all species (as much as possible) as unwitting aspirants to the human species. Fittingly, therefore, Rolston argues that anthropocentrism is a category mistake because it holds nonhuman species up to moral standards similar to those we hold ourselves to, as if it were illegitimate that there should be *different kinds of animals.*

Environmental nonanthropocentrism must then carefully parse the interrelations of the value plurality in nature: "intrinsic animal natures and their ecological places in the world" (Rolston 1989, 134). That is, individual animals (ourselves included) should be seen as governed by behavioral norms that concern both internal interactions with their respective species members (intraspecific relations between conspecifics) and external interactions with members of other species (interspecific relations between heterospecifics). Classical ethics, being focused exclusively on human behavior towards other human beings, seeks to identify good *interhuman* behavior. The anthropocentric mistake is to think that this human behavior set exhausts normative (as opposed to aesthetic) axiology. Interhuman 'morality' (if that term is to be limited to animals which possess 'moral agency' or volition) is but a species of the axiological genus, lying within a larger framework of 'nonmoral'

---

[5]I recognize that species essentialism is highly problematic in the philosophy of biology. However, my argument does not depend on species essentialism being true; it only depends on species nominalism being false. That is, while it is likely that species (and other biological kinds) do not have unchanging essences (otherwise evolution would be impossible!) it is not the case that species (and other biological kinds) are nothing but convenient naming conventions drawn from a contingent cultural repertoire projected onto an arbitrary group of things. Even though species (and other biological kinds) are "thoroughly heterogeneous collections of individuals whose phenotypic properties [change] over time, and [vary] across the population at any given time" (Wilson et al. 2007, 193), radical skepticism about the existence of species does not "do justice to natural kinds as they are studied in biology and other special sciences" (Brigandt 2009, 79). That is, species identification is scientifically convenient for a reason outside simple taxonomic utility. Homeostatic property clustering (stable grouping) of species (and other biological kinds) is something experienced by scientists as external to their own acts of categorization, and as such, possesses sufficient metaphysical reality for my proposed ethic of animal difference to proceed. For a fuller treatment of my view on the metaphysical status of species, see Kowalsky (2012, 129–132).

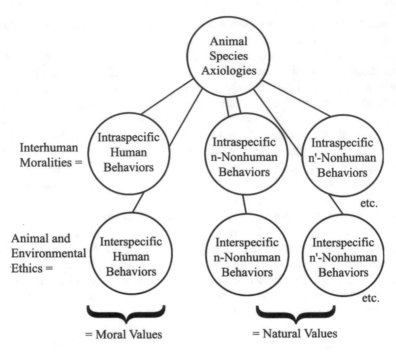

Fig. 5.1 Axiological categorization of species difference

values which relativizes human morality. Animal and environmental ethics, by way of contrast with interhuman ethics, prescribe good *interspecific* behaviors for humans, or our moral duties to nonhuman agents, entities, and systems (Fig. 5.1).

But none of these moral (i.e., good human) behavior sets have anything to do with how nonhuman animal behaviors should be assessed. In Rolston's words, "the appropriate evaluative category is not nature's moral goodness, for there are no moral agents in nonhuman nature. The appropriate category is *one or more kinds of nonmoral goodness*, better called nature's value. Such value is not to be mapped by projection from culture, much less from human moral systems within culture" (Rolston 1992, 252 emphasis mine). The axiological inter/intra distinction can be applied to any species, be it comprised of moral agents or not. There is a set of good intraspecific behaviors for any given species, just as there is a set of good interspecific behaviors for that species. And because species are specific and special, there are often pointed differences between any two species-specific sets of good behaviors. For instance, it is good (though not 'moral') intraspecific behavior for juvenile iguanas to eat the feces of adult iguanas, but familial coprophagy is not very good behavior for most other species, likely including our own. For lions, it is good intraspecific behavior for the newly dominant male to eat the cubs of the previously dominant male, but cannibalizing stepchildren is not very good behavior for many other species, likely including our own.

The same goes for nonhuman interspecific relations. For cuckoo birds, it is good interspecific behavior for them to lay their eggs in the nests of other unsuspecting bird species, leaving non-cuckoos to raise overly large cuckoo chicks which out-compete the surrogate parents' own offspring. Yet cuckoldry is rightly considered bad human behavior, both when it is intraspecific (as goes the dictionary definition of the term) and interspecific (like the legends and tales of Romulus and Remus or Tarzan). Brood parasitism is not very good behavior for most other species either. Finally, for some species, say peregrine falcons, it is good interspecific behavior to consume the flesh and blood of other species, but this does not mean it is good interspecific behavior for other species, say the Ruby-throated Hummingbird, to engage in carnivorous predation.

In common parlance, it is often said that nasty animal behavior 'just is,' as if it cannot be subject to evaluation at all. But with these axiological distinctions in place, we are in a position to capture the intuition of the 'just is' while also avoiding the temptation to see the natural world as lacking any value whatsoever, which often slips into seeing it as a value-neutral repository of material for us to exploit in whatever way we see fit. The best way for humans to assess the natural behaviors of (especially wild) nonhuman animals is to see them as good-in-themselves. It is not our place to say that—because humans are generally not supposed to prey on each other, eat each other, eat shit, or impregnate other people's wives so that another parental pair will raise offspring not their own—the sorts of animals which do exactly these things are behaving badly. Nor will it suffice to say that those behavior patterns are value-neutral, for saying so anthropocentrically denies the conceivability of other-than-human value. Positive value is not the same as 'moral' value; morality is what humans are obligated to do, while positive value is broader than human morality. Each animal has its own set of proper behaviors and thus positive values, and our species' set is not necessarily the same as any other set.

The categorizations established above are generic and as such, empty of content; the scope of this paper permits only preliminary gestures towards their filling. However, it cannot be that humans ought to simply stand back and watch disinterestedly as animals go about their business, for we are not isolated observers. What we observe are interactions, and we are ourselves animals who interact and are interacted with by animals other than us. Indeed, sometimes we are subject to parasitism or even predation by heterospecifics. Rolston (1988, 84–88) offers two ethically naturalistic principles for human treatment of other animals (the principle of the non-addition of suffering, and the prohibition against ecologically pointless suffering), but for him these apply across the board to any and all sentient nonhuman animals, and are not situated within the relativization of interhuman ethics by the larger category of normative species behavior sets.

Elsewhere, Rolston suggests that human treatment of nonhuman animals should be "homologous with nature," i.e. having "functional similarities" (1989, 134). Furthermore, he argues that our animal ethics should take their "cues from the nature of animals and their place in nature and from our animal roots and human ecology… 'Naturally' must apply to the object animal and to the subject human" (Rolston 1989, 135). What I propose, then, is that the right way for humans to treat nonhuman animals will depend on the way those particular animals are naturally treated by both conspecifics and other heterospecifics.[6] We should not want to treat animals in ways that fail to do justice to their constitutive ecological relations. Secondly, humans should treat other animals in ways consistent with our own species-specific natural history and needs. Just because ticks like to infest moose hides doesn't mean we should try to do the same, but likewise, just because ticks don't use moose hides, bones or antlers for clothing and tools doesn't mean we shouldn't be permitted to do so. "Resource use of one animal by another," Rolston says, "is a characteristic of the world humans inhabit (a premised fact), one which they are under no obligation to remake (a concluded ought)" (1989, 134). An ethic of animal difference refines this position by particularizing it: resource use of one animal by another will depend on the kind of animals in question and their respective natural histories.[7]

For example, whether it is ethical for a human to hunt a mule deer will depend, at least, on whether mule deer are typically prey species, whether humans are a typically predatory species, and if mule deer provide goods suitable to their being treated as prey by humans (i.e., meat, hides, sinew, bone, or homologous goods that predators seek through predation). The question would be posed again, and potentially answered differently, with respect to human resource use of grizzly bears, golden eagles, Richardson's ground squirrels, leopard frogs, or what have you. If it is found to be ethical to use an animal on these terms, then Rolston's principle of the non-addition of suffering should come into force: animals should not be subjected to more pain than they would suffer if they were living (and dying) in the wild as undomesticated animals. However, Rolston's prohibition against ecologically pointless suffering (one cannot cause pain in an animal—even if it is less than it might experience, say, in the claws of a hawk—if that pain does not have or resemble an evolutionary function) is made virtually otiose by the naturalization of each animal's constitution and relation with the other, unless the use of the animal is clearly a desecration or dishonorable.[8]

Peter Wenz criticizes Rolston's ethic as "conservative in the worst sense. It papers over difficulties in the status quo that a philosopher should be exposing" (Wenz 1989, 7), and most humane ethicists would likely level the same charge against an ethic of animal difference that is open, in principle, to the killing and use of animals by humans. However, the ethic I am proposing here is more radical than

---

[6]Gruen (2011) allows that "[a]lthough some of the morally relevant facts might be gleaned from species membership, many of them won't be so apparent...the fact[s] that dandelions reproduce asexually or that gibbons are monogamous, don't tell us anything about how we should treat those organisms,...or what obligations or duties we might have towards them in light of such information" (pp. 55–57). To the contrary, I would argue that these facts suggest—at the very least—that humans ought not to attempt to engage in reproductive activities with dandelions and gibbons, and any such attempts by those species towards humans should be rebuffed.

[7]Morally prior to this, of course, is the human duty to maintain ecologically sustainable populations, without which no resource harvest would be permissible.

[8]Besides, Rolston's nature/culture dualism makes virtually any resource use 'cultural' and thus ecologically pointless, making the question "what is natural to humans?" unanswerable on his own terms (1989, 132; cf. Kowalsky 2006).

conservative. While it may permit, in principle, killing an animal for the good of its body, farming an animal for fur or meat may not be permitted. No animals are typically caged species (and few are typically herded by nonhuman heterospecifics) and humans are not a typically caging species (nomadic herding arose contingently a mere 9000 years ago among idiosyncratic cultural groups [Cauvin 2000]). Similar problems beset the use of animals for traction, like horseback riding. It is not even clear how a naturalistic ethic of respect for animal difference could justify animal testing, even for reasons of urgent medical necessity. Rolston (1989, 137) admittedly does not want his ethic to delegitimize "horses, wagons and plows, nomads and camels, cows and milk, chickens and eggs…agriculture…, cities and industry," but the ethic I have outlined here is poised to do just that. To be sure, we will always have to "make some pragmatic compromises" (Rolston 1989, 136)—perhaps for urgent medical necessity, or for the survival of more than seven billion people—but an ethic of animal difference can provide both operative obligations against many present animal cruelties, and aspirational or regulative imperatives which, even as lofty and perhaps unattainable ideals, do not entail colonialist anti-naturalism. While criticizing and revising humane ethics, animal difference can go a long way towards reconciling that field with naturalistic environmental ethics.

5.5 Sameness and the Anthropocene

The Anthropocene, however, is another challenge to the ethic of animal difference that I've proposed. The Anthropocene is the (proposed) name of our current geological epoch, the one wherein human pollution now forms an identifiable layer in the fossil record. The idea that there can be such thing as a 'nature' distinct from the defiling effluents of certain human cultures should be dead, therefore, if it isn't already. Erle Ellis (2011, 40) asserts that the "long trends toward both the intensification of agricultural cultivation and the engineering of ecosystems at increasing scope and scale" are not recent phenomena, but rather began (he thinks) before the Holocene with Paleolithic human fire-drive hunting techniques. There's nothing unique or distinct about the Anthropocene, it would seem, as human beings have always been a geophysical force on the planet. If so, it follows that there is no such thing as animal difference, if by that we mean nonhuman animal behavior sets that are independent of human influence or assessment. The domestication of animals by certain human cultures is at least 10,000 years old, and domesticated lifeforms are the main sort of nonhuman animal encountered by most humans today. With the majority of humans now living in urban areas, most human encounters with wild animals are likely to be in urban settings where such animals are a nuisance at best. The natural habitats of wild animals are fragmented, decreasing in size, and degraded by anthropogenic climate change. The notion of an 'animal' that is other than the 'human' is problematic at best, if not ridiculous, on account of the Anthropocene.

Even though Paul Crutzen proposed the term "Anthropocene" to inspire caution or regret regarding the ways in which anthropogenic effects alter planetary geology, the way the term has been enthusiastically embraced by especially (but not exclusively) humanist scholars suggests anything but caution or regret. As Langdon Winner (2017, 291) notes:

> The basic sensibility that emerges from the notion "Anthropocene"… is one that blends a familiar, threadbare, human-centred worldview, often with lavish infusions of techno-triumphalism, the latest version of a narrative tradition that includes "progress," "development" and "innovation," this time enhanced with austere rituals of hand-wringing.

The hand-wringing is necessary for appearances' sake only, for the normative undertones of "the age in which nature and culture are no longer neatly separable forces or spheres" (Williston 2016, 155) are celebratory: humanity at long last has triumphed in its (supposedly) 200,000 year old war against 'nature.' Even though current rates of anthropogenic ecological change are greater than have been seen for hundreds of millions of years, ecomodernists Michael Shellenberger and Ted Nordhaus (2011, 10) assert that these are changes of "scope and scale, not of kind." If that is so, then there can be no ethically naturalistic critique of the project which finds its culmination in the Anthropocene. All we can and ought to do is adapt to our new Anthropocene conditions, just as nonhuman animals are currently being forced to. The way things are is simply the way things have to be, and there can be no ethical response to it other than acquiescence (and innovation, of course).

This situation, however, is an odd one for normative ethics: if the only response to (so-called) human domination of the planet is acceptance of (so-called) human domination of the planet, then ethics have no normativity vis-à-vis (so-called) human domination of the planet. Ethics in this case simply have no purchase on the orientation of human behavior. The projects of 'our species' (let us pretend, for the moment, that domination of the planet is, in fact, appropriately described as 'human') are entirely naturalized in the sense of being devoid of agency, volition, or freedom. 'Our' tendency to dominate the globe is *itself* a geo-physical law, we are led to believe. If the Anthropocene is what its boosters want it to be, it is the condition for the *impossibility* of an ethical critique of the Anthropocene itself; the Anthropocene narrative is "the rubber stamp [of] the fait accompli" (Charbonneau 2018, 145). This chapter is not the place for a defense of the reality of human moral agency, but if there is anything that is phenomenologically true about the human condition, it is that all of us—regardless of language, culture, color, or creed—make significant choices. We form societies, for instance, and there is a virtually infinite array of societies that we can form. If this is so, then we have to face the possibility that the Anthropocene—even as a geological reality—is also a social construction. It is not simply human; it is the result of a contingent set of some forms of human culture. As such, it can be subject to normative evaluation, and does not function as a natural limit or barrier to normative evaluation. There is no need to accept the Anthropocene as a given, or to see it as characteristically human. Rather, the need is the inverse.

An ethic of animal difference *can* speak to the Anthropocene project critically, therefore, but so can humane ethics. Humane ethics have resources with which to decry the ongoing domestication of animals (Comstock 1992), they critique certain breeding and grooming practices of companion animals, and voice concern for the condition and treatment of wild animals, urban or otherwise. However, all this is merely formal; in content, the humane ethic succumbs to the ironies of progressivism. As if offering a summary of the earlier parts of this chapter, Don McKay (2008–2009, 11) contends that "No less than the technological mindset, Romanticism converts the other into the Same of the human self, but by a soft and seductive path, the generous extension of citizenship rather than violent reduction to utility." Humane ethics model their standards for animal treatment on ethics of human treatment, which is why they are more consistent with an embrace of the Anthropocene than may have originally seemed. At best, humane ethics would offer a reformist balm to the Anthropocene's 'human' domination of all that exists on this Earth. So long as that domination is 'humane,' the colonialist and imperialist projects of both are morally consistent. The anti-naturalism of humane ethics offers nothing but grounds for *accepting* the radical 'humanization' of the planet, which is the Anthropocene per se. Both humane ethics and the humanist celebration of the Anthropocene articulate themselves as fulfilling "the project that has centrally occupied humanity for thousands of years—emancipating ourselves from nature, tribalism, peonage, and poverty…" (Shellenberger and Nordhaus 2011, 11). Just as humane ethics progressively expand the definition of 'person' to include members of marginal human communities and (some) nonhuman animals, the Anthropocene is the progressive expansion of the 'human' to geologically and biologically include *all* of nonhuman nature (animal or vegetable or mineral). The anti-naturalism in humane ethics is of a piece with the Anthropocene's own anti-naturalistic declaration of the end of nature. The Same triumphs over the Other.

It is fitting that this book should be edited by and have so many contributors from within the Dutch context, because the Netherlands is essentially a case study in the Anthropocene. As the saying goes, God made the world, but the Dutch made the Netherlands. The Netherlands is the most densely populated country in Europe, and if the embrace of the Anthropocene becomes writ large across the globe, the Netherlands' levels of population density, land use, and types of animal encounters may become the model for every square meter of the terrestrial surface.[9] Very little of the Netherlands currently counts as 'wild landscape,' and of that which does, most is space reclaimed from the environing 'cultural landscape' of urbanization, industrialization, and agriculturalization. Likewise, the Anthropocene is the radical suppression of wildness, writing domestication and 'civilization' (literally, city-fication) into everything everywhere for all time, from the geological strata to the heady airs of the atmosphere. This is the naturalization which the Anthropocene seeks to achieve: a particular version of human society—broadly speaking, high technology human sedentism (and not necessarily Dutch!)—standardized across time and space, around which all otherness must and will be subordinated (even if some is allowed to remain in isolated pockets for recreational purposes or curiosity's sake).

---

[9]Shellenberger and Nordhaus (2011, 8–9) use the precarious technological gamble of the city of Venice as their metaphor for the Anthropocene, but their comparison is seamlessly applicable to the Netherlands as an entire country.

This is why humane ethics appear best suited for domesticated animals within urban and rural settings. The more oppositional and irreducible differences there are between wild animals and the core context of sedentary agrarian human civilization, the less those animals can be tolerated (in their wild form). Civilization—which as a process, anthropocentrically makes something fit for a city—conceives of anything outside its own ordering as a chaotic threat. Furthermore, to make something a citizen—be it a human or another animal—is to reconstruct it as a member of the city, a place where that thing's wildness is unsuitable. Domestication therefore removes difference from animals by genetically changing them so that they can physically and psychologically tolerate captivity by humans. It forces alignment with the strictures of agrarian human sedentism, being the literal anthropomorphization of wild animals, or a genetic reification of the other to the same. Moral progressivism—be it embodied in humane ethics or the Anthropocene—emanates outward from the agrarian sense of self and generates a barnyard ethic of animal treatment. The progressivist anti-naturalism of both, at base, aim at the triumph of the selfsame which brings order to recalcitrant and repulsive nature with the point of a weapon, if necessary.

However, this particular and contingent social project does not need to be naturalized. Indeed, naturalizing it is both archaeologically and anthropologically false. It is clear that for the vast majority of our species' chronology, we lived without domestication, agriculture, or sedentary civilization (let alone industrialization, mechanization, and mass urbanization), and we were not any less human for lacking it. Foraging—the primary mode of human subsistence for 95% of the human past—does not require the large-scale manipulation of the ecosystem in a manner that sharply contrasts with how the ecosystem would function without human presence (Tudge 1998, 5–7). There's *nothing* universally human about either treating nonhuman animals as "feral permanently retarded human children" (Pluhar 1991, 26) in need of house training, or viewing the Earth Mother as standing in need of geological domestication. *Nor should there be.* If the violent overcoming of the Other by the Same is fundamentally immoral, then both humane ethics and the Anthropocene project are morally suspect at best.

Nor is it an impossible task to respond to animal or geological alterity without antagonism. Most foraging cultures known to anthropology view(ed) wild animals as exemplars of foreign ways of being that humans could not actually participate in—and thus companionable behaviors were seen as inappropriate. In these non-agrarian contexts, wild animals were viewed as both different from humanity and yet as positive and unopposed to humanness, a non-oppositional encounter with alterity which is precisely what humane ethics and the Anthropocene lack. The Anthropocene's geological domination of the planet and humane ethics' (im)moral domination of animals are thus non-natural in the sense of being contingent and unnecessary (not naturalized) and anti-natural in the sense of being opposed to wildness (i.e., the natural evolutionary and ecological state of all animals). Contrary to the progressive narrative, there can (and indeed should) be differentiation between humans and other animals without endemic conflict, and difference within reciprocal relationship.

This is where an ethic of animal difference fits in. In so far as the Anthropocene forces the denaturing of nonhuman animal otherness by taming, domestication, genetic modification, agricultural and urban vilification, habitat destruction, and climate change, an ethic of animal difference will morally condemn the Anthropocene. Such an ethic will provide grounds for resisting those anthropocentric forces which convert the Otherness of nonhuman animals into something more conducive to the Sameness of high-technology sedentary human civilization. That this contingent form of human culture is currently writing itself into the geological record is metaethically irrelevant. The task of philosophical ethics is not to take human cultural constructions for granted, but to subject them moral examination. That is precisely what an ethic of animal difference would do. It is premised on the wild, evolutionary, and ecological otherness of nonhuman animals, and from that vantage point rejects the radical suppression of wild alterity by the Anthropocene (and humane ethics).

What if the Anthropocene cannot be stopped? How then shall humans orient themselves towards the animals which remain after the anthropocentric juggernaut has conquered all the places where both can live? On the one hand, temporary pragmatic compromises can be made. Insofar as domesticated animals are incorporated into sedentary industrial-agrarian social systems, we may apply certain anthropocentric moral standards to their treatment (perhaps alleviation of suffering) as a form of ironic respect for what remains of their wild form's alterity (e.g., allowing chickens to express 'natural' scratching behavior). Insofar as wild animals migrate into urban and rural areas and even speciate in response to anthropogenic pressures, we should allow them to do so, at least in honor of what remains of their eco-evolutionary agency. If they're nuisances in our cities, let them be nuisances as a sign to us of the horror of having brought our cities to the point where wild animals have no other choice but to be a nuisance therein.

But let us not celebrate these new feral beasts or hybrid species as an innovative response to a naturalized Anthropocene. Let them rather be icons of the failure of the currently dominant form of human culture to respond to Otherness without violence. If resistance is the spirit of the compromises we make, then an ethic of animal difference can still issue a moral vocation that transcends the fait accompli of the Anthropocene. Ethics can—without being hamstrung by naturalizing contingent 'realities' like the status quo—offer aspirational or regulative imperatives that provide resources with which to critique the Anthropocene juggernaut, even if it is currently the victor. There is no need to fully collaborate with the colonizer or the imperialist 'human.' Resistance is possible, and resistance is obligatory.

5.6 Conclusion

As with all essays, this chapter can remain only a proposal, and as such its results are indeterminate and open. Much careful work needs to be done to identify humanity's natural intra- and inter-specific behavioral norms, as well as the norms of those species with whom our species most commonly interacts. This is, however, a project worth embarking on. If the vast majority of the Western tradition has been the immoral attempt at conceptually, technologically, and normatively mastering the Other by reducing it to the Same, then the ethical framework proposed in this chapter offers a way forward without perpetuating that colonialist and imperialist agenda. Whether or not it is too late to actually stop the colonialist and imperialist agenda of the Anthropocene is beside the point. What matters is that we recognize its agenda and recover resources with which to oppose it at every step. The Otherness of animals different than ourselves is one such source of grounding. Let us return to the animals themselves![10]

References

Adams, C.J. 1994. *Neither man nor beast: Feminism and the defense of animals*. New York: Continuum.

Brigandt, I. 2009. Natural kinds in evolution and systematics: Metaphysical and epistemological considerations. *Acta Biotheoretica* 57: 77–97.

Callicott, J.B. 1980. Animal liberation: A triangular affair. *Environmental Ethics* 2: 311–338.

Callicott, J.B. 1989. Animal liberation and environmental ethics: Back together again. In *In defense of the land ethic: Essays in environmental philosophy*, ed. J.B. Callicott, 49–59. Albany: State University of New York.

Cauvin, J. 2000. *The birth of the gods and the origins of agriculture*. Trans. Trevor Watkins. Cambridge: Cambridge University Press.

Charbonneau, B. 2018. *The green light: A self-critique of the ecological movement*. Trans. Christian Roy. London: Bloomsbury.

Clark, J. 2014. What is living in deep ecology? *The Trumpeter* 30: 157–183.

Comstock, G. 1992. Pigs and piety: A theocentric perspective on food animals. *Between the Species* 8: 121–135.

Donaldson, Sue, and Will Kymlicka. 2011. *Zoopolis: A political theory of animal rights*. Oxford: Oxford University Press.

Ellis, E. 2011. The planet of no return: Human resilience on an artificial earth. In *Love your monsters: Postenvironmentalism and the Anthropocene*, ed. M. Shellenberger and T. Nordhaus, 37–46. Breakthrough Institute.

Golberg, S.A. 2011. Happy, fat, and meatless: A proposal for a 21st-century vegetarianism. *Table Matters*. http://www.tablematters.com/index.php/plate/vm/vm1. Accessed 11 October 2011.

Goodpaster, K.E. 1979. From egoism to environmentalism. In *Ethics and problems of the 21st century*, ed. K.E. Goodpaster and K.M. Sayre, 21–59. Notre Dame and London: University of Notre Dame Press.

Gruen, L. 2011. *Ethics and animals: An introduction*. Cambridge: Cambridge University Press.

Hargrove, E.C. 1992. *The animal rights, environmental ethics debate: The environmental perspective*. Albany, NY: State University of New York.

Hegel, G.W.F. 1956. *The philosophy of history*. Trans. J. Sibree. New York: Dover.

Jonas, H. 2001. *The gnostic religion: The message of the alien god and the beginnings of Christianity*, 3d ed. Boston: Beacon.

Kemmerer, L. 2011. Theorizing 'others'. In *Theorizing animals: Re-thinking humanimal relations*, ed. N. Taylor and T. Signal, 59–84. Leiden, Netherlands: Brill.

---

[10]Large parts of this chapter are taken from Kowalsky (2016), "Towards an Ethic of Animal Difference" https://doi.org/10.5840/envirophil201692336. The author gratefully acknowledges the permission granted by the Philosophy Documentation Centre to reprint this material here.

Kowalsky, N. 2006. Following human nature. *Environmental Ethics* 28: 165–183.
Kowalsky, N. 2012. Science and transcendence: Westphal, Derrida, and responsibility. *Zygon* 47: 118–139.
Kowalsky, N. 2016. Towards an ethic of animal difference. *Environmental Philosophy* 13: 239–267.
Leopold, A. 1949. *A Sand County almanac: And sketches here and there.* London: Oxford University Press.
McKay, D. 2008. Ediacaran and Anthropocene: Poetry as a reader of deep time. *Prairie Fire* 29: 4–15.
Midgley, M. 1994. The end of Anthropocentrism? In *Philosophy and the natural environment* (Royal Institute of Philosophy Supplements): 36, ed. R. Attfield and A. Belsey, 103–112. Cambridge: Cambridge University Press.
Naess, A. 1989. *Ecology, community and lifestyle: Outline of an ecosophy.* Trans. David Rothenberg. Cambridge: Cambridge University Press.
Nussbaum, M. 2006. *Frontiers of justice: Disability, nationality, species membership.* Cambridge, MA: Belknap.
Nussbaum, M., and C. Faralli. 2007. On the new frontiers of justice: A dialogue. *Ratio Juris* 20: 145–161.
Pluhar, E.B. 1991. The joy of killing. *Between the Species* 7: 121–128.
Raterman, T. 2008. An environmentalist's lament on predation. *Environmental Ethics* 30: 417–434.
Regan, T. 2004. *The case for animal rights.* Updated version. Berkeley: University of California.
Rodman, J. 1977. The liberation of nature? *Inquiry* 20: 83–131.
Rolston, H., III. 1988. *Environmental ethics: Duties to and values in the natural world.* Philadelphia: Temple University Press.
Rolston, H., III. 1989. Treating animals naturally? *Between the Species* 5: 131–137.
Rolston, H., III. 1992. Disvalues in nature. *The Monist* 75: 250–278.
Rolston, H., III. 2012. *A new environmental ethics: The next millennium for life on earth.* New York: Routledge.
Seitz, B. 2010. Hunting for meaning: A glimpse of the game. In *Hunting—Philosophy for everyone: In search of the wild life,* ed. N. Kowalsky, 69–79. Malden, MA: Wiley-Blackwell.
Shellenberger, M. and T. Nordhaus. 2011. Evolve: The case for modernization as the road to salvation. In *Love your monsters: Postenvironmentalism and the Anthropocene,* eds. M. Shellenberger and T. Nordhaus, 8–16. Breakthrough Institute.
Singer, P. 1974. All animals are equal. *Philosophic Exchange* 5: 103–116.
Singer, P. 1975. *Animal liberation: A new ethics for our treatment of animals.* New York: Avon.
Singer, P. 1981. *The expanding circle: Ethics and sociobiology.* New York: Farrar, Straus & Giroux.
Singer, P. 1997. The drowning child and the expanding circle. *New Internationalist,* April. http://www.utilitarian.net/singer/by/199704–.htm. Accessed 23 March 2020.
Singer, P., and J. Mason. 2006. *The ethics of what we eat: Why our food choices matter.* Melbourne: Text Publishing.
Stanescu, J. 2012a. Toward a dark animal studies: On vegetarian vampires, beautiful souls, and becoming vegan. *Journal for Critical Animal Studies* 10: 26–50.
Stanescu, J. 2012b. Species trouble: Judith Butler, mourning, and the precarious lives of animals. *Hypatia* 27: 567–582.
Taylor, N. 2011. Can sociology contribute to the emancipation of animals? In *Theorizing animals: Re-thinking humanimal relations,* ed. N. Taylor and T. Signal, 201–220. Leiden, Netherlands: Brill.
Tudge, C. 1998. *Neanderthals, bandits and farmers: How agriculture really began.* New Haven and London: Yale University Press.
Twine, R. 2014. Ecofeminism and veganism: Revisiting the question of universalism. In *Ecofeminism: Feminist intersections with other animals and the earth,* ed. C.J. Adams and L. Gruen, 191–207. New York: Bloomsbury.
Vera, S. 2008. Apes get legal rights in Spain, to surprise of bullfight critics. *The Times,* June 27. http://purdue.edu/bioethics/blog/?p=171. Accessed 24 November 2011.

Wenz, P. 1989. Treating animals naturally. *Between the Species* 5: 1–10.

Williston, B. 2016. The sublime anthropocene. *Environmental Philosophy* 13: 155–174.

Wilson, R. A., M.J. Barker, and I. Brigandt. 2007. When traditional essentialism fails: Biological natural kinds. *Philosophical Topics* 35: 189–215.

Winner, L. 2017. Rebranding the anthropocene: A rectification of names. *Techné: Research in Philosophy and Technology* 21: 282–294.

Wissenburg, M. 2011. The lion and the lamb: Ecological implications of Martha Nussbaum's animal ethics. *Environmental Politics* 20: 391–409.

Nathan Kowalsky (PhD Katholieke Universiteit Leuven) is an associate professor of philosophy at St. Joseph's College, University of Alberta, Canada. His main areas of research are environmental philosophy, philosophy of religion, and philosophy of culture. Specifically, his interests lie in the interrelationships between wildness, non-agrarian hermeneutics, evil, divine alterity, and critiques of modern technology. He is still plugging away at a collection of short essays (or is a textbook?) on the unusual studies of science and technology in society (STS), and should probably get around to writing his overdue book on natural evil before too long. He is the editor of *Hunting—Philosophy for Everyone: In Search of the Wild Life* (Wiley, 2010) and editor-in-chief of the environmental humanities journal *the Trumpeter*.

Chapter 6
Should the Lion Eat Straw Like the Ox? Animal Ethics and the Predation Problem

Jozef Keulartz

Abstract Stephen Clark's article The Rights of Wild Things from 1979 was the starting point for the consideration in the animal ethics literature of the so-called 'predation problem'. Clark examines the response of David George Ritchie to Henry Stephens Salt, the first writer who has argued explicitly in favor of animal rights. Ritchie attempts to demonstrate—via reductio ad absurdum—that animals cannot have rights, because granting them rights would oblige us to protect prey animals against predators that wrongly violate their rights. This article navigates the reader through the debate sparked off by Clarke's article, with as final destination what I consider to be the best way to deal with the predation problem. I will successively discuss arguments against the predation reductio from Singer's utilitarian approach, Regan's deontological approach, Nussbaum's capability approach, and Donaldson and Kymlicka's political theory of animal rights.

6.1 Introduction

Stephen Clark's article *The Rights of Wild Things* from 1979 was the starting point for the consideration in the animal ethics literature of the so-called 'predation problem' (Dorado 2015, 234). In this article, Clark examines the response of Scottish philosopher David George Ritchie (1853–1903) to Henry Stephens Salt (1851–1939), who is credited to be the first writer to have argued explicitly in favor of animal rights. Ritchie attempts to demonstrate—via *reductio ad absurdum*—that animals cannot

This chapter is a reprint from an article published in the *Journal of Agricultural and Environmental Ethics* (2016) 29: 813–834.

J. Keulartz (✉)
Radboud University, Nijmegen, The Netherlands
e-mail: jozef.keulartz@wur.nl

Wageningen University and Research, Wageningen, The Netherlands

© The Author(s) 2021
B. Bovenkerk and J. Keulartz (eds.), *Animals in Our Midst: The Challenges of Co-existing with Animals in the Anthropocene*, The International Library of Environmental, Agricultural and Food Ethics 33,
https://doi.org/10.1007/978-3-030-63523-7_6

have rights, because granting them rights would oblige us to protect prey animals against predators that wrongly violate the victim's rights.

> In our guardianship of the rights of animals, must we not protect the weak among them against the strong? Must we not put to death blackbirds and thrushes because they feed on worms, or (if capital punishment offends our humanitarianism) starve them slowly by permanent captivity and vegetarian diet? What becomes of the 'return to nature' if we must prevent the cat's nocturnal wanderings, lest she should wickedly slay a mouse? Are we not to vindicate the rights of the persecuted prey of the stronger? Or is our declaration of the rights of every creeping thing to remain a mere hypocritical formula to gratify pug-loving sentimentalists. (Ritchie 2002, 109–110)

Clark argues against this predation *reductio*; he rejects Ritchie's conclusion that, if non-human animals had rights, we should be obliged to defend them against predators. This conclusion, Clark asserts, "either does not follow, follows in the abstract but not in practice, or is not absurd" (Clark 1979, 187).

This article navigates the reader through the debate sparked off by Clarke's article, with as final destination what I consider to be the best way to deal with the predation problem.

I will first argue that the utilitarian approach to the predation *reductio* is ultimately a dead end. Utilitarians can only avoid this *reductio* if they are prepared to reconsider their opinion of predation as an evil that must be eradicated (Sect. 6.2).

As I will argue next, Tom Regan's rights-based approach to animal ethics offers a less gloomy picture of predation and provides a more solid way to escape the predation *reductio* than the consequentialist approach. According to Regan we have no duty to interfere with wildlife to prevent predation because members of both predator and prey species possess a certain 'competence' and are capable of 'using their natural abilities' to survive on their own in the wild. This recourse to the notion of 'competence' could open an avenue for a more balanced view of the predator-prey relationship, in which predator and prey are no longer seen respectively as invincible and defenseless (Sect. 6.3).

To explore this avenue, I then turn to Martha Nussbaum's capabilities approach that centers on the idea that a creature's well-being is dependent on its opportunities to realize some basic natural abilities or competences. However, the considerable conceptual gains that Nussbaum is able to achieve through the introduction of the species-specific norm of flourishing in the discussion of the predation problem are at least partly being undone by the way she compiles a catalogue of innate or 'basic' capabilities relevant to animal species (Sect. 6.4).

I finally turn to Donaldson and Kymlicka's political theory of animal rights. There are important similarities between this theory and Nussbaum's capabilities approach. But there is also a distinct difference: whereas Nussbaum attaches considerable importance to species membership, Donaldson and Kymlicka focus on community membership, thus taking account of the sociopolitical context of animal justice. They succeed in making further headway on the road to a satisfactory solution of the predator problem. However, with their sovereignty model Donaldson and Kymlicka have taken a place-based approach with regard to wild animals that ultimately fails to take sufficient account of the scope and scale of the anthropogenic stress that is inflicted upon these animals during the current stage of the Anthropocene (Sect. 6.5).

6.2   Utilitarianism

In his article *The Rights of Wild Things*, Stephen Clark mentions an important argument that utilitarians usually put forward to avoid the predation *reductio*: that the evil of predation cannot be eliminated without introducing worse ones. "Caribou may be spared the pain of wolves, or Eskimos, but the consequent population explosion will lead to overgrazing, disease, famine, and a population crash" (id., 175).

This argument was introduced in 1973 by Peter Singer in his reply to David Rosinger who asked him if we have a moral responsibility to prevent predation. In answering this question, Singer makes a distinction between domestic pets such as cats and dogs, and wild animals, like the lion. With respect to carnivorous pets, Singer thinks it right to try to raise them on a special vegetarian diet. But as for wild animals, he claims to be fairly sure, "judging from man's past record of attempts to mold nature to his own aims, that we would be more likely to increase the net amount of animal suffering if we interfered with wildlife, than to decrease it... So, in practice, I would definitely say that wildlife should be left alone" (Singer 1973; 1975, 226). Although Singer cautions against interfering with ecosystems because he fears that doing so would cause more harm than good, as a matter of principle, he believes that "if, in some way, we could be reasonably certain that interfering with wildlife in a particular way would, in the long run, greatly reduce the amount of killing and suffering in the animal world, it would, I think, be right to interfere" (Singer 1973).[1]

In order to avoid what he has called a "conceptual absurdity"—that we risk to cause more suffering than we would prevent—Steve Sapontzis has added a proviso to the presumption that we are morally obligated to prevent predation. He claims that we are only committed to stop predation "whenever doing so would not occasion as much or more suffering than it would prevent" (Sapontzis 1984, 31). Sapontzis contends that this reformulation still contains a substantive obligation: it would for instance oblige us to prevent our pets from being predators, something that is also endorsed by Singer. "It would also obligate us to begin exploring other ways in which we could reduce the suffering caused by predation without occasioning as much or more suffering, e.g., in zoos, wildlife preserves, and other areas where we are already managing animals" (ibid.).

---

[1]Fellow-consequentialist Aaron Simmons fully agrees with Singer that saving wild animals "on any large scale would have disastrous ecological consequences" (Simmons 2009, 26). He doesn't believe at all in measures to avoid these bad consequences such as feeding vegetarian diets to predators to prevent starvation, or feeding contraceptives to prey animals to curb overpopulation. If done on any large scale, such measures as proposed by environmental philosophers like Mark Sagoff (1984) would not counter but only compound serious ecological problems.

In a similar vein, Charles Fink has argued that it does not follow from the alleged fact that large-scale interventions in ecosystems would do more harm than good that we should do nothing at all. Fink accuses all those who take an all-or-nothing approach to the predation problem of black-and-white-thinking. He concludes that "it is not *inherently* absurd to suppose that there is an obligation to protect animals from natural predators, even if this obligation has limited practical application" (Fink 2005, 15).

That we run the risk to cause more harm than good is not the only reason why there can be no obligation to defend prey animals against their predators. Another reason, also mentioned by Stephen Clark, concerns the over-demanding nature of such a duty. Any realistic attempt to fulfill a duty to intervene in predation would inevitably be detrimental to our performance of other duties. Our possibilities for positive action are simply limited: "Most of us, not being wandering preachers, can be vegetarians quite easily. Some of us can be vegans. But very few of us can wholeheartedly devote ourselves to the defense of mice" (Clark 1979, 179/180; cf. Hadley 2006).

In connection to this second objection Sapontzis speaks about a "practical absurdity", and he is convinced that this objection is as easy to refute as the objection in the case of the conceptual absurdity. It may be true that we are unable to eliminate predation entirely; but this does not render the obligation to prevent predation meaningless. It can function as a moral ideal that we should work toward and try to approximate ever more closely: "Consequently, it is not practically absurd" (Sapontzis 1984, 32).

6.2.1   Piecemeal Engineering

So, the consensus among utilitarians is that we should intervene in predator-prey relations whenever doing so would not cause more harm than good or be overly demanding and incur costs that significantly outweigh the benefits. In his article *Policing Nature* from 2003, Tyler Cowen identifies a number of policy measures that seem to meet these criteria. Cowen's starting point is that we are already inevitably intervening in nature in massive ways, through agriculture, fishing, industry, building, mining and, of course, through nature conservation. These policies obviously affect predators and prey animals differently. Cowen calls for attempts to shift nature's balance of power to the detriment of predators and to the benefit of their victims. We should, however, do so in a cautious and humble way, without upsetting nature's balance in intolerable fashion. "We should count negative impacts on carnivores as positive features of the human policy, rather than as negative features, as we usually do. Doing so would make us less likely to support the populations of various aggressive carnivores" (Cowen 2003, 174). In order to shift nature's balance of power in the desired manner, we should, at the very least, stop subsidizing the propagation of carnivores and limit or eliminate programs to protect endangered carnivores or prevent their extinction. We should make hunting strictures against killing carnivores less tight than those against killing non-carnivores, or perhaps remove them altogether. And we should be more willing to use carnivores than non-carnivores in laboratory experiments.

6.2.2   The Balance of Nature and the Argument from Ignorance

The utilitarian consensus that the way forward is to proceed in a piecemeal and small-scale manner is based on two interconnected assumptions: that the balance of nature is essentially good for animals, and that we are simply too ignorant to police nature without constantly running the risk to disturb this balance. Both these underlying assumptions have however increasingly come under attack during the past few years.

The claim that the balance of nature is on the overall good for animals has already been contested in a paper from 1995 by Yew-Kwang Ng. But it took until the late 2000s before this paper started to gain real traction (see, e.g., Dawrst 2009; Horta 2010a, b; Tomasik 2015). In his paper entitled *Towards Welfare Biology* Ng argues, on the basis of evolutionary economics and population dynamics, that the natural equilibrium is something quite terrible because all species suffer enormously in this situation. This is due to the prevalence of the reproductive strategy know as 'r-selection', which consists in producing large numbers of offspring per reproductive cycle. The overwhelming majority of animals that follow this reproductive strategy, including fishes, amphibians and reptiles, die shortly after birth, from starvation or by being eaten alive by predators or parasites. But even the tiny minority of animals that follow the other important reproductive strategy know as 'K-selection' will experience their share of suffering and misery as well, because they often also have a large numbers of eggs or offspring, which will be wasted before they reach sexual maturity. Ng concludes that for animals in the wild, their pain and suffering vastly outweigh their pleasure and happiness, so the widely accepted idyllic view that the current balance of nature is overall good for animals is definitely false.

As a consequence of this analysis, Ng and his followers, suggest that we can increase the level of overall animal welfare by lowering the birth-rates and reducing the number of those animals whose lives are not worth living (Ng 1995, 271/275). Consequently, Oscar Horta (2010b) considers species protection and biodiversity conservation as counterproductive to the promotion of animal welfare because this will increase rather than reduce the number of suffering animals. In a similar vein, Brian Tomasik has expressed the hope "that the animal-rights movement doesn't end up increasing support for wilderness preservation and human non-interference of all kinds" (Tomasik 2015, 148).

The second underlying claim, that our understanding is too limited to intervene in nature without causing serious ecological problems, is also far from unproblematic. As Clare Palmer has remarked, this argument from ignorance is not a resilient argument that interfering with wildlife is morally unacceptable in principle. Given the continuous development of sophisticated techniques such as vaccination, radio tracking, and wildlife contraception or sterilization, the argument from ignorance of the consequences will lose its validity in an increasing number of cases (Palmer 2010, 30; 2015, 205).

Sue Donaldson and Will Kymlicka agree with Palmer that the argument from ignorance—they prefer to call it the 'fallibility argument'—seems to miss the target. This argument suggests that if we had the adequate tools and techniques at our disposal, we should start re-engineering the natural world to reduce suffering overall, thereby "turning nature into a well-managed zoo in which each animal has its own safe enclosure and guaranteed food source" (Donaldson and Kymlicka 2011, 164).

6.2.3   Paradise Engineering

To get a glimpse of this future re-engineered world, we might take a look at the work of David Pearce, a British philosopher and co-founder of Humanity+, the international transhumanist organization, whose purpose is the fundamental transformation of the human condition by developing and making widely available technologies to greatly enhance intellectual, physical, and psychological capacities, eventually building a 'Triple S' civilization of Superhappiness, Superlongevity and Superintelligence. Pearce's ideas have inspired a strain of transhumanism called 'paradise engineering', an abolitionist program to achieve nothing less than the elimination of literally all suffering on the planet. He has outlined this program in his 1995 book-length internet manifesto *The Hedonistic Imperative*. In this manifesto, Pearce explains how technologies such as genetic engineering, nanotechnology, pharmacology, and neurosurgery could potentially converge to abolish suffering in all sentient life.

An important part of Pearce's abolitionist program aims to limit or eliminate predation, reducing the suffering of prey animals. He distinguishes two solutions to the 'barbarities' of predation: extinction and reprogramming. The first solution is "to use indiscriminate depot-contraception on carnivores and allow predators rapidly to die out, managing the resultant population effects on prey species via more selective forms of depot-contraception" (Pearce 2009, 6). The second solution concerns the genetic 'reprogramming' or otherwise behavioral conversion of aggressive carnivores into model citizens in our wildlife parks. "With suitable surveillance and computer control, whole communities of ex-predators could be discreetly guided in the norms of non-violent behaviour" (id., 8). Reprogramming and behavioral management can help ensure "the civilised survival of reformed lions and their relatives for human ecotourists to enjoy, if we so choose" (id., 10).[2]

---

[2] Just as David Pearce, White's Professor of Moral Philosophy Jeff McMahan has also embraced "the heretical conclusion that we have reason to desire the extinction of all carnivorous species" (McMahan 2010, 7). McMahan is likewise in favor of selecting carnivorous species for extinction and herbivorous species for survival, and would also support using genetic modification to gradually turn carnivorous species into herbivorous ones, "thereby fulfilling Isaiah's prophecy" (id., 2).

Although Pearce's ideas about future technological developments are highly speculative, it nonetheless appears that utilitarians can no longer hide behind the argument from ignorance, but have to show their colors. They can choose to remain firmly
committed to the view that predation is inherently bad, but then they can no longer
escape from supporting some form of 'paradise engineering' by which nature will be
turned into a well-managed zoo. Or they can choose to avoid this predation *reductio*,
but then they will have to be prepared to reconsider their opinion of predation as an
evil that must be eradicated.

6.3 Rights Theories

Tom Regan's rights-based approach to animal ethics offers a less gloomy picture
of predation and provides a more solid way to escape the predation *reductio* than
the utilitarian approach. Regan is also opposed to interference with nature to protect
prey animals but not because doing so would cause more suffering than it would
prevent. He argues that, although wild animals can certainly harm one another, they
cannot violate one another's rights since, in contrast to human predators, nonhuman
predators are not moral agents, but only moral patients; they do not possess the
relevant capacities to be held morally responsible for their actions. So, we have no
duty "to assist the sheep against the attack of the wolf, since the wolf neither can nor
does violate anyone's rights" (Regan 1983, 285). With respect to animals in the wild,
we have no positive duties of assistance but only negative duties of non-intervention
and are not allowed to confine, torture or kill them. Wildlife managers, Regan claims,
"should be principally concerned with *letting animals be*, keeping human predators
out of their affairs, allowing these 'other nations' to carve out their own destiny" (id.,
357).

Recently, Josh Milburn has convincingly given Regan's account a much-needed
degree of nuance, arguing that the attribution of moral responsibility is not a question
of 'either-or' but of 'less-or-more'. Milburn illustrates this point by the following
example. A wolf killing a deer in isolated woodland does not violate the rights of the
deer, but if the wolf's killing of the deer had taken place in a zoo, then there is some
moral agent who is blameworthy in this situation, namely the zookeeper who placed
the deer in the wolf's enclosure. This example shows that intervention in wildlife is
only morally warranted "in those cases in which morally responsible agents can be
found, and only to the degree that they can be found" (Milburn 2015, 288).[3] Milburn
concludes that the rights of prey do not generally necessitate intervention, because
"the vast majority of predator–prey interactions are not linked to moral agents in
an important way" (ibid.). As much as I appreciate the nuance that Milburn has
introduced in animal rights theory, I don't share this conclusion because, as I will

---

[3] In a similar vein Dale Jamieson has argued that moral evaluation is clearly in order when "predation
is in some way affected by human agency, either because we have structured the encounter or because
the predator is under our direct or indirect control" (Jamieson 2008, 186/7).

argue later, in this stage of the so-called 'Anthropocene', we humans are massively implicated in predator-prey relationships, for better or worse.

There are two kinds of objections that have been raised against Regan's rights-based approach. The first kind of objections concerns the question if it is really morally relevant whether or not perpetrators are moral agents. The second kind of objections is no longer about the perpetrator but concerns its victim, and concerns the question whether it makes any difference if this victim is human or non-human.

6.3.1   Lack of Moral Agency

According to Steve Sapontzis, it is totally irrelevant for our obligation to prevent harm whether that harm is caused by a moral agent or not. He points out that we should separate our moral judgement of an act (as right or wrong) from our moral judgement of the actor (as innocent or culpable). Because we routinely hold parents responsible for preventing their 'innocent' young child from tormenting the cat, we also seem to have an obligation to stop the cat from killing birds (Sapontzis 1984, 27/8).[4]

Dale Jamieson (1990) has also criticized the way in which Regan tries to avoid the predation *reductio* by limiting the duty to render assistance to those animals in need that are the victim of moral agents, the only one's that can commit injustices. He illuminates the problematic character of this limitation by considering five hypothetical cases, in which a man will be crushed by a falling boulder unless I warn him. In the cases 1–3 a woman intentionally or inadvertently causes the boulder to roll toward the man; in the cases 4 and 5 the boulder is set in motion by a wolf and a landslide respectively. On the basis of Regan's theory, we don't have a duty of justice to warn the man in the cases 4 and 5, because neither wolfs nor landslides are moral agents and therefore cannot violate rights (see also Cowen 2003, 176). To avoid this highly counterintuitive conclusion Jamieson argues that we should supplement Regan's theory with a class of nondiscretionary duties that rest on some ground other than justice.

Regan has addressed Jamieson's criticism in his book *Defending Animal Rights* from 2001, and again, in almost identical terms, in his preface to the second edition of *The Case for Animal Rights*, published in 2004. Regan dismisses Jamieson's objection because a careful reading of the relevant passages would reveal that he has never maintained that we owe nothing to those in need who are not victims of injustice. He has only insisted that we do not owe anything to such individuals *on the grounds of justice*. There is nothing in the rights view, Regan contends, that prevents it from recognizing a general *prima facie* duty of beneficence that includes duties of assistance to those in need. "Thus there is nothing in my theory that would preclude

---

[4]This example is quite similar to Milburn's example of the zookeeper who placed a deer in a wolf's enclosure.

recognizing a duty to warn the hiker about the free-falling boulder" (Regan 2001, 51).

Regan has acknowledged that he is "at least partly to blame" for Jamieson's misreading of his texts, because he does nowhere discuss other duties of assistance than those we have to victims of injustice, which he regards as a 'symptom of the incompleteness' of the theory developed in *The Case*. "In hindsight, I recognize that it would have been better had I said more about duties of assistance other than those owed to victims of injustice." (Regan 2004, xxvii).[5]

6.3.2 Non-human Victims

But adding duties of assistance that rest on some ground other than justice does not affect Regan's view that we have no duty to protect the sheep against the wolf. Why Regan holds on to this non-interventionist view will become clear as we shift the focus from predators to prey and look at the second important question raised by Regan's rights-based approach: whether it makes a difference if the victim is a human or non-human animal.

This question was posed by Carl Cohen in an article from 1997, in which he asks us to imagine two cases. In the first case a baby zebra is hunted to death by a lioness. If zebras have a right to live, we ought to intervene, but we usually don't do so. In the second case the lioness is about to attack a human baby, and now we surely will intervene to stop the lioness. So, the question is: what accounts for the moral difference? Cohen's answer is that "animals cannot be the bearers of rights because the concept of rights is essentially *human*; it is rooted in, and has force within, a human moral world" (Cohen 1997, 95). The baby zebra has no right not to be slaughtered by the lioness, nor has the lioness the right to kill that baby zebra, simply because the concept of rights does not apply to animals.

In his article *The Predation Argument* from 2005, Charles Fink has also discussed the question why we should save a human life from predators but not an animal's life when doing so would be equally within our power, but he came to a conclusion diametrically opposed to Cohen's. It may be true, as Regan asserts, that wolves are not moral agents and thus cannot violate the rights of sheep, but it does not follow, Fink believes, that we have no obligation to assist the sheep against the attack of the wolf, considering what our reaction would be if a human being were attacked by a wolf. If we have a duty to protect all members of the moral community from harm, even if this harm is not caused by moral agents, then it would certainly seem to follow that, if sheep are members of the moral community, "there is an obligation to protect them from wolves, whether or not wolves violate their rights" (Fink 2005, 12).

---

[5]However, as Regan has noted "most emphatically", the duty of beneficence has serious limitations. Notably, promoting some one's good should never go at the expense of another one's rights. "In this respect, the demands of justice always take precedence over the claims of beneficence" (ibid.).

In the preface of the second edition of *The Case*, Regan has also addressed Cohen's critical question concerning the moral difference between the case in which a wild animal is threatened by a predator and the case in which the predator is threatening a human child. In his view, we have a duty to save a human child from predators but no such duty with regard to wild animals. The crucial difference between the two cases is that members of both predator and prey species possess a certain 'competence' and are capable of 'using their natural abilities' to survive on their own in the wild, whereas young children do not have the same competence and are unable to survive, in the wild or in the home, without our assistance. We honor this competence of wild animals by just *letting them be*, even if their lives are threatened by predators.[6]

This recourse to the notion of 'competence' could open an avenue for a more realistic solution of the predator problem. It allows for a more balanced view of the predator-prey relationship, in which predator and prey are no longer seen respectively as invincible and defenseless. To explore this avenue, I now turn to Martha Nussbaum's capabilities approach that after all centers on the idea that a creature's well-being is dependent on its opportunities to realize some basic natural abilities or competences.

6.4   The Capabilities Approach

The capability approach differs from the consequentialist approach and the rights approach in one very important respect. Nussbaum rejects the view, taken by both these approaches, that species membership itself is of no ethical and political significance at all. Following James Rachels, Nussbaum calls this view 'moral individualism'.[7] The capabilities approach, by contrast, does in fact attach moral significance to species membership as such. It is based on a species-specific norm of flourishing, that tells us what the appropriate benchmark is for judging whether a member of a species has decent opportunities for flourishing. The capabilities approach has also a strong affirmative character; it "treats animals as subjects and agents, not just as objects of compassion" (Nussbaum 2006, 351), and commits us to support the capabilities of all morally considerable beings, up to some minimum threshold level specific to each species.

However, the significant conceptual gains that Nussbaum would be able to achieve through the introduction of the species-specific norm of flourishing in the discussion of the predation problem are at least partly being undone by the way she compiles

---

[6]As Regan points out, that we have a prima facie duty to assist the child from the lion, does not oblige us to develop general policies "that seek to eradicate every predatory animal under the sun", let alone that we should develop such policies because predatory animals harm their prey (Regan 2004, xxxvii; cf. Donaldson and Kymlicka 2011, 165).

[7]According to Rachels, "moral individualism is a thesis about the justification of judgments concerning how individuals may be treated. The basic idea is that how an individual may be treated is determined, not by considering his group memberships, but by considering his own particular characteristics" (Rachels 1990, 173).

a catalogue of innate or 'basic' capabilities relevant to animal species (Keulartz 2016a). On the one hand, Nussbaum's account of animal capabilities seems to be distinctly pluralist. The capabilities approach is attentive to the fact that each species has a different form of life, and is capable of recognizing a wide range of types of animal dignity, and of the corresponding needs for flourishing. But on the other hand, Nussbaum suggests a one-size-fits-all approach, that has a distinctly anthropocentric character as it applies the same human yardstick to all animal species.[8] Although she fully acknowledges that species-specific entitlements of animals are based upon their various characteristic forms of life and flourishing, she nonetheless wants to use the existing list of human core capabilities "to map out, in a highly tentative and general way, some basic political principles that can guide law and public policy in dealing with animals" (id., 392).[9]

6.4.1 The Other Species Capability

Nussbaum's list of central capabilities includes Life; Bodily Health; Bodily Integrity; Senses, Imagination, and Thought; Emotions; Practical Reason; Affiliation; Other Species; Play; and Control over One's Environment. What seems most problematic, when applied to animals, is the Other Species capability, i.e. the capability or entitlement to be able to live with concern for and in relation to animals, plants, and the world of nature (Cripps 2010, 8). This capability, Nussbaum suggests, "calls for the gradual formation of an interdependent world in which all species will enjoy cooperative and mutually supportive relations with one another. Nature is not that way and never has been. So it calls, in a very general way, for the gradual supplanting of the natural by the just" (Nussbaum 2006, 399). Due to the inclusion of the Other Species capability in her list of central capabilities, Nussbaum's solution to the predation problem is highly ambivalent.

Like most animal ethicists, Nussbaum attaches moral weight to the possibility for animals to enjoy sovereignty. She supports "the idea that species autonomy is part of the good for nonhuman animals" (id., 375). So at first glance, she seems to endorse the view that animals can pursue their own flourishing best when left to their own devices, and that we have no positive duties to support their welfare, providing them with food, shelter and healthcare. Such a "benevolent despotism" of humans over animals might even be perceived as morally repugnant, because part of what it is to

[8]In *Women and Human Development*, Nussbaum argues that the central capabilities "are held to have value in themselves, in making the life that includes them fully human" (emphasis added) (2000, 74).

[9]In her review of Steven Wise's book *Rattling the Cage*, Nussbaum points to an important difference in the ethical evaluation that is involved in preparing capabilities lists: "With the human capabilities, we are evaluating ourselves. If we get it wrong, we are the ones who take the consequences. With animals, we are again the ones performing the evaluation – and there is great danger that we will get it wrong" (Nussbaum 2001, 1542/3).

flourish for animals "is to settle certain very important matters on its own, without human intervention, even of a benevolent sort" (id., 373).

On closer inspection, however, Nussbaum does not fully accept the view that we have no positive duties towards animals in the wild, although she admits that there is "much truth" in this view. The reason is that in today's world it is hardly the case anymore for animals to live sovereign and autonomous lives, unaffected by human interference. The environments on which animals depend for their survival are being increasingly disturbed or destroyed by human activity, and their opportunities for nutrition, shelter, and free movement are in constant decline. Under these human-caused conditions of deprivation, Nussbaum believes that we have a much greater moral responsibility to assist wild animals' flourishing than may at first appear.

But if non-intervention is not a plausible option, the question arises what measures should be taken to assist animals in the wild. More specifically, Nussbaum asks, "Should humans police the animal world, protecting vulnerable animals from predators?" (id., 379). This seems absurd, Nussbaum contends, should it imply that all vulnerable animals or, alternatively, all predators were to be put in 'protective detention', because this would surely do more harm than good. But, like Peter Singer, Nussbaum believes that we should protect prey animals from predation if we can do so without such massive, harm-producing interventions.

Another important question raised by Nussbaum concerns the introduction of 'natural predators' to control animal populations. As an example, she mentions the case of the introduction of wolves to control an overpopulation of elks, something that took place in Yellow Stone in 1995.[10] Nussbaum is opposed to such introductions of predators. She prefers any non-violent method of population control to such a violent method. The "painless predation" of animals through human hunting, she argues, may be an alternative to "other deaths that elks would die, such as starving or being torn apart by wolves" (Nussbaum 2006, 394). In an interview with Carla Faralli, Nussbaum puts it this way:

> Sometimes people think that they have done a great good thing if they make hunting illegal and then, when the deer are reproducing too rapidly and can't find enough to eat, they introduce wolves to tear the deer apart. Actually, I am sure that for the deer the hunter's gun is better than the wolves' jaws, more sudden and less excruciating (Nussbaum and Faralli 2007, 158)

6.4.2 Broadening the Capabilities Approach

Nussbaum's aversion to predation is rooted in her vision of nature. She warns for the danger "of romanticizing nature, or suggesting that things are in order as they

---

[10]With the return of the wolf the elk herd, one of the world's largest elk herds, declined 40% in five years. The wolves prevented elk from overbrowsing willow and aspen near rivers and streams, and this gave rise to a substantial rebound of the beaver, a keystone species that may increase species diversity et cetera. Recently, some doubts have been raised regarding this success story (Mech 2012).

are, if only we humans should stop interfering" (Nussbaum 2006, 367). But she runs the risk of falling into the other extreme, by demonizing nature.[11] Following John Stuart Mill in his essay *Nature*, she portrays predators as vicious criminals, merciless executioners and great monsters, inflicting painful torture and gruesome death on other vulnerable and defenseless creatures. Consequently, Nussbaum maintains that the harm-causing capabilities of predators "are not among those that should be protected by political and social principles" (id., 369). And she also seems to ignore or seriously underestimate the prey animal's natural abilities to evade predators. So, contrary to our initial expectation, Nussbaum's capabilities approach seems far from offering the prospect of a more balanced view of the predator-prey relationship.

However, Nussbaum's version of the capabilities approach is not uncontested; some authors, such as Breena Holland and David Schlosberg, have proposed to re-shape this version because it suffers from a too narrow view of the capabilities necessary for the nonhuman world to function and flourish. Although Nussbaum's approach, by contrast to most other approaches to animal ethics, does attach ethical and political significance to species membership as such, it nonetheless adheres to a liberal individualist framework. Schlosberg notably has argued that the capabilities approach should be broadened to include not only individual animals but also entire species and ecosystems. Such broadening allows us to evaluate the predation problem in a wide ecological context. It sheds new light on the question what it means to flourish as a prey animal: "We need to understand and accept that part of the flourishing of animals is to be the protein for other life forms…To be food for others is the essence of functioning for some beings" (Schlosberg 2007, 151).

Elizabeth Cripps has questioned whether Schlosberg's solution of the predator problem is convincing. Because, even if prey animals as a species benefit from performing the function to be food for other species, it is far from obvious that individual prey animals themselves will flourish when killed for food. To say that it is part of an individual prey animal to be food for another species, "overlooks precisely the concern for the capacity of individual animal lives to go better or worse that Nussbaum wants to recognize" (Cripps 2010, 10; cf. Hailwood 2012).

Cripps suggests that it might be possible to reinvigorate Schlosberg's attempt to make flourishing as a species, which often requires predation, compatible with flourishing as an individual by introducing the notion of 'risk'. As an example to illustrate what she means, Cripps refers to a proposal by a group of scientists to introduce the Old World cheetah as ecological replacement for the extinct American cheetah. This cat has played a crucial role in shaping the astounding speed of the pronghorn antelope, among other traits such as visual acuity. In the absence of this predator, "the pronghorn appears overbuilt today in precisely those traits that make it so distinctive among North American mammals, raising the question of whether a reconstitution of Pleistocene selective pressures warrants consideration" (Donlan

---

[11] Val Plumwood, who has profoundly reflected on the meaning of her experience of being crocodile prey after surviving a crocodile attack in February 1985 in Australia's Kakadu National Park, more or less mockingly remarked that "Predation is often demonised as bringing unnecessary pain and suffering to an otherwise peaceful vegan world of female gathering" (Plumwood 2012, 84).

et al. 2006, 662). According to Cripps, this could indicate that, due to lack of cheetahs, the pronghorn cannot flourish fully because it has no incentive to make full use of its remarkable abilities. "Thus, quite apart from the benefit to the species, it might be in the individual pronghorn's interest to run a risk of being killed by a cheetah" (Cripps 2010, 17).

As we will see in the next section, with this suggestion Cripps anticipates, as it were, the solution to the predation problem that Sue Donaldson and Will Kymlicka have presented in their seminal 2011 book *Zoopolis*.

6.5 Political Theory of Animal Rights

Donaldson and Kymlicka have developed their political theory of animal rights as an alternative to the traditional animal rights theory. Due to its one-sided focus on the intrinsic moral status or standing of animals as the sole basis of our moral obligations towards them, the traditional animal rights theory seems unable to resolve a wide range of pressing issues regarding human-animal interactions, and is thereby at least partly to blame for what Donaldson and Kymlicka perceive as the political and intellectual impasse of the animal advocacy movement. To overcome this impasse, they have made an attempt to shift the debate from the field of moral theory to the field of political theory, focusing on the differential obligations that arise from the varied ways that animals are related to human societies and institutions.

In *Zoopolis*, Donaldson and Kymlicka draw upon the concepts and categories of political theory to illuminate the specific rights and responsibilities we have in our various relationships with animals. They distinguish three types of morally significant human-animal relationships: domesticated animals, such as companion animals and farm animals, should be considered and treated as our co-citizens; wild animals should be recognized as members of separate, sovereign nations, entitled to protection from infringements of their right to self-determination; and, lastly, "liminal" animals, i.e., non-domesticated animals such as rats and raccoons who live among humans, should be designated the status of "denizens".[12]

6.5.1 Similarities and Dissimilarities with the Capabilities Approach

Donaldson and Kymlicka concede that they are sympathetic to Nussbaum's capabilities approach, and that, at the most abstract level, their own citizenship model could be described in broadly capability terms (2011, 95, 275). Like Nussbaum, they treat

---

[12]The group of liminal animals include opportunistic animals, agricultural symbiotics (or niche specialists), feral animals and introduced exotics (see Donaldson and Kymlicka 2011, 219–226).

animals not just as passive victims of human domination and mere objects of compassion but rather as subjects with a clear capacity for agency. And like Nussbaum, they consequently also challenge the one-sided focus of most accounts of animal ethics on negative rights—"thou shall not kill, use, or keep animals" (id., 254). It is the dominant view within animal ethics that the abolishment of animal exploitation and the liberation of animals from enslavement will ultimately rule out virtually all forms of human-animal interaction—"there should be no human-animal relations" (ibid.). According to Donaldson and Kymlicka, this narrow vision of animal rights is at the root of the impasse of the animal advocacy movement because it may discourage all efforts to find out what non-exploitative relations might look like, and what kind of positive obligations we owe to animals, be they domesticated, wild or liminal.

So there are important similarities between Nussbaum's capabilities approach and Donaldson and Kymlicka's political theory of animal rights: both consider animals as moral agents rather than as moral patients, and both aim to complement negative rights with positive rights. But there is also a distinct difference: Nussbaum attaches considerable importance to species membership, whereas Donaldson and Kymlicka focus on community membership, thus taking account of the sociopolitical context of animal justice. They acknowledge that Nussbaum's species norm of flourishing is probably a reasonable standard for animals living in the wild, but they deny that this norm makes sense with respect to domesticated and liminal animals. Another drawback of the preoccupation with species norms is a lack of sensitivity to the associations *between* species and individual variation *within* species.

6.5.2 Competence and Risk

According to Donaldson and Kymlicka, the argument that the flourishing of wild animals would be undermined by interfering with wildlife to prevent predation is "perhaps the most important one, but also the least developed" (2011, 165). This "flourishing argument", as they call it, needs qualification and clarification. Similar to Cripps' critique of Schlosberg, they argue that it is difficult to see how preventing a deer from being killed by a predator is detrimental to her flourishing. And just like Cripps they invoke the notion of 'risk' to address this question. For societies with an interest in self-determination, eliminating the risk of harm or suffering "would involve a terrible curtailment of freedom, including the freedom to fully develop and explore one's capabilities. Individual action to protect a human child at the moment of harm contributes to her flourishing; collective action to prohibit the actions or processes that create the risk of harm is likely to undermine human flourishing. So, too, with animals" (id, 166).

Donaldson and Kymlicka believe that, when it comes to the daily management of the risks of living in the wild, it is reasonable to assume that wild animals are fully competent in general to address the challenges they face: they have the skills to secure and store food, to find or construct shelter, to care for the young, to cover long distances, to hunt, and also to reduce the risk of predation.[13] Because wild animals are competent to manage their own affairs, we are not obligated to systematically intervene to end predation or to control natural food cycles. Respect for the sovereignty of wild animals, in fact, rules out this kind of intervention as it would condemn wild animals to a permanent state of dependency.[14]

6.5.3 Positive and Negative Duties

It would, however, be a mistake to think that respect for sovereignty requires a complete hands-off approach with respect to animals in the wild. Donaldson and Kymlicka mention two broad categories of assistance and intervention that do not threaten but may even promote values of autonomy and self-determination: large-scale interventions to prevent or mitigate natural or human-caused disasters, such as deflecting a large meteor on a collision course with a wilderness zone populated by numerous animals, or halting an aggressive new bacterium which is ready to invade and destroy an ecosystem; and small-scale or micro-scale interventions aimed to aid or rescue individual animals in distress, such as saving an animal who has fallen through the ice from drowning or releasing a beached whale to open water.

In addition to these positive duties to aid, assistance, and intervention, there are also important negative duties that derive from the respect for sovereignty we owe animals in the wild. We should never infringe on the rights to their own territory and to autonomy on that territory, which are key components of the principle of sovereignty. These rights impose, first of all, immediate and drastic restrictions on human expansion into wild animal territories and the ongoing fragmentation and destruction of wild animal habitat. They also impose stringent limits on human actions that have harmful impacts beyond the borders of wild animal territories, such as water contamination, air pollution and the various effects of climate change. To avoid such cross-border impacts we will have to reduce our ecological footprint and replace our environmentally destructive behavior and cost-externalizing practices with fair and sustainable ones.

---

[13] Ethological and ecological research shows that prey species have gained over evolutionary time a stunning array of mechanisms to cope with predators for every stage of their struggle: the avoidance of detection by the predator (such as camouflage, refuge use, nocturnality), the avoidance of attack once detected (mimicking animals with strong defenses, signaling to the predator that pursuit is not worthwhile), the avoidance of capture once attacked (fleeing, bluffing strength), and the avoidance of consumption once captured (playing death or 'thanatosis', sacrificing body parts or 'autotomy').

[14] Donaldson and Kymlicka admit that their competence argument is more compelling in relation to some animals than to others. In fact, they agree with Horta (2013) that members of r-selected species have less scope for 'competent agency' than members of K-selected species. But on balance, they believe that "we should still respect the sovereignty of wild animals, including those for whom there is minimal evidence of competent agency" (Donaldson and Kymlicka 2011, 177; cf. 2013, 154).

6.5.4   The Limits of a Place-Based Approach

Donaldson and Kymlicka contrast their sovereignty model with the 'stewardship' model, often found in environmental science, philosophy and policy. In this model, habitat for wild animals is created in the shape of wild areas such as wildlife refuges, nature reserves, and national parks, where humans enjoy sovereign authority and exercise stewardship over wild animals. The sovereignty model, on the other hand, doesn't grant humans the right to govern wild animal territory, but is based on the principle that sovereign entities are entitled to the same or similar claims to authority, and should deal with each other on an equal footing. This means that when we humans enter wild animal territory, "we do so not in the role of stewards and managers, but as visitors to foreign lands" (*id.*, 170).

Both models, however, have one thing in common—like the classical stewardship model, the sovereignty model has taken a place-based approach with regard to wild animals. Donaldson and Kymlicka argue for an immediate check on the expansion of human settlement, for giving wild animals back control over their own territories, for returning vast areas of land currently devoted to animal agriculture to wild animals, for re-establishing wildlife corridors and migration routes et cetera. But such a place-based approach seems to fall far short of what presently is really required to maintain or restore wild animals' autonomy (Sandler 2012).

Hitherto, place-based or *in situ* conservation is usually given priority over 'out of place' or *ex situ* conservation. The latter is considered to be justified only as a supportive measure to the former. This hierarchical understanding of the relationship between *in situ* and *ex situ* conservation reflects the importance of the place of origin—'wild nature' (Braverman 2015, 33). This understanding is however increasingly being called into question given today's ecological challenges that can be summarized under the denominator of the 'Anthropocene', the current geological epoch in which human activities are so profound and pervasive that humanity itself has emerged as a global geophysical force, at least as important as natural forces (Keulartz and Bovenkerk 2016).

6.5.5   Blurring Boundaries

Under Anthropocenic conditions many wild populations are no longer viable on their own. Mainly due to habitat fragmentation and habitat loss, there is an ongoing conversion of what originally were continuous populations to so-called 'metapopulations': collections of subpopulations, that are spread geographically over patches of habitat. Because these patches are usually small and because the movement of the animals between these patches is restricted for lack of connectivity, an increasing number of subpopulations are declining and are teetering on the edge of extinction. In this situation *ex situ* conservation has a more prominent role to play and is now regarded as equivalent, rather than subordinate to *in situ* conservation.

Because it is no longer considered effective to manage wild and captive populations in isolation from one another, practitioners of species conservation therefore increasingly use the so-called One Plan Approach that was officially proposed to the IUCN World Conservation Congress in 2012. The One Plan Approach promotes the interactive exchange of animals between *in situ* populations (in the wild) and *ex situ* populations (in captivity) for mutual reinforcement, a management approach that is also referred to as *inter situ* conservation (Braverman 2014) or *pan situ* conservation (Minteer and Collins 2013). With animals moving in both directions, the stability and sustainability of wild and captive populations can be greatly enhanced. On the one hand, captive populations can be used for restocking in areas with declining populations or for reintroduction in areas where populations have gone extinct; on the other hand, the demographic and genetic viability of *ex situ* populations can be boosted by supplying genetic founders from wildlife populations (Byers et al. 2013).

With the One Plan Approach captive populations can be used for the conservation of *in situ* populations on the brink of extinction as a result of habitat fragmentation. But what if *in situ* conservation itself is being undermined by that other major environmental stressor—rapid global climate change—, that makes the species' historic indigenous ranges increasingly inhospitable? And when, moreover, populations are not able to move on their own to other areas with more suitable environmental conditions? A conservation measure that may prevent species that are unable to keep pace with rapid climate change from going extinct is assisted migration or assisted colonization, i.e. the intentional movement of 'climate refugees' to new habitats outside their historical range, which they otherwise could not reach (Hoegh-Guldberg et al. 2008). Whereas *inter* or *pan situ* conservation involves the movement of animals from one location to another *within* the species' indigenous range, assisted migration or colonization relates to animal translocations *outside* the species' indigenous range.[15]

The emergence of these new conservation strategies makes it clear that the distinction between classic *in situ* (on-site) and *ex situ* (off-site) conservation is becoming blurred to the point of disappearing entirely. We witness what Braverman (2015, 15) has called a shift "from bifurcation to amalgamation" of *in situ* and *ex situ* conservation: the increased development of hybrid approaches, that integrate the wild and the captive. (Pritchard et al. 2011; Redford et al. 2012; 2013; Minteer et al. 2016).

It is clear that with the ongoing blurring of the boundaries between *in situ* and *ex situ* conservation, placed-based models such as the classic stewardship model but also Donaldson and Kymlicka's sovereignty model, are increasingly rendered meaningless. Under Anthropocenic conditions positive interventions can no longer be limited to providing assistance in the event of natural or human-caused disasters, or to micro-level individual acts of compassion only. Apart from these isolated, incidental cases, Donaldson and Kymlicka believe that we should leave nature to its own devices—"in general, a hands-off principle towards wild animals is a sound one"

---

[15]"The indigenous range of a species is the known or inferred distribution generated from historical (written or verbal) records, or physical evidence of the species' occurrence" (IUNC/SSC 2013, 2).

(2011, 185). But that ship seems to have sailed already.[16] Ironically or not, but today we are morally obligated to systematically interfere with wildlife, not to prevent predation, as consequentialists in particular would have it, but to assist endangered species in maintaining and improving their competences to survive on their own in the wild, including the skills to hunt and to avoid predators, something which is already taking place on the ground.

6.5.6 Learning to Hunt and to Avoid Predators

Especially zoo-based expertise in sustaining small but demographically and genetically sound populations of captive animals has been proven useful for the conservation of small and declining populations in the wild. Zoo-based skills in animal handling may, moreover, be helpful at many of the main stages of animal translocations, from capture, transport, and captive breeding, to pre-release training (Fa et al. 2011, 210). The latter is particularly important because captive animals may lack the behavioral competences needed for survival in the wild, and may thus compromise the ability of captive populations to contribute to the recovery of wild populations. Pre-release training is aimed at maintaining or developing the skills that may have been lost in captivity such as orientation and navigation, finding or building suitable nest sites, hunting and foraging behavior, and predator avoidance (Earnhardt 2010; McPhee and Carlstead 2010).

Predator avoidance training is vital to the success of conservation efforts that rely on captive animals because a substantial number of post-release deaths are due to predators. It usually consists of exposure to live predators or to predator models paired with some aversive or stressful stimulus such as an alarm signal. Given that many animals are so-called 'mesopredators', i.e. animals which both predate and are predated upon, antipredator training has often to be combined with developing predatory skills. A case in point is the black-footed ferret.

In 1986, the population of ferrets had diminished to a mere 18 individuals, but thanks to a captive breeding program, between 500 and 800 now roam the prairie of the US state of Wyoming. The program was not, however, entirely plain sailing. When the kits were released they were far too blasé to make themselves scarce when predators such as eagles, coyotes and badgers arrived on the scene. The researchers tried to resolve this problem by building a mock predator. They attached wheels to a stuffed badger, which would win fame as RoboBadger. The only way the ferrets could escape RoboBadger was to find a burrow. The researchers then tried to increase

---

[16]Donaldson and Kymlicka grossly underestimate the impact of anthropogenic environmental change, that currently takes place at such a fast speed and large scale that it definitely poses a threat to the resilience of the Earth System (Steffen et al. 2007, 2011). They denounce the "fashionable talk" of the Anthropocene, that has had "the perverse effect of making continued human encroachment on, and management of, wild animal habitat seem inevitable" (Donald and Kymlicka 2016).

the ferrets' aversion to RoboBadger by firing rubber bands at them. (McCarthy 2004, 196/7).[17]

But the ferrets have not only to learn how to avoid predators, but also how to locate and kill prairie dogs which make up between 65 to 90 percent of their diet. In addition, they have to learn how to invade and inhabit prairie dogs burrows because they do not build their own burrows. Their preconditioning period lasts for 30 days. During that time the ferrets ideally kill four prairie dogs and live in an actual prairie dog burrow system. The survival rate of these animals is about ten times higher than animals released straight out of the cage (Braverman 2015, 119–123).

6.6   Concluding Remarks

Donaldson and Kymlicka rightly argue that wild animals are fully competent to manage their own affairs, and that we therefore should refrain from protecting prey animals against predation. Without predation, prey animal's possibilities for flourishing will be diminished, because all the amazing capabilities they have gained over evolutionary time to cope with predators might be rendered meaningless. All in all, we can safely conclude that it is counterproductive to extent Nussbaum's Other Species capability to the animal kingdom. Instead of working to ensure that all species will enjoy cooperative and mutually supportive relations, we should respect the natural capabilities of animals, be they predator or prey, without romanticizing or demonizing their agonistic interactions.

But their place-based sovereignty model is inadequate in the light of what has been called the planet's 'sixth mass extinction'. Unlike earlier mass extinctions, the current one is not primarily driven by natural events such as meteorite impacts or volcanic eruptions but by the effects of the activities of *Homo sapiens*. Especially human-caused rapid climate change together with habitat conversion, fragmentation, and destruction have led to a global wave of species and population extirpations and declines in local species abundance.[18]

If we really want to stop or even reverse this so-called 'defaunation' process we can no longer hold on to the idea that species conservation can be accomplished with minimal management by establishing large nature reserves and by creating connections such as corridors and stepping stones between them. Preserving the ecological status quo through such traditional measures increasingly resembles a Sisyphean task. *In situ* conservation (in the wild) is no longer effective without *ex situ* conservation (in zoos and aquariums).

---

[17]This was, by the way, not a great success as became clear when the ferrets started riding on the back of RoboBadger.

[18]Surprisingly enough, nowhere in Zoopolis do Donaldson and Kymlicka even mention the staggering decline in species numbers; they use the notion of 'extinction' only in relation to those proposals from animal rights theorists who call for a complete end to domestication and the extinction of domesticated species.

Donaldson and Kymlicka condemn capturing animals and putting them in zoos, even in the most progressive zoos, as "a violation of their basic individual rights, and a violation of their rights as members of sovereign communities" (2011, 283, cf. 293). Such condemnation only shows how blind they are to the important role that zoos and other *ex situ* institutions have to play under current conditions of anthropogenic stress (see Keulartz 2016b).

References

Braverman, I. 2014. Captive for life: Conserving extinct in the wild species through ex situ breeding. In *The ethics of captivity*, ed. L. Gruen, 93–212. Oxford, UK: Oxford University Press.

Braverman, I. 2015. *Wild life: The institution of nature*. Stanford, CA: Stanford University Press.

Byers, O., C. Lees, J. Wilcken, and C. Schwitzer. 2013. The one plan approach: The philosophy and implementation of CBSG's approach to integrated species conservation planning. *WAZA Magazine* 14: 2–5.

Clark, S.R.L. 1979. The rights of wild things. *Inquiry* 22 (1–4): 171–188.

Cohen, C. 1997. Do animals have rights? *Ethics & Behavior* 7 (2): 91–102.

Cowen, T. 2003. Policing nature. *Environmental Ethics* 25 (2): 169–182.

Cripps, E. 2010. Saving the polar bear, saving the world: Can the capabilities approach do justice to humans, animals and ecosystems? *Res Publica* 16: 1–22.

Dawrst, A. 2009. The predominance of wild-animal suffering over happiness: An open problem. *Essays on Reducing Suffering*. http://reducing-suffering.org/wp-content/uploads/2015/02/wild-animals_2015-02-28.pdf. Accessed 23 March 2020.

Donaldson, S., and W. Kymlicka. 2011. *Zoopolis. A political theory of animal rights*. Oxford, NY: Oxford University Press.

Donaldson, S., and W. Kymlicka. 2013. A defense of animal citizens and sovereigns. *LEAP: Laws, Ethics and Philosophy* 1: 143–160.

Donaldson, S., and W. Kymlicka. 2016. Between wild and domesticated: Rethinking categories and boundaries in response to animal agency. In *Animal ethics in the age of humans*, ed. B. Bovenkerk and J. Keulartz, 225–239. Dordrecht: Springer.

Donlan, J., et al. 2006. Pleistocene rewilding: An optimistic agenda for twenty-first century conservation. *The American Naturalist* 168: 160–183.

Dorado, D. 2015. Ethical interventions in the wild: An annotated Bibliography. *Relations: Beyond Anthropocentrism* 3 (2): 219–238.

Earnhardt, J.M. 2010. The role of captive populations in reintroduction programs. In *Wild mammals in captivity*, ed. D.G. Kleiman, K.V. Thompson, and C.K. Baer, 263–267. Chicago: University of Chicago Press.

Fa, J., S. Funk, and D. O'Connell. 2011. *Zoo conservation biology*. Cambridge: Cambridge University Press.

Fink, Ch. 2005. The predation argument. *Between the Species* 13 (5): 1–15.

Hadley, J. 2006. The duty to aid nonhuman animals in dire need. *Journal of Applied Philosophy* 23 (4): 445–451.

Hailwood, S. 2012. Bewildering Nussbaum: Capability justice and predation. *The Journal of Political Philosophy* 20 (3): 293–313.

Hoegh-Guldberg, O., L. Hughes, S. McIntyre, D.B. Lindenmayer, et al. 2008. Assisted colonization and rapid climate change. *Science* 321 (5887): 345–346.

Horta, O. 2010a. Debunking the idyllic view of natural processes: Population dynamics and suffering in the wild. *Télos* 17 (1): 73–88.

Horta, O. 2010b. The ethics of the ecology of fear against the nonspeciesist paradigm: A shift in the aims of intervention in nature. *Between the Species* 13 (10): 163–187.

Horta, O. 2013. Zoopolis, intervention, and the state of nature. *LEAP: Laws, Ethics and Philosophy* 1: 113–125.

IUCN/SSC. 2013. *Guidelines for reintroductions and other conservation translocations. Version 1.0*. Gland: IUCN Species Survival Commission.

Jamieson, D. 1990. Rights, justice, and duties to provide assistance: A critique of Regan's theory of rights. *Ethics* 100: 349–362.

Jamieson, D. 2008. The rights of animals and the demands of nature. *Environmental Values* 17: 181–199.

Keulartz, J. 2016a. Towards an animal ethics for the Anthropocene. In *Animal ethics in the age of humans*, ed. B. Bovenkerk and J. Keulartz, 243–264. Dordrecht: Springer.

Keulartz, J. 2016b. Ethics of the zoo. *Oxford Research Encyclopedia of Environmental Science*.

Keulartz, J., and B. Bovenkerk. 2016. Changing relationships with non-human animals in the Anthropocene. In *Animal ethics in the age of humans*, ed. B. Bovenkerk and J. Keulartz, 1–22. Dordrecht: Springer.

McCarthy, S. 2004. *Becoming a tiger: How baby animals learn to live in the wild*. New York: Harper Perennial.

McMahan, J. 2010. The meat eaters. *The New York Times*, 19 September.

McPhee, M.E., and K. Carlstead. 2010. The importance of maintaining behaviours in captive mammals. In *Wild mammals in captivity*, ed. D.G. Kleiman, K.V. Thompson and C.K. Baer, 303–313. Chicago: The University of Chicago Press.

Mech, L.D. 2012. Is science in danger of sanctifying the wolf? *Biological Conservation* 150: 143–149.

Milburn, J. 2015. Rabbits, Stoats and the predator problem: why a strong animal rights position need not call for human intervention to protect prey from predators. *Res Publica* 1 (3): 273–289.

Minteer, B., and J. Collins. 2013. Ecological ethics in captivity: Balancing values and responsibilities in zoo and aquarium research under rapid global change. *ILAR* 54 (1): 41–51.

Minteer, B., J. Collins, and A. Raschke. 2016. Between the wild and the walled: The evolution and ethics of zoo conservation. In *Routledge companion to environmental ethics*, ed. B. Hale and A. Light. London: Routledge.

Ng, Y.-K. 1995. Towards welfare biology: Evolutionary economics of animal consciousness and suffering. *Biology and* Philosophy 10 (4): 255–285.

Nussbaum, M.C. 2000. *Women and human development: The capabilities approach*. Cambridge, UK: Cambridge University Press.

Nussbaum, M.C. 2001. Animal rights: The need for a theoretical basis. *Harvard Law Review* 114: 1506–1549.

Nussbaum, M.C. 2006. *Frontiers of justice: Disability, nationality, species membership*. Cambridge, MA: Havard University Press.

Nussbaum, M.C., and C. Faralli. 2007. On the new frontiers of justice. A dialogue. *Ratio Juris* 20 (2): 145–161.

Palmer, C. 2010. *Animal ethics in context*. New York: Columbia University Press.

Palmer, C. 2015. Against the view that we are normally required to assist wild animals. *Relations* 3 (2): 203–210.

Pearce, D. 1995. The hedonistic imperative. http://www.hedweb.com/hedethic/tabconhi.htm. Accessed 23 March 2020.

Pearce, D. 2009. Reprogramming predators. http://www.hedweb.com/abolitionist-project/reprogramming-predators.html. Accessed 23 March 2020.

Plumwood, V. 2012. *The eye of the crocodile*, ed. L. Shannon. Canberra: ANU E Press.

Pritchard, D., J. Fa, S. Oldfield, and S. Harrop. 2011. Bring captive closer to the wild: Redefining the role of ex situ conservation. *Oryx* 46 (1): 18–23.

Rachels, J. 1990. *Created from animals: The moral implications of Darwinism*. New York: Oxford University Press.

Redford, K., D. Jensen, and J. Breheny. 2012. Integrating the captive and the wild. *Science* 338: 1157–1158.

Redford, K., D. Jensen, and J. Breheny. 2013. The long overdue death of the *ex situ* and *in situ* dichotomy in species conservation. *WAZA Magazine* 14: 19–22.

Regan, T. 1983. *The case for animal rights*. Berkeley, CA: University of California Press.

Regan, T. 2001. The case for animal rights: A decade's passing. In *Defending animal rights*, ed. T. Regan, 39–65. Champaign: University of Illinois Press.

Regan, T. 2004. *The case for animal rights*, 2nd ed. Berkeley, CA: University of California Press.

Ritchie, D.G. 2002. *Natural rights: A criticism of some political and ethical concepts*. London: Routledge.

Sagoff, M. 1984. Animal liberation and environmental ethics: Bad marriage, quick divorce. *Osgoode Hall Law Journal* 22 (2): 297–307.

Sandler, R.L. 2012. *The ethics of species: An introduction*. Cambridge, UK: Cambridge University Press.

Sapontzis, S.F. 1984. Predation. *Ethics and Animals* 5 (2): 27–38.

Schlosberg, D. 2007. *Defining environmental justice: Theories, movements, and nature*. New York: Oxford University Press.

Simmons, A. 2009. Animals, predators, the right to life and the duty to save lives. *Ethics and the Environment* 14: 15–27.

Singer, P. 1973. Food for thought, reply to David Rosinger. *New York Review of Books*, 14 June. http://www.nybooks.com/articles/1973/06/14/food-for-thought/. Accessed 23 March 2020.

Singer, P. 1975. *Animal liberation*. New York: Random House.

Steffen, W., et al. 2007. Are humans now overwhelming the great forces of nature. *Ambio* 36 (8): 614–621.

Steffen, W., et al. 2011. The Anthropocene: From global change to planetary stewardship. *Ambio* 40 (7): 739–761.

Tomasik, B. 2015. The importance of wild-animal suffering. *Relations: Beyond Anthropocentrism* 3 (2): 133–152.

Jozef Keulartz is Emeritus Professor of Environmental Philosophy at the Radboud University Nijmegen, and senior researcher Applied Philosophy at Wageningen University and Research Centre. He has published extensively in different areas of science and technology studies, social and political philosophy, bioethics, environmental ethics and nature policy. His books include *Die verkehrte Welt des Jürgen Habermas* [The Topsy-Turvy World of Jürgen Habermas, Junius, 1995] and *Struggle for Nature—A Critique of Radical Ecology* (Routledge, 1998). He is coeditor of *Pragmatist Ethics for a Technological Culture* (Kluwer, 2002), *Legitimacy in European Nature Conservation Policy* (Springer, 2008), *New Visions of Nature* (Springer, 2009), *Environmental Aesthetics: Crossing Divides and Breaking Ground* (Fordham University Press, 2014), *Old World and New World Perspectives in Environmental Philosophy* (Springer, 2014), and *Animal Ethics in the Age of Humans* (Springer, 2016).

Chapter 7
Justified Species Partiality

Ronald Sandler and John Basl

Abstract A core question in practical ethics is 'which entities do we need to consider in our decision-making?' In this chapter we evaluate the justifications and motivations for defending species-membership views of human moral status. These are views on which human beings have a distinctive type of moral status grounded in their being human or possessing some property that almost perfectly correlates with being human. Many ethicists endorse species-membership views on moral status because they believe that moral status differences are needed to support widely held and purportedly well-justified beliefs about species differentiation in consideration and treatment. We argue against the need to adopt a species-membership or human-privilege view on moral status in order to justify species partiality in consideration and treatment. The sort of partiality with respect to consideration and treatment that motivates species-membership views is largely consistent with more egalitarian views about moral status, according to which an entity's moral status depends on its own features, not the biological group to which it belongs. Given the traditional objections to species-membership views, to the extent that justified species partiality is consistent with alternative views of moral status, there is reason to reject the moral status significance of being human.

7.1 Introduction

A core questions in practical ethics—in trying to figure out what we ought to do in a situation—is 'which entities do we need to consider in our decision-making?' If we need to take nonhuman animals into consideration, for example, there are implications for everything from what we ought to eat to whether we ought to have pets (and if so, which ones). In ethical theory this question is often referred to as the question of moral status. If something has moral status, then it needs to be taken into

R. Sandler (✉) · J. Basl
Department of Philosophy and Religion, Northeastern University, Boston, MA, USA
e-mail: r.sandler@northeastern.edu

© The Author(s) 2021
B. Bovenkerk and J. Keulartz (eds.), *Animals in Our Midst: The Challenges of Co-existing with Animals in the Anthropocene*, The International Library of Environmental, Agricultural and Food Ethics 33,
https://doi.org/10.1007/978-3-030-63523-7_7

consideration regarding actions, practices, and policies that could impact it. So, if we want a theory that accurately depicts the ethically relevant features of the world, then we need to determine which entities have moral status as well as the properties in virtue of which they have it.

In this chapter we evaluate both the justifications and motivations for defending species-membership views of human moral status. These are views on which human beings have a distinctive type of moral status that is grounded in their being human or possessing some property that almost perfectly correlates with being human, such as having the biological basis for moral agency. Species-membership views have been challenged by animal and environmental ethicists as being unsupported, ad hoc, arbitrary, and speciesist—i.e. unjustifiably biased against nonhumans. However, species-membership views persist, especially in the bioethics and disability ethics literatures. Part of what motivates species-membership views is that the criteria or principles that guide ethical treatment of humans and nonhuman animals vary widely. For example, standards and protocols for conducting research on human subjects are quite different from those for conducting research on nonhuman animals. Many ethicists endorse species-membership views on moral status because they believe that moral status differences are needed to support widely held and purportedly well-justified beliefs about species differentiation in consideration and treatment.

In this paper we argue against the need to adopt a species-membership or human-privilege view on moral status in order to justify (contextual) species partiality in consideration and treatment. The sort of partiality with respect to consideration and treatment that motivates species-membership views is largely consistent with more egalitarian views about moral status. We discuss several strategies for justifying favorable consideration and treatment of humans over non-human animals—*justified species partiality*—that do not depend on species-membership or some proxy to ground a distinctive type of moral status. Given the traditional objections to species-membership views, to the extent that justified species partiality is consistent with alternative views of moral status, there is reason to reject the moral status significance of being human.

Here, then, is a summary of our core argument:

1. One of the primary motivations for species membership accounts of human moral status is the need to justify differential consideration and treatment between humans and nonhuman animals.
2. If differential consideration and treatment can be justified on other (i.e. non-species-membership) accounts of human moral status, then this motivation for species membership accounts of human moral status is undermined.
3. Differential consideration and treatment between humans and nonhuman animals can be justified on other accounts of human moral status (e.g. species-egalitarian accounts).
4. Therefore, this primary motivation for species membership accounts of human moral status is undermined.

We begin by reviewing the debate over species-membership views, with an emphasis on recent work by disability ethicists critical of species-egalitarian views on moral status. The aim is not to evaluate the legitimacy of the concerns or the validity of the arguments, but to identify their motivations for supporting a species-membership approach to human moral status. We then discuss three strategies for grounding justified species partiality that do not depend on giving special moral significance to species boundaries. One strategy is to show that a highly egalitarian view of moral status and pluralistic conception of moral considerability can support justified partiality. A second strategy is to show that there can be grounds for species partiality within law and policy even if humans and nonhuman animals have equal moral status. A third strategy is to show that the fact that we are in a better epistemic position with respect to understanding human interests than non-human interests justifies considering them differently in some contexts.

These strategies are not mutually exclusive or collectively exhaustive. Partiality can be justified in different contexts for entirely different reasons, though there are limits to the amount, forms, and situations in which species partiality is justified. To the extent that these strategies, and potentially others as well, can support justified (though not unlimited) species partiality, the case for adopting a species-membership view of moral status is undermined. Furthermore, they provide those that are motivated to justify species partiality with new avenues to defend their view, changing the landscape of debates over these issues in, hopefully, productive ways.

7.2 Species-Membership Views of Moral Status

Biological group membership views of moral status are those on which the moral status of an individual is explained at least in part or in some cases entirely by its *biological* features and/or relationships. The view that human beings have a special, unique or differential moral status (or dignity) is a biological group membership account of moral status. It asserts either that (1) being a member of the species *homo sapiens* is itself morally significant (and explains why members of the species have greater worth or are due greater/special consideration), or (2) that the species boundary accurately tracks something that is morally significant (and explains why members of the species have greater moral worth or are due greater/special consideration).

The primary argument against species membership views of moral status is that they are arbitrary and question begging, and that as a result they either unjustifiably exclude individuals from the scope of ethical concern or else unjustifiably reduce the amount of concern due to them (Singer 1989, 1975; Taylor 1986; McMahan 2005, 2008; Rachels 1999). The answer to "why should only humans be regarded as having moral status?" or "why do humans have a special or unique status?" cannot be "because they are human" (Singer 1989; McMahan 2005, 2008). But it is exceedingly difficult to provide any other justification. The reason for this is that *Homo sapiens* species boundaries do not track anything ethically significant—for example, moral agency, autonomy, language, types (or range) of interests, or ability to participate in social relationships. Some human beings are moral agents, highly autonomous, capable of reciprocal concern, and able to participate in complex cooperative arrangements, but not all are. Moreover, some individuals of some nonhuman species—for example, orangutans and dolphins—are as capable of these as are some humans. Similarly, some nonhuman animals have equal or greater psychological capacities in some respects than do some humans, and so have equally or more complex and diverse interests.

Proponents of the greater moral status of all humans might argue that the fact that all healthy or "species-typical" members of *Homo sapiens* have comparable interests and capacities—e.g. moral agency—justifies treating membership in the species as morally special. However, why should co-membership in a group confer the moral status associated with some members of the group to all members, even those that lack the relevant capacities? There are a lot of possible biological groups—e.g. vertebrates, eukaryotes, and mammals. Why should we prioritize one grouping over another when determining moral status? To privilege one biological grouping over others seems arbitrary and question begging. In response, proponents of the moral relevance of the *Homo sapiens* species boundary sometimes appeal to conspecificity (Kittay 2005, 2017). It is not that the species boundary of *Homo sapiens* is itself morally significant or marks something morally significant. It is the fact that it is *our species*. But this response also begs the question. We are part of a lot of possible biological groups—e.g. mammals, vertebrates, and eukaryotes. Why is co-membership in one biological grouping privileged over all the others in moral status determinations, particularly when that grouping does not track anything that is ethically significant, such as interests or capacities? If we found out that what we currently think of as *Homo sapiens* were really two distinct species with indistinguishable capacities—*Homo napiens and Homo mapiens*—we would not have any new reason not to consider the individuals of the other species. Or, at least, no non-question begging and arbitrary one, since the distinction would not track anything other than one among many possible (and often imperfect) biological groupings.

The alternative to a biological grouping account of moral status is an *individualist* and *capacities-based account*. According to such accounts, what matters to whether and how we should consider something's interests is what the individual is capable of, what its interests are, how it can be harmed and benefited, and the relationships that it can have—i.e. its capacities (McMahan 2002, 2005, 2008; Singer 1975, 1989; Regan 1983, 1985; DeGrazia 1996, 2007, 2014; Rachels 1999; Taylor 1986; Sandler 2013). What these views share, qua individualist capacities based view, is that: (1) ascriptions of moral status difference must be explained (in the sense of being justified); (2) they can only be adequately justified by appeal to something about the entities themselves; and (3) the only thing about the entities themselves that could justify a moral status difference is their having differential capacities and interests—e.g. whether they are moral agents, can have positive and negative experiences, can be benefited or harmed, can participate in certain types of relationships, or can set their own ends.

Individualist capacities-based views of moral status have come in for criticism in recent years from some ethicists working in disability ethics because the views deny that all human beings necessarily (or *qua* human being) have equal moral status greater than that of all nonhuman animals. They believe that the views thereby allow that some people—e.g. those with permanent and severe cognitive impairments (hereafter PSCI)—could have less moral status (or be due lesser consideration) than other people, and some psychologically complex nonhuman animals could have the same or even greater moral status (or be due greater consideration) than some people with PSCIs (Kittay 2017; Carlson 2009; Curtis and Vehmas 2016; Jaworska and Tannenbaum 2014). This is troubling to disability ethicists for several reasons. (As discussed above, our purpose here is not to assess either (1) whether capacities based views actually have the implication that some human beings could have lesser worth or be due less consideration than other human beings or some nonhumans, or (2) whether the concerns below are warranted if they do have that implication. Our aim is to establish the first premise in our overarching argument, which is that the belief that capacities-based views have these implications and give rise to these concerns are a primary motivation for proponents of species-membership approaches to human moral status.)

One reason that it is seen as troubling is that it is offensive to assert that someone's child or loved-one are not due full and equal respect because of their impairments or condition, as well as to compare them to (sometimes unfavorably or as having less status or being due less consideration than) dogs and pigs (Kittay 2005).

Another is that it is problematic to hold that the presumption against harming or exploiting someone is weaker because they are more vulnerable, more dependent, due to their condition or disability. This seems to have things backward. Those who are dependent are due greater protection, not less. We have special responsibilities to consider their interests and needs, precisely because of their vulnerabilities. Therefore, any account of moral status on which people with PSCIs have less presumption against harm and exploitation is seen as problematic.

A third concern is that giving up the idea that all human beings have the same moral status is likely to lead to further exclusions and mistreatment. As Eva Kittay (2017, 31) puts it, "The claim that humans are not equal threatens to plunge us backward. The idea that 'all men were created equal' was hard won and it has taken centuries to make all 'men' include women, racial, ethnic, and sexual minorities, and people with disabilities. To claim that any humans are of unequal value is to let the camel's nose inside the tent. The Nazi's first victims were those with mental disabilities."

A fourth concern is to do with the empirical inadequacy of comparisons between people with PSCIs and psychologically complex nonhuman animals, and the ways in which those inaccurate comparisons are pernicious. The lives, emotions, preferences, perspectives and abilities of people with PSCIs are various and nothing like those of pigs and apes. As Kittay (2017, 25) puts it: "Respectable contemporary philosophers have, for instance, spoken of the radically or severely mentally impaired as unable to recognize familiar people in their lives, as having cognitive abilities comparable to those of a dog, as always remaining at the mental age of an infant, although it is often unclear whether they are speaking of actual people or a hypothetical case."

Kittay has been particularly critical of the way in which "marginal cases"—human beings who lack the "full set" of cognitive capacities of healthy adult humans—have been used in arguments for elevating the moral status of nonhuman animals. When animal ethicists engage in "leveling by intrinsic properties" (Kittay 2017), they are not only elevating nonhumans, in her view they are often diminishing the moral status of people with PSCIs "to the level of the raised status of those nonhuman animals possessing such putatively comparable intrinsic properties" (Kittay 2017, 30).

Perhaps the most influential positive argument in support of the view that all human beings have equal and full moral status, without respect to their individual capacities, is a *reductio ad absurdum* from cases. Proponents offer a range of intuitively objectionable, or downright repulsive things, that they believe would be permitted to do to people with PSCIs if one accepts the individualist and capacities based approach—for example, that it would be permissible to sacrifice them for their organs, do invasive and harmful experiments on them, or euthanize them if resources could be better used elsewhere. If people with PSCIs lack full moral status because of their different capacities, they either do not have as great of protections against these sorts of things or their interests are not weighed as heavily as non-impaired people (and nonhumans with full status). Therefore, to avoid these unacceptable implications, it is necessary to locate people with PSCIs in the category of entities with full and equal status. On their view, this means rejecting the view that individual capacities are fully determinative of moral status, and it leads theorists to try to bring species membership back into the moral status picture (Grau 2010; Kittay 2005, 2017; Curtis and Vehmas 2016; Rothhaar 2019; Kipke 2019).

For example, Grau (2010, 2016) suggests that intuitions about such cases and that all humans have full moral status are sufficient to motivate the 'speciesist' option. Curtis and Vehmas (2016) argue that confidence in the belief that all human beings have equal moral status greater than that of all nonhumans is itself sufficient to warrant belief. And that this is so even in the absence of a positive argument for the view and the absence of counterarguments against views (such as capacities based individualism) that entail that the equal and greater moral status view is false. Kittay (2005, 2017) argues that there are relationships we can enter into with members of our own species and not members of other species (2005), and that "*We have moral obligations to other human beings for the simple reason that we find ourselves in relation to them.* We cannot be the sorts of creatures we are except by being in relation to other human beings" (2017, 36, emphasis original).

But none of these is satisfactory. They amount to just asserting that a view is warranted or else are based on false differences. For example, we cannot be the sorts of creatures we are except by being in relation to a lot of other things, nonhumans included. We can enter into many of the sorts of (nonbiological) relationships with individuals of other species that Kittay highlights as being so important and valuable among humans, including people with PSCIs (Townley 2010). As a result, the discourse seems to be at something of an impasse. On the one hand, there are what seem to be strong philosophical arguments in favor of individualist capacities-based views of moral status and against group membership views. On the other hand, there are strong concerns about the ways in which those views have been developed and presented, as well as about some of their implications.

Both animal ethicists and disability ethicists have the aim of expanding our moral horizons beyond (and removing the prejudices in our ethical theories in favor of) "paradigm" moral subjects—i.e. healthy adult humans. Is it possible to hold that species membership is not a morally relevant property and accept an individualist capacities based approach to moral status—thereby bringing nonhumans more fully into the domain of ethical concern—without giving rise to the worries about the marginalization of and implications for people with PSCIs? That is to ask, is it possible to defend a view of justified species partiality even given moral status egalitarianism? In what follows we argue that it is possible.

7.3   Strategy One: Moral Status Equality and Moral Considerability Diversity

One concern raised by disability ethicists against proponents of leveling moral status by intrinsic properties (or capacities) is that the strategy is empirically inadequate. It fails to recognize how very different are the capacities, interests, desires, perspectives and lives of people with PSCIs and those of nonhuman animals like dogs, pigs and dolphins, which are themselves very different from each other. They do not seem comparable or "like" in the ways that leveling arguments suppose. Another concern is that the focus on intrinsic properties as the basis for moral status excludes other important ethical considerations, such as familial and care relationships (Kittay 2005, 2017; Francis and Norman 1978; Gunnarsson 2008).

Emphasizing the moral significance of relationships and the moral importance of attending to difference have strong analogs in environmental and animal ethics. The core idea is that moral considerability is underdetermined by an entity's capacities. To take an example from Clare Palmer (2010), compare a pet dog and a wild coyote. They have similar cognitive capacities. However, how one should consider their interests is very differently. There is a responsibility to promote the interests of our own pets—to feed them and give them medical care—that we do not have to wild animals. The reason is not that they have different interests (though they sometimes do). It is because there is a history of dependency, shared experiences and emotional engagement with one's pet. Wild animals, in contrast, should not be harmed unnecessarily, but there is not a positive responsibility of beneficence to them. On many environmental ethics, trying to help wild animals is even prima facie ethically problematic (Taylor 1986; Palmer 2010; Sandler 2007; Everett 2001). The reason for this is that it fails to appreciate the significance of their wildness and their relationships within ecological systems. If this is right, then the capacities a nonhuman animal has might tell us something about their moral status, but there is a lot that it does not tell us, particularly regarding consideration and treatment.

The same is true with respect to the moral considerability of human beings. Capacities based accounts of moral status might convey some information about consideration, but it is not the whole or even most important part of the story. Familial relationship, being part of a community, having a shared history, being in a particular role/position, and forms of dependency and vulnerability are, among many other things, also crucial to consideration and treatment (Hursthouse 2006). To be clear, the claim here is not that relationships and contextual factors change an entity's moral status. The view is that they change how we ought to consider and respond to entities with moral status. The fact that a parent's child is their child is an ethically relevant property. It justifies relating to her in ways, taking responsibility in ways, prioritizing her in ways, and being emotionally invested in her wellbeing in ways that it would not be appropriate for them to do with other children or for other adults to do with their child (Williams 2012). But this does not mean that their child has a moral status that other children do not have, or greater or different moral status from them. The same is true of wild coyotes and domesticated basset hounds.

Suppose now that an inclusive egalitarian account of moral status is correct, and that people with PSCIs, cognitively complex nonhuman animals, and healthy adult humans all have equal moral status. This would tell us some very general things about the need to take their interests fully into account or treat them as an end and not a mere means, for example. However, it would not tell us very much about what their interests are, how we should consider them, and how we should treat them. After all, they are the same with respect to having full status, but they are different with respect to consideration, interests and treatment (Singer 2009). Getting from knowledge that something has moral status to a meaningful account of how we ought to take them into consideration, let alone treat them, requires being attentive to their lives, their capacities, their perspectives, their experiences, and their relationships. The more things that are included as having moral status—the more diversity and variety there is within the group—the less moral status ascription can substantively convey about the morally salient features of the individuals and their lives, and about how we ought to respond to them (Hursthouse 2006; Sandler 2013).

Given this, if ethical theory is ultimately about mapping the normative terrain—i.e. the features of the world that are relevant to how we ought to consider others and make decisions about what we ought to do—then moral status must play a more minor role within ethical theory than is often supposed. Moreover, adopting a minimalist conception of moral status helps move beyond the impasse described earlier between disability ethicists and those who advocate for capacities based individualism. It becomes possible to agree that everything with certain intrinsic capacities has equal or full moral status, while recognizing that what is owed to those that have equal or full moral status varies widely. This is possible because we can distinguish between *moral status* (which things have interests that we ought to care about), *moral considerability* (how we ought to consider them), and *treatment* (how we ought to act regarding them). It enables a view on which all human beings and nonhuman animals have the same or full moral status, such that their interests are fully considerable in all the ways that are appropriate to them. But the ways that are appropriate to them differ on the basis of their capacities and relationships. That is, they are differentially considerable. Compassion is due to sentient animals, but not to nonsentient ones. Respect (in the hands-off sense) is due to wild animals but not household pets. Friends and family are due reciprocity and loyalty, strangers are not. All of this can be made sense of in ways that do not involve status differentiation. Indeed, as we have seen, status differentiation does not explain the differential consideration and treatment—the relevant relationships and capacities do.

Moreover, things are only obscured by introducing degrees or levels of status. As already mentioned, the problem is one of inclusion and differentiation. The more inclusive an account of moral status, the more it has to be possible to differentiate the implications of having moral status. Wild animals, household pets, children, people with PSCIs, family members, and strangers can all have moral status, but we should consider and respond to them and their interests in different ways. We certainly should not treat them the same. So 'having moral status' under-explains how we should consider and treat anyone. And the issue here is not one of prioritization. The issue is not how should we rank people and nonhuman animals in some ordering of who to harm first (or how much justification we need to harm them). The ways of responding to them are different in kind. Respect for autonomy is appropriate to other adults, but not to infants. Loyalty is appropriate to friends, but not to wild animals. This is why appeals to degrees or levels of moral status are theoretically and practically unhelpful. Is respect or loyalty a "higher" degree of moral status? The question does not make sense. It looks for a scalar comparison where there is none. There is a plurality of forms or ways of appropriately responding to the interests of individuals—e.g. respecting, promoting, prioritizing, and acknowledging—and there is no strict ordering among them (Hursthouse 1999, 2006; Sandler 2007; Warren 1997).

What generates the controversy over moral status between animal and disability ethicists is the belief that moral status ascriptions convey significant information regarding consideration and treatment. However, when moral status ascriptions are made for a wide range of types of entities the ascriptions can do little practical work, since consideration and treatment are enormously informed by not only whatever the qualifying capacities are for moral status, but also by other capacities that an entity has, as well as by its relationships and the relevant contextual features (Bovenkerk and Meijboom 2012). (For example, a parent-daughter relationship might justify favoritism in some contexts but not others.) Moreover, there are diverse ways of taking individuals with moral status into consideration and responding to them. Therefore, we ought to adopt a minimalist conception of moral status—to have moral status is just to have directly considerable interests—and embrace (non-lexically ordered) pluralism in forms of consideration. On this view, there are not paradigm cases or marginal cases, just cases.

Furthermore, on this view, people with PSCIs are no more or less like dogs or pigs with respect to moral status than are any other people. Each has full moral status. However, context-specific partiality is justified in some cases. Just as Kittay argues, the fact that a person with PSCIs is someone's child and stands in social relationships within a community are justifications for partiality—i.e. differential consideration and treatment. Moreover, the view is sensitive to the differences in capacities among people and nonhuman animals with moral status, so is not empirically inadequate. Getting consideration and treatment right requires being attentive to an individual's particular relationships and capacities.

7.4 Strategy Two: Equal Moral Status Without Equal Political Status

Even if we accept that different individuals are due different forms of consideration and treatment despite having equal moral status, there are cases where it seems difficult to reconcile claims of equal moral status and radical asymmetries in how humans and nonhumans are treated or considered. For example, consider the protections afforded to human research subjects compared to those afforded to nonhuman research subjects. Research involving human subjects is constrained by principles that we typically associate with deontological normative theories, such as respect for persons, which is operationalized in terms of requirements of informed consent and special protections for the vulnerable. Research involving nonhuman animals is subject to entirely different ethical standards. In the US, for example, research on vertebrate animals is constrained by principles that are associated with consequentialism. Animals are not to be used in a wasteful manner and they are not to be caused unnecessary harm, although what constitutes a 'necessary' harm is given by the aims of the research project. If an experiment promises to generate sufficiently useful knowledge, then the harms necessary to animals to complete the experiment are sanctioned.

It might seem as if this asymmetry in the consideration and treatment of humans and nonhuman animals in research (not to mention, in food production) is inconsistent with an individualist capacity-based view of moral status, even given significant differences in their capacities and relationships. If humans and nonhuman animals have full and equal moral status, how can anything like this be justified? (Of course, many proponents of individualist capacity-based views of moral status have been concerned that non-human animals used in research are not considered or treated in a way that is commensurate with their moral status (Singer 2009; Francione 2009; Gruen 2011; Donaldson and Kymlicka 2011; Engel 2012)).

One possibility is that there are important differences in the criteria used to assess whether some action, practice, or institution is unethical and whether some law or regulation that allows for those actions, practices, or institutions are *legitimate* (Cohen 1997; Basl and Schouten 2018). For example, while being unfaithful to one's partner is unethical, it doesn't follow that laws that coerced fidelity would be legitimate. The distinction between ethics, on the one hand, and political legitimacy on the other opens up the possibility that even if humans and nonhuman animals have the same moral status, the laws and regulations that govern animal use in research may be legitimate even though they differ radically from the laws and regulations that govern the use of human subjects. In other words, there is a potential basis for justified species partiality in law that does not track a difference in moral status.

On the basis of this distinction, Basl and Schouten (2018) have suggested that proponents of individualist capacity-based views of moral status make a mistake when they claim, for example, that animal experimentation should be legally abolished on grounds that it is inconsistent with the moral status of nonhuman animals. In order to establish that these practices should be abolished, it must be shown that such legal intervention would be legitimate (and that the current practices are illegitimate), that it would constitute a justified use of the coercive power of the state to prohibit or restrict animal use and experimentation in those ways. On their view, justifying the coercive power of the state requires showing either that it is "authorized by citizens through democratic or majoritarian processes" (Basl and Schouten 2018, 635) or that it is necessary to protect a "political interest", which is an interest necessary to allow citizens to form their conception of their good (Basl and Schouten 2018, 638).

While there has been debate about whether and how to bring animals into the political fold (see, e.g., Nussbaum 2009; Abbey 2007, 2016, Cochrane 2009, 2012; Donaldson and Kymlicka 2011, 2012; Garner 2012, 2013; Meijer 2016), the "political turn in animal ethics" has primarily focused on whether animals are properly subjects of justice, whether they are citizens in the sense they are owed duties of a special class. However, just as we can and should recognize that questions of justice are distinct from questions about other sorts of duties, so too should we recognize that questions of legitimacy are distinctive. The question of whether we, collectively, have duties of fair treatment or distribution toward animals is distinct from whether the state may legitimately coerce compliance with those duties.

Basl and Schouten argue that neither of the above conditions for legitimacy are met with respect to nonhuman animal research. For example, while there is a trend of increasing concern for nonhuman animals among the US public (Gallup 2015), there is little evidence that the public at large is opposed to current practices and policies that govern the use of nonhuman animals in research, or that they are willing to give up the goods that these practices make possible (Donaldson and Kymlicka 2011; Basl and Schouten 2018). This could change, and there are perhaps modifications that could be made to existing animal care and use policies that would be legitimate. However, the fact that there are changes to the current oversight regime that would be authorized by democratic processes does not show that the current oversight regime is illegitimate. Moreover, citizens that are opposed to nonhuman animal research on ethical grounds do not have a political interest in ending the practice. The existence of nonhuman animal research as currently practiced does not limit their ability to form and live according to their conception of the good. Nor does it disrespect their status as a full and equal citizen. Furthermore, nonhuman animals do not themselves meet the criteria for having political interests. They lack the capacities such that they are to be recognized as citizens to whom justifications for coercive interventions are owed.

In order to deploy this strategy in defense of an asymmetry in the consideration of humans and non-human animals, two things must be established. First, as discussed above, it must be shown that there *is not* a legitimate basis for extending certain protections to non-human animals. Second, it must be shown that that there *is* a legitimate basis for extending those protections to *all* humans. What are the prospects for defending this second claim?

There are at least three approaches available to defend the extension of certain protections to all humans as politically legitimate in the sense discussed above. First, it is possible to argue that there exists a majoritarian consensus or other democratic process that licenses the extension of those protections. So long as there is broad general support for equal protections of all humans (and assuming that such protections don't conflict with fundamental political interests), then there is a legitimate basis for protections for all humans. Even absent broad public support for a specific policy extending certain legal protections to all humans, there might be other democratically licensed processes by which such protections are extended in this way. For example, there might be broad public support for extending the relevant protections to nearly all humans coupled with a recognition that it will be difficult, in terms of writing a law or policy, or due to concerns about a slippery slope, to distinguish between those for which there is support and those for which there is not. This could serve as the basis for extending legal protections to all humans.

This approach faces some difficulties. Even though there now may be a majoritarian consensus about the scope of, for example, protections for all human research subjects, this might not always have been the case. Those seeking to justify species partiality are likely to want a less contingent basis for such partiality. The second and third approaches avoid this sort of contingency by showing that extending some protections to all humans is essential to protecting the political interests of citizens.

The second approach is to show that all humans, though not all animals, are or should be seen as citizens, individuals with political interests. On the traditional Rawlsian account of citizenship, citizens are those that have the capacity (or potential capacity) to form and revise their conception of the good and be held responsible or accountable for the way they go about acting according to their conception of the good (Basl and Schouten 2018, 639). Non-human animals fail to meet this standard (though see Meijer 2013, 2017 for a dissenting view). It seems that at least some humans will fail to meet these conditions as well. However, as we discuss in the next section, in conditions of uncertainty it might be that we should, in the case of humans, err in favor of ascribing them capacities or interests even if we aren't sure they have them. This leaves open the possibility of arguing that all humans should be seen as citizens.

The third approach, and perhaps the most promising for justifying species partiality, is to argue that even if only a subset of humans are citizens in the sense relevant to legitimacy, extending legal protections to all humans is essential to protecting the political interests of those that are citizens. Basl and Schouten consider this as a route to justifying some legal protections of animals that would otherwise not be legitimate. For example, it could be that while abolishing factory farming would be otherwise illegitimate, given the role that this practice plays in anthropogenic climate change, abolition is justified on the grounds that it contributes to protecting the political interests of citizens (640, fn. 23).

In the context of thinking about differential treatment and consideration of humans and nonhuman animal research subjects, there are differences between humans and nonhuman animals that might be used to justify species partiality. This is because humans stand in different relationships to one another than do humans to animals. Every human, whether or not they meet the technical conditions for Rawlsian citizenship, is a relative of some that do. This is not true of nonhuman animals, in particular the animals used in animal research. Given the nature of these relationships, there is room to make the case that extending equal consideration and treatment to all humans, in certain contexts, is important to protecting the political interests of citizens. For example, perhaps it is the most efficient mechanism available to protect citizens from certain forms of abuse in scientific research.

The above are overviews of how one might justify species partiality in a particular context by appeal to political legitimacy. They are intended to indicate spaces for proponents of justified species partiality to make their case without a commitment to species membership views of moral status. All the pieces of the above argument are subject to challenge. For example, while the view of political legitimacy used in support of the above argument draws from a fairly standard and widely-held conception of political liberalism (for defense and discussion see Rawls 2005; Ackerman 1980, 1994; Larmore 1987, 1996; Quong 2011; Schouten 2019), it is not uncontroversial. Furthermore, the conditions for legitimacy discussed above are necessary but not sufficient conditions for justifying the coercive intervention of the state. It must also be shown that such interventions are, for example, an efficient use of power compared to other legitimate alternatives.

However, even if one disagrees about the details of the approaches described above, the general lesson still holds. The criteria for what is unethical or unjust, including when those involve moral status claims, are not the same as the criteria for what is politically legitimate or required. As a result, there is room for the proponent of species partiality to argue for limited species partiality without adopting a species-membership view. This is important because the above grounds for limited species partiality, while up for philosophical debate, may be less questionable than species-membership views.

7.5   Strategy Three: Differential Epistemic Position

A third strategy for justifying species partiality in some contexts, even given moral status egalitarianism, appeals to differential uncertainty about the relative capacities, and thereby interests and strength of those interests, between humans and nonhuman animals. Our epistemic state as it concerns inferences or knowledge about the preferences, desires, and other mental states of humans is substantially different than that concerning nonhuman animals. In particular, we often should be more confident in our judgments about the mental life of humans than that of nonhuman animals (Allen 2006; Allen and Bekoff 2007). The reason for this is that other humans are more physiologically and evolutionarily similar to us than are nonhuman animals.

As a result, we are not only in a better position to assess which capacities other humans have, but are in a *relatively* good position to simulate the mental life of other humans and make inferences about their desires and preferences given other information we have. With respect to nonhuman animals, our confidence in such inferences should diminish as we consider animals that are more evolutionarily and physiologically distant from us. This is not to say we should be skeptical that nonhuman animals have a mental life. We can have good physiological, evolutionary, and behavioral evidence that they do (Varner 2002; Godfrey-Smith 2016), without being as confident about its content as we are for other humans.

This difference in epistemic position can have implications for consideration and treatment. For example, imagine having to make a decision about whether to continue life support for a family member that has not left explicit instructions about their desires as compared to making a similar decision about a family pet. In the case of the family member, we are likely to try to determine whether they would have a preference regarding being kept on life support. In the case of the pet, we are not likely to consider whether the pet has such preferences, since it is reasonable to believe that our pets either lack well-formed preferences about how they are to be treated when they are unconscious or that we could not infer with any confidence what their preferences would be if they did have them.

In the above case, the differential epistemic situation justifies or explains differential consideration and treatment of two individuals. This sort of strategy can be deployed to explain or justify common views about tradeoff cases. For example, lifeboat cases that force a tradeoff between a human and a non-human animal, where one or the other must be sacrificed, can be used to motivate species-membership views. If we think that we should, all else equal, favor humans in such cases, it seems that this must be simply *because* all humans have greater status than all nonhumans. However, even if, holding all else equal, on a capacity-based view such favoritism would be unjustified, it may be that in such choice contexts we would actually not be in a good position to judge that all is actually equal. Given the evidence we have available to us, it might be justified to favor humans because we are justified, for epistemic uncertainty reasons, in assuming they have capacities that in some cases it turns out they do not in fact have.

This epistemic uncertainty strategy also supports the second strategy based on the difference between what is ethical and what is politically legitimate. Some of the arguments deploying that strategy depend on being able to show that all or nearly all humans have some capacity or property that nonhuman animals lack. This is difficult, in part because it seems plausible that there are some humans very alike in terms of the relevant capacities to some nonhuman animals. For example, whatever capacities ground political interests or citizenship, it seems either that some humans will lack those capacities or some nonhuman animals will have them. However, when it comes to judging which humans lack those capacities, we might be in an epistemic position where it is most justified to make assumptions (or defining inclusion) on the basis of species membership—i.e. to assume that all humans (but not all nonhumans) have them.

Again, whether this strategy is ultimately successful depends on the details and context in which it is deployed. However, as with the other strategies, it opens up additional space to defend context-dependent partiality in a way that track species membership but does not commit one to thinking that species membership itself is a morally relevant property.

7.6 Conclusion

The primary justification in favor of the special moral status of humans—i.e. the view that all humans have equal moral status greater than that of all nonhumans—is that it is needed to support the differential consideration and treatment due to humans in comparison with nonhuman animals. We have argued that this justification fails. Differential consideration and treatment of humans, what we have called justified species partiality, is consistent with an egalitarian account of moral status. We have shown that (contextual and limited) justified species partiality can be warranted by pluralism in the forms and bases of moral considerability, the distinction between the considerations relevant to ethics and those relevant to policy, and our differential epistemic position with respect to humans and nonhumans.

Admittedly, our approach does not get proponents of species-membership accounts of human moral status everything they want either theoretically or practically. It is not the case that all human beings have equal moral status greater than that of all nonhuman beings. Nor is it the case that all human beings take priority over all nonhuman beings in every case. However, our approach avoids their primary concerns about disvaluing human beings. It also allows for contextual and situational partiality in favor of humans (sometimes humans in generally, sometimes particular humans). Moreover, it does not allow that any human beings have lesser moral status or are due lesser consideration than any nonhumans. For these reasons, justified species partiality in consideration and treatment does not require rejecting capacities-based accounts of moral status.

References

Abbey, R. 2007. Rawlsian resources for animal ethics. *Ethics and the Environment* 12 (1): 1–22.
Abbey, R. 2016. Putting cruelty first: Exploring Judith Shklar's liberalism of fear for animal ethics. *Politics and Animals* 2 (1): 25–36.
Ackerman, B. 1980. *Social justice in the liberal state.* Yale University Press.
Ackerman, B. 1994. Political liberalisms. *Journal of Philosophy* 91 (7): 364–386.
Allen, C. 2006. Ethics and the science of animal minds. *Theoretical Medicine and Bioethics* 27 (4): 375–394.
Allen, C., and M. Bekoff. 2007. Animal minds, cognitive ethology, and ethics. *The Journal of Ethics* 11 (3): 299–317.
Basl, J., and G. Schouten. 2018. Can we use social policy to enhance compliance with moral obligations to animals? *Ethical Theory and Moral Practice* 21 (3): 629–647.
Bovenkerk, B., and F. Meijboom. 2012. The moral status of fish: The importance and limitations of a fundamental discussion for practical ethical questions in fish farming. *Journal of Agricultural and Environmental Ethics* 25: 843–860.
Carlson, L. 2009. Philosophers of intellectual disability: A taxonomy. *Metaphilosophy* 40: 552–566.
Cochrane, A. 2009. Do animals have interests in liberty? *Political Studies* 57 (3): 660–679.
Cochrane, A. 2012. *Animal rights without liberation: Applied ethics and human obligations.* Columbia University Press.
Cohen, G.A. 1997. Where the action is: On the site of distributive justice. *Philosophy and Public Affairs* 16: 3–30.

Curtis, B., and S. Vehmas. 2016. A Moorean argument for the full moral status of those with profound intellectual disability. *Journal of Medical Ethics* 42: 41–45.

DeGrazia, D. 1996. *Taking animals seriously: Mental life and moral status.* Cambridge: Cambridge University Press.

DeGrazia, D. 2007. Human-animal chimeras: Human dignity, moral status, and species prejudice. *Metaphilosophy* 38 (2–3): 309–329.

DeGrazia, D. 2014. On the moral status of infants and the cognitively disabled: A reply to Jaworska and Tannenbaum. *Ethics* 124 (3): 543–556.

Donaldson, S., and W. Kymlicka. 2011. *Zoopolis: A political theory of animal rights.* New York: Oxford University Press.

Donaldson, S., and W. Kymlicka. 2012. Do we need a political theory of animal rights? In Minding Animals international conference, Utrecht, The Netherlands.

Engel, M., Jr. 2012. The commonsense case against animal experimentation. In *The ethics of animal research: Exploring the controversy*, ed. J.R. Garret, 215–236. Cambridge, MA: MIT Press.

Everett, J. 2001. Environmental ethics, animal welfarism, and the problem of predation: A Bambi lover's respect for nature. *Ethics & the Environment* 6 (1): 42–67.

Francione, G. 2009. *Animals as persons: Essays on the abolition of animal exploitation.* New York: Columbia University Press.

Francis, L.P., and R. Norman. 1978. Some animals are more equal than others. *Philosophy* 53: 507–527.

Gallup. 2015. In U.S., more say animals should have same rights as people. Gallup.com. http://www.gallup.com/poll/183275/say-animals-rights-people.aspx.

Garner, R. 2012. Rawls, animals and justice: New literature same response. *Res Publica* 18: 159–172.

Garner, R. 2013. *A theory of justice for animals: Animal rights in a nonideal world.* New York: Oxford University Press.

Godfrey-Smith, P. 2016. *Other minds: The octopus, the sea, and the deep origins of consciousness.* New York: Farrar, Straus and Giroux.

Grau, C. 2010. Moral status, speciesism, and Liao's genetic account. *Journal of Moral Philosophy* 7 (3): 387–396.

Grau, C. 2016. A sensible speciesism? *Philosophical Inquiries* 4 (1): 49–70.

Gruen, L. 2011. *Ethics and animals: An introduction.* Cambridge: Cambridge University Press.

Gunnarsson, L. 2008. The great apes and the severely disabled: Moral status and thick evaluative concepts. *Ethical Theory and Moral Practice* 11: 305–326.

Hursthouse, R. 1999. *On virtue ethics.* Oxford, UK: Oxford University Press.

Hursthouse, R. 2006. Applying virtue ethics to our treatment of other animals. In *The practice of virtue*, ed. J. Welchman, 136–155. Indianapolis, IN: Hackett Publishing.

Jaworska, A., and J. Tannenbaum. 2014. Who has the capacity to participate as a rearee in a person-rearing relationship? *Ethics* 125 (4): 1096–1113.

Kipke, R. 2019. Being human: Why and in what sense it is morally relevant. *Bioethics*, https://doi.org/10.1111/bioe.12656.

Kittay, E. 2005. At the margins of personhood. *Ethics* 116: 100–131.

Kittay, E. 2017. The moral significance of being human. *Proceedings and Addresses of the American Philosophical Association* 91: 22–42.

Larmore, C. 1987. *Patterns of moral complexity.* Cambridge: Cambridge University Press.

Larmore, C. 1996. *The morals of modernity.* Cambridge: Cambridge University Press.

McMahan, J. 2002. *The ethics of killing.* New York: Oxford University Press.

McMahan, J. 2005. Our fellow creatures. *The Journal of Ethics* 9: 353–380.

McMahan, J. 2008. Challenges to human equality. *Journal of Ethics* 12: 81–104.

Meijer, E. 2013. Political communication with animals. *Humanimalia: A Journal of Human/Animal Interface Studies* 5 (1): 28–51.

Meijer, E. 2016. Interspecies democracies. In *Animal ethics in the age of humans*, ed. B. Bovenkerk and J. Keulartz, 35–72. Cham, Switzerland: Springer.

Meijer, E. 2017. Interspecies encounters and the political turn: from dialogues to deliberation. In *Ethical and political approaches to nonhuman animal issues*, ed. A. Woodhall and G. Garmendia da Trindade, 201–226. Cham, Switzerland: Palgrave Macmillan.

Nussbaum, M. 2009. *Frontiers of justice: disability, nationality, species membership.* Harvard University Press.

Palmer, C. 2010. *Animal ethics in context.* New York, NY: Columbia University Press.

Quong, J. 2011. *Liberalism without perfection.* New York: Oxford University Press.

Rawls, J. 2005. *Political liberalism.* New York: Oxford University Press.

Rachels, J. 1999. *Created from animals: The moral implications of darwinism.* New York: Oxford University Press.

Regan, T. 1983. *The case for animal rights.* Berkeley, CA: University of California Press.

Regan, T. 1985. The case for animal rights. In *In defense of animals*, ed. P. Singer, 13–26. New York, NY: Basil Blackwell.

Rothhaar, M. 2019. On justifying arguments of species membership. *Bioethics*, https://doi.org/10.1111/bioe.12657.

Sandler, R. 2007. *Character and environment: A virtue-oriented approach to environmental ethics.* New York: Columbia.

Sandler, R. 2013. *The ethics of species.* Cambridge: Cambridge University Press.

Schouten, G. 2019 *Liberalism, neutrality, and the gendered division of labor.* New York: Oxford University Press.

Singer, P. 1975. *Animal liberation: A new ethics for our treatment of animals.* New York, NY: New York Review.

Singer, P. 1989. All animals are equal. In *Animal rights and human obligations*, ed. T. Regan and P. Singer, 148–162. Upper Saddle River, NJ: Prentice Hall.

Singer, P. 2009. *Animal liberation: The definitive classic of the animal movement.* New York: HarperCollins.

Taylor, P. 1986. *Respect for nature: A theory of environmental ethics.* Princeton, NJ: Princeton University Press.

Townley, C. 2010. Animals and humans: Grounds for separation? *Journal of Social Philosophy* 41 (4): 512–526.

Varner, G.E. 2002. *In nature's interests?* Oxford University Press.

Warren, M.A. 1997. *Moral status: Obligations to persons and other living things.* Oxford: Oxford University Press.

Williams, B. 2012. *Moral luck.* Cambridge: Cambridge University Press.

Ronald Sandler is a professor of philosophy, Chair of the Department of Philosophy and Religion, and Director of the Ethics Institute at Northeastern University. His primary areas of research are environmental ethics, ethics and emerging technologies, and ethical theory. Sandler is the author of *Environmental Ethics* (Oxford), *Food Ethics* (Routledge), *The Ethics of Species* (Cambridge), and *Character and Environment* (Columbia), as well as editor or co-editor of *Ethics and Emerging Technologies* (Palgrave), *Environmental Justice and Environmentalism* (MIT), and *Environmental Virtue Ethics* (Rowman and Littlefield).

John Basl is associate professor of philosophy at Northeastern University. He works in normative and applied ethics, focusing primarily on issues in the ethics of emerging technologies, environmental ethics, and on issues of moral status. His most recent book *The Death of the Ethic of Life* was published in 2019 by Oxford University Press.

Chapter 8
Humanity in the Living, the Living in Humans

Michiel Korthals

Abstract Recent studies in biology, ecology, and medicine make it clear that relationships between living organisms are complex and comprise different forms of collaboration and communication in particular in getting food. It turns even out that relations of collaboration and valuing are more important than those of aggression and predation. I will outline the ways organisms select and value specific items in their network of living and non-living entities. No organism eats everything; all organisms prefer certain foods, companions, and habitats. Relations between organisms are established on the basis of communication, exchange of signs, actions and goods, through mutual learning processes on all levels of life. Micro, meso and macro organisms participate in this process of valuing and communication. Animals and plants therefore show features that were traditionally attributed only to humans, like selfless assistance. The usual distinction between humans and other living beings on the basis of human's sensitivity for altruism, language and values crumbles down due to the circumstance that also non-human living beings are prone to selfless assistance, communication and valuing.

8.1 Introduction: Animals, Plants and Humans

In the past the relations between different species and between species and human beings were seen as quite limited. Studies analysed the enduring competition of non-human living beings with each other in their struggle for survival, and members of different species didn't mix. Eating and being eaten was seen as the main principle. Moreover, it was conceded that whereas humans can act altruistically, other living beings cannot. It was proven that humans had a sensitivity for beauty, but other living animals did not. Anthropomorphism, projecting human-like values as motives for animal behaviour, was seen as a big scientific sin. As a consequence, living entities were conceptualised either from a narrow behaviourist, or from a survivalist and

M. Korthals (✉)
University of Gastronomic Sciences, Pollenza/Bra, Italy

© The Author(s) 2021
B. Bovenkerk and J. Keulartz (eds.), *Animals in Our Midst: The Challenges of Co-existing with Animals in the Anthropocene*, The International Library of Environmental, Agricultural and Food Ethics 33,
https://doi.org/10.1007/978-3-030-63523-7_8

objective point of view: members of different animal species don't mix, don't show altruism, and don't have a sensitivity for beauty. Animal relations were studied from a survivalist, 'objective' approach: animals of a species and species were always acting according to their own advantage, survival of the fittest meant an enduring struggle with each other in a race towards fitness. Although evolution was acknowledged as the important dynamic force, when a species had established itself and its competitors had died off, it kept developing in trying to establish fitness with its environment. For example, bacteria present in the gut system of animals and plants were seen as egocentric agents (driven by selfish genes), in the best case not harming the host, but in worse case killing it. These bacteria were seen as parasites, as for example by Wilson (2014): "Almost all species of plants and animals carry parasites. Which by definition are other species that live on or inside their bodies and in most instances take some little part of the hosts without killing them" (idem, p. 180).

I will argue that what Wilson calls a parasite should count as a worthy collaborator, or even as a separate organ. Cows, squirrels, all animals and plants have microbial collaborators, which ensure food digestion and stimulation of their immune system and even determine their cognitive capabilities. More and more biologists are studying these forms of cooperation (e.g. Bray 2019; Valles-Colomer et al. 2019; Cryan et al. 2019).

Not only values like cooperation and altruism were victims of the behaviourist, survivalist or genetic outlook on nature's processes, but also a value like beauty. It was crazy to argue that beauty played a substantial role in the relations between animals or between plants and animals. Geneticists allowed beauty only as a genetic advantage. However, I will argue that in light of recent research these differences between humans and other living beings collapse. The exceptional place of humans in nature does not hold, not because humans are more like non-humans, but on the contrary, because non-human beings are more like human beings, at least like the ideal human being. They embrace values; they act altruistically and they are sensitive to beauty. I therefore fully agree with the proposition of Harry Kunneman (2017), that humans participate in animality and animals and plants participate in humanity. Moreover, these values comprise a kind of non-egotistic agency: organisms are beings, and this means that in looking for food, assisting others or appreciating beauty they are not ego's acting in an objective world. They are, just like the early Heidegger (1963) or Sartre would say, modes of being: being in the web of food or being in the assistance or in the web of beauty.

First, the role of food will be discussed, secondly the role of values in animal interactions will be tackled. Next, I will discuss various types of relationship, the misleading metaphor of the tree of life and the difficulties anthropocentrism causes in analysing these interactions. Finally, I will argue for caution in deriving practical recommendations of changing these complex connections between beings and their world.

8.2 Food Makes the World Go Around

Food processes are an illuminating focus to find out how members of a species and a species in general are embedded in altruistic relations and value altruistic actions. Looking for food, eating and digesting are for all living beings time consuming and daily activities. Nutrition also establishes an intrinsic relationship between dead entities such as earth, and living organisms, or better: between non-living and living processes. Water, air and minerals play a decisive role in producing the right quality of food. Animals (including humans) are important links in this. Animals live on water but also on sunlight, sugars and oxygen via plants. And animals also feed on minerals and deadly gases made digestible through plants, such as nitrogen. Carbon users such as plants in turn need animals for their reproduction and food production: mammals, insects and birds pollinate flowers and plant shrubs and trees and therefore maintain forests. Plants and trees warn each other for predators through the emission of odours and other signals (Kohn 2013; Wohlleben 2015; Mancuso 2015). Worms and bacteria make the soil fertile for crops and trees. Trees provide food (sugars, water) for other plants and so indirectly for plant eating animals. The fact that all living beings need food, and can feed themselves, assumes that they are assisted by other living beings in providing them their food either directly or indirectly, via their contribution to the establishment and the maintenance of an useful food-context.

Darwin had an idea of these processes in his analysis of the activities of worms in *The Formation of Vegetable Mould, through the Action of Worms*, from 1881: "When we behold a wide, turf-covered expanse, we should remember that its smoothness, on which so much of its beauty depends, is mainly due to all the inequalities having been slowly levelled by worms. It is a marvellous reflection that the whole of the superficial mould over any such expanse has passed, and will again pass, every few years through the bodies of worms. (…) Some other animals, however, still more lowly organised, namely corals, have done far more conspicuous work in having constructed innumerable reefs and islands in the great oceans; but these are almost confined to the tropical zones" (p. 313).

Darwin makes it clear that worms prepare the soil by making it a fertile context of nutrients for plants and other living beings. Worms do this probably unintentionally by digesting soil debris, and as a consequence bringing air, nitrogen, and improving nutrient availability for plants in general. Here they work in their own advantage assisting other living beings.

Another relationship is established when two or more parties evolve in reaction to each other to their reciprocal advantage. The coevolution of flowers and pollinators like bees and certain birds is a very good example. I can only shortly raise the issue of intentionality here. My main point here is that organisms have intentions (for example to select, eat, exchange or give food), however they differ in how many consequences their intentions take into account.

However, living beings can also assist other living beings without pursuing their own advantage. More radical thoughts on this have been proposed and are verified (Simard 2018; Mancuso 2015; Wohlleben 2015). Many individuals of a species assist other individuals of different species in finding their food or even in producing their food without any advantage for the assisting actor. Oaks for example provide with their large underground root system sugars to smaller trees, in particular in times of scarcity.

8.3   Values in Animal Plant Interactions

Organisms cherish a lot of different values, like consoling and assisting others, playing, and enjoying beauty. For example, mammals can give consolation to others when they are sad and apparently are grieving their misfortune. Taking the position of the other and offering others their preferential food stuff or improving their food context happens quite regularly, in particular in parent child relations, but also in inter- and intra-species relationships when for example an elderly animal perceives helpless young being. The case of a gorilla trying to comfort a young child that has fallen in its compound is an emotional example (Goodall 2016). Between species and within species play of elder members with young ones and between young ones is a quite common spectacle.

Susanne Simard (2018) analysed forests and identified what she called a hub tree, or "mother tree". Mother trees are the largest trees in forests that act as central hubs for vast below-ground mycorrhizal networks. A mother tree supports seedlings by infecting them with fungi and supplying them with the nutrients they need to grow. In this way birch and firs communicate by exchanging nutrients. She discovered that Douglas Firs provide carbon to baby firs. She found that there was more carbon sent to the baby firs that came from a specific mother tree, than random baby firs not related to that specific fir tree. It was also found that mother trees change their root structure to make room for baby trees.

Crows cause fires to keep forests open for birds and other animals. The Aboriginals talk about the 'Karrkkanj' (hawk falcon) that brings a burning branch to a forest to set it on fire and catch small fleeing animals. Haviksevken steal bread and throw it in a river to catch fish. Scientific research has confirmed these processes (Bonta et al. 2017). Red squirrels start forests in their diligence to make stocks.

Last but not least, many animals are sensitive to beauty, i.e. which is a value that is not directly related to utility or survivalist motives. Birds and fishes are motivated by beautiful looking members of their species. Perceiving and tasting something as pleasant and as delicious, a phenomenon that regularly happens to animals, cannot directly be connected with a struggle for survival or an advantage for fitness (Prum 2017). Just as with humans, the sensitivity of animals for beauty and something delicious is at least clearly influenced by social context and upbringing. These aesthetic preferences for certain types of food or partners don't seem to have any adaptive benefit. Birds with certain feathers can attract partners and their offspring can have a mix of the best genes, but it is not always the case. Strong evidence seems not to support the good-genes hypothesis (Noble 2016). Colored skin or feathers are both signs of evolutionary progress and of a capacity for beauty of valuing animals (Ryan 2017).

8.4 Do They Communicate with Each Other?

No individual or species arises, lives and dies by itself. Animals are core and junction points in processes of exchange of substances that are food. Animals in this sense work together with plants to produce food and drink, as well as oxygen and other chemicals. All are sensitive to signs other organisms are showing to their own and other species. Demand and response are a general feature of their collaboration.

For many biologists it is increasingly clear that animals, plants and microbes are dependent on a network, where some species jointly develop (in co-evolution), others have the lead, and others just emerge or are parasitic. The cooperation between living beings has remarkable connections. Everyone knows the intensively inter-locked evolutionary developments of flowers and those of pollinators (insects, birds). Deeper chalices were accompanied by longer tongues; some plants even suggest fake flowers to attract certain insect species (Peeters 2015).

But there is a lot more to it: for example, animals, including humans, cannot live without the enormous numbers of bacteria and viruses in their gastrointestinal tract (Enders 2013). The human brain is also inspired by these food processing and producing gut microbes (a fact that was denied by science ten years ago, Sonnen-burg and Sonnenburg 2015). Every animal species, including humans, consists of a combination of various different species (Margulis 1998). Communication, i.e. the exchange of information and of claims and preferences via instinctual, but also cognitive, learned, processes play a role in all these processes. There is learning, communicating and threatening. And there is collaboration.

8.5 Collaboration as a Mechanism of Co-evolution

Darwin has done extensive research on co-evolution in orchids and their pollinators. In a centuries-long developmental process these organisms constantly adapt to each other. But co-evolution goes even further than Darwin's views on orchids. All parties in a co-evolutionary network exert pressure on each other to specify, that is, to look for a niche that has not yet been used in the food supply and more broadly, in the ecosystem. This selective pressure is not often a form of competition or even war, because the goal is different for all parties. The pressure can go in all directions, with often not directly clear consequences. The birds that build forests by burying seeds from trees, or even by burning forests, provide their own food supplies. But they also create open spaces at the same time, with the effect of the germination of other, new seeds (Wall et al. 2013).

The selective pressure on species does not necessarily imply a war for their own survival. The warning systems and defence mechanisms that plants use to warn their own species of predators are sometimes also intended for other species. Some plants ward off predators by luring the enemies of those predators and they warn other species. For example, some plants emit odour signals when they are affected by spider mites. In this way they attract the natural enemies of spider mites, namely predatory mites.

8.6   Tree of Life or Network?

In fact, Darwin has underestimated the full meaning of collaboration and co-evolution due to his idea that evolution can be represented by a tree with branches of species (Kropotkin 1902 was the first to mention this). According to Darwin the general development of species is supposed to form a tree of life, a Great Chain of Being (Peeters 2015). Once species have become a separate branch of this tree, they develop on their own. The branches represent the classifications of species of living beings. Traditionally, it is assumed that species once separated from their ancestors follow their own path separated from other species. But recent evolutionary genetic research shows something completely different. Molecular researchers such as Eric Bapteste, an evolutionary biologist, show that hybridisation (crossbreeding between species) and symbiosis are important factors in evolution (Margulis 1998). Bapteste (2013) says about the tree:

> Ever since Darwin, a phylogenetic tree has been the principal tool for the presentation and study of evolutionary relationships among species. A familiar sight to biologists, the bifurcating tree has been used to provide evidence about the evolutionary history of individual genes as well as about the origin and diversification of many lineages of eukaryotic organisms. Community standards for the selection and assessment of phylogenetic trees are well developed and widely accepted. The tree diagram itself is ingrained in our research culture, our training, and our textbooks. It currently dominates the recognition and interpretation of patterns in genetic data.

However, this metaphor of the tree of life has its limitations. Organisms from different branches can transfer genes to each other—and often do; branches are not separated from each other with an origin in a single main stem.

The transfer of properties (now we would say: genes) between species takes place much more often than Darwin originally thought. There are hardly any separate branches on the tree of life, and the species do not evolve on their own, but they also integrate foreign DNA. Microbes do this to a large extent, but also animals such as mice and rats. Frogs generally have at least ten percent of foreign DNA. Horizontal gene transfer happens quite often, and therefore genomes are not completely isolated within species boundaries. In nature different species can exchange genetic information, as it has been shown for rice to millet (Diao et al. 2006). Even gene transfer takes place between very different species from different kingdoms, as from Agrobacterium to sweet potato (Kyndt et al. 2015). Reproduction between different species does not always lead to infertility, as the dogma still reads. The metaphor of the tree of life must be replaced by a kind of network of constant fusion and differentiation.

The tree of life with its classifications is also problematic in another way. Central here is the overall classification of living organisms in various 'kingdoms', such as Linnaeus distinguished between animals and plants. However, new additions have to be made each time, often on the basis of the invention of new detection methods, such as DNA and epi-genetic research methods. That is why this overall classification has now been replaced by eight 'kingdoms'. Moreover, a major problem of these distinctions is that there are always intermediate forms that are difficult to classify. Every classification remains a human invention, which always tries to capture complexity and renewal, but never completely succeeds.

The 'tree of life' concept misinterprets the mutual cooperation between species. Through this cooperation, species can develop in response to each other (selective pressure) and to the symbiotic networks and processes that organisms are part of. Scholars come up with all sorts of terms for this biosocial collaboration, such as co-evolution, bio-socialities (Rabinow 1996), naturecultures (Haraway 2003), entanglement of matter and meaning (Barad 2007). Barad argues: "matter and meaning cannot be dissociated, not by chemical processing, or centrifuge, or nuclear blast. Mattering is simultaneously a matter of substance and significance, most evidently perhaps when it is the nature of matter that is in question, when the smallest parts of matter are found to be capable of exploding deeply entrenched ideas and large cities."

8.7 Symbiosis, Symbionts, Holobionts and Place

As can be seen from the foregoing, symbiosis is one of the most important mechanisms of evolution. This includes selective pressure, place and co-evolution. Most individuals, or species are mixed forms (holobionts). They consist of different species (symbionts) and yet are ecological units (Margulis 1998). People are also holobionts, which bring symbiotic expression to both their own genome and that of other species, the symbionts. Plants work together with microbes, and feed on inorganic material and thus establish a link between dead matter and life and via-via with live animals (Hansen 1993).

Because of these mutual connections in a locally developed ecosystem, the locally evolved relationships are extremely important. The place is a breeding ground. The mutual symbiotic adjustments of the symbionts in a holobiont are disturbed when one species is removed. Biologists call this preference for locally evolved relationships 'Home-field advancement' (Rúa et al. 2016). The geographical location and the specific soil and aboveground condition is the place where adaptations and co-evolution with micro-organisms take shape. Manure from animals that have eaten local plants and that is broken down by bacteria and insects belongs to one network. Local manure fertilizes the soil better than manure from elsewhere.

Ultimately, the entire web of relationships is carried by microbes (Rosenberg et al. 2016). For example, in humans, communication between the brain and the stomach is just as fundamental to capabilities and behavior as the brain. These stomach-brain connections apply to all mammals. One of the most famous researchers in this area, Emeran Mayer, writes in *The Mind-Gut Connection: How the Hidden Conversation Within Our Bodies Impacts Our Moods, Our Choices, and Our Overall Health* (2016), that the gut and the inhabitants of the gut, the microbes, think for the brain. That is why he calls the bowels the 'second brain'. The bacteria in the gastrointestinal tract play a central role: the presence of certain types of bacteria and their products have a major influence on the willingness to take risks, on thought processes and on moods such as apathy and depression. In addition, the digestion of foods is largely provided by intestinal bacteria, as is the stimulation of the immune system. This microbiome is a second brain.

8.8   Different Types of Relations Inter- and Intra-species

Categorising these different types of relations is a little pretentious (giving the complexity), nevertheless can be useful for further study. Kunneman (2017) has done an interesting job in outlining at least four relations. Relations can be either beneficial or advantageous to all parties, or only to one, or not harmful to the other, or harmful. Advantage means here improving fitness, or even improving quality of life. Short term and long term are often difficult to distinguish. The best known and earliest analysed relation between species that is distinguished is that of parasitism. A relation where only one party profits at the cost of the host can be called parasitic. For example, striga is a plant that is totally dependent on grain species, and after a certain time the host is so exhausted that it dies. Probably inspired by the dominant nine-teenth and twenty century societal ideas about outcasts and beggars, biologists were eager to analyse 'parasitic behaviour' in plants and animals and neglected symmetrical relations. A more mutually beneficial relation happens when both organisms improve their chance for survival. Famous and fundamental example are the mycorrhizal relations between roots of plants with fungi and bacteria, that provide plants with inorganic compounds and trace elements and plants that provide the fungi and bacteria with sugars and other chemical compounds. Many plants have these relations and they are therefore crucial for ecosystems. Seed dispersal by animals which gives them food advantages is also a very good example of mutual symbiosis. Without animals dispersing seeds of trees, shrubs and other plants couldn't move to other places than that of the mother plant.

Cooperation is happening with members of the same species, for example when dolphins chase after fishes, or wolves try to catch a deer. However, other animals are often also involved, and cooperation can therefore be an inter species affair. Altruistic types of symbiosis are also well-known.

8.9 Matter and Meaning; Philosophical Questions

When humans are dependent on other species to such an extent, especially microbes, to what extent do the microbes control them, rather than the people themselves? Did microbes start wars? Do organisms determine how we think? These are difficult questions. Jared Diamond (2000) seems to suggest a positive answer: people are driven by microbes. I think we still do not know enough to say something sensible about this.

There are proposals to answers to these questions. One of these is that we have so many health and psychological problems due to the fact that our modern food is not adapted to our gut system. We have to return to the wild food that our microbiome has received over the past millions of years (thus type of food changed considerably since the agricultural revolution about 10,000 years ago). Jeff Leach (2015) explicitly states that people (not just animals) must also become wild, and therefore he likes to crawl through mud in Central Africa to get rid of health problems.

According to many, the emphasis on hygiene, especially since Pasteur showed the influence of bacteria on food, is having a detrimental impact on health. The 'hygiene hypothesis', for example, states that the excess of hygiene is the cause of a weak immune system, so that people are more susceptible to diseases and allergies (Blaser 2014; Okada 2010). Bacteria function as impulses to strengthen the immune system and when they are no longer present, the immune system does not develop or develops insufficiently.

Incidentally, in the tree and plant world it has been much more common to anticipate the role of fungi (microbes) in the way roots feed (Rainer 2015; Hansen 1993). The roots give sugars through those leaves through photosynthesis, and the roots through the fungi and their fungal threads have a very wide network that provides water and inorganic minerals. Fungi also protect the roots against infections. When plants root between the right fungi, mycorrhiza (i.e., symbioses between roots and fungi) develop and they improve their growth processes. Almost all plants and trees take part in mycorrhiza.

Anyhow, the notions of value, cognition and communication also merit recalibration: animals and plants value their contexts, they remember their predators and develop defence mechanisms. They also warn their fellow species. I hereby join the research of Frans de Waal (2016), who uses a broad concept of cognition, "a wide range of cognitive mechanisms", such as memory and information processing (2016, 282). Values enable to select the information that an individual and a group believe to be relevant.

8.10 Barriers: Classifications, Anthropocentrism and Hubris

Thinking and learning about evolution and cooperation between species is hampered by a number of rationalistic principles. Western rationalism, from Descartes to Heidegger, places strong emphasis on the individual ego that is in a world full of things with qualities, but also on the view that humans are superior to other living beings (Korthals 2018). The view that man is a symbiotic being contradicts this idea of individuality. Men as symbionts means there is no I, there is everywhere a (widespread) we-process, composed of different species, namely symbionts of life and death.

Because of our inadequate senses, our unilateral communication ability through spoken language and the growth of our brains, people believe that other beings do not communicate. That is why it took so long to discover communicative and cognitive skills in other species (see also Meijer and Bovenkerk in this volume). The evolutionary achievements of the brain have simultaneously equipped man with very deficient senses. The brain has shrunk the senses (Wilson, 2014, 48) Language is an extensive network of useful communication with the world for people and therefore the specific possibilities the senses can realize are diminished. With spoken language and a thinking brain, people see themselves as separate beings, separate from plants, animals and microbes. Man is secluded at the top of evolution, or sees himself as a world-shaping against all those other world-blind beings.

As a result of this anthropocentrism, man is blind to important, life-feeding interactions and communications between other animals, plants and dead matter. Due to the great emphasis on language as a superior communication system, other communication systems are not covered. But animals and plants have other, equally effective communication systems that enable co-evolution. Slowly we become a little smarter in research into how living beings live, communicate and, above all, feed themselves and others (De Waal 2016).

In western philosophy, many barriers have been raised against the elaboration of this idea. In particular, the view that man is an exceptional being due to rationality or consciousness makes it difficult to see that animals and plants also think, feel and communicate in a certain way.

This anthropocentrism can be found in leading philosophers such as Descartes, Kant or Wittgenstein. Heidegger also has a strong anthropocentric view in his own way: man and animal differ according to him because man is 'world-meaning', the animal is 'world-poor' and the plant is 'without world'. He claims: "Throughout the course of its life the animal is confined to the environmental world, as it is within a fixed sphere that is incapable of further expansion and contraction" (1995, p 198). That is why he says: "The animal is poor in the world" (idem, p. 186). Plants and animals are locked up in their own environment (Umwelt) and they are never in the open space (Lichtung) of being. "In its essence, language is not the utterance of an organism; nor is it the expression of a living thing. Nor can it be thought in an essentially correct way in terms of its symbolic character, perhaps not even in terms of the character of signification. Language is the clearing-concealing advent of being itself" (idem, p. 248).

However, it is quite naive to single out language as the distinctive feature between man and animals. There are so many other important features that some animals possess and humans do not possess. As I have argued, all living beings have preferences, and communication systems with respect to these preferences. They value their contexts of life and their own life. Here I agree fully with Korsgaard, who in her newest book (2018), starts her reasoning with the acknowledgement that animals (in the quote a rabbit) value items, and then concludes in this way:

> For even if the rabbit's life is not as important to her as yours is to you, nevertheless, for her it contains absolutely *everything of value*, all that can ever be good or bad for her, except possibly the lives of her offspring. The end of her life is the end of all value and goodness for her. So there is something imponderable about these comparisons. (p. 65)

8.11 Philosophical Challenges: Pandora's Box Versus New Skills

The philosophy that symbiosis and mutualistic relationships through eating and excretion are central to life confronts philosophy with countless new issues. It is over with the 'superior man', who, as lord and master of nature, dominates all the living beings. Human rationality, thinking, communication and feeling are not unique. Man does not stand alone. The interwovenness with dead and living processes necessitates attention to local and long-term processes.

At the same time, a new temptation arises: to make this insight useful, and to open a Pandora's box. "Symbiotic relationships are fundamental for all forms of life on the planet and for our own existence, if we could learn to redirect some of them the results could be spectacular, for example if we could transform the symbiotic relationships of plant and nitrogen fixing bacteria from just the legumes to all crops we could change the face of agriculture for ever" (Mancuso 2015, 93).

What Mancuso points out is an important issue: the technological impact of this view. The temptation strikes: technologies are being developed and applied that intervene in holobionts, in microbiomes, such as the introduction of fungi, bacteria and other organisms into the soil. But also by means of intervention in animal or human stomachs and intestines with 'psychobiotics', biotechnologists want to bring about certain health effects. A Pandora's box opens with the knowledge about symbiotic relationships. Countless commercial companies propagate the development and use of these psychobiotics. Regulating bodies do not know how to give these psychobiotic interventions a legal framework (Green et al. 2017).

The question here is whether we know enough about these complicated and subtle relationships, and whether we do not seriously disrupt them with simple interventions. The risks to food supply for animals and people are great, as are those for health. The search for the right combination of, for example, plants and soil bacteria, of human behavior and gastrointestinal bacteria cannot be concluded with a few association studies, some longitudinal studies and a clinical practice.

Philosophically these developments could be seen as challenges to find the right way in which we could and should deal with this process-based embedding of people. This embedding in a context is the biological equivalent of the hermeneutical concept of 'place' and the pragmatic concept of 'practice' (Keulartz et al. 2002). Place is where people (and living organisms in general) grow up and live; it is the starting point for pluralism of opinions, and of practices. Partly because of their place of embedding, plants, animals and people differ more or less from each other.

The other philosophical challenge is how we deal with pluralism and with the differences that are so much needed for the symbiotic networks. For the many who strive for unity that pluralism is a stumbling block. Ethically dealing with symbioses also requires a substantive valuation of the processual and local relationships in a globalizing world society, where place is increasingly difficult to determine and respect. But the place of embedding matters. That is why place has an enormous significance for symbionts and holobionts and it should be a central point of attention. Endless mobility of people, plants, animals and even entire habitats has its clear limits.

There are therefore various epistemic and moral skills to formulate from the idea that people live in the midst of animals and animals in the midst of people. I speak here about 'skills' in the sense of Sen's 'capability theory' (2009). Epistemic skills should focus on complexity and on setting priorities in the enormous amount of relevant connections. Moral skills should focus on responsive interactions with totally different organisms and on the responsibility for the interventions they want to do and their consequences. The question is not: can people govern microbes? But: what does a responsible symbiotic biopolitics look like? Caution and respect are required. One small change and one whole ecosystem changes completely, or even collapses. Mancuso's desire for profound changes in mycorrhiza networks bringing about a new agriculture should be postponed for the time being.

8.12   Conclusion

All living beings communicate with each other and value elements of their ecological contexts, be it as food, partner, companion or enemy. Values like beauty, solidarity and sociality play an important role in living their life. A neutral objective, utility perspective on living organisms neglects these values and the role they play in ecological processes. Although the exact motivation to value for example feathers, colours, and certain types of food is unclear, this cannot only be determined by a genetic disposal (Noble 2016). Kunneman (2017) is right about his idea that animals participate in humanity. Moreover, these values comprise a kind of non-egotistic agency: organisms are beings embedded (located) in a broader food web or web of assistance or a web of beauty. The misleading metaphor of the tree of life causes difficulties in analysing these interactions because it rules out coevolution and symbiosis. Anthropocentrism erect a barrier between humans and other living organisms and therefore denies the wide variety of processes of communication, valuing and solidarity with non-human animals. Finally, I argue for caution in deriving practical recommendations in changing these complex connections between beings and their world.

References

Bapteste, E., et al. 2013. Networks: Expanding evolutionary thinking, Trends. *Genetics* 29 (8): 439–441.

Barad, K. 2007. *Meeting the universe Half-way: Quantum physics and the entanglement of meaning and matter*. Durham, NC: Duke University Press.

Blaser, M. 2014. *Missing microbes: How killing bacteria creates modern plagues*. London: Oneworld Publications.

Bonta et al. 2017. Intentional fire-spreading by 'fire hawk' raptors in northern Australia. *Journal of Ethnobiology* 37 (4).

Bray, N. 2019. The microbiota–gut–brain axis. NatureResearch. https://www.nature.com/articles/d42859-019-00021-3.

Cryan, J.F., K.J. O'Riordan, C.S.M. Cowan, et al. 2019. The microbiota-gut-brain axis. *Physiological Reviews* 99 (4): 1877–2013.

Darwin, Ch. 1881. *The formation of vegetable mould, through the action of worms*. London: J. Murray.

De Waal, F. 2016. *Zijn we slim genoeg om te weten hoe slim dieren zijn?* Amsterdam: Atlas-Contact.

Diao, X., M. Freeling, and D. Lisch. 2006. Horizontal transfer of a plant transposon. *PLoS Biology* 4: e5. https://doi.org/10.1371/journal.pbio.0040005.

Diamond, J. 2000. *Zwaarden, Paarden en Ziektekiemen*. Utrecht: Spectrum.

Enders, G. 2013. *Gut*. London: Greystone.

Goodall, Jane. 2016. Memo RE: Shooting of Harambe at Cincinnati Zoo and Botanical Gardens. janegoodall.org.

Green, J.M., M. Barratt, M. Kinch, and J. Gordon. 2017. Food and microbiota in the FDA regulatory framework. *Science* (July): 39–40.

Hansen, R., and F. Stahl. 1993. *Perennials and their garden habitats*. Portland: Timber Press.

Haraway, D. 2003. *When species meet*. Minneapolis: University of Minnesota Press.

Heidegger, M. 1963. *Sein und Zeit*. Tubingen: Mohr.

Heidegger, M. 1995. *The fundamental concepts of metaphysics: World, finitude, solitude*. Bloomington: University Indiana Press.

Keulartz, J., M. Korthals, M. Schermer, and T. Swierstra. 2002. *Pragmatist ethics for a technological culture*. Dordrecht: Springer.

Kohn, E. 2013. *How forests think*. Berkeley: University of California Press.

Korsgaard, C. 2018. *Fellow creatures*. Oxford University Press.

Korthals, M. 2018. *Goed Eten: Filosofie van Voeding en Landbouw*. Nijmegen: Van Tilt.

Kropotkin, P. 1902. MutuaL Aid. https://theanarchistlibrary.org/library/petr-kropotkin-mutual-aid-a-factor-of-evolution.lt.pdf.

Kunneman, H. 2017. *Amor complexitatis*. SWP Utrecht.

Kyndt, T., D. Quispe, H. Zhai, R. Jarret, M. Ghislain, Q. Liu, et al. 2015. The genome of cultivated sweet potato contains Agrobacterium T-DNAs with expressed genes: an example of a naturally transgenic food crop. *Proceedings of the National Academy of Sciences of the United States of America* 112: 5844–5849. https://doi.org/10.1073/pnas.1419685112.

Leach, J. 2015. *Rewild*. London: CreateSpace.

Mancuso, S., and A. Viola. 2015. *Brilliant green: The surprising history and science of plant intelligence*. New York: Island Press.

Margulis, L. 1998. *Symbiotic planet: A new look at evolution*. Basic Books.

Meyer, E. 2016. *The mind-gut connection: How the hidden conversation within our bodies impacts our moods, our choices, and our overall health.* New York: Harper.

Noble, D. 2016. *Dance to the tune of life.* Cambridge University Press.

Okada, H., C. Kuhn, H. Feillet, and J.F. Bach. 2010. The 'hygiene hypothesis' for autoimmune and allergic diseases: An update. INSERM U1013, Necker-Enfants Malades Hospital, Paris, France.

Peeters, N. 2015. *Botanische Revolutie.* Zeist: KNNV.

Prum, R. 2017. *The evolution of beauty.* Doubleday.

Rabinow, P. 1996. *Essays on the Anthropology of Reason.* Princeton.

Rainer, Th., and C. West. 2015. *Planting in a post-wild world: Designing plant communities for resilient landscapes.* London: Timber Press.

Rosenberg, E., and I. Zilber-Rosenberg. 2016. Microbes drive evolution of animals and plants: The hologenome concept. *mBio* 7(2) (March–April): e01395-15. Published online March 31, 2016 . https://doi.org/10.1128/mbio.01395-15.

Rúa, M., et al. 2016. Home-field advantage? Evidence of local adaptation among plants, soil, and arbuscular mycorrhizal fungi through meta-analysis. *BMC Evolutionary Biology* 16: 122.

Ryan, M. 2017. *A taste for the beautiful.* Princeton University Press.

Sen, A. 2009. *The concept of justice.* London: A. Lane.

Simard, S.W. 2018. Mycorrhizal networks facilitate tree communication, learning and memory. In *Memory and learning in plants*, ed. F. Baluska, M. Gagliano, and G. Witzany. Springer.

Sonnenburg, J., and E. Sonnenburg. 2015. *The Good Gut.* London: Penguin.

Valles-Colomer, M., G. Falony, Y. Darzi, et al. 2019. The neuroactive potential of the human gut microbiota in quality of life and depression. *Nature Microbiology* 4: 623–632.

Wall, R., D. Bardgett, V. Behan-Pelletier, J.E. Herrick, T. Hefin Jones, J. Six, D.R. Strong, and W.H. Van der Putten. 2013. *Soil ecology and ecosystem services.* Oxford: Oxford University Press.

Wilson, E. 2014. *The meaning of human existence.* New York: Norton.

Wohleben, P. 2015. *Das Geheime Leben der Baume.* Ludwig.

Michiel Korthals is currently Professor Philosophy, University of Gastronomic Sciences, Pollenza/Bra (Italy) and emeritus Professor Applied Philosophy, Free University and Wageningen University. Korthals started his career in the tradition of Critical Theory. Inspired by the work of Horkheimer and Habermas he published about social philosophy and education. In the second phase of his career he addresses problems of agriculture, animals, nature and food production and consumption. His main thesis is that the modern regime of food production and consumption produces alienation of food and nature and disregards food capabilities. Main publication: *Before Dinner* (Springer, 2004).

Chapter 9
Comment: The Current State of Nonhuman Animal Agency

Joost Leuven

Daily at schools all over the world, human children rebel against their teachers. When they are forced to make assignments, they don't want to do, children often resist and try to challenge authority. This shows us that agency, the ability to make choices regarding their own lives, matters to them, at least to a certain extent. Good teachers try to accommodate the students' desire for agency by creating lessons and assignments in which there is room for students to make personal choices, like giving them the freedom to decide the specific topic for an essay or classroom presentation. Children don't lose this desire for agency once they grow up though and the desire for agency and the tendency to rebel against attempts to take that agency away, can be seen in human adults as well.

Human beings are not the only ones actively rebelling when their agency is curtailed. As we enter the geological era that some have fittingly named the Anthropocene (due to the increasing impact of human activity on ecosystems and geology), the examples of nonhuman animals trying to rebel against the increasing impediment of their agency also seem to become more numerous and frequent. We can see nonhuman animal rebellion in small and relatively harmless situations, like companion animals who refuse to be disciplined to use the litter box, but also under more serious circumstances, like livestock animals refusing to fall in line when entering a slaughterhouse (Palmer 2001). It is difficult to overstate the extent to which human beings nowadays control the lives of nonhuman animals. For the ones we keep in captivity, we control their movement, when and what they eat and even who they mate with, while the ones still living in the wild have to face the consequences of climate change, deforestation and the destruction of natural habitats in general. The existence of cases of rebellion forces us to consider the moral importance of agency in the lives of nonhuman animals.

J. Leuven (✉)
The Hague, The Netherlands

© The Author(s) 2021
B. Bovenkerk and J. Keulartz (eds.), *Animals in Our Midst: The Challenges of Co-existing with Animals in the Anthropocene*, The International Library of Environmental, Agricultural and Food Ethics 33,
https://doi.org/10.1007/978-3-030-63523-7_9

Especially in an era in which the overall relationship of humans with nonhuman animals is one of near total domination of human beings over nonhuman animal lives, it is necessary that moral philosophers look critically at the topic of self-realization in nonhuman animals. By reflecting on the attempts by the authors in this section of the book to do just that, this comment aims to make sense of nonhuman animal agency in the Anthropocene.

9.1 Changing Perspectives Within Animal Ethics

Traditionally the academic debate on animal ethics has focused predominantly on the question of whether nonhuman animal interests matter and how these interests should be taken into account. Authors such as Tom Regan (1983), Peter Singer (1975) and David Degrazia (1996) have contributed importantly to this debate, making cases against speciesism and in favor of animal rights and welfare. However, in recent years the debate has moved towards more complex questions and authors have now begun to challenge the idea that nonhuman animals are simply passive recipients of rights or care and argue that they are much more than that. Instead they should be seen as actors in a social (Gruen 2015), political (Donaldson and Kymlicka 2011) or moral (Monsó et al. 2018) sense. In their chapter, '*Taking Animal Perspectives into Account in Animal Ethics*', Eva Meijer and Bernice Bovenkerk explore this relatively new idea of animal agency and the ethical ramifications of accepting that nonhuman animals too have an interest in shaping their own lives.

Meijer and Bovenkerk identify and review two opposites on a spectrum of approaches to conceptualizing animal agency. The first they call propositional agency, which links agency to capacity for propositional thought. The second they call materialist agency, which links agency to the capacity to move something else. Both views have major problems, either being too limiting and anthropocentric (as human forms of rationality are taken as the standard) or too broad to be of practical use (as a framework to make normative and political judgments is missing). Meijer and Bovenkerk instead propose an alternative working definition of agency, based on a relational approach to ethics that doesn't take an anthropocentric view of agency as a starting point and one that takes the unique perspective of nonhuman animals into account. In their view agency should be seen as the capacity of a subject to influence the world in a way that expresses their desires and will. Their paper highlights the advantages and disadvantages of a relational approach to agency and makes a case for philosophers to engage differently with nonhuman animals in order to develop an interspecies ethics.

Charlotte Blattner's chapter, '*Turning to Animal Agency in the Anthropocene*', makes a similar case, but from a judicial perspective. Blattner demonstrates how the current legal system is unable to provide justice for nonhuman animals. While there are laws that claim to protect the welfare of nonhuman animals, Blattner argues these laws in practice serve only to silence nonhuman animals and inhibit their agency by cementing the moral and legal status of nonhuman animals as human property.

Changes are therefore necessary to make the legal system fair and just towards nonhuman animals. This can only happen once we acknowledge the difficulty of objectively evaluating competing interests of human and nonhuman animals (due to the fact that, as humans we are the ones making the laws while also being the ones who benefit immensely from exploiting nonhuman animals). To overcome this issue, Blattner convincingly argues in favor of honest, unbiased and open-ended research into nonhuman animal agency to empower animal agency and work towards a multispecies polity.

The importance of removing an anthropocentric perspective from the way we think of nonhuman agency is further emphasized in Nathan Kowalsky's chapter, *'Animal Difference in the Age of the Selfsame'*. Kowalsky critically reflects on the flaws of the 'growing circle of ethics' perspective, put forward by Peter Singer. He sees an historical development of growing concern from only caring for one's own well-being towards also caring for other humans and eventually caring about nonhuman animals as well. Instead of determining the moral worth of a nonhuman animal by measuring them to the human standard, we should judge them solely on their own merits. By challenging the inherent anthropocentrism in how we decide moral considerability, Kowalsky also touches on a type of nonhuman agency that the previous authors did not discuss in depth, namely moral agency.

9.2 The Problem of Predation

Ever since people started making the case for animal rights and questioned the ethics of killing them for food, the question of how to judge predation is one that has been a challenge for philosophers. Especially when one acknowledges the moral agency of nonhuman animals, accepting that their actions might be good or bad, it becomes difficult to ignore the question of whether we have a duty to educate animals in right moral behavior and if we should punish them for behaving in ways we consider deplorable. How should we relate to the state of nature, where life is nasty, brutish and short? Should we try to intervene and police wild animals, e.g. save the gazelle from the hungry lion?

In his chapter, *'Should the Lion Eat Straw Like the Ox? Animal Ethics and the Predation Problem'*, Jozef Keulartz explores this issue and evaluates the way different philosophers have tried to tackle this problem from different ethical perspectives. He reviews Peter Singer's utilitarian approach, Tom Regan's deontological approach, Nussbaum's capability approach and the political animal rights approach of Donaldson and Kymlicka. Singer argues that the evil of predation cannot be eliminated without introducing worse suffering, while Regan tries to argue it is not our duty to intervene and that nonhumans animals should simply be left alone. Keulartz shows the problem with both approaches and he goes on to discuss Nussbaum's capability approach, which seems promising at first, but ultimately fails to offer a more balanced view of the predator-prey relationship. In the end the political animal rights theory of Donaldson and Kymlicka, which argues that wild animals should be recognized as members of separate, sovereign nations, seems the most promising. However, Keulartz identifies problems with their approach as well, arguing that they underestimate the large destruction that is taking place globally, due to climate-change and human activities. Simply giving wild animals back control over their own territories won't save them from the mass extinction that is taking place in the Anthropocene. Keulartz's paper forces us to ask ourselves whether treating wild animal populations as sovereign entities is really the best way to respect nonhuman animal agency.

In the end, it seems that as philosophers we still struggle with how to evaluate the (moral) agency of animals in the context of predator-prey relationship. While it might be clear what the current state of these (often violent) relationships between nonhuman animals *is*, it is not yet clear what they *can* and *ought* to be.

9.3 Human and Nonhuman Animals

It's not strange that, as the academic field of animal ethics matures and other interdisciplinary fields like animal studies emerge as well, scientists and philosophers become more sensitive to the ways humans and nonhuman animals are alike. The chapter by Michiel Korthals, '*Humanity in the Living, the Living in Humans*', highlights this as it shows how difficult it is to maintain a clear boundary between humans and other animals, based on altruism, language or certain values.

Furthermore, as an academic field we should also become more aware of the interconnectedness and similarity in the ways different groups are oppressed. Only by being mindful of all manners of oppression can an ethical theory be developed that respects the agency of human and nonhuman animals alike. The aim of '*Justified species partiality*', the chapter by Ronald Sandler and John Basl, should be seen in this light. While I think that animal ethicists in general might be more aware of and sensitive to the interests of humans with certain impairments or conditions than Sandler and Basl give them credit for, the paper does show how important it is for ethicists to look critically at how their words can be interpreted or (mis)understood. While the famous 'argument from marginal cases' might still be logically valid, no one likes to be called 'marginal', which is another good reason to rename the argument to the 'argument from species overlap', as some philosophers have argued for (Horta 2014).

9.4 The Future of Agency

The state of nonhuman animal agency is dire. Human domination unforgivingly curtails opportunities for nonhuman self-realization, the human legal system works against them and mass extinction due to human-caused climate change threatens their continued existence. The discussed authors not only successfully make clear where the theoretical debate on nonhuman animal agency now stands and what the remaining conceptual hurdles are, but they also make the societal and moral urgency of more research in this area abundantly clear. As Meijer and Bovenkerk discussed in their paper, philosophers and scientists will need to rethink the way they approach this research, inventing new innovative ways of empirical research that engage with nonhumans animals and gives them the opportunity to reshape our relationship with them and determine their place in our shared society.

References

DeGrazia, D. 1996. *Taking animals seriously: Mental life and moral status*. Cambridge: Cambridge University Press.

Donaldson, S., and W. Kymlicka. 2011. *Zoopolis: A political theory of animal rights*. New York: Oxford University Press.

Gruen, L. 2015. *Entangled empathy: An alternative ethic for our relationships with animals*. New York: Lantern Books.

Horta, O. 2014. The scope of the argument from species overlap. *Journal of Applied Philosophy* 31: 142–154. https://doi.org/10.1111/japp.12051.

Monsó, S., J. Benz-Schwarzburg, and A. Huber. 2018. Animal morality: What it means and why it matters. *The Journal of Ethics* 22 (1): 283–310. https://doi.org/10.1007/s10892-018-9275-3.

Palmer, C. 2001. 'Taming the wild profusion of existing things'? A study of Foucault, power and human/animal relationships. *Environmental Ethics* 23 (4): 339–358.

Regan, T. 1983. *The case for animal rights*. Berkley: University of California Press.

Singer, P. 1975. *Animal liberation*. London: Pimlico.

Joost Leuven studied philosophy and cultural sociology at the University of Amsterdam. He wrote his master theses on the political turn in animal ethics and the 'abolitionist vs welfarist' debate within the contemporary animal rights movement. Between 2014 and 2018 he worked as a dual city councillor for the Party for the Animals in the municipality of The Hague. Currently he works as a high school teacher, teaching philosophy and social science.

Part II
Domesticated Animals

Chapter 10
An Introduction to Ecomodernism

Hidde Boersma

Abstract Land use change has detrimental impacts on the planet. It is not only a major cause of biodiversity loss, through habitat destruction and fragmentation, but also an important driver for climate change, through deforestation and peat oxidation. Land use change is mainly driven by food production, of which meat production comprises the major share. Ecomodernists therefore feel reduction of the impact of meat production is paramount for a sustainable future. To achieve this, ecomodernists focus on intensification of the production process to produce more on less land, both through the closing of global yield gaps and through the development of integrated indoor systems like agroparks. On the demand side, ecomodernists feel a diverse strategy is needed, from the development of meat substitutes and lab meat, to the persuasion of consumers to move from beef to monogastrics like pork or chicken.

10.1 Introduction

With the 2005 essay *The Death of Environmentalism* Ted Nordhaus and Michael Shellenberger introduced the world to Ecomodernism, and with that the start of a new green movement. The first 10 years they build their organization and philosophy, culminating in 2015 in *The Ecomodernist Manifesto*, a document, written by experts from different backgrounds, where the basic principles of the new movement where laid down. The writers felt the classic Green environmentalists lost sight of saving the planet, amidst political interests. Ecomodernists generally agree they owe a great deal to the environmental movement from the 1960s and 1970s, but felt increasingly frustrated by the lack of results of the last decades of the twentieth century. It was time for the old movement to die, for a new one to thrive.

Ecomodernists, as a principle, strive to reduce mankind's impact on the planet by concentrating its activity on as little land as possible. The more humans intensify their activities, the more space there is for nature to thrive. At this moment, humans

H. Boersma (✉)
Amsterdam, The Netherlands

© The Author(s) 2021
B. Bovenkerk and J. Keulartz (eds.), *Animals in Our Midst: The Challenges of Co-existing with Animals in the Anthropocene*, The International Library of Environmental, Agricultural and Food Ethics 33,
https://doi.org/10.1007/978-3-030-63523-7_10

use more than 70% of the ice-free land for the build environment, infrastructure, energy, but mainly for agriculture (IPCC 2019). And although the earth's population grows from 7,5 billion to 10 billion in 2100 and at the same time will get more prosperous, ecomodernists feel it is possible to reduce our need for land to a maximum of approximately 30% of the total ice-free land.

As a consequence of this principle, ecomodernists favor nuclear energy as an energy source for combating climate change, over renewables like wind and sun, as nuclear reactors provide the most dense form of energy, and thus save space for nature. Ecomodernists also support the ongoing urbanization trend. Currently, half of the population, 3,5 billion people, live in cities thereby occupying only 3% of the globe. Estimations are that in 2100 more than 70% of the people will live in cities (UN DESA 2018). Finally, on agriculture, ecomodernists believe intensive agriculture is the most sustainable way of combining a well fed and green planet in 2050.

As to the stance on agriculture, the focus point of this essay, ecomodernists owe a big deal to the work of Ben Phalan and Andrew Balmford, both from the University of Cambridge, UK, and their work on the land sharing versus land sparing debate. In a nutshell, this debate revolves around the question whether for biodiversity it is better to combine high yielding agriculture with setting away large swaths of the planet for nature, or go for wildlife-friendly farming, with lower yields and thus requiring more land, but with more biodiversity on the farm. The results of Phalan and Balmford, but also of others, unequivocally show that land sparing in most cases saves more biodiversity than land sharing, currently, but also in scenarios where in 2050 70% more calories need to be produced (Phalan et al. 2011; Balmford et al. 2015; Hodgson et al. 2010; Egan 2012). This is true for birds, mammals, trees, plants and insects, and can simply be explained by the fact that animals or plants don't like agricultural fields, no matter if they are extensively or intensively managed. Especially rare organisms need 'wild' nature to thrive.

The aim to reduce land use, fits with another core principle of ecomodernism: saving nature by not needing it. In stark contrast with the classic green motto of wanting to live more in harmony with nature as a way to save the planet, ecomodernists feel that dependency on nature for one's survival leads to its detriment. There are some striking historical examples where nature was saved by not needing it anymore. The population of whales, whose oil was used for lighting, was saved by the discovery of fossil oil, and the same accounted for forest all around the world: by switching from wood to oil and coal many forests around the world were saved from the axe. Moreover, the rise of synthetic rubber, saved forests from turning into rubber plantations, and, more recently, wild fish stocks are being saved by the development of fish farms (Asafu-Adjaye et al. 2015). The less we need nature for its ecosystem services, its resources or its land, the more it can follow its own dynamic, and the more wild nature there is for us to enjoy. Nature should be saved for its intrinsic value, not for its use for humankind, as then it is bound to lose.

So how do animals fit in this picture? Most ecomodernists don't oppose eating animals from a moral point of view, but naturally feel those animals should be treated well. They however also realize that it is often hard to exactly pinpoint what exactly comprises a good life for animals and how to measure it, without falling in the trap of anthropomorphizing. Ecomodernists often question our meat consumption from an efficiency point of view: how much land can we save if we eat less meat, and how do we produce meat in the least harmful way for the planet, with the lowest possible greenhouse gas emissions and eutrophication potential.

10.2   The Optimal Role of Animals in Our Food System

From an environmental point of view, eating meat generally is a very inefficient and wasteful way of producing food. Life Cycle Analyses (LCAs) show that meat production over the whole range uses more land and comes with more greenhouse gas emissions per kilo protein then a plant-based diet (FAO 2006). There is however a large difference between the various meat sources regarding their environmental impact, mainly through their different feed conversion ratios. While cows on average need around 6 kilo feed to produce 1 kilo of meat, chicken need less than 2 kilo.

This means that to produce a kilo of beef, 100 m^2 of land is necessary while greenhouse gas (ghg) emissions reach 300 kilo CO_2-equivalent for that same kilo. A kilo of pork is produced on 13 m^2, poultry on 8 m^2. Similar calculations can be made for milk and eggs, using respectively 4 and 5 m^2 per kilo protein. For comparison: the production of 1 kilo of proteins from pulses costs on average 1 m^2 (Nijdam et al. 2012).

This seems like a clear call for eating less meat, but it is not as simple as that. Animals serve an important role in the feed system, as they are able to feed on feed not suitable for human consumption, like leaves and stems from crops, co-products of the production system like spent grain of breweries and even food waste. Furthermore cows feed on grass, non-digestible for humans, which often grows on soils not suitable for crops. Historically, it is in this function animals roamed on the premises of the small scale Dutch farms of the nineteenth century and before: by upcycling waste, the system, which was characterized more by scarcity then by the current abundance, was made more circular and less nutrients were wasted (Bieleman 2010).

During the twentieth century, agriculture professionalized, leading to specialization of the various tasks. Farmers specialized in crops or cattle, specialized butcheries were erected. But animals still somewhat serve the same functions as before: more than half of the feed of Dutch pigs consists of residual feed, for instance from the starch industry. Over the years however, several residual feeds, like swill and animal meal have become off-limits, because of disease risks. This made circular agriculture harder, and sentenced a lot of food waste to the incinerator.

Recently, Dutch researchers from Wageningen University calculated the optimal amount of animal protein consumed per day per capita for a system with as few food waste as possible, to be between 9 and 23 grams (Van Zanten 2019). The large range can be explained by legislation, which allows for residual streams, and by the types of animals used for upcycling. As humans need approximately 60 grams of protein per day, only a third of it should be of animal origin, the rest should come from either plants or seafood. Currently Dutch consumers consume 104 grams of protein on average, of which 74 are of animal origin, on a daily basis, which implies a drastic reduction of meat consumption is crucial for an optimal and sustainable future agricultural system (Dagevos et al. 2019).

It is important to notice that this analysis largely applies to rich countries in North America and Europa, as this is where most meat is consumed. Where American citizens consume 120 kilo meat per year, which translates into more than 400 kilocalories per day, Indians only devour 4 kilo and China around 60. In many developing countries increasing meat consuming improves overall health, as meat is an easy and dense source of nutrients, much more than vegetables. Ecomodernists therefore accept UN predictions that global meat consumption will continue to rise, which begs the question what the most sustainable way is to raise animals.

10.3   The Case for Intensification

Under pressure of population growth, technological progress, globalization and scarcer land, meat production in the Netherlands, but also in other western countries changed during the twentieth century from extensive grazing or roaming, to intensive production systems. This was especially true for pork and poultry, as they cannot live on grass and became more and more 'land independent', fed by feed sometimes from far away. As a result of this intensification process, during the twentieth century the number of pigs in the Netherlands grew from 3 million in 1960 to 12,5 million in 2018, while the numbers of broiler chickens grew from 2,4 to 48 billion. Cows showed a far more modest growth (CBS). Similar intensification trends are currently underway in China and other countries like Brazil, where the economy and the population are growing (FAO 2006).

The process of intensification has increased food security but led to a series of environmental problems, ranging from eutrophication of streams and waters due to overuse of fertilizer, to land degradation and deforestation through the increasing need for land for grazing and for the production of soy and maize for feed (ibid.). It furthermore led to rising greenhouse gas emissions, raised concerns about animal welfare and made people question the impact on the landscape.

Lately, this has led to an increasing interest in meat raised in extensive systems. In the US sales of grass-fed beef is on the rise, as opposed to feedlot-finished beef (Hayek and Garrett 2018). Similarly, sales of organic produce is growing in the Netherlands, often motivated by both environmental and animal welfare concerns (Bionext Trendrapport 2018). Environmental organizations like Friends of the Earth (2017) call for buying locally and organically produced meat, next to reducing consumption in general.

Ecomodernists very much question whether a move to a more extensive meat production system is the right direction. As Marian Swain (2017), fellow by the American thinktank the Breakthrough Institute notices in her essay on The Future of Meat, intensive livestock farming produces more meat, more quickly, with fewer animals. The controlled environment and formulated feed promote optimized growth and reduce losses. As most of the greenhouse gas emissions come from the need for land for feed production, and, in the case of ruminants, the production of methane through burping, it are these characteristics which make intensive cattle farming more environmentally friendly. The direct use of fossil fuels in those intensive system is small, usually less than 20% of the total greenhouse gas emissions of agriculture.

Several studies endorse the notion that intensification is more sustainable: Pelletier et al. (2010) for instance find that in the US pasture beef almost takes 20 kilo of CO_2-equivalents per kilo to produce, while the more intensive feedlot-finished beef costs a little more than 15 kilo. Dutch intensive agriculture is even more efficient, with 10,9 kilo of CO_2-eq emissions per kilo beef. In poultry, it takes 39,2 kilo CO_2-eq emissions to create a per kilo protein in intensive production systems, compared to 48,7 in extensive systems. Only for pork, extensive systems have lower greenhouse gas emissions: 47,6 over 52,0. With total combined yearly greenhouse gas emissions of 1609 million ton the production of chicken and pork is dwarfed by the 5024 million ton which is yearly emitted by the production of beef (FAO 2017).

Similar calculations can be made for land use, where in the US feedlot-finished beef needs only half of the land compared to its grass-fed equivalent. In 2016, the American agronomist Jason Lusk from Oklahoma University calculated that if there had been no intensification and technological progress since the 1950s, 15,3 million more beef cows would be needed to produce the amount of meat we produce now. Also, 228 million more acres of corn would be needed, as well as 101 million more acres of soybeans (Lusk 2016). Thus, intensification saves nature from being ploughed under, a phenomenon also known as the Borlaug hypothesis, named after Norman Borlaug, the famous plant breeder and Nobel Peace prize laureate, also known as the father of the Green Revolution.

Recently, a team of scientists endorsed the importance of sparing land by introducing the concept of carbon benefit. Currently, life cycle analyses of food production often do not take into account the carbon storage opportunity of fields taken out of production, but the concept of carbon opportunity costs in this study solves this lacuna. The consortium of scientists shows that intensification of production, and thereby freeing land for forest is a powerful and efficient climate change mitigation strategy (Searchinger et al. 2018a). In a similar vein, Lamb et al. (2016) two years earlier calculated that by intensifying agriculture in the UK on the best suitable places, and returning marginal land to nature, the UK would be able to produce the same amount of food, whilst at the same time achieving the goals of the 2015 Paris Agreement, without changing its energy portfolio.

The Dutch Scientific Council for Government Policy (WRR) was an early advocate of intensification. Already in 1992 the council calculated that for the European Union it is possible to produce a similar amount of food on a quarter of the current area, by concentration and optimizing production in the most fertile places. In this way, 75% of all European fields, could utopically be rewilded, used for climate change mitigation or have any other beneficial function.

Similar calculations can be made for the Netherlands, where at the moment already 80% of economic value in agriculture is made on 20% of the land (Van de Klundert

2012). By intensifying these areas further, it is possible to put marginal lands like the poor sandy soils in the northern province of Drenthe out of production. Those areas can be turned into reserves for meadow and farmland birds, which are an important part of the identity of the Dutch landscape and people. It also opens up the opportunity to stop oxidation of peatlands in the west of the country by flooding these areas and turning them into biodiverse wetlands, instead of using them for milk cows to graze upon.

Opponents of such land sparing strategies often point to the so called rebound effect, also known as Jevons Paradox. Historically, gains in efficiency often do not lead to environmental wins but ramp up consumption by lowering prices, thereby increasing the pressure on earth and its resources. William Jevons already in 1865 showed that improved technology in the coal industry led to more, not less fuel consumption. Applied to agriculture, it means that a farmer who intensifies his production most often will not spare land, but expands his business either because technology allows him, or because low prices force him.

Ecomodernists agree this is a serious problem. Especially in the age of rapid economic growth, combined with a rising population, intensification and agricultural expansion often go hand in hand. Ecomodernists therefore believe that intensification only works when accompanied by strict zoning policies from the government, which allow for the setting aside of large swaths of nature (Boersma et al. 2018). Ecomodernists even consider the erection of an IPCC-like body for global land use (Ellis and Mehrabi 2019). Without strict land use policies, it will be very hard to save biodiversity and stabilize the climate, whatever the agricultural system in place.

10.4 How History Shapes the Way We Think About Animal Farming

Making the case for intensification isn't popular. This stems not only from misconceptions about what sustainability actually compromises or from sincere concerns about animal welfare, but also has historical and sociological components. What scientifically counts as sustainable, often isn't considered beautiful or culturally correct, and this makes for a murky debate. People often make inconsistent, and sometimes even contradictory demands.

Such contradictions is what the German philosopher Jürgen Habermas (1985) calls the conflict between the worlds of subjectivity, objectivity, and intersubjectivity, or, more simply put, the worlds of facts, social norms and individual experiences. As an example, in animal agriculture this means that from a scientific, rational point of view it is best to keep cows indoors, with high productivity and low emissions under tightly controlled circumstances. But from a normative point of view, those animals belong in the meadow, as this is how it should be, this is what we see on paintings of famous Dutch artists of the seventeenth to nineteenth century like Paulus Potter and Willem Maris. Grazing cows are part of the Dutch identity. Finally, from an individual point of view, the sight of a cow slogging in a field at 35 °C, might make an onlooker think it is better to take the animal inside. Such conflicting views slow down transitions, as they lead to contradicting views and thus conflicting policy recommendations. Transitions often only come to fruition when the three worlds come together.

In the discussion about the future of animal agriculture, environmental organizations have a peculiar position. Environmentalists generally are highly educated, live in urban environments and live a cosmopolitan lifestyle. They optimally benefit from the globalized, industrialized and interconnected world and live a wealthy life, mostly shielded from the hardship of the normal world, in what the German philosopher Peter Sloterdijk (2005) calls a Crystal Palace.

But when it comes to the landscape in their direct surroundings, they expect the exact opposite. They don't like the globalized and highly connected, technological food chain, and what it has done to the landscape. For them, farmers should move to a way of farming resembling the past: more extensive and in harmony with nature. The conflicts between peoples own lifestyles, in this case cosmopolitanism, and preferences on how others should live their lives, show how hard it is to align the world of subjectivity, objectivity, and intersubjectivity.

To whether animals are better off outside, the evidence is mixed. In intensive systems, disease rates are often lower, as are animal losses, but in extensive systems animals are able to display more natural behavior. For cows, living outside in hot summers can be stressful. With their sophisticated metabolism of grass, their bodies generate a lot of heath, and cows therefore dislike temperature above 16 °C. Recently, Von Keyserlingk et al. (2009) from the University of British Columbia showed that cows in temperate climates like their pastures, but only at night when it is cooler. In more tropical regions, like India and Brazil, from a welfare stance, it might be better for cows to always stay inside. For pigs, Hötzel et al. (2004) showed that indoor pigs displayed more aggressive and unnatural behavior then their outdoor living equivalents. If held indoors, efforts should be taken to mimic outside circumstances. Finally, chicken by nature are forest animals, and refrain from going far in open fields.

As for global human health, indoor systems are preferred. Diseases like the various versions of the avian flu, which can also infect humans, are more easily spread in open, outdoor systems, sometimes through contact with wild relatives like ducks. Similarly, wild boars more easily spread swine fever to pigs roaming outside, then to those living in closed systems (Brown and Bevins 2018).

10.5   The Future of Animal Farming

How do ecomodernists feel animal farming should develop the coming decades? First of all, efforts should focus on making best practices common practice. In crop agriculture, closing the yield gap between maximum and actual yields, is a major strategy in solving hunger and malnutrition while saving nature (Floey et al. 2011), and a similar effort should be made in animal agriculture. Research shows there are large gaps in efficiency and productivity between countries, but also between farms within countries. Especially in beef production the gap between the best and the worst performing farms is large. Nijdam et al. (2012) show that greenhouse gas emissions vary from 9 kilo per kilo beef for the most efficient intensive system, to 129 kilo for the worst performing pastoral system. Land use varies from 7 m^2 per kilo for the best to 420 m^2 per kilo for the worst performer. For pigs and chickens the gaps are smaller, but the best performing systems still use half the land of the more unproductive ones.

The country of Brazil serves as an example where closing the gap can make a difference. Although soy plantations get the most attention as a driver of deforestation of the Amazon and the Cerrado (a vast tropical savanna), 80% of the fields which used to be pristine forest are used for grazing, in a very extensive way, often with only one cow per hectare (Pendrill et al. 2019). The introduction of feedlot-finishing alone will already drastically lower land demand.

For the places where intensification already is in place, ecomodernists feel that environmental gains can be made by transforming farms into so-called agroparks. Agroparks are integrated, concentrated facilities where multiple actors in the food chain come together, from breeding to processing, and where residual flows can be re-used on site. Agroparks aim to make as efficient as possible use of space, scale, distance and waste (Smeets 2011).

Agroparks come from the field of industrial ecology and were first mentioned at the turn of the century. They were born out of the idea that a strict separation of functions would increase livability of the countryside: agroparks for agriculture, the landscape for nature. Agroparks are inspired by the mixed farms which were common in the Netherlands in 1900, especially on the poorer, sandy soils of the east and south of the country. Those farms kept a couple of animals, mainly for the manure, grew the feed themselves and even slaughtered the animal on site. The discovery of artificial fertilizer, the growing global market and better infrastructure made mixed farms to inefficient and led to specialization. This went so far that whole regions began to focus on the production of one type of product, like vegetables under glass in the area around the cities of Rotterdam and The Hague, and pigs in the province of Noord-Brabant.

However, technological progress has made it possible to operate mixed farms that are actually competitive enough in the global market. Mixed farms are more sustainable as they are easier to make circular by re-using residual streams on site. They furthermore avoid transporting animals, and thus lowering stress levels, as an agroparks can have their own butcher. Key to making this work is scaling up: butcheries are only profitable on farms with more than 200.000 animals, and agroparks make this possible, in an animal-friendly way.

Agroparks build on the success of the famous greenhouses in the Netherlands. By controlling all circumstances in a closed circuit, Dutch horticulturists are able to make products with minimal input. Where a kilo of tomatoes in an open field in Spain need 60 L of water to grow, Dutch growers work with only 8. Furthermore, yields per hectare are significantly higher than elsewhere in the world, which means the Dutch are making efficient use of space. The advantages of indoor systems are reflected in the price. The costs of a kilo chicken in a closed system add up to 2,30 euro, in an open system to 3,70 (ibid.).

Size and proximity in agroparks also make horizontal integration possible. This means for instance that slaughter waste can be processed into useful resources, which can be used to grow crops. The energy and heat of this process can be used on site, or returned to the net as green energy. Wastewater from the greenhouses can serve as feed for algae, fish or mollusks. As animals live in a closed off area, it is easier to separate feces and urine, and reuse nutrients as phosphate and nitrogen. Technical improvements to make such things work, are often expensive, and are only cost-effective in large systems.

In 2018, Peter Smeets, researcher at Wageningen University and Research, retired geographer Steeph Buijs and me developed a redesign of the Dutch landscape, based on increasing urbanization and the use of agroparks for the production of food (ibid.). We were inspired by the plan of the Chinese government to create so-called metropolitan regions, like the greater-Beijing area, which will grow to 130 million people (Johnson 2015). To house them, the government allocated an area as big as 200.000 km^2, which means every inhabitant has on average 1500 m^2 at her disposal, 1,5 times as much as people living in the Randstad, the urbanized area in the west of the Netherlands. The Chinese plan to make this area self-sufficient for fruit, vegetables, meat, eggs and dairy. Staple crops and feed will be imported from the hinterland of from other countries.

The Netherlands should also be developed like a metropolitan region. For that, it is essential to think across borders. City planners already for a long time call the Netherlands, and more specifically the Randstad, an empty city, and the same counts for surrounding areas like the Ruhr in Germany, the Brussels-Antwerp axis in Belgium and the area around Lille in France. This whole area houses 35 million people, and Dutch agriculture at the moment produces mainly for this region: 70% of all products are sold here, which makes the area largely self-sufficient. This forms the perfect basis for the development of a metropolitan region.

With the clustering of agricultural activities there is finally room to really improve nature and biodiversity in the Netherlands. The last couple of decades the Dutch landscape has become cluttered, and it has proven hard to create large areas for nature and recreation (PBL 2017). In the 1990s the Dutch government planned the Ecological Main Structure, in an effort to connect Dutch nature areas, but it largely failed to materialize due to political inertness in the following decades (LNV 2018). To improve biodiversity, interconnected reserves are crucial, as many scientists have shown in different parts of the world (Kuussaari et al. 2009).

In our plan, agroparks are built or expanded in areas which are already leading the way. This means more greenhouses around Rotterdam and The Hague, and more in the north of the province of Noord-Holland, where Seed valley is located. Poultry agroparks will be built around the Veluwe, in the middle of the country, while de Peel in the province of Noord-Brabant is perfectly suited for agroparks for pigs. We furthermore anticipate a continuing urbanization, and thus expanded housing areas around cities. Through this concentrating of housing and food production, large swaths of land can be returned to nature, most notably the peatlands of the 'Green Heart', the area in between the cities of Utrecht, Rotterdam and Amsterdam, which will be completely turned into wetlands, both for nature purposes and to increase water storage potential, in the wake of climate change.

10.6 The Future of Animal Eating

Clearly, the production side isn't the only lever to pull when it comes to making animal agriculture more sustainable; there is also the demand side. As stated above, from an environmental point of view, reduction of meat consumption is paramount. Although a full plant-based lifestyle isn't necessarily the best, as long as many people eat more than their fair share of meat, every extra vegetarian should be welcomed.

Like every major behavioral change, getting people to eat less meat is hard. Notwithstanding years of promotion of ideas like meatless Monday, average meat consumption in the Netherlands has stagnated at 77 kilo per year, down from a peak of 79 kilo in 2010. The last two years no change in consumption has been detected. Similarly, the percentage of vegetarians is relatively stable at approximately 4% (Dagevos et al. 2019). To lower the impact of the consumption of meat, several strategies have to be set in motion simultaneously. People have different values when it comes to diet and eating meat in particular, and to get people to change their diet, means catering to those different values.

As full vegetarianism is hard to adhere to, one important strategy might be to get people to swap their beef for chicken or pork. The World Resource Institute in December 2018 calculated that swapping one third of ones beef for pork or poultry already reduces ones greenhouse gas emissions by 14% and land use by 13% (Searchinger et al. 2018b). The transition from beef to chicken is already in place in most western countries, as Linus Blomqvist (2019) notices in his essay *Eat Meat, Not Too Much, Mostly Monogastrics*: "In the US, beef consumption declined from 169 kcal/capita/day at its peak in 1976 to just 100 kcal/capita/day in 2013, a drop of over 40 percent."

Another strategy is the development of meat substitutes. The last couple of years there have been many improvements and new introductions, and substitutes now resemble real meat in taste, structure and look. A prominent example is the Impossible Burger, which uses a plant heme, produced by modified yeast, to let its burger 'bleed' like a really burger, thereby mimicking its mouthfeel. The pea-based Beyond Burger uses beet juice, to do the same. Both are available in restaurants and shops around the world.

A third important strategy is the development of lab meat.[1] In 2013 Dutch pharmacologist Mark Post from Maastricht University presented the first burger created from stem cells in a petri dish (Jha 2013). Back then, it costed 250.000 euro to create, but since then various companies have optimized and standardized the process and got

---

[1] See also the chapter by Cor van der Weele in this Volume.

the price down to 1000 euro per kilo. In a few years lab meat will probably be afford-able for the masses. Lab meat presumably will have the biggest impact replacing bulk meat like nuggets, minced meat and shawarma, as these are the simplest form of meat, containing only muscle tissue. As these meat products constitute a major chunk of total consumption, lab meat might make a big difference. Several studies already show lab meat being superior from an environmental point of view, espe-cially with regard to land use (Tuomisto and Teixeira de Mattos 2011; Smetana et al. 2015).

Ecomodernists recognize that in the end, a sizeable part of the people still prefer the real deal over meat substitutes. To tackle this, in the Netherlands, food writer Joel Broekaert (in Hertzberger et al. 2018) introduced the concept 'eating less meat by eating more meat', in which he advocates for eating the whole animal instead of only parts. For him, there are four major upsides, namely (1) people will get more knowledgeable about the food chain and eat their animal with more respect; (2) farmers will be able to capitalize on the whole animal, instead of having to sell leftover parts for lower prices to the processing industry; (3) people's quality of life will rise, as the parts we seldom eat actually are the most tasteful; and lastly (4) when people recognize what good meat really tastes like, it makes it easier to swap tasteless chicken in other dishes for substitutes like tofu.

10.7   Conclusion

For a sustainable future, ecomodernists aim to minimize humanity's footprint primarily by shrinking the area humans use to live and produce. Where humans now use more than 70% of the ice-free land, ecomodernists aim to reduce that to around 30%. As cattle, pigs and chickens tread heavily on the earth, our meat consumption and production serve as an important lever. Ecomodernists believe that a pragmatic combination of intensification and demand reduction are the most important ways to lower impact. The former consists of a combination of closing the yield gap between the best and the worst performers, along with the development of innovative concepts like agroparks; the latter needs a combination of the (further) development of meat substitutes and lab meat, combined with a move away from beef to pork or chicken, and an increased use of the whole animal instead of only a few parts.

References

Asafu-Adjaye, J., et al. 2015. *An ecomodernist manifesto*. http://www.ecomodernism.org/manifesto. Accessed 23 March 2020.
Balmford, A., R. Green, and B. Phalan. 2015. Land for food & land for nature? *Daedalus* 144 (4): 57–75.
Bieleman, J. 2010. *Five centuries of farming: A short history of Dutch agriculture 1500–2000*. Wageningen Academic Publishers.

Bionext Trendrapport. 2018. Bionext. https://files.smart.pr/f1/176a685def40eb83f93864f9ccfbce/TRENDRAPPORT-BIOLOGISCHE-SECTOR-2018.pdf. Accessed 23 March 2020.

Blomqvist, L. 2019. Eat meat, not too much, mostly monogastrics. Breakthrough Institute, January 29.

Boersma, H., et al. 2018. *Feeding the city: Farming and the future of food*. Amsterdam, The Netherlands: Van Gennep Publishers.

Brown, V.R., and S.N. Bevins. 2018. A review of African swine fever and the potential for introduction into the United States and the possibility of subsequent establishment in feral swine and native ticks. *Frontiers in Veterinary Science* 5 (11): 42–59.

Dagevos, H., et al. 2019. *Vleesconsumptie per hoofd van de bevolking in Nederland, 2005–2018.* Wageningen Economic Research Report.

Egan, J.F. 2012. A comparison of land-sharing and land-sparing strategies for plant richness conservation in agricultural landscapes. *Ecological Application* 22 (2): 459–471.

Ellis, C.E., and Z. Mehrabi. 2019. Half Earth: promises, pitfalls, and prospects of dedicating Half of Earth's land to conservation. *Current Opinion in Environmental Sustainability* 38: 22–30.

Floey, J.A., et al. 2011. Solutions for a cultivated planet. *Nature* 478: 337–342.

Food and Agriculture Organization (FAO). 2006. *Livestock's long shadow: Environmental issues and options.* http://www.fao.org/3/a-a0701e.pdf. Accessed 23 March 2020.

Food and Agriculture Organization (FAO). 2017. *Global Livestock Environmental Assessment Model (GLEAM).* http://www.fao.org/gleam/results/en. Accessed 23 March 2020.

Friends of the Earth. 2017. *What is better meat?* https://friendsoftheearth.uk/food/what-better-meat. Accessed 23 March 2020.

Habermas, J. 1985. *Theory of communicative action*, vol. 1. Boston: Beacon Press.

Hayek, M.N., and R.D. Garrett. 2018. Nationwide shift to grass-fed beef requires larger cattle population. *Environmental Research Letters* 13 (8).

Hertzberger, R., et al. 2018. *Je bent wat je leest*. Nederland Leest Geschenkboek.

Hodgson, J.A., et al. 2010. Comparing organic farming and land sparing: Optimizing yield and butterfly populations at a landscape scale. *Ecology Letters* 13: 1358–1367.

Hötzel, M.J., et al. 2004. Behaviour of sows and piglets reared in intensive outdoor or indoor systems. *Applied Animal Behaviour Science* 86 (1): 27–39.

Intergovernmental Panel on Climate Chane (IPCC). 2019. *Special report on climate change, desertification, land degradation, sustainable land management, food security, and greenhouse gas fluxes in terrestrial ecosystems.* https://www.ipcc.ch/srccl/. Accessed 23 March 2020.

Jevons, W.S. 1865. *The coal question: An inquiry concerning the progress of the nation, and the probable exhaustion of our coal-mines.* Macmillan.

Jha, A. 2013. First lab-grown hamburger gets full marks for 'mouth feel'. *The Guardian*, August 6.

Johnson, I. 2015. As Beijing becomes a supercity, the rapid growth brings pains. *NY Times*, July 29.

Kuussaari, M., et al. 2009. Extinction debt: A challenge for biodiversity conservation. *Trends in Ecology & Evolution* 24 (10): 564–571.

Lamb, A., et al. 2016. The potential for land sparing to offset greenhouse gas emissions from agriculture. *Nature Climate Change* 6: 488–492.

Lusk, J. 2016. Why industrial farms are good for the environment. *NY Times*, September 25.

Ministerie van Landbouw, Natuur en Voedselkwaliteit (LNV). 2018. *Eindrapportage groot project Ecologische Hoofdstructuur.* https://www.rijksoverheid.nl/documenten/rapporten/2018/10/15/eindrapportage-groot-project-ecologische-hoofdstructuur. Accessed 23 March 2020.

Nijdam, D., et al. 2012. The price of protein: Review of land use and carbon footprints from life cycle assessments of animal food products and their substitutes. *Food Policy* 37 (6): 760–770.

Nordhaus, T., and M. Shellenberger. 2005. The death of environmentalism. *Grist Magazine*, January 13.

Pelletier, N., et al. 2010. Comparative life cycle environmental impacts of three beef production strategies in the Upper Midwestern United States. *Agricultural Systems* 103: 380–389.

Pendrill, F., et al. 2019. Deforestation displaced: Trade in forest-risk commodities and the prospects for a global forest transition. *Environmental Research Letters* 14 (5).

Phalan et al. 2011. Reconciling food production and biodiversity conservation: Land sharing and land sparing compared. *Science* 333 (6047): 1289–1291.

Planbureau voor de Leefomgeving (PBL). 2017. *Tussenbalans van de Leefomgeving* (nr. 2908). https://themasites.pbl.nl/balansvandeleefomgeving/wp-content/uploads/pbl-2017-tussen balans-van-de-leefomgeving-2908.pdf. Accessed 23 March 2020.

Searchinger, T.D., et al. 2018a. Assessing the efficiency of changes in land use for mitigating climate change. *Nature* 564: 249–253.

Searchinger, T.D., et al. 2018b. *Creating a sustainable food future*. World Resources Institute (WRI) Synthesis Report. https://www.wri.org/publication/creating-sustainable-food-future. Accessed 23 March 2020.

Sloterdijk, P. 2005. *Im Weltinnenraum des Kapitals*. Frankfurt a/M: Suhrkamp.

Smeets, P.J.A.M. 2011. *Expedition agroparks: Research by design into sustainable development and agriculture in the network society*. Wageningen, The Netherlands: Wageningen University Press.

Smetana, S., et al. 2015. Meat alternatives: Life cycle assessment of most known meat substitutes. *The International Journal of Life Cycle Assessment* 20 (9): 1254–1267.

Swain, M. 2017. The future of meat. *Breakthrough Institute*, May 18.

Tuomisto, H.L., and M.J. Teixeira de Mattos. 2011. Environmental impacts of cultured meat production. *Environmental Science & Technology* 45 (14): 6117–6123.

United Nations, Department of Economic and Social Affairs (UN DESA). 2018. *The world's cities in 2018*. Data Booklet. https://www.un.org/en/events/citiesday/assets/pdf/the_worlds_cities_in_ 2018_data_booklet.pdf. Accessed 22 March 2020.

Van de Klundert, B. 2012. *Expeditie wildernis. Ervaringen met het sublieme in de Nederlandse natuur*. Zeist, The Netherlands: KNVV Uitgeverij.

Van Zanten, H.H.E. 2019. The role of farm animals in a circular food system. *Global Food Security* 21: 18–22.

Von Keyserlingk, M.A.G., et al. 2009. The welfare of dairy cattle—Key concepts and the role of science. *Journal of Dairy Science* 92 (9): 4101–4111.

Wetenschappelijke Raad voor Regeringsbeleid (WRR). 1992. *Grond voor keuzen: Vier perspectieven voor de landelijke gebieden in de Europese Gemeenschap*. Den Haag, The Netherlands: Sdu Uitgeverij.

Hidde Boersma is a freelance publicist with in Ph.D. in soil microbiology. He writes mainly about ecomodernism, agriculture and biotechnology. He is the co-author of 'Ecomodernisme', 'Feeding the City' and More!—why abundance brings prosperity and sustainability. In 2017 he debuted with the documentary Well Fed, on genetic modification in developing countries, he now works on the follow up Bite Me on gmo-mosquito's to combat malaria. Boersma is a 3 times TEdx-speaker.

Chapter 11
Place-Making by Cows in an Intensive Dairy Farm: A Sociolinguistic Approach to Nonhuman Animal Agency

Leonie Cornips and Louis van den Hengel

Abstract Based on recent ethnographic fieldwork at an intensive dairy farm, this chapter examines the usefulness of posthuman critical theory for developing a new sociolinguistic approach to nonhuman animal agency. We explore how dairy cows, as encaged sentient beings whose mobility is profoundly restricted by bars and fences, negotiate their environment as a material-semiotic resource in linguistic acts of place-making. Drawing on the fields of critical posthumanism, new materialism and sociolinguistics, we explain how dairy cows imbue their physical space with meaning through materiality, the body and language. By developing a non-anthropocentric approach to language as a practice of more-than-human sociality, we argue for establishing egalitarian research perspectives beyond the assumptions of human exceptionalism and species hierarchy. The chapter thus aims to contribute towards a new understanding of nonhuman agency and interspecies relationships in the Anthropocene.

11.1 Introduction

Human thinkers in the western philosophical tradition have long relied upon the silencing of nonhuman animal others to confirm the exceptionalism of their own species. Since Aristotle, philosophers and scientists have defined "man" as a "rational animal" distinguished from other animals by his—and, more recently, her or their—capacity for a special kind of thinking, variously described as self-consciousness, reason, or representational thought (Cull 2015, 19). If, as Eva Meijer asserts (2016, 73), in this tradition "humans are viewed as radically different from other animals," then language is commonly seen as "one of the main ways in which this difference is

L. Cornips (✉)
NL-Lab, Humanities Cluster (KNAW), Amsterdam, The Netherlands
e-mail: leonie.cornips@maastrichtuniversity.nl

L. Cornips · L. van den Hengel
Faculty of Arts and Social Sciences, Maastricht University, Maastricht, The Netherlands

B. Bovenkerk and J. Keulartz (eds.), *Animals in Our Midst: The Challenges of Co-existing with Animals in the Anthropocene*, The International Library of Environmental, Agricultural and Food Ethics 33,
https://doi.org/10.1007/978-3-030-63523-7_11

expressed." The idea that language is what makes us human, or more precisely, that the possession of language allows humans to separate themselves from nonhuman nature, including their own animality, is indeed a key component of philosophical humanism and its exclusionary conceptions of individual and collective personhood. The philosopher Giorgio Agamben (2004, 33) uses the term "anthropological machine" to refer to the process by which the human is defined over and against what is nonhuman or animal, thus dividing the human subject from more-than-human forms of sentience, sociality, intelligence and communication. The strict identification of the human with language, or what Agamben (2004, 38) calls an articulation between "speaking being" and "living being," is central to the functioning of the anthropological machine, that is, the ways in which humans ought to continually create themselves as speaking political beings by creating hierarchies between human, animal, vegetable and mineral species. The traditional humanist understanding of the human as a unique creature, one that rises above the natural world of animals, plants and the physical environment, thus rests on a fundamental denial that nonhuman animals might be capable of language and other forms of complex symbolic communication.

In this chapter, we wish to move beyond the assumptions of human exceptionalism and species hierarchy in order to advance an understanding of language that displaces the centrality of the human subject. Specifically, we will explore how dairy cows, as caged living beings or what sociologist Rhoda Wilkie (2010, 115) has called "sentient commodities," negotiate their environment as a material-semiotic resource in the production of a meaningful world. While their physical mobility is profoundly restricted by bars and fences, we will examine how dairy cows enact social and linguistic agency through complex assemblages formed by human and nonhuman bodies, materials and environments. Starting from the assumption that, within the context of dairy farming, the subjectivities of cows and humans are continuously co-produced, we want to highlight how recognizing the linguistic agency of dairy cows may allow us to resist anthropocentric understandings of interspecies relationships and to formulate a new perspective on language as a social practice of human-nonhuman interaction. The central aim of this chapter, therefore, is to elaborate a radically post- or non-anthropocentric sociolinguistic approach that may help foster more egalitarian relationships to and between different species, or, as Agamben (2004, 83) phrases it, to bring to a "standstill" the anthropological machine that has historically articulated humanity and animality through their mutual exclusion.

The chapter has four sections. The first section examines how traditional humanist conceptions of language have structured dominant philosophical and linguistic understandings of human-animal relationships. The second section discusses the cognitive, emotional and social capacities of cows as sentient and intelligent beings, and proceeds to argue for the usefulness of posthuman critical theory for expanding the linguistic research agenda to include the study of nonhuman animal languages. This sets the stage for the third section, which discusses recent fieldwork at an intensive dairy farm in order to explore how cows, as social actors, engage in processes of linguistic place-making. Drawing, on the one hand, on recent work at the intersection of critical posthumanism and applied linguistics (Pennycook 2018) and, on the other, on new materialist conceptions of agency as a distributed phenomenon (Bennett 2010), we will elaborate a non-anthropocentric approach to human and nonhuman language practices. In the last and concluding section, we consider some of the implications of our findings for negotiating, or renegotiating, contemporary questions of nonhuman animal agency. As a whole this chapter argues that acknowledging nonhuman linguistic agency is essential for thinking through and responding to the specific conditions and challenges of the Anthropocene, where the advent of the human as a global geophysical force has muddled conventional distinctions between culture and nature, human and nonhuman, self and other. If, as Donna Haraway suggests, nonhuman animals "are not here just to think with," but rather they are here to "*live with*" (Haraway 2003, 5, emphasis added), then we must indeed embrace modes of inquiry suited to the task of confronting human *and* nonhuman acts of language, sociality and world-making.

11.2 Language and the Politics of Human Exceptionalism

The view that nonhuman animals have no speech, and therefore cannot establish themselves as ethical, juridical and political subjects, goes back at least to the ancient Greeks. In a well-known passage of his *Politics*, Aristotle associates the formation of political community with the supposedly unique human capacity for reasoned speech. Aristotle insists that the capacity for speech informed by reason (*logos*) is what separates "man", as a political animal, from the mere beasts who do not speak but simply produce sound (*phonè*):

> And so the reason why man is a political animal more than any bee or any gregarious animal is clear. For nature, as we often say, does nothing in vain; and man alone of the animals possesses speech (*logos*). The mere voice (*phonè*), it is true, can indicate pain and pleasure, and therefore is available in the other animals as well (for their nature has been developed so far as to have feelings of pain and pleasure and to signify them to one another), but speech, for its part, is designed to express the useful and the harmful and therefore also the just and the unjust.[1]

Because nonhuman animals are incapable of speech, Aristotle argues, they cannot express the civic and moral virtues that he considered essential for the wellbeing of the household and the city-state. Aristotle, as Derrida explains in *The Beast and the Sovereign* (2009, 343–349), thus makes a categorical distinction between human and nonhuman animals to posit an inextricable link between language and the political sphere. His philosophy therefore not only delimits political agency to certain privileged human beings—the free adult male citizens that make up the *polis*—but also actively excludes nonhuman animal voices from the definition of language itself.

Many later philosophers have followed Aristotle in rejecting the linguistic and cognitive abilities of nonhuman animals (Meijer 2016, 75–76). Descartes, for example, argued that animals do not think because they cannot speak: he regarded

---

[1] Aristotle, *Politics*, 1253a. Translation by Louis van den Hengel.

nonhuman animals as machines, governed by the laws of physical matter alone and hence devoid of mind and self-awareness.[2] Although he did recognize that animals such as magpies and parrots can utter words, and that dogs make noises that might resemble speech, Descartes maintained that other animals "cannot speak as we do: that is, they cannot show that they are thinking what they are saying" (Descartes 1985, 140). Heidegger, despite his phenomenological critique of Descartes, also claimed that nonhuman animals are incapable of language and therefore lack access to what he called "world-formation", that is, the ability to form true and conscious relationships with others and with their environment. According to Heidegger, nonhuman animals are unable to apprehend the world *as such*—that is to say, in the world-forming ways that language, understood as *logos*, allows for—because they are captivated by their instincts and bound to their environments. Animals do have access to the world—they are not, like stones and other inert objects, what Heidegger calls "world-less" (*weltlos*)—but their relationship to it is an impoverished one: Heidegger calls nonhuman animals "poor in world" (*weltarm*), whereas humans are deemed to be world-forming (*weltbildend*). Insofar as an animal is essentially absorbed in its environment (*Umwelt*), it cannot truly act in relation to the world (*Welt*) as such, or, as Heidegger puts it (1995, 239), an animal "behaves within an environment but never within a world."

Although Heidegger repeatedly observes that "the relation between poverty in world and world-formation does not entail hierarchical assessment" (1995, 192), he does nonetheless privilege human beings to the extent that, in his view, only language, by which he means *human* language, is capable of disclosing the world as an intelligible and meaningful place. "Language alone," he writes, "brings what is, as something that is, into the Open for the first time. Where there is no language, as in the being of stone, plant, and animal, there is also no openness of what is, and consequently no openness either of what is not" (Heidegger 1971, 73). In fore-grounding human language as central to the practice of world-formation, Heidegger not only seeks to demarcate human from nonhuman animals, but also postulates a distinction between language and communication that echoes Aristotle's assumption of a fundamental difference between speech and sound, between *logos* and *phonè*, as well as Descartes' view on animals as mindless machines. For Heidegger, language is "not only and not primarily an audible and written expression of what is to be communicated" (ibid.), but rather it serves to manifest the world *as such* as a field of significance: an open space of possibilities, as opposed to an animal's instinctual captivation. Language, in this view, is thus what separates human being-in-the-world, or *Dasein*, from the being of other animals, which, as Heidegger writes, "has nothing to do with the selfhood of the human being comporting him- or herself as a person" (Heidegger 1995, 238–239).

Linguistics, no less than philosophy, has long reiterated the humanist understanding of language as a fundamental dividing line between human and nonhuman

---

[2]Most commentators attribute to Descartes the concomitant view that animals, because they cannot think, have no feelings and do not suffer pain, yet some scholars (Harrison 1992; Cottingham 2008, 163–173) have sought to contest this interpretation.

animals, thus reinforcing the predominant view of language as the essence of human personhood. Aristotle's distinction between *logos* and *phonè* is still holding ground in concepts of language as either a mental or social construct in two dominant contemporary linguistic theories, namely generative grammar (Chomsky 2002, 2006) and (variationist) sociolinguistics (Labov 1994, 2001). The generative framework, advanced since the late 1950 s by Noam Chomsky and others, theorizes language as a human mental construct where processes of thinking and knowledge about abstract symbols are generated: a cognitive system or "inner mental tool" (Berwick and Chomsky 2016, 164) that works independently from phonetics or the speaking voice (*phonè*), which is assigned to language-in-use. This view consolidates the older assumption that the primary function of language is for human thought rather than for external communication. Even though language can, of course, be used to communicate with others, most of speech is inner speech, or, as Chomsky (2002, 148) puts it: "almost all the use of language is to oneself." In suggesting that language, defined in the narrow sense of an abstract computational system for thought, does not occur beyond the human brain, Chomsky gives a new inflection to the Cartesian understanding of language as essentially disembodied and non-social, while at the same time reinforcing the anthropocentric idea that language constitutes "a yawning chasm between what we [humans] can do and what other animals cannot" (Berwick and Chomsky 2016, 110).[3]

Language, in generative linguistics, is thus seen as a species-specific ability that sets human beings apart from nonhuman animal others: "When we study human language, we are approaching what some might call the 'human essence', the distinctive qualities of mind that are, so far as we know, unique to man (sic)" (Chomsky 2006, 88). Even though, in a paper coauthored for *Science*, Chomsky has acknowledged that "available data suggest a much stronger continuity between animals and humans with respect to speech than previously believed" (Hauser et al. 2002, 1574), he nonetheless maintains his faith in a uniquely human property of language, located either in the capacity for recursion—the ability to "generate an infinite range of expressions from a finite set of elements" (ibid.)—or in what he calls the "creative aspect" of language use, that is, the "distinctively human ability to express new thoughts and to understand entirely new expressions of thought" (Chomsky 2006, 6).[4] And while Chomsky, as Donna Haraway notes, has been cautious enough to present the idea of linguistic uniqueness as "a testable hypothesis, not an assumption rooted in premises of human exceptionalism" (Haraway 2008, 373 note 44), there is no doubt that the tradition of anthropocentric thought assumed in the generative

---

[3]In his *Fundamental Concepts of Metaphysics*, Heidegger in a similar manner suggests that nonhuman animals are "separated from man by an *abyss*" (1995, 26, emphasis added). Because nonhuman animals lack language, they cannot apprehend other beings conceptually, *as* beings: for Heidegger, only humans are capable of grasping that which is *as such*.

[4]In this view, the linguistic ability to *innovate*—to form new statements that express new thoughts appropriate to but not directly caused by their immediate contexts—is considered a fundamental factor that distinguishes human language, seen as free from control by any detectable stimuli, from nonhuman animal communication, which is assumed to occur only in response to an external environment or to internal drives.

framework has been a serious obstacle to investigating the linguistic, rather than communicative, abilities of nonhuman animals.

In contrast to Chomsky's non-social view of language, sociolinguistic research has built on the pioneering work of William Labov and others to theorize language as both dependent on cognition and interconnected with the workings of society and culture.[5] In the sociolinguistic framework, language is understood as a social construct stemming from the need to contextualize how humans use language in interaction with others, aiming to find out how and why languages vary and change, and how (groups of) speakers employ linguistic resources to shape individual and collective identities, communities and social hierarchies. This view converges with recent approaches to language as an embodied (Bucholtz and Hall 2016), multimodal (Müller et al. 2013) and multisensory (Pennycook and Otsuji 2015) phenomenon that includes not only verbal speech but, among others, bodily gestures and facial expressions, actions, movements, sensorial practices of meaning-making through tasting, touching, seeing and smelling, as well as the mediation of embodiment by material objects, spaces and environments. Encompassing a wide range of research areas, including the social meaning of different language varieties, the role of stylization in language use, the construction of social identity categories like class and gender through language practices, bi- and multilingualism, and social norms and attitudes towards linguistic diversity, sociolinguistics has opened valuable new avenues for researchers interested in the manifold relationships between language, identity and power.

Although much sociolinguistic research remains faithful to the human as the most important user of language—in fact, the very notion that humans may *use* certain linguistic skills and resources is in no small part dependent on liberal humanist conceptions of choice and agency—this framework is nevertheless promising for a linguistics that wishes to be inclusive of human *and* nonhuman actors (Cornips 2019). By approaching language as both embodied and embedded in a variety of interactive social practices, contexts and environments, sociolinguistic studies challenge the anthropocentric understanding of a language as an exclusively verbal, decontextualized object that is completely autonomous, inaccessible from outside the mind, and therefore somehow fixed in character. In this perspective, embodiment is central to the production and interpretation of language as a form of social practice, while bodies, in turn, are themselves part of the semiotic landscape as they are "imbricated in complex arrangements that include nonhuman as well as human participants, whether animals, epidemics, objects, or technologies" (Bucholtz and Hall 2016, 186).

Broadening up the concept of language (grammar) to include multimodal and multisensory practices of meaning-making allows us to foreground nonhuman semiotic capacities, including specific sensorial abilities such as olfactory ones for cows and dogs, as language-specific grammatical means. It thus provides a useful framework to analyze differences between and among human and nonhuman animals

---

[5]Neither generative nor sociolinguistic theory has questioned the legitimacy of each other's discipline, yet attempts to integrate both have not been successful (Cornips and Gregersen 2016).

in terms of grammatical possibilities and expressions instead of simply ascribing deficiencies to the latter (Kulick 2017, 373). For example, if a cow in an indoor dairy farm steps back and withdraws her face through the iron bars when humans approach her, this bodily movement combined with head positioning, gaze direction and the sound of the moving iron bar may be analyzed equally to how human animals phrase *negation* as in the sentence *do NOT approach me* (Cornips and van Koppen 2019). Further below, we will demonstrate how recent scholarship produced at the intersection of sociolinguistic theory and critical posthumanism will allow us to take the study of language beyond the speaking human subject and into the more-than-human material world. But first, let us discuss the linguistic abilities and communicative competence of cows in more detail. What can a non-anthropocentric approach to language contribute to our understanding of the ways in which cows speak to each other and to humans? And how can this understanding, in turn, help us confront, and respond to, the enabling and constraining conditions under which dairy cows, as speaking beings, participate in the formation of a meaningful world?

11.3 Cows as Social and Linguistic Beings

Human thinkers, as we have seen, have produced the idea of language as a uniquely human trait by categorically marginalizing nonhuman animal speakers, denying them recognition as linguistic subjects. But, as Eva Meijer suggests, learning about how other animals use language "can help us understand them better, and build new relations with them; challenging an anthropocentric view of language can help us see animals of other species, and their languages, differently" (Meijer 2016, 74). Recent research into how different animal species communicate, ranging from birds and bees to whales, apes and cephalopods, indeed suggests that there may not be a "sharp divide between human language and nonhuman communicative systems" (Evans 2014, 258; see also Meijer 2019). This does not mean that human and nonhuman forms of communication are the same, but, as Alastair Pennycook (2018, 82) notes, "it *is* an argument against human exceptionalism." In this chapter, we take the view that both human and other animals create meaning through language conceptualized as a social, spatial and artefactual resource. By theorizing language in terms of material-semiotic assemblages and spatial repertoires (Pennycook 2017), we wish to avoid an anthropocentric definition of language that not only *a priori* excludes nonhuman animals, but also neglects all other aspects of language beyond the "distinctive qualities of mind" so often privileged in philosophy, linguistics and cognitive science.

Cows, including domestic dairy cows, have distinct personalities and stable personality characteristics and have a clear capacity to lead rich and socially complex lives. Measured assessments of cows' cognitive, emotional and social abilities provide scientific support for what people familiar with cows already know, namely that cows demonstrate intelligence, experience a range of emotions and display a high level of social complexity, including social learning, in ways that human animals can recognize (Marino and Allen 2017; Colvin et al. 2017). When given the opportunity, cows form strongly bonded social groups, with mother cows and calves sharing an especially powerful emotional connection that, in part, depends upon the possibility for the mother to be able to lick her child for several hours after birth (Marino and Allen 2017, 484). Cows are competent learners and possess both short- and long-term memories: they are capable of discriminating between different objects, colors and geometric shapes, and are able to learn and recognize individual differences among humans, as well as conspecifics, under a variety of circumstances.[6] These abilities show that cows do not merely respond to external stimuli but engage in the formation and categorization of mental concepts (Colvin et al. 2017, 7). Moreover, cows display emotional reactions to their own learning and in response to each other's feelings, which has been suggested to reflect sophisticated levels of psychological capacities such as self-awareness and empathy (Hagen and Broom 2004; Marino and Allen 2017, 482–483).[7]

As social mammals, cows depend on each other for interaction and emotional support; social isolation therefore inflicts great stress on them, as does the immediate, and life-long, postpartum separation of mothers and calves in intensive dairy farming. In commercial settings, where human-cow relations are deeply instrumentalized and commodified, the possibilities for cows to express species-appropriate behavior are severely compromised by periods of confinement in indoor housing, health problems due to higher milk yields and distress caused by various forms of social separation. As caged living beings, with little or no opportunity to escape their exploitation by humans, cows raised for food in factory farms experience "unnatural conditions from birth to slaughter" (Marino and Allen 2017, 474), including procedures that cause severe pain and suffering such as dehorning and disbudding. In these circumstances, where young calves are raised individually and cows are killed before their time, social bonding formation is extremely difficult to establish and maintain (McLennan 2013), which has devastating consequences for their well-being and welfare. It is decidedly problematic, then, that most research into the lives of cows is done within the framework of their use as "livestock" for human consumption. As the scientific literature on cow psychology and behavior is dominated by an applied science perspective mainly relevant to human practices of intensive farming (e.g. training cows to use automatic feeders) (Marino and Allen 2017, 475), there is a felt need to understand, and relate to, cows on their own terms.

---

[6]In one study, cows have been demonstrated, within a few training sessions, an ability to discriminate photographs of different cows' faces from faces of other species. A later study has shown that heifers can differentiate between two-dimensional facial images of familiar and unfamiliar cows, treating these images as mental representations of real individuals (Marino and Allen 2017, 478–479). So far, there is no knowledge yet on social learning from humans or the use of human-given cues in cattle (Nawroth et al. 2019, 5).

[7]Emotional reactions to learning in cows have to do with "the positive emotions and excitement that go with realizing one is controlling a situation" (Marino and Allen 2017, 482). This does not merely show that cows understand the causal relation between accomplishing a task and receiving a reward, but rather it suggests that they learn to experience task solving as intrinsically rewarding by adopting "an emotional perspective on their own agency" (Hagen and Broom 2004, 212).

While it is undeniable that dairy cows are always already caught in the anthropological machine of industrial animal production—an "apparatus" (Despret 2008) that essentially prevents them from experiencing a full quality of life—we do believe that an inquiry into how dairy cows make use of language, conceptualized in a non-anthropocentric manner, can help human animals to get to know cows better and to understand them as "the someones they actually are" (Colvin et al 2017, 3). This will, in turn, allow us to respond to the question of nonhuman animal agency in new ways that not only serve to challenge established structures of species hierarchy, but also entail a fundamental rethinking of how agency is enacted in and through language as a practice of human-nonhuman sociality. In the context of what has been termed the "cage age," it is routinely assumed that the restrictive and monotonous captive environments in which domesticated animals usually live, will "limit the frequency and diversity with which [their] agency is expressed" (Špinka and Wemelsfelder 2011, 34). Yet, as we will demonstrate below, these same restrictive conditions can, paradoxically, also give rise to new modes of linguistic agency and resilience, revealing the copious ways in which dairy cows, as speaking beings, orient themselves towards the world.

Dairy farmers bring their own perspectives on how cows, as social and sentient beings whose freedom of movement is nevertheless severely restricted, give meaning to their physical environment and negotiate their housing conditions. A female dairy farmer based in the south of the Netherlands recently provided this chapter's first author with a hand-written letter with some of her thoughts in preparation of an interview addressing how cows and farmers communicate with each other.[8] She wrote:

> *A true story*: Cows are herd animals and they have a leader who will inform the others what to expect. Mientje was always the first waiting by the fence for the farmer to collect them [from the meadow] to be milked. The cows would first be treated to snacks in the barn which is a feast. They might become so impatient as children, and Mientje always watched carefully how the farmer would unlock the fence. An iron slide bolt. For days she would be licking that bolt and the farmer assumed she liked the taste of it, but actually she was practicing how to accomplish that [unlocking the bolt] by sliding it across with her tongue bit by bit long enough, and yes the farmer stayed away for too long and she opened the fence by herself, moved a bit backwards so that the fence could open further and so she managed to steer all the cows to the barn where there were no snacks present since it was no milking time yet. The barn was an overshitted barn that first had to be cleaned with very restless cows back in the meadow. The blacksmith made a new bolt.

Mientje, in the narrative above, is clearly positioned as an actor, even though her actions arise from within a state of unfreedom that makes it difficult, if not impossible, to draw sharp boundaries between action and passion, between doing and suffering. Seemingly functionless activities such as repetitive licking and/or biting of non-food objects, including bars and fences, are common stereotypic behaviors in captive ungulates and are caused by the frustration of natural behavior patterns

---

[8]The interview anticipated in the letter took place at the second farm (Farm 2) in the south of the Netherlands where the first author conducted her field research which is further discussed below. This interview took place on 15 February 2019 and is not discussed further in the current chapter.

or by repeated attempts to deal with some problem (Bergeron et al. 2006). Tongue rolling, object licking and biting at fences—important indicators of compromised animal welfare—are especially prevalent among intensively housed cattle, as they are routinely deprived of the freedom to pursue natural patterns of grazing and rumination (Moran and Doyle 2015, 47). Nevertheless, it is also clear that Mientje's actions are not at all inconsequential or meaningless. On the contrary, by unlocking the bolt, leading the herd to the barn in the expectation of finding some snacks and by shitting the barn when they find nothing there, Mientje and the other cows not only spur their humans into action (cleaning the barn, producing a new bolt) but also engage in linguistic acts of place-making by transforming their shared living space into a site for negotiating, or renegotiating, the semantics of power, resistance and belonging.

A place, in the sociolinguistic sense, is not simply a fixed geographic location but rather a changeable site of symbolic meaning as well as a material assemblage of objects or things that mediate social processes and relationships (Johnstone 2011; Cornips and de Rooij 2018a; Peck et al. 2019). Place-making, then, involves the assigning, through interaction and other forms of connectivity, of social meanings to physical (and, increasingly, digital) spaces, thereby "creating places that are perceived as the basis of belonging" (Cornips and de Rooij 2018b, 7–8). In contrast to other branches of linguistics where languages are seen as "naturally" anchored to specific spaces—a view that only holds if a language is conceptualized as a monolithic and identifiable object detached from real-time practices—a sociolinguistics of place takes a practice-based approach focused on speakers and their activities. This shifts the focus from the linguistic system or structure to a whole range of situated practices in which speech is produced, so that what is typically labelled as *a* language is reconceived as a linguistic resource that only becomes socially meaningful in combination with other material-semiotic resources distributed across people, places and environments (Pennycook 2017).

Although previous sociolinguistic research has conceptualized place-making primarily or even exclusively in terms of human practices and institutions, we suggest that other animals, like cows, also engage their senses, thoughts and emotions in the material-semiotic production of the world as a meaningful place. This entails a clear break away from the previously discussed humanist conceptualizations of language as a computational system located exclusively within the human mind—a view on language which, as we have seen, is itself informed by a desire to place the human above all other animals—and steers us towards an understanding of language as a *distributed* phenomenon, an emergent property deriving from the interactions and interrelations between human and nonhuman actors, including spatial resources and things usually seen as inanimate (Cowley 2011; Pennycook 2017). This shift in thinking corresponds to the critical posthumanist "turn" that has been put on the linguistic research agenda recently by Alastair Pennycook, who urges us "not just to broaden an understanding of communication but to relocate where social semiosis occurs" (Pennycook 2016, 446). Once we acknowledge that, as Pennycook (2018, 51) notes, "linguistic and other semiotic resources are not contained in someone's head, nor just choices available within a speech community, but are spatially distributed," we can begin to explore how dairy cows, such as Mientje, engage in linguistic place-making in relation to other cows, farmers, fences, iron bars and spaces such as barns and meadows, as well as through embodied acts of looking, smelling, licking, walking, eating, defecating, playing and listening.

11.4 Linguistic Place-Making in an Intensive Dairy Farm

In the previous section, we suggested that cows, as sentient and intelligent beings, engage their cognitive, emotional and social abilities in practices of linguistic place-making. Just as for people, we assume that the formation of meaningful bonds between cows and a place is "a powerful factor in social life… and is often based on the social relationships that are enacted in a place" (Schieffelin 2018, 35). In this section and the next, we will examine in more detail how intensively housed cows engage in place-making through language, understood as a distributed phenomenon emerging from within "material webs of human and nonhuman assemblages" (Pennycook 2017, 279). Drawing on recent fieldwork at an intensive dairy farm, we seek to demonstrate how in this context linguistic place-making occurs through multimodal and multilingual repertoires where human and nonhuman bodies, materials and environments come together in co-shaping motion. We will pay special attention to the questions of material and nonhuman animal agency, not merely because "processes of place-making and place itself are always sensible to power dynamics and asymmetries" (Schieffelin 2018, 34), but also because these questions are crucial for thinking through the challenges of human-nonhuman coexistence in the current context of the Anthropocene.

First, a cow becomes connected to her place as a "territory of knowledge" (Schieffelin 2018, 30, citing Århem 1998) through her verbal practices. While cattle vocalizations are often proposed as indicators of animal welfare, scientific analysis of naturally occurring contact calls produced by crossbred beef cows and their calves have provided insight into the acoustic structure and information encoded in these vocalizations. One study showed that calf calls encode age, but not sex, and are produced (F0 = 142.8 ± 1.80 Hz) when separated from their mothers and preceded suckling (Padilla de la Torre et al. 2015, 58). Also, indoor housed calves produce individually recognizable calls to their mothers and vice versa whereas indoor housed cows signal verbally that they are hungry, sexually aroused, and experience milking delay in distinctive ways (Jahns 2013, 247). Thus, although cow sounds may be meaningless to most human animals, they constitute meaningful signs recognizable by mother cows and their calves, as well as by fellow cows as a sociolinguistic community of practice.

Further, cows establish place-making through visual, auditory, olfactory, gustatory and tactile practices, as well as through creative behavior such as play. Sight is a cow's most dominant sense, with a field of vision of at least 330° and a fine eye for details. Cows pay more attention to moving objects than ones that remain still, such as bars, and they are often "spooked" by sudden movements. A cow's hearing is better than that of horses, but she is less able to locate sounds compared to goats, dogs and humans. She has an acute sense of touch, which enables her to enjoy some forms of tactile contact, such as scratching behind the ears, but it also means that the conditions of industrial farming cause her considerable pain. Olfaction plays an important role in cows' social lives, and there is evidence that they can detect the scent of stress hormones present in the urine of fellow cows (Marino and Allen 2017, 475–476). Cows engage in all forms of play found in mammals, including gamboling and running, playing with objects such as balls and social play with members of other species. While play is an important indication of an animal's pleasure, curiosity and capacity to innovate, and as such it "forms the basis for complex object-related and social abilities" (Marino and Allen 2017, 481), play behavior in captive animals is also dependent on their housing conditions; for example, being released from confinement will increase the frequency of movement-based forms of play such as galloping and bucking. In what follows, we will discuss specific examples from field research to demonstrate how dairy cows can mobilize these structural constraints to imbue their environment with linguistic meanings and thus negotiate their positioning within an anthropological machine that is, by and large, designed to deprive them from the opportunity to speak.

11.4.1   The Fieldwork Site

From May 2018 through 15 February 2019, this chapter's first author conducted fieldwork in three dairy farms in the south and west of the Netherlands and in one small dairy farm in Norway. The observations presented in this chapter are based on data collection at Farm 1 in the south of the Netherlands, where the first author spent several weeks during her holidays in May and which subsequently became the site for three days of ethnographic observation, including two days of gathering audio- and video recordings.[9] The dairy farm counts about 150 adult cows, heifers and calves. The dairy cows are milked by robots, while an automatic feeder takes care of pushing the food towards them, minimizing embodied practices between farmers and cows. A small camp site and some holiday apartments accompany the farm, as so often in the south. Many tourists, children in particular, seek contact with the newly born calves, as well as with the older calves and heifers to be discussed below, petting them and speaking to them. Feeding the newly born calves, who are housed individually in fiberglass cages outside the barn, is an especially popular activity among the human

---

[9]The fieldwork took place on 25 July (observation), 26 July and 17 August 2018 (audio- and video recordings). A written and signed consent form by the owners of Farm 1 was obtained abiding by the guidelines for research as stated in the protocol of the Ethics Assessment Committee Humanities of the Radboud University Nijmegen and adopted by the Royal Academy of Arts and Sciences (KNAW). From the perspective of establishing egalitarian research methods for interspecies collaboration, there is a need to examine how to receive permission from the nonhuman animals under study, while at the same time one should interrogate how the bioethical framework of "informed consent" is set up through human-centred discourses of rational agency and choice.

visitors. During the on-site observations, the cows would often reach through the bars and fences to touch the farmers, tourists and field worker through licking and nuzzling. They would also establish contact through nonverbal interactions such as eye gaze and body positioning, as well as by using language in the form of rumbling, calling, hooting, sniffing and coughing. These practices would happen frequently, even though neither humans nor cows were able to traverse to sharing space with each other directly.

Farmers usually assign dairy cows to fixed places in artificial groupings based on their age without male peers, revealing extreme power asymmetries between cows and humans. In the farm under observation, cows are assigned to eight distinct places differentiated by age: new-born calves, older calves up until a few weeks of age, young heifers, older and oldest heifers, dry cows (pregnant cows), and dairy cows. As noted, new-born female calves are separated from their mothers immediately after birth and isolated in fiberglass enclosures, so-called "igloos," for about three weeks. In Farm 1, these igloos are placed in the open air facing the dairy cows in the open barn. After this period, the somewhat older calves are housed together with their age mates in igloos holding up to four or five animals, positioned sideways to the open barn so that visual contact with the older cows is much more restricted. Growing older, the calves are placed in the so-called *jongveestal* (young cattle barn) in four different age groups (see Fig. 11.1). The dry cows are housed in a separate space and the dairy cows reside in the large open barn that also contains three milking robots. In spring and summer, the dry cows and dairy cows can graze in the meadow during the day and, when it is very hot, during the late evening and night. The assignment to specific physical places in distinct housings prevents the calves, heifers and cows from forming a natural herd that would include a matrilineal social structure with strongly clustered networks and many non-random attachment and avoidance relationships (Marino and Allen 2017, 488). It also prevents the younger ones from engaging in processes of cognitive and social learning, and deprives them from being comforted by older conspecifics, including mother cows.

The fieldwork took place in the *jongveestal*, where audio and video recordings were made while observing the calves and heifers. The *jongveestal* is an oblong building, about twenty by ten meters, with half bowed windows, touching the house where the farmer's family lives. It is the oldest barn on site with a main entrance in the middle of the long front side and a full opening at one of the short sides. The cattle stay indoors: their day includes some combinations of eating, lying and

←———————— 20 meters ————————→			
Section 1	Section 2	Section 3	Section 4
n=12	n=9	n=7	n=8

←———————————————————————————————→

Calves (older than three weeks)   Heifers (about one year)

Fig. 11.1 Jongveestal ("young cattle barn"), housing 36 calves and heifers (Friday 17 August 2018, 2.30–6 p.m.)

standing. The oldest heifers were about one year old during the fieldwork.[10] In May, swallows would fly in and out to take care of their new-born in the many nests they had fabricated under the beams of this old building. The floor of the *jongveestal* consists of cubicle divisions for calves and heifers to lie down and stand up (see Picture 11.3 below), while in-between the cubicles, they can stand or walk on discrete beams where feces and urine pass.

In the *jongveestal*, the calves and heifers ($n = 36$) were spatially positioned in four sections divided by iron bars, as illustrated in Fig. 11.1. The calves and heifers in the *jongveestal* are thus profoundly restricted in their mobility—much more so than the adult cows who are able to graze in the meadow, but less than the new-born calves confined to the small igloos. Consequently, from birth onwards throughout their lives, calves, heifers and dairy cows—either individually or with same-sex and age mates—are confined to human-made physical spaces. How, then, do they manage to assign their own meanings to the restricted environment in which they are placed?

11.4.2 Place-Making Through Practices of Sociality and Multilingualism

The housing conditions of the *jongveestal* not only restrict mobility but also limit the visual, auditory, olfactory, gustatory and tactile practices that calves and heifers may display under less restrictive conditions. This significantly affects their modes of sociality and processes of belonging: calves and heifers cannot touch and/or allogroom each other cross-sectionally; a lack of daylight hinders optimal vision and the walls obstruct a far vision; the sound of tractors may penetrate; calves and heifers are dependent on the farmer for how to lie down as well as for when, what and how to eat (with no attention for individual food preferences); ventilation is often not optimal so that calves and heifers, whose sense of smell is far superior to that of humans, deal with omnipresent scents of ammonia, carbon dioxide, methane and hydrogen sulfide (Vallez 2013, 12); and the beams on the floor, often slippery due to feces and urine, impede playing and running (see Picture 11.3). Limited space allowance furthermore makes it difficult to maintain a preferred distance to neighbors with whom the individual likes to bond or not. The spatial distance that cows establish between each other is affected by their relationship and proximity might indicate the existence of a social bond (McLennan 2013, 26). Under more natural conditions, cows seem to engage with particular individuals with whom they prefer to spend their time, creating voluntary bonds while grazing and lying together in close proximity (McLennan 2013, 49–50). In the captive environment of the *jongveestal*, however,

---

[10]During her holidays in May, the first author frequently visited the *jongveestal* since there was one heifer who was very much looking for contact with human animals. She was positioned near the main entrance and would nearly jump towards the author to put her head on her shoulder. This extravert expression of contact seeking behavior might be interpreted as indicative of a willingness to engage in interspecies collaboration.

a calf or heifer who is a non-preferred partner may stand, lie down or eat in closer proximity than would naturally occur, which may lead to feelings of uneasiness and has been suggested to have negative consequences for animal welfare (McLennan 2013, 52).

How, if at all, do calves and heifers in the *jongveestal* create sociality under these conditions? And how is this sociality mobilized in and through material-semiotic practices of place-making? Fieldwork observations show that the calves and heifers in their cubicle divisions (see Fig. 11.1) may not show any sign of interaction or connectivity, thus is it not self-evident for individuals who are placed in a restricted space to construe it together as a social place. Picture 11.1 shows an example: four older calves stand in the second section of the *jongveestal*. Although they share a restricted physical space, their body positioning does not reveal any form of co-shaping the act of standing together. The calves position their bottoms to each other, taking diverging positions, avoiding eye contact and body contact. Although the calves in the right corner seem to align sideways, there is no form of interaction. Their bodies don't touch and while the calf in the middle bows her head, the calf in the right corner is rubbing her chin at the iron bar and wooden demarcation while establishing eye contact with the fieldworker, as shown in Picture 11.2.

Both pictures also show that in the second section of the barn four cows are standing on their feet whereas two cows are lying down in cubicle divisions. The calf to the right in Picture 11.2 was headbutted by another for about two seconds when trying to move over to the most right-hand section of the barn (not visible in the picture). Picture 11.3, however, shows calves in section 2 mirroring each other's body positions when lying down in the cubicles. Although their bodies do not touch and they are not able to lie down in a circle as less restrictive settings, they are able to choose to lie down all together in the same way at the same time. The two calves in the cubicles in the back are facing each other whereas the two shown in the foreground

Picture 11.1   Standing in the jongveestal (Farm 1, 17 August 2018)

Picture 11.2 Calf is rubbing her chin at the fence while establishing eye contact with the fieldworker (Farm 1, 17 August 2018)

do not. Note that the younger calf in section 1 has decided to look out of the window instead of synchronizing with the others.

As a material-semiotic resource, bodily synchronizing can be seen as part of what Frans de Waal has described as "identification" with the other, a process of "bodily mapping the self onto the other (or the other onto the self)" which not only relates to a capacity for shared neural representation, but also forms "a precondition for imitation and empathy" (De Waal 2012, 123). During the fieldwork, a clear practice of bodily synchronizing—which we consider here as a social form of meaning-making typical for encaged dairy cows—emerged during the communicative event of feeding (by the farmer) and eating (by the cows). The farmer feeds the calves twice a day by putting upside down a wheelbarrow loaded with food on the ground before them. An iron feed fence separates the human and nonhuman animals during this event while at the same time it mediates the meanings that arise from their mutual interaction. The farmer provides the food from one side of the fence whereas the calves on the other side need to position themselves before individual openings and put their heads through the bars in order to reach the food below. Since there are only as many openings as calves, every individual has to touch her neighbor to secure a place (see Pictures 11.4a, b).

As the calves put their heads through the iron bars they simultaneously bow their heads forward and downward to pick up the food from the ground, taking a slightly more upward head position in order to chew. Within this joint "embodiment of movement," a collective form of sociality in which each calf will instantaneously "follow and lead" (Argent 2012, 120),[11] the calves engage one another and the space around them in bodily acts of identification that articulate the *jongveestal* as a shared social place. Thus, although for encaged calves feeding is clearly a habitual and

---

[11] Argent writes about synchronizing between horses and riders.

Picture 11.3 Bodily synchronizing in cubicle divisions in the jongveestal (Farm 1, 17 August 2018)

Picture 11.4 a The calves positioning themselves at the feed fence (wheelbarrow on the right);
b The calves are synchronizing during their eating practice

routinized practice, from a sociolinguistic perspective it also entails creative acts of place-making through what Argent calls "kinesic, haptic, and proxemic communication modes" (Argent 2012, 119). The synchrony of movement that occurs in and through the spacing of interactional distances not only orients calves to group living, but also enables them to imbue their restricted environment with meaning in the form of social bonding and thus constitute themselves as linguistic agents.

Specifically, we argue that nonhuman practices of place-making in an intensive dairy farm can be seen as a form of bi- or multilingualism peculiar to the context of industrial animal production. The iron bars make sound when the synchronizing calves put their heads through them and move their faces up and down in co-motion to reach for the food on the ground during eating practice. These sounds are not meaningless or arbitrary but constitute a semiotic resource for calves to reinforce social bonding, specifically since the sounds of the iron bars and bodies co-shape each other acoustically. Eating practices in the *jongveestal*, then, are acts of place-making where calves do not only use their vocal tract to produce language, but also establish themselves as linguistic beings through the rhythmic clattering of the iron bars that shapes the synchronizing bodies into socially meaningful sounds. In other words, these calves engage in a process of nonhuman place-making not only by producing *one language* with their own bodies, that is, the words or vocalizations for "greeting" and "hunger" which are inextricably combined with multimodal and multisensory ways of meaning-making through body positioning and visual, auditory, olfactory, gustatory and tactile practices, but also by producing a *second language* with the material-semiotic means that both compose and transcend their restricted environment.

Being socialized into the environment of an industrial dairy farm, then, for cows implies being or becoming bilingual, where bilingualism is to be understood as a "complex set of practices" (Heller 2007, 15) which draw on linguistic resources that belong to two codes which are structurally maximally divergent (see Auer, forthcoming, 8), in this case one code produced by vocal tract and one code produced by synchronizing bodies and iron bars. The latter language is more context-dependent than the former because of its restriction to the practice of feeding in conditions of captivity. These two codes thus reveal structural constraints in the linguistic sense: they can combine together in a multimodal way but cannot be mixed. The observations do not show that calves alternate between the two codes within one single discourse—the social practice of eating together—even if both codes are part of the broader material-semiotic assemblage through which place-making is established. In the sociolinguistic framework, the two codes might be said to correspond to different social functions and identities, since different languages, or language varieties, are associated with diverging "processes of construction of social difference and social inequality" (Heller 2007, 15). The "bars and bodies" code will be associated primarily with encaged individuals, suggesting that this form of bilingualism is specific to the complex network of connections among human and nonhuman agents that constitutes daily life at an intensive dairy farm.

Crucially, the material presence of the feed fence, which both enables and constrains the expression of linguistic agency in this context, should be understood not merely as a demarcation of physical space, but as belonging to the spatial repertoire through which language is produced as a "distributed effect of a range of interacting objects, people and places" (Pennycook 2017, 278). As an embodied and embedded practice, linguistic place-making in the *jongveestal* is thus not simply a conditioned response to an unresponsive environment, nor does it arise from the individual communicative competence of calves and heifers; rather, it emerges from within a complex assemblage of material-semiotic resources distributed across human and nonhuman subjects, artefacts and environments, including the means of confinement by which humans seek to restrict the freedom of other animals. In other words, the conditions of captivity in an intensive dairy farm are not external but intrinsic to how cows engage in acts of place-making that we, in Heideggerian terms, might understand as linguistic practices of world-formation. This view, as we will conclude, has significant consequences for how we may conceptualize the expression of nonhuman animal agency, in particular linguistic agency, in the troubling context of the Anthropocene.

11.5   Conclusion

From a traditional humanist perspective, domesticated captive animals are doubly barred from entering into a meaningful relationship with the world: not only are nonhuman animals, in this view, by nature captivated in their environment (as Heidegger and many others have suggested), it is also assumed that confinement in cages does nothing but further limit their natural instincts and capabilities. This view effectively renders nonhuman animals, cows in particular, mute and dumb, while at the same time it reinforces a traditional mechanistic worldview where both nature and matter are considered to be passive and inert, available for manipulation by humans and exploitable for profit (Merchant 1992, 48–55). The critically posthumanist perspective developed in this chapter, by contrast, not only acknowledges cows as the social and intelligent speaking beings that they are, but also approaches their material encagement—the bars and fences meant as barriers to prevent calves and heifers from freely going wherever they want—as a social and spatial artefactual resource for building a meaningful world. Paradoxically, then, it is their state of unfreedom that allows dairy cows to open up the restricted environment of industrial animal farming, exemplified here by the young cattle barn, as a *linguistically* meaningful place.

In drawing attention to the linguistic agency of dairy cows, we do not wish to reiterate the familiar observation that "agency is intrinsic to the way animals behave" (Špinka and Wemelsfelder 2011, 34), nor are we suggesting that the capacity of captive animals to act is somehow "expanded" or "curtailed" through acts of linguistic place-making. In the context of industrial dairy farming, where categorical boundaries between humans and other animals, as well as between organisms and machines, have collapsed—a condition exemplifying the "implosion of nature and culture" (Haraway 2003, 16) that marks the Anthropocene—neither agency nor language can be understood as a property of individual persons or collectivities. Rather, we must account for how different forms of agency, including linguistic agency, emerge from within what political theorist Jane Bennett (2010, 107) has called "agentic assemblages," that is, networks of human and nonhuman actors living together in relations of systemic inequality. In this chapter, therefore, we have tried to show how a non-anthropocentric approach to linguistic place-making, understood as a practice of more-than-human sociality, can help us reckon with the question of nonhuman animal agency in new ways.

Assemblages, as Pennycook (2017, 278) notes, "describe the way things are brought together and function in new ways" and as such they provide a way of thinking about agency as a distributed force, much like we described language and cognition as spatially distributed. Bennett, indeed, suggests that we think of agency as "distributed across an ontologically heterogeneous field, rather than being a capacity localized in a human body or in a collective produced (only) by human efforts" (Bennett 2010, 23). Linguistic agency, then, is not an individual or collective competence that can be "mastered" or "possessed" but should rather be seen as a processually emergent quality arising from multiple assemblages of human and nonhuman elements, including material things, artefacts and spaces. This conception of agency, rooted in what Bennett (2010) calls a "vital materialism" and what Pennycook (2018) describes as a "posthumanist applied linguistics," disturbs the traditional understanding of agency as the capacity for self-willed action, linked especially to human subjectivity and intentionality, as well as the corollary presumption that the more-than-human material world—including other animals and the physical environment—is essentially passive, inert and predetermined in its operations.

Throughout this chapter, we have sought to demonstrate the usefulness of a non-anthropocentric approach to language and language practices in light of a long history of human exceptionalism that has routinely denied nonhuman animals the freedom and ability to speak. We have elaborated a posthumanist conception of language as a distributed effect of multiple interacting bodies in order to foreground the fluidity through which a cow, a calf, calves, a wheelbarrow, a farmer, an iron feed fence, a lock, the clattering of bars, sounds of chewing, sounds of puffing, sounds of urinating, the smell of food, urine, feces, other bodies in proximity or distance, movements up and down, become relationally entangled with one another and, crucially, with the anthropological machine of industrial animal production. Furthermore, we have shown how rethinking nonhuman animal agency in terms of material-semiotic assemblages, as an equally distributed effect of linguistic interactions and social processes, allows us to break away from the idea of lifeless matter, including the Cartesian understanding of nonhuman animals as mindless machines, an idea which has shaped the pervasive modes of human exceptionalism and instrumentalism that have traditionally characterized the humanist agenda and which continue to inform ideas about the "muteness" and "bruteness" of nonhuman creatures today. In this way, we hope to contribute to a greater recognition among humans of other animals, not only as sentient living beings, but as intelligent, social, speaking beings, linguistic agents who even under poor conditions form rich and complex relationships with the world to make it a meaningful place.

Acknowledgements The first author is very grateful to the participating farmer for all support, the opportunity to make recordings, and time.

References

Agamben, G. 2004. *The open: Man and animal*. Stanford: Stanford University Press.

Argent, G. 2012. Toward a privileging of the nonverbal: Communication, corporeal synchrony, and transcendence in humans and horses. In *Experiencing animal minds: An anthology of animal-human encounters*, ed. J.A. Smith and R.W. Mitchell, 111–128. New York: Columbia University Press.

Århem, K. 1998. Powers of place: Landscape, territory and local belonging in Northwest Amazonia. In *Locality and belonging*, ed. N. Lovell, 78–102. London and New York: Routledge.

Auer, P. forthcoming. "Translanguaging" or "doing languages"? Multilingual practices and the notion of "codes". In *Language(s): Multilingualism and it consequences*, ed. J. MacSwann. Bristol: Multilingual Matters. Retrieved from https://www.researchgate.net/publication/332 593230_'Translanguaging'_or_'doing_languages'_Multilingual_practices_and_the_notion_of_ 'codes'.

Bennett, J. 2010. *Vibrant matter: A political ecology of things*. Durham and London: Duke University Press.

Bergeron, R., A.J. Badnell-Waters, S. Lambton, and G. Mason. 2006. Stereotypic oral behaviour in captive ungulates: Foraging, diet and gastrointestinal function. In *Stereotypic animal behaviour: Fundamentals and applications to welfare*, 2nd ed., ed. G. Mason and J. Rushen, 19–57. Wallingford and Cambridge: CABI.

Berwick, R.C., and N. Chomsky. 2016. *Why only us: Language and evolution*. Cambridge and London: The MIT Press.

Bucholtz, M., and K. Hall. 2016. Embodied sociolinguistics. In *Sociolinguistics: Theoretical debates*, ed. N. Coupland, 173–197. Cambridge: Cambridge University Press.

Chomsky, N. 2002. *On nature and language*. Cambridge: Cambridge University Press.

Chomsky, N. 2006. *Language and mind*, 3rd ed. Cambridge: Cambridge University Press.

Colvin, C.M., K. Allen, and L. Marino. 2017. *Thinking cows: A review of cognition, emotion, and the social lives of domestic cows*. https://www.farmsanctuary.org/wp-content/uploads/2017/10/ TSP_COWS_WhitePaper_vF_web-v2.pdf.

Cornips, L. 2019. The final frontier: Non-human animals on the linguistic research agenda. *Linguistics in the Netherlands* 36 (1): 13–19.

Cornips, L., and F. Gregersen. 2016. The impact of Labov's contribution to general linguistic theory. *Journal of Sociolinguistics* 20 (4): 498–524. https://doi.org/10.1111/josl.12197.

Cornips, L., and V. de Rooij (eds.). 2018a. *The sociolinguistics of place and belonging: Perspectives from the margins*. Amsterdam: John Benjamins.

Cornips, L., and V. de Rooij. 2018b. Introduction: Belonging through linguistic place-making in center-periphery constellations. In *The sociolinguistics of place and belonging: Perspectives from the margins*, ed. L. Cornips and V. de Rooij, 1–16. Amsterdam: John Benjamins.

Cornips, L., and M. van Koppen. 2019. Embodied grammar in Dutch dairy cows. Poster presented at the workshop Animal Linguistics: Take the Leap!, L'École normale supérieure, Paris, June 17.

Cottingham, J. 2008. *Cartesian reflections: Essays on Descartes' philosophy*. Oxford: Oxford University Press.

Cowley, S.J. 2011. *Distributed language*. Amsterdam: John Benjamins.

Cull, L. 2015. From *homo performans* to interspecies collaboration: Expanding the concept of performance to include animals. In *Performing animality: Animals in performance practices*, ed. L. Orozco and J. Parker-Starbuck, 19–36. Basingstoke: Palgrave Macmillan.

De Waal, F. 2012. A bottom-up view of empathy. In *The primate mind: Built to connect with other minds*, ed. F. de Waal and P.F. Ferrari, 121–138. Cambridge and London: Harvard University Press.

Derrida, J. 2009. *The beast and the sovereign: Volume 1*, trans. G. Bennington. Chicago and London: University of Chicago Press.

Descartes, R. 1985. Discourse on the method. In *The philosophical writings of Descartes: Volume 1*, trans. J. Cottingham, R. Stoothoff, and D. Murdoch, 111–175. Cambridge: Cambridge University Press.

Despret, V. 2008. The becomings of subjectivity in animal worlds. *Subjectivity* 23: 123–139. https://doi.org/10.1057/sub.2008.15.

Evans, V. 2014. *The language myth: Why language is not an instinct*. Cambridge: Cambridge University Press.

Hagen, K., and D.M. Broom. 2004. Emotional reactions to learning in cattle. *Applied Animal Behaviour Science* 85: 203–213.

Haraway, D. 2003. *The companion species manifesto: Dogs, people, and significant otherness*. Chicago: Prickly Paradigm Press.

Haraway, D. 2008. *When species meet*. Minneapolis and London: University of Minnesota Press.

Harrison, P. 1992. Descartes on animals. *The Philosophical Quarterly* 42 (167): 219–227. https://doi.org/10.2307/2220217.

Hauser, M.D., N. Chomsky, and W.T. Fitch. 2002. The faculty of language: What is it, who has it, and how did it evolve? *Science* 298 (5598): 1569–1579. http://dx.doi.org/10.1126/science.298.5598.1569.

Heidegger, M. 1971. *Poetry, language, thought*, trans. A. Hofstadter. New York: Harper & Row.

Heidegger, M. 1995. *The fundamental concepts of metaphysics: World, finitude, solitude*, trans. W. McNeill and N. Walker. Bloomington and Indianapolis: Indiana University Press.

Heller, M. 2007. Bilingualism as ideology and practice. In *Bilingualism: A social approach*, ed. M. Heller, 1–22. Basingstoke: Palgrave Macmillan.

Jahns, G. 2013. Computational intelligence to recognize animal vocalization and diagnose animal health status. In *Computational intelligence in intelligent data analysis*, ed. C. Moewes and A. Nürnberger, 239–249. Berlin: Springer.

Johnstone, B. 2011. Language and place. In *The Cambridge handbook of sociolinguistics*, ed. R. Mesthrie, 203–217. Cambridge: Cambridge University Press. https://doi.org/10.1017/CBO9780511997068.017.

Kulick, D. 2017. Human-animal communication. *Annual Review of Anthropology* 46: 357–378. https://doi.org/10.1146/annurev-anthro-102116-041723.

Labov, W. 1994. *Principles of linguistic change. Volume 1: Internal factors*. Oxford and Cambridge: Blackwell.

Labov, W. 2001. *Principles of linguistic change. Volume 2: Social factors*. Oxford and Cambridge: Blackwell.

Marino, L., and K. Allen 2017. The psychology of cows. *Animal Behavior and Cognition* 4 (4): 474–498. https://dx.doi.org/10.26451/abc.04.04.06.2017.

McLennan, K.M. 2013. *Social bonds in dairy cattle: The effect of dynamic group systems on welfare and productivity*. Doctoral thesis, The University of Northampton. http://nectar.northampton.ac.uk/6466/.

Meijer, E. 2016. Speaking with animals: Philosophical interspecies investigations. In *Thinking about animals in the age of the Anthropocene*, ed. M. Tønnessen, K. Armstrong Oma, and S. Rattasepp, 73–88. Lanham: Lexington Books.

Meijer, E. 2019. *Animal languages: The secret conversations of the natural world*. London: John Murray.

Merchant, C. 1992. *Radical ecology: The search for a livable world*. London and New York: Routledge.

Moran, J., and R. Doyle. 2015. *Cow talk: Understanding dairy cow behavior to improve their welfare on Asian farms*. Melbourne: CSIRO Publishing. http://www.publish.csiro.au/ebook/7929.

Müller, C., J. Bressem, and S.H. Ladewig. 2013. Towards a grammar of gestures: A form-based view. In *Body—Language—Communication: An international handbook on multimodality in human interaction*, ed. C. Müller, A. Cienki, E. Fricke, S.H. Ladewig, D. McNeill, and S. Tessendorf, 707–733. Berlin and Boston: De Gruyter Mouton.

Nawroth, C., J. Langbein, M. Coulon, V. Gabor, S. Oesterwind, J. Benz-Schwarzburg, and E. von Borell. 2019. Farm animal cognition: Linking behavior, welfare and ethics. *Frontiers in Veterinary Science* 6 (24): 1–16. https://doi.org/10.3389/fvets.2019.00024.

Padilla de la Torre, M., E.F. Briefer, T. Reader, and A.G. McElligott. 2015. Acoustic analysis of cattle (*Bos taurus*) mother-offspring contact calls from a source-filter theory perspective. *Applied Animal Behaviour Science* 163: 58–68. http://dx.doi.org/10.1016/j.applanim.2014.11.017.

Peck, A., C. Stroud, and Q. Williams (eds.). 2019. *Making sense of people and place in linguistic landscapes*. London and New York: Bloomsbury.

Pennycook, A. 2016. Posthumanist applied linguistics. *Applied Linguistics* 39 (4): 445–461. https://doi.org/10.1093/applin/amw016.

Pennycook, A. 2017. Translanguaging and semiotic assemblages. *International Journal of Multilingualism* 14 (3): 269–282. https://doi.org/10.1080/14790718.2017.1315810.

Pennycook, A. 2018. *Posthumanist applied linguistics*. London and New York: Routledge.

Pennycook, A., and E. Otsuji. 2015. Making scents of the landscape. *Linguistic Landscape* 1 (3): 191–212. https://doi.org/10.1075/ll.1.3.01pen.

Schieffelin, B.B. 2018. Language socialization and making sense of place. In *The sociolinguistics of place and belonging: Perspectives from the margins*, ed. L. Cornips and V. de Rooij, 27–54. Amsterdam: John Benjamins.

Špinka, M., and F. Wemelsfelder 2011. Environmental challenge and animal agency. In *Animal welfare*, 2nd ed., ed. M.C. Appleby, J.A. Mench, I.A.S. Olsson, and B.O. Hughes, 27–43. Wallingford and Cambridge: CABI.

Vallez, T. 2013. *Huisvesting van jongvee op een melkveebedrijf*. Master thesis, Ghent University. https://lib.ugent.be/fulltxt/RUG01/002/062/712/RUG01-002062712_2013_0001_AC.pdf.

Wilkie, R.M. 2010. *Livestock/deadstock: Working with farm animals from birth to slaughter*. Philadelphia: Temple University Press.

Leonie Cornips is senior researcher at the NL-Lab, Humanities Cluster (KNAW) in Amsterdam and professor *Languageculture in Limburg* at Maastricht University, the Netherlands. Her research deals with sociolinguistic issues. Current topics are national and regional identity construction through language practices, bidialectal child acquisition, multilingualism and intra- and inter-species interactions, in particular, dairy cow language and communication. She received her Ph.D. from the University of Amsterdam. Her homepage is https://pure.knaw.nl/portal/en/persons/l-cornips/publications/.

Louis van den Hengel is Assistant Professor of Gender and Diversity Studies at Maastricht University, the Netherlands. His research deals with the potentiality of contemporary art performances for activating new socio-political imaginaries and for envisioning an affirmative ethics of difference for the twenty-first century. In other words, he is interested not only in how performance negotiates new relationships between contemporary art and the cultural politics of gender, sexuality, race, class, and species difference, but also in how it may help actualize new modalities of being in, and becoming with, the world. His most recent publications include contributions to journals such as *Criticism: An Interdisciplinary Quarterly* and to edited volumes such as *Doing Gender in Media, Art and Culture* (Routledge, 2018) and *Entanglements and Weavings: Diffractive Approaches to Gender and Love* (Brill, 2021).

Chapter 12
The Vanishing Ethics of Husbandry

Paul B. Thompson

Abstract The ethics of food production should include philosophical discussion of the condition or welfare of livestock, including for animals being raised in high volume, concentrated production systems (e.g. factory farms). Philosophers should aid producers and scientists in specifying conditions for improved welfare in these systems. An adequately non-ideal approach to this problem should recognize both the economic rationale for these systems as well as the way that they constrain opportunities for improving animal welfare. Recent philosophical work on animal ethics has been dominated by authors who not only neglect this imperative, but also defeat it by drawing on oversimplified and rhetorically overstated descriptions of the conditions in which factory farmed animals actually live. This feature of philosophical animal ethics reflects a form of structural narcissism in which adopting a morally correct attitude defeats actions that could actually improve the welfare of livestock in factory farms to a considerable degree.

12.1 Introduction

Bernard Rollin has argued that when university programs in animal husbandry began to relabel themselves as programs in animal science, there was an accompanying shift in ethics. The changeover occurred during the 1970s, as the agricultural sciences generally began to adopt a more positivist ethos (Johnson 1976). Rollin's claim is that while animal husbandry had both implied and encouraged an ethic of caring for livestock and consideration of their interests, the turn to science discouraged empathy and substituted a headlong pursuit of efficiency in its place (Rollin 2004). Husbandry had "vanished" from the curriculum of students training for animal agriculture, as well as in the organization of veterinary research. Rollin's thesis has been developed as a vehicle for both exploring and reforming practices in industrial animal production

P. B. Thompson (✉)
Michigan State University, East Lansing, MI, USA
e-mail: thomp649@msu.edu

(Harfeld 2011). In this paper, I will explore a very different sense of vanishing. My focus will be on the way that husbandry ethics are missing from the discourse of philosophers working on animal issues.

For both Rollin and myself, husbandry ethics consists of norms and standards for the care of animals in livestock production settings. Such standards give rise to philosophical puzzles, conundrums and even paradox. The conceptual work needed to develop and implement husbandry ethics is especially important in the Anthropocene because climate change promises to exacerbate already-existing deficits of animal welfare in industrial production systems. The highly influential report *Livestock's Long Shadow* from the Food and Agriculture Organization (FAO) of the United Nations has been widely cited by philosophers and animal advocates for documenting how methane emissions from animal production contribute to the greenhouse effect. In this philosophical literature, the FAO report is often cited as a supporting argument for condemnation of industrial animal production and in support of ethical vegetarianism (Ilea 2009; De Bakker and Dagevos 2012). Yet the report itself argues for *more* use of intensive animal production systems owing to their greater efficiency of emissions per unit of consumable animal protein when compared to traditional pasture-based production (Steinfeld et al. 2006).

Both Rollin and I (as well as a handful of other philosophers such as Peter Sandøe and several contributors to this volume), have undertaken philosophical analyses of the challenges that attend a functional and practical ethics of husbandry. It is not as if there is absolutely *no* philosophical research on these questions. However, I contend that this work remains marginalized in mainstream academic philosophy. This chapter extends an argument made in my 2015 book, *From Field to Fork: Food Ethics for Everyone.* I compared three ways in which the ethics of livestock production might be structured differently. First, one can ask whether vegetarianism is ethically mandatory. This is an old question with a distinguished philosophical pedigree dating back to Ancient Greece. Second, one can ask whether industrial animal production is ethically acceptable. This question typically presumes a negative answer to the first, but acknowledges the potential for housing and treatment of livestock species that fails to respect animal interests in a morally significant way. A negative answer to this question might, then, lead to a third: How should industrial animal production be reformed to improve animal welfare? Almost all philosophers who have taken the trouble to ask have concluded that industrial animal production is *not* ethically acceptable, but very few have been interested in the third question. Their philosophical curiosity is satisfied by finding some alternative, morally acceptable source of meat, milk or eggs (Thompson 2015, 134–137).

This lacuna in the philosophical literature is how I will understand the vanishing ethic of husbandry. Why is it that philosophers who are interested in animal ethics are so incurious about what counts as improving the lives of livestock? Any defensible answer to this broad question would require consideration of many themes, some of which will have little philosophical relevance. A narrower thesis is explicitly normative: The approach that philosophers are taking to livestock exhibits a form of narcissism that deserves critique. This narcissism is not limited to philosophers, but reflects a broader cultural movement evident in other dietetic disciplines. While I will touch briefly on this trend, my focus in this paper will be limited to the philosophical community, on the one hand, and to animals and animal products, on the other. In the concluding section, I will link it to emerging applications of extreme biotechnology that are attempting to decouple consciousness from meat production, entirely. This component of my argument will connect with my previous work on "the opposite of human enhancement" (Thompson 2008). I begin, however, with a frank (and probably unpopular) statement on the state of animals currently housed in industrial production systems.

12.2 Industrial Animal Production

Production of meat, milk and eggs—the primary food products derived from live-stock—went through a dramatic transformation over the course of the twentieth century. Circa 1900, livestock farmers throughout the industrialized world raised their animals in comparative small groups on pasture, with occasional confined housing during inclement weather. By the year 2000, concentrated animal feeding operations (CAFOs) or "factory farms" had displaced a large percentage of this extensive production. CAFO amass much larger herds and flocks, often in large industrial barns, to facilitate mechanical delivery of feed and water, as well as auto-mated milking, collection of eggs and herd or flock management, including manure disposal. A comprehensive overview of CAFO systems for each agricultural species would exceed the remit of the present essay, but a number of reasonable summaries are available (see Rollin 1995; Norwood and Lusk 2011; Mench 2018).

Although CAFOs pose both animal welfare and environmental challenges, they are not going away soon. Global demand for animal products is growing. Total global meat production increased from 71.36 million tons in 1961 to in 317.85 million tons in 2014 (Ritchie and Roser 2017). The FAO projects that "Between 1997/99 and 2030, annual meat consumption in developing countries is projected to increase from 25.5 to 37 kg per person, compared with an increase from 88 to 100 kg in industrial countries (FAO 2003). Comparable percentage growth in consumption of milk products and eggs is also predicted. Whatever the moral case for reducing or eliminating the consumption of animal products from one's diet, the economic drivers for increasing production remain strong. Absent an almost unimaginable upsurge of political support for regulations that would constrain demand or regulate production, it would appear that livestock will continue to be produced for human consumption of animal products for the foreseeable future.

What is more, for reasons already foreshadowed, more and more of this production will occur in CAFOs. Although the capital costs for an intensive animal feeding facility are high, they are distributed over a large number of salable units. When the cost of production per unit of product is viewed over the usable life-span of these facilities, they are economically competitive. When combined with the feed, labor and management efficiencies of scale, as well as market advantages derived from being able to reliably supply a high volume of product, CAFOs are economically attractive investments for producers who are focused solely on the monetary bottom line (Mench et al. 2008; Norwood and Lusk 2011). The original FAO report on climate impacts of livestock production argues CAFOs also limit the environmental impact of producing meat, milk and eggs when environmental costs are computed on a per unit basis (Steinfeld et al. 2006). As we move deeper into the Anthropocene, the case for using these industrial systems grows stronger, not weaker. Intensive animal feeding facilities introduce the potential for greater efficiencies in landscape impact from animal production (Capper 2012). Adjustment of feed rations in CAFOs facilitates additional means for limiting climate forcing emissions (Hristov et al. 2013). Cost efficiency coincides with environmental efficiency, yet the economic and environmental rationales for CAFOs appear to be on a collision course with animal welfare.

Authors from Ruth Harrison (1964) to Peter Singer (Singer and Mason 2007) have pilloried factory farming for neglecting animal interests. It is important to temper these criticisms by recognizing that for some producers, at least, improving animal husbandry was a motivation for moving toward more industrialized production methods, in the first place. Jim and Pamela Braun are Iowa hog farmers. They describe how up until 1969, pigs on their family farm had been raised in an extensive (e.g. open field) system. In an attempt to limit MMA (*mastitis metritus agalactia*) infections that were becoming difficult to control in their pasture-based farming system, Jim Braun's father shifted to a totally confined, indoor system. They report, "Each stall was its own self-contained sow hotel, with an automatic feeder, waterer and manure removal system. We farrowed year round, and the sows could not run from their shots, thereby helping to ensure the health and safety of the piglets" (Braun and Braun 1998, 40–41). The Brauns are not arguing for the welfare of their pigs in this article; they simply take that for granted. They go on to criticize vertical integration in the pork industry and the subsequent loss of control by family farmers that would allow them to make changes based on animal welfare.

Of course, it is possible that the Braun's decision was a mistake, especially when welfare impacts beyond MMA are included in the evaluation. My point is not to defend any particular model of industrial production, but only to show that some producers saw confinement systems as beneficial to their animals. As noted already, there is little doubt that CAFOs led to a dramatic change in the economics of livestock farming, just as the Braun's claim. While it might have been reasonable for an old-school animal producer to assert that their personal economic interests were (at least roughly) consonant with the health and well-being of their animals, that was largely because the animals themselves represented a large share of the farmer's total capital investment. The large barns, automated feeders, watering systems and mechanisms for manure disposal or retrieval of milk and eggs changed that. In many cases, maximizing return on investment in equipment required accepting the reduced yield in per-animal production of the salable commodity that accompanied rising rates of herd or flock morbidity and mortality (Norwood and Lusk 2011; Bennett and Thompson 2018). Indeed, recognition of the welfare deficits associated with CAFOs motivates the ethics of husbandry: How should we reform these systems?

12.3 Reforming Husbandry in Industrial Animal Production

The economic structure of animal production provides a clue to one of the most important philosophical features of husbandry ethics: In industrialized economies, husbandry ethics must be addressed collectively. An individual producer acting alone cannot adopt many of the changes that husbandry ethics recommends. The farming approach that maximizes capital returns will be the most competitive in a market economy. To the extent that animal products are pure commodities, with one example fully substitutable for another, price will be the dominant factor in consumer decision making, and the system that is most efficient in its utilization of capital will be the system that can offer products at the lowest price. Producers who fall too far short of this efficiency standard will not be able to recover the cost of their investments, and will eventually fail. Farmers must recover moneys expended on buildings and equipment just as much as they must recover the costs of feed and labor. As buildings and equipment become an ever-larger share of the livestock producer's expenditures, there is a downward spiral in which *only* producers who are willing to exploit animals remain in the industry.

There are several possible responses to this situation, each with respective strengths and weaknesses. First, collective action can take place at the level of the state by regulating production systems based on animal welfare. This approach has been taken throughout Europe. While it is philosophically satisfying, it suffers from three main problems. First, regulation does not necessarily entail compliance. Implementation of European rules has been slow and there is evidence that compliance is highly variable (see Thompson 2015, 154–156). Second, regulations tend to be quite inflexible, meaning that they can actually retard change in a production practice when new science and technology becomes available. When the replacement of poor systems requires large capital investments, farmers are deterred from taking action to improve welfare. The sheer cost of a new barn is itself a form of deterrence, but a producer must also be confident that these large investments will continue to comply with regulations throughout the useful life of the facility. The result is a vicious circle. Producers do not improve for fear that they will not comply with the rules, while regulators do not revise rules for fear that producers will be financially unable to comply. Finally, the existence of regulations may encourage moral complacency. Producers and consumers alike presume that once regulations are in place they no longer ask the compelling ethical questions implied by a husbandry ethic. There is thus some risk that a too strict regulatory environment can actually undercut the motivation for continued work on husbandry ethics.

Along with Canada, Australia and New Zealand, the United States relies almost exclusively on collective action taken by producers themselves. This has taken the form of husbandry guidelines and industry standards that voluntarily bar certain problematic practices. For example, the United Egg Producers, the principal trade organization for shell eggs in the U.S., has promoted a standard that ends forced molting and institutes minimum space requirements for hens in the facilities operated by its members. Tail docking in pig and milk production has also been significantly curtailed, and there have been improvements in welfare prior to slaughter in processing facilities. Although these are voluntary standards, they have significantly improved the lives of many thousands of animals. Producer action does achieve substantial compliance and it has the advantage of flexibility. It is much easier to implement incremental changes when producers are directly involved. However, voluntarily developed standards are often quite low and some other American commodity groups have failed to take any kind of meaningful action at all (Mench et al. 2011). As such there is a continued need for documentation of remediable deficits in animal welfare—a key activity of husbandry ethics.

Finally, there is the potential for decommodification of animal products by enabling and encouraging consumers to choose meat, milk and egg based foods certified to meet higher standards of welfare. The popularity of this approach has grown in Europe and America alike, but there are two weaknesses. One is that in depending on consumer willingness to pay, the best that animal welfare certification can achieve is improvement for a subset of farmed animals. Ethical meat, milk and eggs appeal to niche markets. Commodity production standards will still apply in many production systems (Kehlbacher et al. 2012). Second, animal welfare labels are, in effect, marketing devices. They are subject to all of the distortions and obfuscations that we typically associate with advertising. This means that, on the one hand, consumers are skeptical that welfare claims are true, while on the other they can be misled by anthropomorphic images of animals used to promote these products. Inconsistency in the various schemes currently used to measure animal welfare may also undercut consumer confidence (Main et al. 2014).

Ethical inquiry into what actually improves the lives of farmed animals operates in the logical space circumscribed by these options, which are not necessarily exclusive of one another. Husbandry ethics must be open to the possibility that reform of CAFOs might call for doing away with them altogether. The arguments cited above notwithstanding, this possibility is reinforced when the environmental impact of CAFOs are taken into consideration (Ilea 2009; Fairlie 2010). Yet an honest concern for animal welfare should take note of the fact that millions of animals currently live in the CAFO environment, and that CAFO-like systems are rapidly displacing extensive animal production in Asia, Latin America and Africa (Thornton 2010). To the extent that improving the quality of life for these animals is a moral priority, there are compelling reasons to undertake husbandry ethics, even if these CAFOs cannot be ethically justified. This means that animal yhusbandr is a form of non-ideal ethics: Inquiry into the welfare of animals aims to make morally compelling improvements in quality of life. It does not presume that improvements in welfare justify the continuation of these systems, on either animal welfare or environmental grounds. This feature of the husbandry ethic holds for CAFOs and for more traditional, extensive systems alike. Many arguments for veganism, for example, hold that *no* form of animal agriculture is morally acceptable, but this does not logically vitiate the question of how the lives of animals living in these systems could be made better.

12.4   Philosophers and Animal Husbandry

Given the preceding discussion, one might think that there would be a robust philosophical discourse on how animal welfare could be improved in industrial systems.
This discourse might probe when or under what circumstances practices that compromise animal welfare are truly unavoidable or unnecessary. It might investigate trade-
offs between animal welfare deficits and benefits to humans, especially those on
limited budgets. Most fundamentally, it would take up the deep philosophical questions that arise in drawing up diverse and sometimes logically contradictory indicators
of welfare to make a justifiable evaluation of the comparative merits of alternative
systems for improving welfare (Fraser 1999). Ethologists and veterinary researchers
are accumulating a large body of empirical research on the condition of animals
raised in CAFOs, and one might think that philosophers would take some interest in
the value dimensions of this work.

Of course, some philosophers have done precisely that (Rollin 1995; Appleby
et al. 2014). Yet in what follows I will take a more polemical turn, focusing on what
I take to be the dominant strands of thinking by philosophers writing on the animals
amongst us. There is, I submit, an archetypical mode of address toward industrial
animal agriculture among mainstream philosophers. It consists of a few sentences
(or a paragraph at most) reciting the horrific conditions in CAFOs, followed by a
blanket statement of moral condemnation. This generally appears quite early in the
analysis, from which the author moves on to consider their favored philosophical
topic. For example, Alistair Norcross begins his widely read paper "Puppies, Pigs
and People: Eating Meat and Marginal Cases," by sketching the thought experiment
of Fred, who tortures puppies in order to attain sensory pleasure in his consumption
of chocolate. Norcross motivates the significance of this thought experiment with
the following:

> No decent person would even contemplate torturing puppies merely to enhance a gustatory
> experience. However, billions of animals endure intense suffering every year for precisely
> this end. Most of the chicken, veal, beef, and pork consumed in the US comes from intensive
> confinement facilities, in which the animals live cramped, stress-filled lives and endure
> unanaesthetized mutilations. (Norcross 2004, 230–231)

Norcross provides no peer-reviewed literature in support of these empirical claims,
though he does reference several rabble-rousing critiques of industrial agriculture.
The rest of his article takes up a variety of well-known philosophical questions,
including the extent to which "marginal cases" (e.g. humans suffering cognitive
deficits of various kinds) challenge our intuitions. He makes no further references
to practices in industrial agriculture beyond noting the 8 billion chickens slaughtered in 1998, calling them "the most cruelly treated of all animals raised for human
consumption, with the possible exception of veal calves" (Norcross 2004, 232). He
ends thusly: "I conclude that our intuitions that Fred's behavior is morally impermissible are accurate. Furthermore, given that the behavior of those who knowingly
support factory farming is morally indistinguishable, it follows that their behavior is
also morally impermissible" (Norcross 2004, 244).

Norcross thus draws a moral conclusion that references animal agriculture while providing no discussion of any specific husbandry practice (whether in industrial *or* traditional extensive systems). His normative thesis is an attempt to shame his readers, who he has characterized as mindlessly supporting factory farming. He is, I submit, mobilizing intuitions widely shared about "factory farming". One could undertake a critique of the broad claims that Norcross makes about the condition of animals in CAFOs,[1] but it is more important to stress that none of the ethical problems that actually arise in CAFOs really concern Norcross. His argument does not depend on whether opportunities available for ameliorating factory farming's deleterious effects on an animal's quality of life are required by regulation, adopted through producer cooperation or supported by consumers hoping to support more humane production systems. He is deploying a pre-existing intuition about the "torture" animals endure in these systems to stimulate interest in philosophical problems that have no bearing on an animal's quality of life, at all.

It is easy to find instances of this archetype in the work of contemporary philosophers. Rosalind Hursthouse notes how Peter Singer's *Animal Liberation* created awareness of "how horrible the factory farm conditions were" (Hursthouse 2011, 117) and then implies that anyone who is informed about animal ethics "knows, as I do, that in regularly eating commercially farmed meat we are being party to a huge amount of animal suffering" (Hursthouse 2011, 129). Yet Hursthouse's interest lies in exploring how virtue theory compares with Singer's utilitarianism in offering a philosophical analysis of animal ethics, and there is nothing in her discussion that takes up ways in which the suffering she notes could be reduced. Jeff McMahan motivates a consequentialist analysis of animal death with a single sentence on industrial agriculture: "An increasingly common view among morally reflective people is that, whereas factory farming is objectionable because of the suffering it inflicts on animals, it is permissible to eat animals if they are reared humanely and killed with little or no pain or terror" (McMahon 2016). It would rapidly become tedious to recite instance after instance in which this archetype occurs in the philosophical literature on animals. The point is not to suggest that these authors should be taking up husbandry questions in lieu of the philosophical issues that they do investigate. Rather, it is note how the irredeemable nature of factory farming is so thoroughly engrained in the philosophical literature that it can be cited in a passing comment and without support from any factual discussion the actual conditions that animals in these systems endure.

---

[1]For example, unanesthetized surgical procedures (e.g. mutilations) are common in traditional animal production. This is not a practice that distinguishes production in CAFOs from all forms of livestock farming (or, indeed from things done to pets or mutilations that human parents practice on their children). Stress is also common in traditional systems, though as the empirical literature shows, stress is not always detrimental to welfare (Moberg 2000). CAFO production does involve crowding, but the ethology literature suggests that this is much less problematic for chickens (who have a flocking instinct) than it might be for humans. The lack of opportunities for perching and nesting, and the impact of a large flock size on feather pecking are almost certainly much more serious issues from the chicken's perspective (Lay et al. 2011). In all these respects, the claims that Norcross makes to elicit the intuition of cruelty in factory farming are misleading or ill informed.

To cite just one more pieces of evidence, *The Oxford Handbook of Animal Ethics* includes some 35 essays on various topics, including 11 on practical questions. Although the *Handbook* is nearly 1000 pages long and putatively covers the full range of topics in animal ethics, no chapter takes up husbandry ethics. Three of the "practical question" chapters do mention livestock farming. Elizabeth Harmon uses my archetypal strategy, claiming blandly that "factory farming involves subjecting animals to intense suffering" (Harmon 2011, 727). She then moves on to her chief concern: arguing that killing animals is itself morally wrong. David DeGrazia includes a short discussion of harmful impacts on livestock in his article of animal confinement before concluding, "*I contend that wherever the term "factory farming" is properly applied the conditions of confinement are so intensive that they render the animals' lives not worth living*" (DeGrazia 2011, 757; italics in the original). The only extended discussion of animal production in the *Oxford Handbook* occurs in a chapter entitled "Vegetarianism." Here the extensive peer reviewed literature on the welfare of animals in contemporary livestock systems is ignored in favor of treatments intended to shock readers into support for vegan diets (Rachels 2011). The *Handbook* editors have not thought to include any treatment of the philosophical issues that arise in the practice of animal husbandry.

To sum up, a small cadre of philosophers *do* work alongside veterinarians, cognitive ethologists and animal producer groups to fashion better husbandry methods for confined *und* unconfined livestock production. Nevertheless, this topic is simply not on the radar screen of mainstream philosophers writing on animal issues. Most philosophers writing on animal issues hold university appointments and as such might be expected to rely on (or at least be informed about) studies by their peers in science. However, when philosophers do make empirical claims about industrial agriculture, they do not consult the extensive literature in peer-reviewed journals such as *Animal Welfare*, the *Journal of Applied Animal Welfare Science* or in animal science and veterinary outlets such as *Poultry Science* or *The Journal of Animal Science*. Instead, they rely exclusively on reports from journalists or animal activists.

12.5 Animal Husbandry and Animal Activism

The fact that philosophers ignore the peer-reviewed literature on animal welfare should not be taken to imply that what they say about welfare in CAFOs is false. The most frequently cited source in the "Vegetarianism," article just discussed is Peter Singer and Jim Mason's *The Ethics of What We Eat*. As far as I can tell, virtually everything that Singer and Mason say in this book is either true or was true at one time. Other philosophers have built their impressions of industrial agriculture by reading materials published by animal protection advocacy groups such as the Humane Society of the United States (see McPherson 2016). Much of what they say is also true because activist groups *do* generally rely on the peer-reviewed literature from animal welfare science in making their claims. Does this imply that philosophers who pass over detailed discussions of the conditions that obtain in CAFOs are justified in doing so? I argue that it does not. There is a gap between the peer-reviewed literature crucial to husbandry ethics and the literature summaries that are prepared by animal protectionists. Philosophers should be more mindful of this gap than they typically are.

The empirical side of husbandry ethics involves documenting the state of welfare in animal production. There is now an extensive peer-reviewed literature on the welfare of livestock, and as noted, animal protection groups are avid consumers of this literature. However, there are important sources of implicit bias that emerge when the scientific findings are summarized in critiques of industrial production systems. First and most obviously, peer reviewed literature usually weighs both benefits and costs to welfare, while the summaries mention only costs. More subtly, welfare claims in scientific studies are qualified because the data are far from complete. In the scientific literature, classic Humean skepticism combines with lack of statistical power, leading scientists to inject a note of skepticism into all of their claims. There is also the fact that data from actual production environments is extremely scarce, and that producers may lack both the skill and the motivation to remedy this situation over the near term. Activist tracts report claims about the state of welfare in production systems as simple assertions. Accounts of harm to welfare are stripped of any qualification by noting offsetting benefits or acknowledgement of uncertainties. The critics are accurately reporting what the literature states, but unqualified declarative sentences imply both more certainty and more sweeping generality than one would find in the scientific literature.

Second, there are systemic availability biases in the scientific literature. We know quite a bit more about the welfare of animals in CAFOs than we do about the welfare of animals raised in traditional systems, simply because those systems are harder to study. This is because there are hundreds or thousands of small producers, requiring study methods that standardize and aggregate data. In contrast, one can collect data on thousands, tens and hundreds of thousands of animals from a single industrial farm. It is even more difficult when those small producers are in far-flung rural locals in Africa, Asia or Latin America, where travel distances, language barriers and field conditions make data collection expensive and difficult. Activists report findings on CAFOs because that is what gets studied, but it is at least logically possible that welfare on traditional farms and ranches is as bad or worse. The implication that CAFOs are the main problem is an artifact of reporting what we know and remaining silent about what we don't know. The inference that animals in CAFOs endure significantly more suffering than animals raised by small farmers may be plausible, but it is not supported by data.

What is more, activists are trying to motivate change, and producers are trying to forestall it. Both have a tendency to cherry pick such data as *is* available. Ethological studies have made enormous progress since Thomas Nagel speculated that we can't really know what it is like to be a bat in the 1970s (Nagel 1974). Experimental studies have done much to reveal what conditions, needs or amenities appear to be of most importance to animals of different species (Mench 1998). Animal welfare science is replete with studies quantifying maladies from bone breakage to mortality, and the activists love to cite it. But behavioral and physiological studies have also given us strong reason to think that chickens are much more concerned with perches than with crowding, and to suspect (here we're less sure) that pigs are less stressed by confinement than by worries about whether the boss pig will get their food (discussed in Thompson 2015). Where husbandry ethics sees a puzzle in such findings, the activist looks for findings that will help the campaign. What gets reported are claims about injury, pain and stress, but difficult questions about how to limit injury, relieve pain or limit stress are omitted.

Finally, welfare scientists often (I would say generally) care about animals and hope that their work will have uptake among animal producers. When they are able to identify a cost-effective method for improving the condition of animals, they want producers to use it. Especially when producer organizations are in the lead on encouraging change (as in the U.S.) there is thus an incentive to adopt a posture of working with them to encourage change. Even when welfare scientists work with state agencies, they will be dealing with ministries and departments organized for the governance of agricultural production. Officials in these agencies will have little interest in campaigns designed to put livestock farmers out of business. With a few important exceptions, activist organizations have taken a political stance of opposing the interests of animal producers, especially in the United States. Organizations (such as People for the Ethical Treatment of Animals) advocating vegetarianism of any kind are acting in direct opposition to producers' economic interests. At the same time, animal welfare scientists are aware that they need pressure from activist groups to incentivize change by producers. The tension that emerges as advocates of husbandry ethics (as I'm now claiming that many animal welfare scientists are) try to occupy a middle ground between producers and advocates can be seen in the rhetoric. At one juncture standards for animal welfare are weakened in hopes that producers can be enticed to adopt them, while at another juncture they are overstated in hopes of increasing pressure for political or market reform.

There is a philosophical complement to these observations. It is that cognizance of the implicit biases and tensions of working to improve animal welfare should become part of animal ethics itself. An ethic that sides with activists is just as problematic as one that sides entirely with producers, who have their own reasons for resisting change. Mainstream philosophers who adopt the archetypical approach seem to be unaware of this problem. A philosopher should aim to expose and articulate what makes husbandry reform intellectually and politically challenging. Once this work is done, it may be possible to articulate the case for or against a particular practice, or to encourage the average consumer to pay more attention to one set of welfare product claims, over and against another. The mainstream philosophical community seems to lack any appreciation of this context. They may be selling the animals who live in these systems short, as a result.

12.6   The Eclipse of Husbandry and the Rise of Narcissism

Following the analysis from my 2015 book, I interpret the literature in animal ethics as preoccupied with one of two dietary questions. Either one should be a vegetarian (probably a vegan) or failing that, one should avoid eating meat, milk or eggs from industrial farms. Nothing that I have said in this paper can be construed as a rebuttal to either of these dietary claims. I have simply not engaged them. Instead, I have argued that preoccupation with these dietary questions has prevented mainstream philosophers from engaging questions that could lead to significant improvement in millions of animal lives. Unlike David DeGrazia, I believe that the lives of animals in CAFOs *are* worth living, but like the majority of animal welfare scientists contributing empirical findings for husbandry ethics, I believe that their lives could be significantly better than they currently are.[2]

In light of this, I hope it will not be considered too impolite for me to suggest that there is a thread of narcissism in mainstream animal ethics. Narcissism is, of course, a philosophical and psychological phenomenon with a complex history, influenced significantly by the thought of Friedrich Nietzsche. The term has been used to critique forms of absorption with the self that frustrate both social attachment and political engagement, but this should not imply that all forms of self-reflection have these outcomes (Gendlen 1987). Richard Rorty wrote that narcissism is simply a pejorative way of observing that our situated humanity pervades all of our observations, going on to claim that he was proud to be a narcissist (Rorty 1979). Given my professed affiliation with pragmatism, Rorty's statement might be taken as indicative. Yet Jeffery Stout insists that even given this position, Rorty can (and should) maintain a commitment to the potential for objective truth. Our situatedness is neither an excuse for adopting indefeasible views nor does it make all viewpoints equally narcissistic (Stout 2007). My sentiments are with Stout. Rather than taking a deeper dive into the philosophical literature on narcissism, my approach will be to offer a series of characterizations that emerge directly from the subject at hand.

A strong claim of narcissism might go like this: At bottom, mainstream animal ethics is less about the animals than it is clean hands. There are indications of this strong narcissism in the philosophical literature. Norcross writes that the ethical issue is one of not "supporting" the torture of animals, not about undertaking actions that would make them better off. This might be taken to mean that a person's attitude

---

[2]Indeed, many of the ills DeGrazia notes in the lead-up to his sweeping conclusion have been targeted by husbandry ethics. When bone-strength characteristics are included in the index of traits used by poultry breeders, problems decrease, but leg problems increase with selection based on fast growth (González-Cerón et al. 2015). It is also possible to reduce injuries through feed additives and behavioral management. It is an open question whether reforms could ever make factory farming ethically defensible on quality of life grounds. Nevertheless, there are measurable improvements that can be and have been made. DeGrazia does not do enough to show that we should simply dismiss opportunities to improve the welfare of these animals on the ground that their lives are not worth living, (see Thompson 2020). The argument might be persuasive to someone who is wondering about their own "support" of livestock farming, but it is irrelevant to the question of whether and how systems should be changed.

to torture is more important from the moral perspective than the effects of torture. Yet strong narcissism overstates the case. Dietetic ethics intersects with economic markets, and it quite reasonable for a vegetarian to think that their refusal of animal foods lowers the demand curve for these products, reducing a farmer's incentive to produce them (see Norwood and Lusk 2011). This is a more reasonable interpretation of what Norcross means by support. As previously stated, I have no desire to argue against the dietetic approach to animal ethics. My claim is only that it is seriously incomplete, and that animals themselves are the losers.

Efforts to promote human gustatory satisfaction by displacing the experience of an animal entirely exhibit a more nuanced sense of narcissism. This was the problem considered in my "blind chicken" scenario (Thompson 2008). It is exceedingly unlikely that blind chickens actually have better welfare than sighted ones. My point was to bring our discomfort with extreme genetic manipulation as a solution to animal welfare problems into the foreground (see also Thompson 2010). While most of the response to this paper has suggest that I was not forceful enough in articulating objections to genetic transformation (see Ferrari 2012; Bos et al. 2018), some have argued that the lack of realism in the blind chicken thought experiment disguises the insight that what matters really is welfare, after all (Sandøe et al. 2014). In the decade since my paper was published, the totally insentient animal organism has become a reality and the mainstream animal ethicists love it. A growing literature documents the enthusiasm for cellular production of animal products, eliminating animal minds entirely (Chiles 2013; Welin 2013). Concerns that manipulation of stem cells and genetically engineered heme and other biologics might stimulate concern about the extreme instrumentalization of food have been raised (Thompson 2014), but there is little evidence that they will quell animal advocates' excitement over the prospects of eating a meat product that did not come from a sentient being.

Are proponents of cellular protein production advocating on behalf of animals? If one truly thinks that the lives lived by livestock today are not worth living, it is feasible to think that future generations of human beings will be doing a favor to future animals by not bringing them into existence at all. Claire Palmer called attention to paradoxical elements in my original analysis of animal disenhancement by interpreting it as an instance of the non-identity problem. As described initially by Derek Parfit, the problem is a radical discontinuity between the identity of the individuals being harmed (or benefited) and the identity of the individuals that actually eventuate, given the intervention under discussion (Palmer 2011). In the original example, a genetic intervention leads to an animal that suffers less (perhaps because of a reduced capacity for pain) than the one that would have resulted if the intervention had not taken place. But there is a problem in thinking that one is either benefiting the animal that does not come into existence, or harming the one that does. In the case of cellular meats, it is potentially millions of animals that never come into existence, but if Palmer is correct in claiming that we would be making a metaphysical error to think that we are benefiting the animals that *do not* come into existence, what possible benefit could there actually be? The most straightforward answer that I can see is that people who eat cellular meat are benefiting themselves.

They are satisfying a gustatory desire, while assuring themselves that they are not "supporting" the suffering endured by livestock being raised in confined settings.

There are also environmental reasons for not eating meat, but here, too, the interests of animals that never exist do not really come into play. Expanding the outlook on dietary ideals into environmental ethics does not resolve the problem of narcissism with respect to animal interests. Broadening even further, Christina Van Dyke has reviewed a number of emerging food practices advocated under the heading of "food ethics". Her focus has been on the analogy between dietetics and traditional forms of spiritual askesis, or ascetic practice. Like traditional religious spirituality, these practices combine social formation and personal redemption, albeit defined in terms of health, when we are talking food. Van Dyke argues that on any philosophically secular account, these dietetic regimes qualify as genuine spiritual practices. However, like the spiritual practices of religious extremists, dietetics become pathological when absorption with one's individualistic salvation overwhelms the social aspect of spirituality and the pursuit of conviviality (Van Dyke 2018). In other words, dietetics cease to function as properly ethical practices of spirituality when they become narcissistic.

Donna Haraway has written convincingly on the role of interspecies relationships in framing normative networks. She has characterized thinking that one is doing a creature a favor by making sure it never exists as a form of exterminism, linking it to genocide (Haraway 2008). Haraway does not mention cellular meat, but she is targeting what she characterizes as extreme vegan views that would call for the total elimination of animal production. Her claim here is a little vague. She is clearly claiming that human-animal relationships are constitutive of moral situatedness. This claim might be developed through a feminist care ethics that emphasizes the maintenance of interdependencies and network bonds (see Noddings 2013). However, the meaning of Haraway's reference to genocide is less straightforward. She seems to imply that in thinking that food animals would have been better off not to have existed at all, the extreme animal ethicists arrogate to themselves a standpoint capable of determining the ultimate value of another being's life. Deciding who should live and who should never be born is genocidal exterminism, even if it is not yet genocide, because it is only a half-step away from deciding who should live and who should die.

This reading of the desire for cellular meat is narcissistic in that one elevates oneself to a Godlike standpoint to decide the fate of other creatures. One might, of course, claim that livestock breeders have themselves taken on that role long, long ago. There is certainly a germ of truth in this worry. On this view, the problem with advanced breeding, including stem cell technologies, gene editing and cellular techniques is that they exacerbate a germinal trope that was, indeed, present in conventional breeding, but that was held in check by the limitations of technique. Breeders did pursue self-regarding roles in selecting which animals would reproduce. Yet breeders were unable to sever themselves from relational responsibilities to the progeny that resulted from their activity, and this limitation had a morally salvific socializing effect. It blocked the complete instrumentalization of the animal, and situated the breeder's instrumental goal within a situated network calling for attentiveness to animal needs. The imperative of husbandry, of a moral regard for the animal itself, thus also blocks total realization of the narcissistic instinct present in all self-interested action.

Advocacy for cellular production of animal proteins does not fully satisfy the conditions of genocidal exterminism outlined by Haraway. Yet when this advocacy is coupled with the archetypical caricature of family farming that I have critique above, the resulting dismissal of husbandry ethics in CAFOs alleviates the need for empathy or attention to the animal's experience. There is, then, a kind of *cultural narcissism* that emerges throughout the scholarly practice of philosophy. The repetition of the archetype and the subsequent failure to actually consider the condition in which animals live reproduces (if not also encouraging) a pattern of normative practice. The lives of animals in factory farms are repetitively characterized as involving extreme suffering, so much so that in DeGrazia's words, their lives are not worth living. Engaging substantively with this suffering is taken to be both pointless, and even problematic from a moral perspective. Animal suffering in CAFOs engages no philosophical interest, because the lives of these animals are without value. It would be better if these animals did not exist; we should pursue strategies that eliminate them. With no value, these lives cannot generate any ethical response other than disengagement of one's self from the nexus in which these worthless lives are embroiled. Individual philosophers may not *feel* like they are ignoring the interests of animals, and might take umbrage at the suggestion that they do not care about how animals actually fare in factory farms. Yet by ignoring the questions of ethical husbandry, they replicate a pattern of disengagement that can be observed in other forms of structural injustice. Our overweening concern with our own consumption reinforce institutions that militate against improving the lot of the animals themselves.

12.7 Conclusion

I have tried to sketch the contours of an argument that would hold mainstream philosophers accountable in part for the lack of movement toward improving the condition of animals living in industrial production systems. Those who do discuss these systems fall prey to implicit biases associated with activism for animal causes. A non-ideal theory would excuse activists for using whatever tools are available to motivate change, at least insofar as they resist outright falsehoods (after all, that's what activists do). But non-ideal theory would hold that, like animal welfare scientists themselves, those who occupy the social position of a philosopher or a scholar have a duty to present a more nuanced and complex account any controversy on which they report (see Pielke 2007). In making this case, I have emphasized a set of questions that emerge *within* husbandry ethics: the need to address collective action, the trade-offs between distinct welfare indicators, the matter of how far our ability to reform a system is really constrained. I have gestured at an archetype that I find too commonplace with most philosophers who have taken up animal ethics, and I have gone on to illustrate how this archetype oversimplifies the tasks of husbandry ethics, even when the claims it makes about industrial agriculture are strictly true.

The archetype conjoins with philosophers' penchant for relying on activist reports for their empirical understanding of what happens on industrial farms. Although activist reports are often factually accurate, they overstate what is known about the condition of animals in CAFOs in two ethically significant respects. First, by simply listing known welfare deficits, they fail to contextualize these deficits within a larger and comprehensive understanding of animal welfare, one that would include not only comparative discussions of traditional farming and wildlife, but that would also acknowledge what we *do not* know about how animals fare in all of these settings. Second, they fail to convey what ethologists and veterinary specialists *have* learned about ways in which the welfare of animals may depend on features that livestock species do not share with human beings. While anthropomorphism has its place, these statements promote a pernicious form of anthropomorphism that fails to respect ethically important differences in animal lives.

As an interest in the ethics of husbandry has vanished from the philosophical discourse, it is almost certainly disappearing from the consciousness of the average person woefully disconnected from the production of food. It is, thus, not surprising that radical responses to the suffering of animals in these systems advocate disappearance of the animals themselves. I argue that this mode of thought exhibits narcissism in several forms. In the extreme, it is a concern for *my* involvement that erases interest in what might be done for the animals themselves. It emerges more subtly in the view that biotechnology could resolve the factory farm issue by doing away with animal consciousness altogether. This thought conjoins with Christina Van Dyke's analysis of dietetic spirituality and finds further reinforcement in Donna Haraway's discussions of genocidal exterminism. In the end, however, a more modest *structural* form of narcissism may be the most appropriate diagnosis. Here narcissism is a reflection of the cultural institutions that block understanding, leaving us to think that monitoring of our own personal conduct is a morally adequate response to circumstances of structural injustice. This view of narcissism as a cultural form owes more to Nietzsche than to Freud. In contrast, some serious philosophical dialog with people who are trying to mitigate the suffering in factory farms is a better estimate of what the profession of philosophy owes to the animals in our midst.

References

Appleby, M., D. Weary, and P. Sandøe (eds.). 2014. *Dilemmas in animal welfare*. Wallingford, UK: CAB International.

Bennett, R.M., and P. Thompson. 2018. Economics. In *Animal welfare*, 3rd ed, ed. M.C. Appleby, I.A.S. Olsson, and F. Galindo, 335–348. Wallingford, UK: CAB International.

Bos, J.M., B. Bovenkerk, P.H. Feindt, and Y.K. Van Dam. 2018. The quantified animal: Precision livestock farming and the ethical implications of objectification. *Food Ethics* 2: 77–92.

Braun, J., and P. Braun. 1998. Inside the industry from a family hog farmer. In *Pigs, profits, and rural communities*, ed. K.M. Thu and E.P. Durrenberger, 39–56. Albany, NY: State University of New York Press.

Capper, J.L. 2012. Is the grass always greener? Comparing the environmental impact of conventional, natural and grass-fed beef production systems. *Animals* 2: 127–143.

Chiles, R.M. 2013. If they come, we will build it: in vitro meat and the discursive struggle over future agrofood expectations. *Agriculture and Human Values* 30: 511–523.

De Bakker, E., and H. Dagevos. 2012. Reducing meat consumption in today's consumer society: Questioning the citizen-consumer gap. *Journal of Agricultural and Environmental Ethics* 25: 877–894.

De Jonge, J., and H.C.M. van Trijp. 2013. Meeting heterogeneity in consumer demand for animal welfare: A reflection on existing knowledge and implications for the meat sector. *Journal of Agricultural and Environmental Ethics* 26: 629–661.

DeGrazia, D. 2011. The ethics of confining animals: From farms to zoos to human homes. In *The Oxford handbook of animal ethics*, ed. T.L. Beauchamp and R.G. Frey, 738–768. New York: Oxford University Press.

Fairlie, S. 2010. *Meat: A benign extravagance*. White River Junction, VT: Chelsea Green Publishing.

FAO (Food and Agriculture Organization of the United Nations). 2003. *World agriculture: Towards 2015/2030: An FAO perspective*. Rome: FAO.

Ferrari, A. 2012. Animal disenhancement for animal welfare: The apparent philosophical conundrums and the real exploitation of animals. A response to Thompson and Palmer. *NanoEthics* 6: 65–76.

Fraser, D. 1999. Animal ethics and animal welfare science: Bridging the two cultures. *Applied Animal Behaviour Science* 65: 171–189.

Gendlin, E.T. 1987. A philosophical critique of the concept of narcissism: The significance of the awareness movement. In *Pathologies of the Modern Self: Postmodern Studies on Narcissism, Schizophrenia, and Depression*, ed. D.M. Levin, 251–304. New York: New York University Press.

González-Cerón, F., R. Rekaya, and S.E. Aggrey. 2015. Genetic analysis of bone quality traits and growth in a random mating broiler population. *Poultry Science* 94: 883–889.

Haraway, D.J. 2008. *When species meet*. Minneapolis, MN: University of Minnesota Press.

Harfeld, J. 2011. Husbandry to industry: Animal agriculture, ethics and public policy. *Between the Species* 13 (10), Article 9.

Harmon, E. 2011. The moral significance of animal pain and death. In *The Oxford handbook of animal ethics*, ed. T.L. Beauchamp and R.G. Frey, 726–737. New York: Oxford University Press.

Harrison, R. 1964. *Animal machines: The new factory farming industry*. London: Vincent Stuart Publishers.

Hristov, A.N., J. Oh, C. Lee, R. Meinen, F. Montes, T. Ott, J. Firkins et al. 2013. Mitigation of greenhouse gas emissions in livestock production: A review of technical options for non-CO2 emissions. FAO Animal Production and Health Paper No 177. Rome: FAO.

Hursthouse, R. 2011. Virtue ethics and the treatment of animals. In *The Oxford handbook of animal ethics*, ed. T.L. Beauchamp and R.G. Frey, 119–143. New York: Oxford University Press.

Ilea, R.C. 2009. Intensive livestock farming: Global trends, increased environmental concerns, and ethical solutions. *Journal of Agricultural and Environmental Ethics* 22: 153–167.

Johnson, G.L. 1976. Philosophic foundations: Problems, knowledge, solutions. *European Journal of Agricultural Economics* 3 (2–3): 207–234.

Kehlbacher, A., R. Bennettand, and K. Balcombe. 2012. Measuring the consumer benefits of improving farm animal welfare to inform welfare labelling. *Food Policy* 37: 627–633.

Lay, D.C., R.M. Fulton, P.Y. Hester, D.M. Karcher, J.B. Kjaer, J.A. Mench, B.A. Mullens, et al. 2011. Hen welfare in different housing systems. *Poultry Science* 90: 278–294.

Main, D.C.J., S. Mullan, C. Atkinson, M. Cooper, J.H.M. Wrathall, and H.J. Blokhuis. 2014. Best practice framework for animal welfare certification schemes. *Trends in Food Science & Technology* 37: 127–136.

McMahon, J. 2016. The comparative badness of animal suffering and death. In *The ethics of killing animals*, ed. T. Višak and R. Garner, 65–85. New York: Oxford University Press.

McPherson, T. 2016. Why I am a vegan (and you should be one, too). In *Philosophy Comes to Dinner: Arguments about the Ethics of Eating*, ed. A. Chignell, T. Cuneo, and M. C. Halteman, 73–91. New York: Routledge.

Mench, J.A. 1998. Why it is important to understand animal behavior. *ILAR Journal* 39: 20–26.

Mench, J. A. (ed.). 2018. *Advances in agricultural animal welfare*. Cambridge, MA: Woodhead Publishing, Elsevier.

Mench, J.A., E.A. Pajor, H. James, and P.B. Thompson. 2008. *The Welfare of Animals in Concentrated Animal Feeding Operations*. Technical Report to the Pew Commission on Industrial Animal Production. http://www.pcifapia.org/_images/212-7_PCIFAP_AmlWlBng_FINAL_REVISED_7-14-08.pdf. Accessed 23 March 2020.

Mench, J.A., D.A. Sumner, and J.T. Rosen-Molina. 2011. Sustainability of egg production in the United States—The policy and market context. *Poultry Science* 90: 229–240.

Moberg, G.P. 2000. Biological response to stress: Implications for animal welfare. In *The biology of animal stress: Basic principles and implications for animal welfare*, ed. G.P. Moberg and J.A. Mench, 1–21. Wallingford, UK: CAB International.

Nagel, T. 1974. What is it like to be a bat? *The Philosophical Review* 83: 435–450.

Noddings, N. 2013. *Caring: A Relational Approach to Ethics and Moral Education*. Berkeley: University of California Press.

Norcross, A. 2004. Puppies, pigs and people: Eating meat and marginal cases. *Philosophical Perspectives* 18: 229–245.

Norwood, F.B., and J. Lusk. 2011. *Compassion by the Pound: The Economics of Farm Animal Welfare*. New York: Oxford University Press.

Palmer, C. 2011. Animal disenhancement and the non-identity problem: A response to Thompson. *NanoEthics* 5: 43–48.

Pielke Jr., R.A. 2007. *The honest broker: Making sense of science in policy and politics*. New York: Cambridge University Press.

Rachels, S. 2011. Vegetarianism. In *The Oxford handbook of animal ethics*, ed. T.L. Beauchamp and R.G. Frey, 877–905. New York: Oxford University Press.

Ritchie, H., and M. Roser. 2017. Meat and seafood production and consumption. *Our World in Data*. https://ourworldindata.org/meat-and-seafood-production-consumption. Accessed 23 March 2020.

Rollin, B.E. 1995. *Farm animal welfare: Social, bioethical and research issues*. Ames, IA: Iowa State University Press.

Rolllin, B.E. 2004. Annual meeting keynote address: Animal agriculture and the emerging social ethic for animals. *Journal of Animal Science* 82: 955–964.

Rorty, Richard. 1979. *Philosophy and the Mirror of Nature*. Princeton, NJ: Princeton University Press.

Sandøe, P., P.M. Hocking, B. Förkman, K. Haldane, H.H. Kristensen, and C. Palmer. 2014. The Blind Hens' Challenge: Does it undermine the view that only welfare matters in our dealings with animals? *Environmental Values* 23 (6): 727–742.

Singer, P., and J. Mason. 2007. *The ethics of what we eat: Why our food choices matter*. Emmaus, PA: Rodale Press.

Steinfeld, H., P. Gerber, T.D. Wassenaar, V. Castel, M. Rosales, and C. de Haan. 2006. *Livestock's long shadow: Environmental issues and options*. Rome: FAO.

Stout, J. 2007. On our interest in getting things right: Pragmatism without narcissism. In *The New Pragmatists*, ed. C. Misak, 7–31. New York: Oxford University Press.

Thompson, P.B. 2008. The opposite of human enhancement: Nanotechnology and the blind chicken problem. *NanoEthics* 2: 305–316.

Thompson, P.B. 2010. Why using genetics to address welfare may not be a good idea. *Poultry Science* 89: 814–821.

Thompson, P.B. 2014. Artificial meat. In *Ethics and Emerging Technologies*, ed. R.L. Sandler, 516–530. London: Palgrave Macmillan.

Thompson, P.B. 2015. *From field to fork: Food ethics for everyone*. New York: Oxford University Press.

Thompson, P.B. 2020. Philosophical ethics and the improvement of farmed animal lives. *Animal Frontiers* 10 (1): 21–28.

Thornton, P.K. 2010. Livestock production: recent trends, future prospects. *Philosophical Transactions of the Royal Society B: Biological Sciences* 365: 2853–2867.

Van Dyke, C. 2018. Eat Y'self Fiiter: Orthorexia, health and gender. In *Oxford Handbook of Food Ethics*, ed. A. Barnhill, T. Doggett, and M. Budolfson, 553–571. New York: Oxford University Press.

Welin, S. 2013. Introducing the new meat: Problems and prospects. *Etikk i praksis-Nordic Journal of Applied Ethics* 1: 24–37.

Paul B. Thompson is the W. K. Kellogg Professor of Agricultural, Food and Community Ethics at Michigan State University. Much of his work on animals has been advisory and practical, serving in an advisory capacity to a number of U.S.-based organizations. Thompson sits on welfare advisory committees to the American Humane Associations and the United Egg Producers, and has acted as a consultant to the National Pork Producers Council. As a founding member of both the Agriculture, Food and Human Values Society and the European Society for Agricultural and Food Ethics, Thompson has been active in philosophical reflection on agricultural production methods for nearly 40 years. Thompson's most recent book is *The Spirit of the Soil: Agriculture and Environmental Ethics* 2nd edition, a thorough revision of the 1995 original.

Chapter 13
Reimagining Human Responsibility Towards Animals for Disaster Management in the Anthropocene

Andreia De Paula Vieira and Raymond Anthony

Abstract Animals, like human beings, are prone to suffering harms, such as disease, injury and death, as a result of anthropogenic and natural disasters. Animals are disproportionately prone to risk and adversely affected by disasters, and thus require humane and respectful care when disasters strike, due to socially situated vulnerabilities based on how human communities assess and value their moral standing and function. The inability to integrate animals into disaster risk and management practices and processes can sometimes be associated with a lack of understanding about what animal ethics and animal health and welfare require when designing disaster management programs. This chapter seeks to reimagine human responsibility towards animals for disaster management. The pervasiveness of disasters and their impacts on animals, human-animal and animal-environment relationships underscore the importance of effective animal disaster management supported by sound ethical decision-making processes. To this end, we delineate six ethically responsible animal caretaking aims for consideration when developing disaster management plans and policies. These aims, which address central vulnerabilities experienced by domesticated animals during disasters, are meant to be action-guiding within the disaster management context. They include: (1) Save lives and mitigate harm; (2) Protect animal welfare and respect animals' experiences; (3) Observe, recognize and promote distributive justice; (4) Advance public involvement; (5) Empower caregivers, guardians, owners and community members; (6) Bolster public health and veterinary community professionalism, including engagement in multidisciplinary teams and applied scientific developments. To bring about these aims, we offer a set of practical and straightforward action steps for animal caregivers and disaster management teams to ensure that animals' interests are systematically promoted in disaster

A. De Paula Vieira
Animal Welfare Scientist and One Health Researcher, Curitiba, Brazil
e-mail: apvieirabr@gmail.com

R. Anthony (✉)
Department of Philosophy, University of Alaska Anchorage, Anchorage, AK, USA
e-mail: rxanthony@alaska.edu

© The Author(s) 2021
B. Bovenkerk and J. Keulartz (eds.), *Animals in Our Midst: The Challenges of Co-existing with Animals in the Anthropocene*, The International Library of Environmental, Agricultural and Food Ethics 33,
https://doi.org/10.1007/978-3-030-63523-7_13

management. They include: (1) Respect and humane treatment; (2) Collaboration and effective disaster communication; (3) Strengthening systems of information sharing, surveillance, scientific research, management and training; (4) Community outreach and proactive contact; (5) Cultural sensitivity and attitudes check, and (6) Reflection, review and reform.

13.1 Introduction

We are living through the Anthropocene, an epoch defined by the fact that human activities have touched nearly every aspect of life on Earth,[1] including accidental or inadvertent pollution from industries that result in the deaths of fish and pigs,[2] as well as intentional projects such as damming rivers that flood and drown a plethora of animal species and rapid, widespread urban development that contributes to wildfires that consume wild animals' habitats.[3] Further, the effects of climate change and environmental degradation have left humans and animals vulnerable to drought, food shortages and lack of habitability. In addition, the rise of emerging infectious disease outbreaks has been connected to industrial agriculture, environmental destruction and habitat loss (FAO 2017; Johnson et al 2020; Hiko and Malicha 2016).

Animals are constantly vulnerable to disasters and are not equally protected when they occur. For example, at the time of drafting this chapter, the world is gripped by two major disasters—the Australian bushfires and a novel coronavirus pandemic. In the first instance, conservative estimates point to upwards of 800 million mammals, reptiles and birds affected by the New South Wales fires.[4] A viral Internet video from Adelaide of a koala approaching a group of cyclists and climbing on one of the bicycles to get a drink has become an iconic image during this calamity.[5] It and other images of injured or charred animals have ushered in an overabundance of concern, including handmade goods and medical supplies from across the globe to help the animals injured in the heatwave and wildfires. Could the Australian animals' vulnerabilities have been reduced and many lives spared? What landscape management disaster plans were in place and were they designed to safeguard the wildlife populations and their habitats and/or shepherd human behavior to care for the animals during an anticipated climate-induced crisis? How might real-time sentinel mapping of wildlife populations have mitigated these negative effects? What disaster strategies were in place to evacuate animals in vivariums as well as in research and shelter facilities in case the fires reached these places? In order to prevent similar future disasters,

---

[1] The authors are grateful to Clemens Driessen for helpful comments in an earlier draft.

[2] https://www.bbc.com/news/world-us-canada-48911918 and https://www.theguardian.com/environment/chinas-choice/2014/apr/17/china-water, respectively.

[3] https://www.nytimes.com/2019/03/21/climate/missouri-river-flooding-dams-climate.html   and https://www.bbc.com/news/world-us-canada-46178230, respectively.

[4] https://www.abc.net.au/news/2020-01-31/fact-check-have-bushfires-killed-more-than-a-billion-animals/11912538.

[5] https://www.abc.net.au/news/2019-12-28/thirsty-koala-fed-by-cyclist-in-adelaide/11830276.

what have decision-makers learned and implemented about effectively integrating animal, human and environmental health and welfare? What disaster regulations should be enacted?

In the second instance, the SARS-CoV-2/COVID-19 outbreak,[6] which originated in China, has led to a global pandemic that had infected more than 25 million people by 5 September 2020. Animals can be major reservoirs of zoonotic diseases, which can jump to humans and vice versa, especially when basic public health measures are not vigilantly observed in relation to animals and the environment. These measures include prevention, detection, monitoring and eliminating outbreaks and epidemics through sanitation and epidemiological surveillance. Initial reports speculate that the COVID-19 outbreak was caused by a spillover infectious virus that surfaced at a live wholesale seafood and wet market in Wuhan, China. Differently from the bushfires in Australia, perhaps due to delayed confirmation and notification of the outbreak by the local authorities, little was said by field investigators and researchers about the number of animal deaths and the impact of the outbreak on the health and welfare of infected and non-infected animals. It is unclear what has happened to the tens of thousands of animals that would have been sold in Wuhan in conjunction with the Lunar New Year celebrations after authorities banned the trade of live animals. Have they been slaughtered (if so, how) or have some been abandoned and what was the main motivation for doing so?

The questions associated with these examples highlight significant ethical challenges posed by disasters. While a host of difficult choices must be made during a disaster, our ethical commitments to animals will frame how they count morally and how disaster planning, together with improved emergency-response capacity, should be designed and deployed to prevent and reduce risks to both humans and animals. Thus, further research, regulations and practices in animal disaster management should consider what outcomes are intended for animals in specific disaster events, how are they justified, and what ethical and scientific blind spots exist when it comes to how the substance and effects of human activities, such as better animal welfare and care and husbandry practices, influence regard for animals and their welfare.

Our focus in this chapter is on the plight of domesticated animals—those with whom we have direct or proximate contact.[7] We begin by defining "disaster" and discuss the ethical biases that result in many animals, by and large, still being left out of or minimized in disaster management plans. Next, we discuss the importance of improving disaster management for animals in the Anthropocene. We argue that animal health and welfare perspectives, together with an emphasis on human-animal-environment relationships should be strengthened in disaster risk reduction and management strategies, together with measures traditionally considered. We

---

[6] https://www.who.int/news-room/detail/30-01-2020-statement-on-the-second-meeting-of-the-int ernational-health-regulations-(2005)-emergency-committee-regarding-the-outbreak-of-novel-cor onavirus-(2019-ncov).

[7] Some animals that are recovered from disasters for rehabilitation by qualified animal health and welfare professionals become domesticated if they cannot be returned to the wild.

discuss six ethically responsible caretaking aims for animal disaster management, and in concert with these aims, we end by offering practical and straightforward recommendations to increase the visibility of animals' interests during a disaster. These recommendations are meant to catalyze further engagement and strengthen policies and practices on the subject.

13.2　Animal Disaster Ethics: Developing Disaster Frameworks

Disasters are emergencies endured by people and animals and can be induced by anthropogenic or natural agents. Anthropogenic cum technological disasters include fires, environmental contamination, toxicological or chemical events, and disasters due to human negligence or abuse, conflict, criminal activity or terrorism. Meanwhile natural disasters fall under four broad categories: (a) Hydro-meteorological-climatological: floods, wave surges, storms, hurricanes, cyclones, landslides, avalanches, fire, droughts and climate change; (b) Geophysical: tsunamis, earthquakes and volcanic eruptions; (c) Biological: pandemic diseases, epidemics and insect infestations; and (d) Extraterrestrial: asteroids, meteoroids, and comets that alter interplanetary conditions that affect the Earth's magnetosphere, ionosphere, and thermosphere (EM-DAT 2020; Heath 1999). Disasters can be international, national or local in scope. The onset of a disaster can be sudden/rapid (fire, flood, avalanche, mudslide and earthquake) or slow (disease, biosecurity breach).

A disaster occurs when the ability to anticipate and reduce risk to natural or anthropogenic hazards overtakes standard health and well-being accommodations and the conventional capacity to cope is destabilized. Disaster management is necessary when the scale, timing and unpredictability of events threatens to overrun routine capabilities of civic and public health systems, communities and individuals to address the emergency (Nelson et al. 2007). Disaster management activities include risk communication, regulating environmental conditions, minimizing and detecting disease threats and outbreaks, planning for emergency medical and public health response capacities, and preventing secondary emergencies following a disaster (Salinsky 2002). Disaster management teams must address a complex emergency situation in the most humane and respectful way possible for all the parties involved—human, animal and environment (Murray and McCutcheon 1999). The experience and skills of the social, behavioral and health sciences, prevention and surveillance, risk communication, data gathering, architecture and planning, environmental sciences, engineering, and public safety are commonly required in traditional disaster management.

However, disaster management is also a poignant animal issue. Indeed, animal disaster management is a "wicked problem" (Glassey 2020), marked by the confluence of increasing human dependence on animals for survival (Delgado et al. 1999) including nutrition, food security, health, safety and livelihood. The challenges posed by climate change and extreme natural events, population growth and urban sprawl, emerging and reemerging diseases, and global political and economic instability also bring focus on human ethical commitments towards animals. The capacity for human communities to recover after a disaster is inextricably linked to how animals fare.[8] According to the international Sendai Framework for Disaster Risk Reduction (United Nations 2015), countries should enhance their disaster preparedness and recovery efforts, strengthen governance and invest in disaster risk reduction since there are significant economic, social, environmental and public health and safety benefits in doing so. A focus on understanding the complex interconnections between health and welfare at the intersection of human-animal-nature conflicts can lead to preventative and mitigation measures that reduce the number of deaths, injuries, disabilities and losses in economic, physical, social, cultural and environmental assets. The Sendai Framework includes strategies for stakeholder engagement and dialogue to develop implementable community guidelines and financing intervention, dependable surveillance, strategic planning that enhances clear lines of governance and authority for decision-making in veterinary and public health emergencies, early warning systems, coordinated and reliable risk analysis, equitable triage protocols for animals during medical support and rescue, and practicable policies for landscape planning and infrastructure (e.g. evacuation centers and temporary housing) that reflect scientific advice and the most recent evidence-based information.

How a disaster is framed is key in successfully preparing for and responding to it. In framing a disaster in terms of its management aims, the disaster management team reveals their ethical commitments. This involves making explicit the priorities, values and moral assumptions, and reasons underpinning crisis policies and actions while fostering coordination at all levels to manage an all-encompassing crisis (Institute of Medicine 2007). Disaster management aims can highlight the adequacy of the infrastructure involved in advancing equity, inclusion and community relationships, which will be necessary in mobilizing political will. Further, this framing provides a window into the people, devices, systems, procedures and methods necessary to realize significant community ends during a disaster, including constraints such as existing laws, regulations and public policies. A disaster calls for specialized communications and surveillance systems, adequate equipment, trained responders and deployment of professionals who can provide quick and appropriate response to the threat (Institute of Medicine 2003; O'Toole et al. 2002). Furthermore, adequate disaster preparedness involves a well-prepared community to ensure vulnerable populations are well-integrated into an existing infrastructure (for example, see

---

[8] A case in point is the January 2019 collapse of the Feijão dam in Brumadinho, which has been billed as Brazil's worst industrial accident. The incident not only killed at least 248 people, but also engulfed nearby farms, thus affecting the environment on which local and regional communities built their economies. Numerous farm animals were terminated on humanitarian grounds per the directive of the Federal Council of Veterinary Medicine (CFMV) (https://www1.folha.uol.com.br/cotidiano/2019/01/animais-agonizando-sao-mortos-a-tiros-em-brumadinho.shtml). In addition, the response team also rescued more than 400 animals (https://crmvpb.org.br/a-atuacao-da-brigada-veterinaria-no-resgate-de-animais-em-brumadinho/).

AVMA Emergency Preparedness and Response Guide 2012; Itoh 2018; Murray and McCutcheon 1999; Vinícius de Souza 2018; Powers 2016). Government and animal industries' investments in capacity building and personnel training, together with practical operation and maintenance records allow for better governance, since they prevent mistakes in operating pre-established contingency and biosecurity plans. For example, previously designed action plans to address a pandemic like COVID-19 were not executed due to under-resourced facilities and a lack of personal protective equipment for frontline animal caregivers (Winders 2020). Since an essential objective of disaster management is minimizing the vulnerability of affected populations, animal disaster management plans should include detailed information about mitigation and prevention such as housing, husbandry and waste management standards for animals, the built environment, and social, political, environmental and economic structures around animals, including those in animal facilities. These include zoos, vivariums, sanctuaries and concentrated animal feeding operations to minimize both the loss of animal lives and poor welfare conditions during recovery from a disaster.

Particularly urgent in the Anthropocene are holistic framings which can serve as a foundation for governments, civic and public health systems, disaster management professionals, animal health and welfare, veterinary emergency care, surveillance and public health, private sector stakeholders, animal-related organizations and facilities, university researchers and communities to investigate specific risk reduction strategies, develop guidelines for disaster management and provide effective messaging during outbreak response.

The One Health Framework,[9] which is gaining popularity in zoonotic disease control, brings the connectivity of human-animal-environmental health and welfare issues into sharp focus when dealing with animal, environmental and public health crises (ECDC 2018; Rist et al. 2014; Stauffer and Conti 2014). One Health seeks "to promote, improve and defend the health and well-being of all species and the ecosystem, by enhancing cooperation and collaboration between physicians, veterinarians, other scientific health and environmental professionals and by promoting strengths in leadership and management to achieve these goals" (http://www.one healthinitiative.com/mission.php). The One Health resolution marks the first time a holistic definition was formally agreed upon to address the interconnections between human-animal-ecosystem health, and it resulted in greater public visibility for the well-being of animals (Zinsstag et al. 2011). Here "health is a state of complete physical, mental and social well-being and not merely the absence of disease or infirmity" (https://www.who.int/about/who-we-are/constitution). Within this framework, field investigations and scientific and technological disaster developments would seek to clarify the effects of past disasters on climate and environmental variability, welfare and disease occurrence in order to predict and plan for future disasters. Further, because disasters challenge the welfare of the agent-environment-host triad, as in the case of a pandemic, the framework may be applied to ongoing challenges to shed light on changes in the intensity of disease outbreaks in humans and animals, the

---

[9] A competing framework, One Welfare, has also been gaining traction in disaster management as an alternative to One Health (Pinillos 2018).

access of pathogens to new landscapes, the relationship between previous disaster variables and the epidemiology of diseases and interventions, as well as the effects of biological development, emergence and re-emergence of infectious diseases. The OIE (World Organisation for Animal Health), a One Health initiative partner, has taken a leadership role in the development of animal disaster guidelines and in identifying the current state of disaster management and risk reduction processes (Dalla Villa et al. 2017). The OIE supports the Veterinary Services of member countries to enhance their resilience and strengthen their disaster management capacity, reduce risks at the global level and promote close collaboration among emergency services and all other agencies involved in disaster management. The OIE also provides wide access to the epidemiological information that public veterinary institutions and organizations are called to collect at national and international levels through the OIE Information System WAHIS, the World Health Organization (WHO), and the Animal Disease Notification System (ADNS-EC) in Europe.

Considering animal disaster management through the lens of One Health can broaden current knowledge and provide new ways to minimize harms, such as revealing how animals cope immunologically with infections or respond in search and rescue missions. A One Health perspective can offer an important set of orienting questions to guide disaster management. For example, could improved welfare practices boost animals' immunity and reduce the spread of infectious diseases amongst animals and humans? To what extent have climate mitigation or adaptation strategies been designed to reduce harms to animals? Could the transmissibility of recent outbreaks have been minimized if the health and welfare of animals were given priority?[10] Could scientists demonstrate the possible link between human and animal welfare in terms of the social, economic and political complexity of emergency planning, response and recovery involving animals? Could animal welfare science improve search and rescue missions often performed by animals? While One Health is a promising candidate for animal disaster management, the questions outlined here reveal various anthropocentric biases, animal ethics and considerations as well as other moral, political and budgetary priorities and commitments (Van Herten et al. 2019).

13.3 Animal Disaster Ethics: Revealing Animal Vulnerabilities

As more disasters are emerging, some of unpredictable scale and magnitude, it becomes clear that the Anthropocene has heightened animals' vulnerability. While

---

[10] For example, Britain's Foot and Mouth Disease outbreak (2001), avian influenza outbreaks across China and Asia, the highly pathogenic A(H5N8) strain of avian influenza (HPAI) epidemic that occurred in 29 European countries between 2016–2017, the porcine reproductive and respiratory syndrome (PRRS) outbreak, and the H5N2 outbreak that ravaged poultry systems in more than 20 US states in 2015–2016.

concern for some animals is palpable (e.g., companion animals and those that capti-
vate the human imagination such as koalas and polar bears), what constitutes an
adequate response to the needs and interests of other animals is not universally
consistent (e.g., livestock). Since Hurricane Andrew in 1992 (Heath 1999), emergency and evacuation plans and early warning systems in the United States have
started to address the importance of contingency plans to save animals; however,
after 25 years, as demonstrated in the responses to Hurricanes Harvey, Irma, and
Maria, these contingency plans are still not widespread and lack the breadth and
depth to technically, scientifically and responsibly address animals' issues before,
during and after a disaster. Other countries prone to natural disasters, such as Brazil
(from drought, landslides and flooding) and Japan (from tsunamis, earthquakes, landslides and flooding), also continue to struggle to address the welfare of animals during
emergencies (Itoh 2018; Vinícius de Souza 2018). When a disaster strikes, human
considerations tend to take precedence or are still considered independently from
animal considerations (consider the two examples that began this chapter). Consequently, the necessary infrastructure, methods and capacity to adequately address
animal-related issues is absent in many cases (photographer Yasusuke Ota's depic-
tion of this omission in the context of the aftermath of Tohoku earthquake and tsunami
in Japan is an excellent example of a lack of aesthetic, moral and evidence-based
engagement in disaster management).[11]

Another challenge that pushes for better animal disaster management plans is
the problem faced by animal owners who leave companion animals and livestock

[11] The authors thank Clemens Driessen for alerting us to this exhibition. https://www.aestheticama
gazine.com/yasusuke-ota-the-abandoned-animals-of-fukushima-amsterdam.

behind during a disaster. It is still common that first responders have neither clear outcomes for minimizing welfare harms to the fewest number of animals nor the capacity to rescue them because the resources or equipment are not available to save people, animals and property. While animals are considered property under the law in most places, given the intimate relationships people have with their animals (e.g., as members of families, and as sources of nutrition, food security and livelihoods) (Sawyer and Huertas 2018), they are increasingly being granted more social consideration (Meijboom and Stassen 2016). Further, there is every indication that animals' welfare and lives will continue to be a major issue affecting disaster management and rescue in the future (LEGS 2015). The COVID-19 pandemic, for example, has raised fear and anxiety among pet and companion animal owners, livestock producers, zoos, shelters and consumers of animal products. There has been little guidance on how to ensure the welfare of feral, wild and community animals during mandatory stay-at-home orders for those on whom they depend or of livestock when meat and milk processing cannot occur. Local community values and practical constraints in tandem with technical and scientific information should be factored into decision making about how animals are managed during a disaster and what constitutes desired outcomes for animals vis-à-vis disaster response.

Disasters may present opportunities to explore protections for animals in preparation for future ones. In the US, the Pet Evacuation and Transportation Standards (PETS) Act was passed shortly after Hurricane Katrina in 2006 to mitigate loss of life of some animal species during a disaster. As a federal law, the PETS Act mandates that in order for states, cities, and counties to receive federal funding for disaster relief plans, those plans must "account for the needs of individuals with household companion animals, pets and service animals before, during, and following a major disaster or emergency."[12] The Act allows the Federal Emergency Management Agency (FEMA) to provide funding to states and localities for the creation, operation, and maintenance of pet-friendly emergency shelters, along with other disaster emergency actions for companion and service animals.[13] Rescuing and rehabilitating animals need not be in conflict with promoting human welfare and agency, and can serve to soften the human-animal divide. For example, during Hurricane Irma in the US (2017), Florida's Governor Rick Scott urged hotels to waive their no-pet policies for pet owners seeking refuge from the hurricane. The University of Florida's College of Veterinary Medicine (UFCVM), part of the state's disaster response system, also set up pet-friendly shelters so that whole families could stay together. "Do not leave your pet behind," was the refrain from the UFCVM since, "If it's not safe for you, it's not safe for your pet." Similarly, the George R. Brown Convention Center in Houston, Texas, permitted survivors to bring their animal companions with them. Some caregivers and owners will not leave their homes unless they know their animals can accompany them or that their animals will be saved. Owners who do not relinquish their animals during a disaster have made it harder for first responders to evacuate people—their target survivor group—which can also inadvertently sabotage rescue

---

[12] See p. 1 of https://www.congress.gov/109/plaws/publ308/PLAW-109publ308.pdf.

[13] https://www.congress.gov/109/plaws/publ308/PLAW-109publ308.pdf.

efforts to save animals, especially if the owners put themselves and the animals at risk when trying to save the latter. Furthermore, when animals are left behind, the trauma of abandoning them can haunt both rescuers and owners. From a public health and biosecurity perspective, however, it must be made clear to owners when it is necessary to practice physical distancing from their pets; quarantined animals may be carrying a pathogenic agent, especially if a natural disaster occurs concomitantly with an animal or public health emergency (WSAVA 2020). In the case of veterinary and public health emergencies, reference laboratories in every country should be considered in disaster management plans and work in tandem to investigate the efficacy of interspecific transmission and the manifestation of a disease in animals so that subsequent diagnostic, therapeutic and prevention interventions can be developed and deployed effectively. Local and international laboratories should be involved early on and interactions should happen often for proper technical collaboration. An example of a multi-nation concerted effort to improve animal disaster management is the OIE Reference Laboratories and Collaborating Centres that have the objective of harmonizing and exchanging data, and sharing information and reference materials to improve disease surveillance, control and veterinary emergencies worldwide.[14]

There are many anthropocentric reasons to provide for animals during a disaster—for example, humans have an acquired responsibility due to animals' membership in our homes, the human-animal bond, their health and welfare, psychosocial and emotional trauma, the potential for environmental degradation, and savings in time, labor and financial expense if animals are neglected during an emergency. While we have a duty to plan and prepare well ahead of a disaster for our own benefit, recent disasters have made clear that the heavy loss of animal lives and their poor welfare due to disease and injury constitutes a moral harm in terms of their injustice and inhumaneness. Therefore, there are also non-anthropocentric reasons for providing aid to animals during a disaster. The weighty effects of our continued domestication of animals in the modern age signal the need to carefully consider the ethical aspects of animal disaster management and to incorporate ethical considerations involving animals into emergency planning activities. For animals, first responders will most likely be their immediate caretakers. Here, it will be incumbent upon the organizations that engage first responders to develop disaster plans that include evacuation (also taking into consideration the capacities of certain species of animals to fend for themselves and the health status of both owners and animals) and having contingency plans if the animals cannot be removed or can be a hazard to humans and vice versa, such as wildlife.

Disaster management is still largely defined by the interests of human communities. Members of the public and elected officials are hardly surveyed to discern their commitment to protecting human and animal lives prior to and during a disaster, as well as its impact on the ecosystem, including their perceptions and the relative weight placed on human-animal lives and how resources should be allocated to mitigate future disasters.

---

[14] See https://www.oie.int/scientific-expertise/collaborating-centres/reference-centre-networks/.

Disasters, however, remind us that human beings and animals share an ecological landscape. In a sense, many of our uses of animals themselves may constitute a hazard, exposure or vulnerability for animals when not properly cared for or managed. Human beings have much work to do to ensure conditions for mutual coexistence and alter activities and projects in ways that minimize human-animal conflicts in order for effective interspecies relationships to flourish. Disasters are occasions that can draw people into caring for animals in extraordinary situations and to reconsider the "norms of normality" by rethinking our existing uses of animals and practices that give rise to their vulnerabilities in the first place.[15]

The vulnerabilities experienced by animals take many forms and manifest themselves in different ways. For example, reducing animals' vulnerability during emergencies include mitigation and prevention strategies that prepare for possible evacuation, redesign of animal housing, and handling waste pollution and carcass disposal effectively to minimize public health hazards. Animals also experience socially situated vulnerabilities, that is, how human communities assess and value the moral standing of animals and their function, and how a lack of understanding about what constitutes good animal welfare during disasters may impact their consideration.

Arluke and Sanders (1996) suggest a sociozoologic scale to assign relative moral worth to humans and animals. The socially situated vulnerability that follows species lines and/or our traditionally cultivated uses of animals and that forms the basis of a deep-seated cultural hierarchy of valuing animals, influences how we make decisions about animals in disasters. Irvine (2009) argues that one's species status on the scale influences one's relative moral considerability and the extent to which resources will be devoted to save one's life. The scale is complicated by other considerations such as economic value, function and types of relationships and liability.

Although human lives tend to have priority during an emergency, increasingly animals matter. Our close relationship with dogs and cats and the recognition that the human-animal bond is a significant feature during rescue and evacuation, and has propelled these "near-person" (Varner 2012) companion animals into the "also victims" arena during a disaster. In the US, the PETS Act provides accommodation for companion animals and their caregivers. Livestock and research animals tend to be more vulnerable than companion animals due to the position that livestock occupy on the sociozoologic scale. They suffer disproportionately in the wake of a disaster, especially if no disaster management plan is in effect. Historically (and because of their high stocking densities and paucity of hazard mitigation strategies), livestock experience more injuries, disease and loss of life/termination. Farmers and producers responsible for their care, when faced by delivery failures at processing plants, may not be able to sell their animals even though they care for them (FAO 2020). Rescuing large animal populations (for example, a herd of cattle, wildlife translocation, research animals) is a tremendous effort compared with rescuing a small number of family pets. Whereas a family can often bring their pets in their personal vehicle, moving large numbers of animals requires many transport vehicles

---

[15] The authors are indebted to Clemens Driessen for this insightful skein of thought.

and adequate shelter facilities, which may not be possible during an emergency situation because this type of priority is not included in disaster management plans. These conditions make euthanasia, depopulation or culling very likely to minimize harms to these animals. However, we argue that alternative strategies that also prioritize animal welfare should be developed and implemented in the field through practical guidelines that include indemnity procedures for loss of animals and mitigation of animal suffering. During Hurricane Katrina (2005), millions of farm animals in the United States died. Hurricane Sandy claimed the lives of tens of thousands of research animals because there was no conceived contingency plan for them.[16] Meanwhile, more than 3 million chickens and 5,000 pigs died during Hurricane Florence (2018). The US Department of Agriculture (the primary agricultural regulator) does not have the resources to address animal welfare and mortality. The chronic effects of disasters may influence animals and can predispose those that are already health or welfare compromised to infectious and non-infectious diseases due to low immunity that leads to distress, behavioral maladaptation and negative affective states (FAO 2020). In disaster emergency sites, feed and water quality and quantity may be severely lacking and common management practices such as moving manure, moving feed and stock, and automated activities that rely on energy supplies can be subverted due to power outages.

Ethical judgments are implicit in all decisions and recommendations made about how to conceptualize a "disaster" (AVMA 2012; Irvine, 2009), its impact on animal welfare (Anthony 2004; AVMA 2020a, b; Sawyer and Huertas 2018), or which disaster management framework to deploy (e.g., One Health). The current COVID-19 crisis provides many examples about the ethical decisions associated with animal care during a global pandemic. Some farmers in North America have had to dump milk following lockdown and social distancing restrictions when processing plants and institutional buyers shut down (Splitter 2020). Meanwhile, supply chain disruptions meant that some poultry farmers were required to depopulate their animals (i.e., the rapid, large-scale destruction of multiple healthy animals in the most efficient way possible) (Kevany 2020). With fewer slaughterhouse spaces to process the market animal surplus, farmers and farm workers are forced depopulate them, resulting in food waste and animal welfare issues when depopulation or euthanasia go awry. Terminating animals before they are able to go to market has a significant emotional and financial toll on farmers. Ethically, depopulation due to a lack of operational processing plants during COVID-19 is entirely different from depopulation necessary to curtail disease spread within a herd or flock or to society. Another significant animal welfare problem during the COVID-19 pandemic is limiting animals' access to feed and water in an attempt to slow their growth (AVMA 2020a).

Animals are also put at risk differentially by their physical or housing conditions. The way animal facilities are organized and the magnitude of animals housed can result in disastrous consequences for animals and humans alike. For example, erecting concentrated animal feeding operations (CAFOs) in floodplains that are in

---

[16] https://www.businessinsider.com/hurricane-sandy-killed-tens-of-thousands-of-research-ani mals-2012-11).

the path of hurricanes or storms (with little option for animals to evacuate on their own), or encouraging stocking densities that incubate and exacerbate animal diseases without clear disaster management strategy are cases of negligent planning. These forms of 'human-induced' hazards fly in the face of acknowledging the independent moral value of animals. They are examples of difficult conversations we must have regarding how we value certain animals. In the Netherlands, thousands of animals have succumbed to stable fires because farmers did not want or could not afford to invest in fire alarms and management systems.[17] Further, being cognizant of the 'carrying capacity' of a particular geographical locale—that is, the number of animals in a particular location and the location's susceptibility to certain kinds of hazards— is essential for both human and animal well-being and to minimize environmental degradation (Irvine 2009). The proximity of animal farms to human communities and precarious ecological entities, and management of farm waste and pollution, continue to have negative implications for human, animal and environmental health as experienced since the Australian Bushfires or Hurricanes Floyd and Dennis (1999) and Florence (2019), despite being almost 20 years apart.[18]

In summary, an ethical conclusion about whether the interests of animals are regarded morally is largely contingent on the type and magnitude of disaster facing a community, how animals are viewed relative to human interests and priorities, and what disaster management plans are in place to attend to animals during a crisis. High and low income countries should take steps to consider the impact of disasters on both humans and animals since their fates are often inextricably bound together. Acknowledging our 'solidarity' with animals during a disaster can serve as an effective and equitable basis for mitigating harm to all affected parties. Where animals are more directly tied to peoples' livelihoods (and thus, cannot be easily replaced), early disaster interventions for animals need to reduce disaster damage (e.g., animal suffering, mortality, morbidity, displacement, asset damage) and indirect losses, in order to promote overall economic recovery and owners', producers' and communities' and veterinary professionals' psychological and social well-being (Campbell and Knowles 2011; FAO 2020; Knowles and Campbell 2014; Martin et al. 2020; Rollin 2011).

---

[17] The authors are grateful to Bernice Bovenkerk for this addition and link: https://www.verzekera ars.nl/media/5048/20180705-actieplan-brandveilige-veestallen-definitief.pdf. To date, a new action plan has been agreed upon by Dutch farmers, the fire department, an animal protection organization and an insurers' organization in order to create safer barns.

[18] At the time of drafting this chapter, citizens in Mozambique, Malawi and Zimbabwe are facing floodwaters and waterborne disease outbreaks (like cholera) in the aftermath of the category 2 Cyclone Idai. Early reports suggest that the storm claimed nearly 1000 human lives and countless livestock lives. Also, floodwaters in the US Midwest have meant that farmers who subscribe to conventional forms of agriculture are contending with economic losses due to lost stockpiled grain and diseased animals and dead livestock (https://www.pbs.org/newshour/nation/for-midwest-far mers-floodwaters-threaten-millions-in-crop-and-livestock-losses).

13.4   Animal Disaster Management: A Reimagining

Animal disaster ethics is a distinctive component of disaster management activity. It asks a central question: "How are animals regarded during a disaster?" The ethical aims orienting this activity involve a societal component—the responsible caretaking of vulnerable animals and groups. It is a systematic social activity governed by norms and motivated by core values to minimize human and animal harm and protect public interests. It aims to bring about welfare outcomes for animals commensurate with their interests and needs. It also obligates communities—not just individuals—to promote these outcomes for the common human-animal good. As indicated above, not all animals are equally protected during disasters and some may be subjected to harms more than once during a disaster (e.g., laboratory animals conscripted in the fight against SARS-CoV-2/COVID-19).

The core ethical problems in disaster management all apply in the case of animals: an unprepared public, limited resources, special responsibilities to vulnerable populations, special obligations of health and veterinary professionals, lack of capacity building and training, community engagement and involvement in all disaster management phases, effective disaster communication, and barriers to gathering relevant evidence to guide interventions. Disaster animal managers need to act rapidly and decisively on the basis of incomplete knowledge. Ensuring public trust and confidence are an essential part of a robust disaster management program involving animals, which can include developing response mechanisms regarding triage care, separation measures such as quarantine, isolation, and physical distancing, and measures to prevent animal-to-human transmission from companion, laboratory, livestock or wild animals (adapted from Heath 1999; Jennings and Arras 2016; OIE 2016). Rapidly growing imbalances regarding supply and demand of essential resources and services during a disaster will require clear ethical guidance on rationing scarce resources and sound triage principles, including implementation procedures that are executable, transparent, equitable, inclusive and engender public trust. Knowledge of animal behavior and of the capacity of different species to cope in different disasters are also crucial.

Transparency and direct links to community and stakeholder involvement will also ensure that public health decision-making related to animal welfare will be effective, humane and just (Vroegindewey 2012), especially if large numbers of animals must be destroyed through depopulation. In the event that depopulation is necessary (such as the highly pathogenic Avian Influenza outbreak in 2014–2015 in the US[19]), adherence to strong ethical standards and procedures, and state and federal laws and regulations should take precedence to ensure that as much consideration as possible is given to the welfare of affected animals.

By and large, animal disaster managers straddle two differing worlds. They are challenged to extend the humanitarian impulse directly to animals within the constraints of a human-centric world. It is imperative, then, that those working in this realm appreciate the vulnerabilities and social and economic positioning of animals

---

[19] https://www.ers.usda.gov/webdocs/publications/86282/ldpm-282-02.pdf?v=3994.

within the risk and the emergency scenarios at the intersection of animal welfare and ethics. Here, sensitivity to the moral considerability placed on human-animal relationships should be observed alongside respect for the interests of animals. Assisting animals and their caregivers can ensure greater survivability and better long-term outcomes for the whole community (Sawyer and Huertas 2018; Vinícius de Souza 2018). The perspectives of those interested in good outcomes (e.g., clinical, behavioral and affective) for animals such as animal owners and caregivers, farmers, the public, first responders, veterinarians, industry agents, aid agencies, policy makers and public health officials, and affected communities, should be considered when deciding the fate of animals during a crisis. Without the support of these groups, the public and animal caregivers may reproach governments for their disregard for animal welfare, a sentiment that could frustrate the disaster management process. According to Sawyer and Huertas (2018), common barriers to effective animal protection from disasters include insufficient knowledge of animal needs in emergencies and a lack of animal management skills; absence of resources for veterinary emergencies within the disaster cycle; lack of recognition to protect animals despite a high dependency between people and their animals; responsibility for veterinary emergencies (nationally and internationally) is either unassigned or ineffective; absence of integration (people and animals) in emergency management; lack of organization amongst subsistence livestock owners making emergency management of animals very difficult (pp. 2–3). The current COVID-19 pandemic reminds us that disasters are rapidly evolving situations and can be experienced differently by different communities. Some communities may be better prepared than others and have contingency plans in place. Disaster management plans should have clear decision-making matrices to outline when animals should be quarantined, depopulated, slaughtered in alternative facilities, sent to a shelter and so on. Furthermore, disasters can put extraordinary and sustained demands on essential community services and public health systems, and frontline workers, veterinarians and those caring for animals, leading to compassion fatigue.

13.5 Animal Disaster Management: Humanitarian Impulse and Animal Welfare Science

How might disaster managers and responders sharpen their sensitivity and judgment regarding animals and their interests before and when disaster strikes?

During a disaster, human beings and animals experience atypical and urgent need of rescue and protection. This is a time of shared vulnerability and solidarity. During an emergency, no one is self-reliant and animals in confined settings are dependent on human beings for their rescue, evacuation and care (e.g., during euthanasia and depopulation; when planning management of animal facilities, shelters and so forth; when performing translocation, rehabilitation and release practices; during triage and clinical treatment; when developing scientific and technological prevention and mitigation strategies). The focus of animal disaster ethics is to understand animals' needs within the context of the humanitarian impulse to aid animals in need, which includes reducing pain, suffering, and loss of life. This humanitarian impulse is at the core of the discipline of animal welfare science, and in the case of disaster management it is recast as respect for animals, ensuring humane treatment, and minimizing harm and vulnerability (i.e., protecting animal welfare), with a view toward the well-being of the entire community and the common good. Animal loss or poor animal welfare prior to, during or after a disaster have devastating implications for owners, caregivers and communities. Communities who rely on animals for social and economic well-being, food security, health, and livelihoods are most in need of community disaster innovations (LEGS 2015; Sawyer and Huertas 2018). The experiences of citizen and animal advocacy groups who self-organized in the wake of Hurricane Sandy underscores the need to consider the collective wisdom and agency of a stricken community. In the aftermath of the storm, animal advocacy organizations like the ASPCA and PETA, local residents, government agencies, FEMA, veterinarians, Petsmart Charities, Iams, and Del Monte Foods banded together as part of a broad though unintegrated coalition to assist the region's animal survivors. Aid came in the form of search-and-rescue operations, food and veterinary services and care, temporary emergency shelters for lost companion animals found during the storm, and use of social media to reunite animals with their families. It is necessary for animal shelters and other animal facilities to establish practical disaster management and planning, and that policies and personnel are prepared and have prompt access to necessary infrastructure.[20]

As animals and their interests gain moral significance in disasters, a deliberate, comprehensive and systematic disaster management system will require sufficient input from central stakeholder groups and planning (i) to prevent disasters through reliable scientific evidence, technology and surveillance sentinel systems, and (ii) to mitigate and prevent potential hazards and strengthen response practices (e.g., by deploying evacuation plans that include animals, since they are evacuated together with their owners or are part of search and rescue operations).[21] The goals of a comprehensive and systematic planning system should be to reduce animal suffering, loss of life, and exposure to agents, venomous and synanthropic animals, chemical and contaminants, contaminated water, and to limit the scale of depopulation and improve recovery initiatives. Research is also needed on the effects of disasters on animal diseases that are not vector-borne and on the impacts of social and economic factors on the consequences of disasters for animals in different parts of the world. Here, animal welfare science is important in determining the research trajectories to pursue.

---

[20] https://www.thedailybeast.com/how-pets-survived-hurricane-sandy?ref=scroll.

[21] This aspect of the chapter is currently being pursued by the authors through a grant-funded research project with colleagues in Alaska, Brazil and Japan. Technological solutions (e.g., robots, artificial intelligence and monitoring devices, easy escape housing) that augment animals' capacity to be self-reliant during an emergency may help animals evacuate or seek shelter quickly.

The examples in the introduction and many others highlighted in this chapter, as well as captive animals abandoned in conflict zones and animals harmed by floods and earthquakes, highlight the invisibility or lack of attention to animal care at the population level when disaster strikes. Reducing negative impacts to animals requires advanced planning, and prevention and mitigation strategies. Animals are subject to varied impacts during a disaster; some prosper in the absence of humans while others to suffer. Since disasters that affect animals are likely to become more common in the Anthropocene, much work needs to be done to ensure that animals' needs and interests become part of the established norms in disaster management. Moreover, disaster management programs should lay out practical and executable guidance that considers all phases of a crisis. This involves examining how ethical processes and animal welfare science apply to and are implemented in the content of policies and processes associated with both specific and integrated disaster events (i.e., compound extreme events) that occur alongside natural and anthropogenic calamities. Examples of multiple simultaneous crises that add a further layer of complexity to an already difficult response include the Australian bushfires, a dengue fever outbreak in Singapore, wildfires in California, and inclement meteorological events in Southern Brazil, all of which are happening due to sudden temperature changes and are concurrent with the COVID-19 pandemic. Special preparations are necessary to address such layered disaster events. Responding to local catastrophes during a global disaster highlights the personal decisions each citizen must make and also the stress placed on different systems (e.g., health, food production). Compound extreme events accentuate the need for trustworthy disaster communicators to demonstrate empathy when framing key questions and answers about personal and interspecific threats. Doing so can ensure public acceptance of recommendations regarding how to navigate a human-animal relationship and equitably allocate scarce resources (FAO 2020; OIE 2016).

Animal welfare science (AWS) can be characterized as the rigorous use of scientific methods to study the quality of life of animals, including companion animals, wildlife, research animals, and those farmed for food. AWS, however, is also borne out of ethical concern for animals (Fraser et al. 1997; Fraser and Weary 2003) and while there is still some conceptual disagreement about what constitutes animal welfare and how to assess it (Weary and Robbins 2019), the field of study can inform deliberations about practices involving animals with a view towards animals' perspectives or animal-based measures. AWS integrates ethological or behavioral, psychological, physiological, environmental, and health measures or indicators to identify whether life is going well or poorly for animals in different contexts (De Paula Vieira and Anthony 2020; Fraser 2008; De Paula Vieira et al. 2008). AWS can broaden how veterinarians and other disaster management professionals consider what is important to animals during an emergency, including highlighting human activities and built environments that lead to vulnerabilities for animals, developing frameworks to set desired outcomes for species, and evaluating the likelihood of success of a contingency plan (Allen and Taylor 2014; Anthony 2004; OIE 2018). AWS will also be essential in informing the development of evidence-based assessments in concert with ethical objectives to minimize harm to the fewest numbers of animals, including when and how euthanasia or depopulation should proceed for the affected species.

In a disaster, AWS can promote good outcomes for animals, through offering technical, systematic and species-specific guidance to manage animals as well as strategies to minimize suffering and loss of life. For example, the welfare impacts of toxins on animals' behavior, physiology and affective states, aversive handling and depopulation techniques, identification of measurable species-specific harms, and effective/positive human-animal-technology interactions in a crisis. AWS training can provide first responders with the knowledge and skills to approach, handle and terminate animals with the least harm and—in the absence of trained responders— empower laypersons who lack specialized knowledge. Training in AWS can help responders recognize when animals are distressed and what constitutes poor and good animal welfare, and to take effective steps to address welfare harms. Further, AWS may also help responders and managers identify animals' natural capacities that might help them cope during rescue or evacuation as well as inform the design of housing systems that can increase animals' chances of survival. AWS can provide, for example, an understanding of population dynamics, animals' affective states and of social behaviors when coping with disasters, provide strategies for curbing zoonoses, animal handling, translocation/relocation management, and the assessment of the effectiveness of a rescue procedure or depopulation techniques at the species level. Through systematic scientific evaluation, AWS can identify indirect harms to non-target animals as a result of ecological or social group disruptions or use of a depopulation technique.

Addressing significant ethical and animal welfare aspects are important to ensuring public support and inclusion of diverse social, cultural, practical and normative perspectives regarding animals in developing a strong, well-functioning disaster management system. Animal disaster management plans should consist of properly trained and well-equipped individuals (e.g., veterinarian animal health and welfare services, animal welfare experts, wildlife service managers, epidemiologists, vaccination administrators, and strike teams) to respond to welfare considerations as well as to the link between humans and animals (Sawyer and Huertas 2018). These management plans should have clear outcomes for animals and their owners in an emergency situation to minimize unnecessary termination of animals' lives through depopulation (AVMA Guidelines on the Depopulation of Animals 2019 Edition). As part of emergency preparedness and response, disaster response teams must decide whether an animal/animals can be saved and what constitutes a good death for the animal/animals given exigent circumstances. Members of the team (which typically will include veterinarians and animal behavior specialists) can advise animal owners, research institutions and animal industries to form an emergency operation plan to minimize welfare harms and the loss of animal life during a disaster as well as promote effective and responsible communication to society and professionals.

Disaster management should consider a cycle of processes that need to be assessed dynamically and continuously, engaging different sectors and actors. In all processes, animals should also be taken into consideration. Disasters bring into focus the practical intervention, welfare, public health, civil defense and protection, biose-curity and scientific challenges associated with each phase of the disaster manage-ment cycle (FAO 2020) as well as the inevitable normative decisions and choices reflecting ethical values that must be made through judicious deliberation regarding our responsibilities to animals (Mepham 2016; Schwartz 2020; van Herten et al. 2020). Below we exemplify common aspects of the disaster management cycle that should be considered by multi-professional teams when devising disaster manage-ment plans aimed at maximizing animal health and welfare (adapted from EM-DAT 2020; OIE 2016; Heath and Linnabary 2015, pp. 174–190; Sawyer and Huertas 2018, pp. 20–23).

(a) Planning: Planning is central to all phases. A community-centered disaster oper-ations plan should be concrete and implementable and consist of contingency and action plans. It should incorporate the needs of animals and their owners. Such plans should identify and prioritize realistic threats and delineate the response mission, goals, capabilities and any gaps to meet them such as through the law or descriptive epidemiology, environmental and other disaster-specific data sets.

(b) Prevention: Prevention is necessary to avoid harms. It should consider the existing infra-structure for animals in rural and urban areas and include geographic information regarding distributions of animal populations and loca-tions, etc. Prevention strategies and funding allocation should include a defined exit strategy involving removal of threats, conservation efforts, epidemiological data of populations through passive and active surveillance, destination or relo-cation of animals to alternative sites to avoid droughts or flooding and for vector control. Prevention includes mapping risks and vulnerabilities to animals, such as susceptibility of certain populations to landslides and infectious diseases.

(c) Mitigation: Mitigation involves interventions aimed at minimizing the impact or costs of disasters to vulnerable animals ahead of their occurrences through anticipation measures. It includes identifying what legislation, regulations and their enforcement are needed and strengthening commitments to resource availability when disaster strikes to protect animals and their welfare. Miti-gation reduces animals' vulnerabilities to physical, behavioral and psycho-logical harms. Specific disaster technologies and scientific developments can help to mitigate harms for animals and should be encouraged. Other examples are strengthening animal shelters and building structures in low-risk zones or constructing physical barriers to prevent flooding or the effects of hurricanes, typhoons or tornadoes.

(d) Preparedness: Preparedness planning involves all threats that cannot be elimi-nated through prevention or mitigation, but must be executed in order to strategi-cally organize and plan the response when disaster strikes. It involves educating and training community members and professionals. Roles of participating animal health and welfare organizations and other officials and stakeholders should be clearly defined. Vulnerable areas and threats to animals should be identified and a network of operational and public communication strategies, including simulation exercises that would consider animals in all steps and evacuation plans should be devised. Essential ingredients include credentialing responders, bolstering public awareness of animals' issues during a disaster and strengthening caretaker capacity to address both human and dependent animals' needs.

(e) Alert: Predictive models can warn a population to evacuate before a disaster strikes. This phase relates to disaster prediction through monitoring the level of animal risk via specific technologies and current scientific data, such as seismology networks, hydrometeorology sensors, and cameras, alongside welfare-friendly animal training and signals/cues that would be essential for an evacuation, such as during an earthquake.

(f) Response/emergency relief: Response/emergency relief focuses on minimizing morbidity, mortality and protecting goods, and sets the stage for helping communities bounce back in the recovery phase. This stage involves the execution of preparedness plans (action and contingency) in concert with different disaster management professionals and organizations. It involves search and rescue, veterinary services and care, evacuation and temporary shelter, and safety and protection. The welfare of animals working in search and rescue operations should also receive specific attention. When disaster triage for animals needs to be performed, first responders and related professionals should have the ethical decision-making tools to maximize the use of resources in order to save the most animals and minimize risks to responders. Training in triage care should advance systematic and immediate assessment to treat critically ill or injured animals and rehabilitation.[22]

(g) Recovery/Rehabilitation: This phase involves activities that center around a vision of a desired future or to restore a community to a pre-disaster status quo as best as possible, including reinstating basic services. Here health, genetic tests, psychological and behavioral rehabilitation practices can be intensified in proper animal facilities to prepare for release and essential monitoring of animal populations and wildlife, post-release. The recovery phase is the longest and most expensive and can take several months or years (e.g., the impact of the Exxon Valdez oil spill on animals and the environment in Alaska's Prince William Sound). The recovery phase gives decision makers a unique opportunity to improve the animal health/welfare infrastructure.

(h) Reconstruction: Financial resources to cover material damages or reconstruct animal facilities or to indemnify losses are important assets in this phase. Inter-institutional coordination and implementation of new legislation and practices also underscore this phase.

Unfortunately, current disaster responses adopted globally expose significant gaps and challenges in disaster management. The COVID-19 outbreak, for example, has uncovered a lack of attention to the risks that infected animals and humans pose to public, animal and environmental health. In particular, the FAO Guidelines

---

[22] For example, see http://veterinarynews.dvm360.com/hurricane-lessons-four-things-we-learned-harvey-and-irma.

(2020) emphasized that disasters not only impact the supply chain (e.g., resulting in animal losses, reduced slaughtering and processing capacity as well as misconceptions regarding animals and animal products being hosts or vehicles of zoonosis that can infect humans) but also the prevention and control capacity of common animal health and welfare services. These include labor shortages disrupting common animal health and welfare practices adopted by farmers and food processors, delays and reduced testing and diagnostic capacities as well as animal disease surveillance and reporting due to restrictions in testing for animal diseases. Additionally, the COVID-19 pandemic underscores the importance of including animals in disaster management practices in the future. Further, the pandemic stresses the need to engage the public and all relevant stakeholders to develop concerted response mechanisms to be followed by practical and systematic knowledge supported by disaster simulation models and the best available scientific evidence, including from animal welfare science.

Incorporating AWS and ethics into epidemiological and environmental studies could illuminate our current understanding of the natural history of disease and of epidemic processes by considering characteristics of the agent, host and environment together with animal care and husbandry (e.g., success of immune transfer, epigenetic effects, level of pathogens in the environment, pre-clinical and clinical signs, local commitment to animal welfare, effects of the human-animal bond, ability to perform species-specific behaviors, and experience of positive and negative affect). Such characteristics could be used not only in observational and experimental studies, but also in predictive epidemiological models when deciding on criteria such as parsimony, goals and data "best fits." It is paramount that current integrated models of epidemiological population projections (e.g., cohort component models, Bayesian probabilistic projections) begin to include current animal welfare science data and expert opinions in order to enhance understanding of animals and how to improve their welfare in the short, middle and long term. These models would, for example, reflect cutting-edge animal welfare and health knowledge in disease outbreaks, thus allowing veterinary epidemiologists to better represent animals' realities and coping mechanisms under professional frameworks (e.g., the One Health initiative [De Paula Vieira and Anthony 2020]). Information technologies can also be used to improve the quality, completeness, and speed of information obtained in field investigations and the speed and sophistication of reports that can be generated from that information at the individual or aggregate level. An example of a decision support system used for emergency planning, response, and recovery that facilitates decision-making when veterinary services are involved in crisis situations is the Veterinary Information System for Non-Epidemic Emergencies (SIVENE), that includes a database, web application, mobile app, and Web Geographic Information Systems (GIS) component. SIVENE provides Italian Veterinary Services (local health units and national and regional veterinary services) with an emergency management tool for disaster management. The data is maintained within its database and converted into real-time information (Possenti et al. 2020).

13.6 Animal Disaster Management: Aims and Recommendations for Ethically Responsible Caretaking

Disaster or emergency ethics is oriented to promote the public interest. In the human case, it has become a systematic field of study (O'Mathuna et al. 2014; Zack 2009). However, animal disaster ethics has yet to catch on as a systematic field, but its time may be ripe (see Heath and Linnabary 2015; Itoh 2018; Meijboom and Stassen 2016; Mepham 2016; Sawyer and Huertas 2018; Vinícius de Souza 2018). As a practical matter, we have a responsibility to domesticated animals simply by virtue of their dependence upon us. Domesticated animals (including wildlife rescued for emergency treatment and rehabilitation) should not be left to fend for themselves during a disaster. The aims of animal disaster ethics should be to minimize the vulnerability of animals to physical hazards and their social stations. For example, minimizing animals' risks to hazards in the first place, providing humane treatment of animals until they are terminated, selecting and using termination methods that are swift, efficient and humane, minimizing negative psychological and emotional tolls on animal caregivers, owners and the public, and mitigating harm (e.g., spread of disease) to adjacent animals (Meijboom and Stassen 2016; AVMA 2013; Rollin 2009).

Determining obligations to animals during a disaster requires having a method of ethical assessment and decision-making that explores the various dimensions of both the hazard and decision on animal life and welfare, and which weighs the considerations that will impact the relevant parties according to specific objectives. Practicing responsible caretaking[23] emphasizes the dynamic and multidisciplinary nature of disasters involving animals and the need for a problem-posing approach to animal disaster management that prioritizes concrete problems and reveals inequities. Animal welfare and One Health considerations serve as an appropriate orientation for action-guidance. The fluid, all-encompassing, unpredictable and uncertain nature of disaster ethics requires practicable and operational guidance for veterinary, public health, civil defense and protection services and other interested professionals embedded in disaster management who must act quickly and decisively.

A centerpiece of animal disaster management is saving lives and ensuring that every effort has been taken in the planning and response phases to ensure the humane treatment of animals. Accordingly, improving critical disaster management issues involves identifying and reflecting on ethical principles, values and inherent biases relevant to disaster management and the plight of animals during disasters.

The nature and complexity of the task of animal disaster management suggest that a one-size-fits-all formula is inadequate and that ethical assessment, analysis and

---

[23] Haynes (2008) distinguishes an ethics of *caretaking* from one of *caregiving*. The latter is more appropriate to primary caregivers of animals, such as farmers and ranchers. In the case of institutional responsibility such as stewardship of the food system, citizen-consumers, policy-makers and industry agents have a collaborative role to inculcate and express virtues of *caretaking* in the design, development, and maintenance of the industrial food system.

deliberation (involving impact on animals) should occur continuously and at several levels (Mepham 2016; Zack 2009). General principles such as a lifeboat scenario do not provide appropriate guidance in the wake and aftermath of a disaster, for they are largely academic and removed from the realities of a crisis situation. For one thing, tens or hundreds of thousands of animals may be involved in a situation over which human beings have little control. Secondly, the nature of disaster management requires proportionality, flexibility, and patience to allow events to evolve and clarify. In real-world contexts, disasters are marked by the pressure of time, interrupted communications and coordination, constant recalibration in response to uncertainty, imperfect knowledge and inadequate equipment and supplies, and legal sanctions and enforcement. Disaster managers and first responders must also contend with unforeseeable developments, huge financial losses and emotional distress, containment of harm (e.g., zoonotic disease) to the health and well-being of the human public, long recovery time and adequate capacity/resources, and safety of responders and strike team members.

Management and response decisions are context-dependent and reflect the social and cultural norms and prioritization of ethical factors (e.g., analysis of beliefs, values and interests) of various stakeholders, as well as legal, economic and practical constraints and considerations. For example, responses to flooding will be very different depending on a community's capacity to mobilize assistance swiftly. The community's capacity is influenced by whether the disaster is connected to overall readiness, human culpability and whether legal fault can be assigned (this will impact who will pay for response or recovery), geography, local political and economic factors like resource allocation and wealth distribution and how and which ethical issues are recognized, deliberated, weighted and prioritized. Also, different communities may place different importance on human life and livelihoods, protection of property, risks and harms to human and animal safety, suffering and loss of life, and community resilience. These differences can impact the objectives and desired outcomes of a response and rescue. For example, in the case of the collapsed Feijão dam in Brazil, the objective was to rescue human lives first. The rescue efforts were hampered by the scale of the disaster, a lack of overall response readiness, unavailability of equipment and confusion about culpability. By the time a local veterinary group (CRMV-MG) was mobilized, decisions to kill animals that were assessed to be in irrevocable distress were made strictly on technical grounds. However, this decision may have been influenced by a relatively weak network for prioritizing animals' interests that is not yet deep or widespread in Brazil. A seemingly clinical or scientific determination about whether to rescue animal survivors may also be impacted by social, economic and ethical factors, such as who is responsible for the long-term care costs for animals, the possibility of reunification with owners, and the cost of rehabilitation or relocation.[24] In contrast, greater attention and preparation were given to pets and companion animals in the Hurricane Sandy case due

[24] https://emais.estadao.com.br/noticias/comportamento,sobe-para-57-o-numero-de-animais-res gatados-em-brumadinho-mg,70002701999; and https://www.bbc.com/news/world-latin-america-48935651.

to the availability of reliable weather forecasts, and a deeper network of existing frameworks to mobilize people and resources to rescue animals

Not all communities attribute the same moral status to animals. Further, different communities will subscribe to different risk analyses. These differences will impact investment in local capacity to address large-scale and abrupt onset of a disaster. Hence, the ethical aims of emergency preparedness and response for animal disaster ethics is up against deeply embedded background conditions that result in elevated risks or vulnerabilities for animals during a disaster. While it will take time to dismantle the deep-seated cultural hierarchy of valuing animals, there are specific opportunities for both ethics and science to help reimagine good outcomes for animals in emergency situations. As alluded to above, AWS and frameworks such as One Health can offer evidence-based support to minimize welfare harms to the fewest number of animals and unnecessary euthanasia and depopulation.

The lack of attention to the needs and interests of some animals as a function of their position on the sociozoologic scale is a preexisting inadequacy in the disaster management (including veterinary and public health emergencies) infrastructure and delivery of aid. While not all animals can be saved due to resource scarcity and a stressed response system, it is important to address institutional or systemic biases about the needs of animals and organizational roles in disaster management. A first step in emergency preparedness is to take into account the population of isolated persons and animals in a given area who might have special vulnerabilities. For example, we should not neglect the experiences of rural or farming communities. Next, by addressing implicit and explicit institutional biases around how we talk about or experience animals in our mixed communities (Midgley 1983), we might begin to appreciate their different meanings and see different ways to value them in order to provide effective solutions in times of crisis.

Disaster management activities should protect public safety, and promote health and welfare to produce desired outcomes consistent with a community's social values. In the Anthropocene, animal interests intersect with human interests. Thus, disaster management activities should minimize the extent of death, injury, disability and suffering during and after the emergency. The disaster management objective to reduce morbidity and mortality of isolated individuals also includes protection and promotion of the health and welfare of the human-animal-environment community with a view to the interest of the common good (adapted from Jennings and Arras 2016; OIE 2016; Heath 1999). With these objectives in mind, we propose six ethically responsible caretaking aims involving animals in a disaster:

1. Saving lives and mitigating harm: Disaster management activities should be respectful of and humane towards animals, individuals and groups, with a view toward public safety, health, and welfare and animal care during and after an emergency. These activities should also include confronting structural factors and deep systemic prejudices that give rise to preventable anthropogenic vulnerabilities endured by animals (for example, livestock).

2. Protect animal welfare and respect for animals' experiences: Disaster management activities should be mindful of standard veterinary clinical measures and

veterinary services, the functioning capacities and behavioral needs of different species of animals and their affective states, and how they are coping during the emergency and its aftermath. Well-trained professionals with knowledge of the capacities and behavior of each species and effective handling should be emphasized.

3. Observe recognition and distributive justice: Disaster management activities should ensure that animals and their interests do not remain invisible during a disaster and that the benefits and burden imposed on the population by the emergency are shared as equitably and fairly as is practicable. As COVID-19 reminds us, the health and welfare of animals should not be ignored during a disaster. Research infrastructure and resources to identify the natural history of emerging diseases and spillover events from animal health and welfare perspectives should be strengthened through animal health and well-being disaster reference centers.

4. Advance public involvement: To maintain public trust, disaster management activities should be grounded in and include decision-making processes that are equitable, inclusive, transparent, and accountable. This basis can help to identify participation and knowledge gaps that should be addressed with appropriate systematic ethics and scientific assessment, outreach and education models. In normal times, an open process of community engagement informed by frank and full consideration of the relevant animal health and welfare science and ethical assessment of community values and interests should be encouraged as part of disaster management governance.

5. Empowerment of caregivers, guardians, owners and community members: Disaster management activities should strive to empower animal caregivers, guardians and owners and community members through education, training and mutual communication exchange as part of community vigilance, responsibility, solidarity and resilience, and developing capacity to provide effective animal care during and after the disaster.

6. Public health and veterinary community professionalism: Disaster management activities should recognize and enhance the skills and competencies of, and coordination among, public health and veterinary professionals. It should also include protective and coping strategies that can help minimize unnecessary mental and emotional distress on both the affected animals and the disaster management professionals.

13.7 Recommendations

In the Anthropocene, we—individuals, communities, governments, businesses, and professionals in disaster management—bear a moral responsibility to identify where the barriers to ethically responsible animal disaster management are likely to occur and to take appropriate steps to rectify them in order to prevent or reduce harm to animals. The collective interests embodied in disaster management measures should also include those of animals.

The foregoing discussion highlights that disaster management solutions in the Anthropocene have to be at the intersection of human-animal-environmental touch-points and cannot be amended by simply attending to human interests. The list of unanswered questions for animal disaster management in the Anthropocene is long. How can we empower first responders to be resilient under the chronic stress of a zoonosis and natural disaster? How should individuals and communities prepare for layered disaster events? How could animal welfare scientists catalyze the engagement with the public and other professionals to come up with funding, science-based poli-cies and technologies that benefit and maximize resources benefiting animals during disasters? What sort of public engagement, risk communication and early warning systems can improve uptake so that individuals, governments, organizations and communities have feasible and effective intervention strategies at their disposal to act ethically to advance animal care during a disaster or a compound extreme event? While the One Health framework can provide a foundation for guiding collective attention, ethical inquiry that actually improves the lives of all animals during a disaster requires government regulation based on animal welfare, voluntary commer-cial schemes (e.g., standards and guidelines) to minimize the vulnerability of animals and commitment by private sector stakeholders, communities and individuals towards disaster preparedness and caretaking activities.

Towards advancing the six ethically responsible caretaking aims we need a publicly accountable set of operating procedures or action steps that can empower immediate caregivers of animals and disaster management teams to ensure that animals' interests are systematically promoted in disaster management. They are (not an exhaustive list):

1. Respect and Humane Treatment: Animals should not be considered a "prob-lem" or an afterthought for disaster management. Indeed, animals can be a valu-able resource for emergency planning, mitigation and response (e.g., animals as sentinels of danger and vehicles to assist in the evacuation of human beings). Disaster management and implementation should recommend a strategic frame-work for deliberation and action and strive for humane outcomes for animals in a crisis situation. Disaster management should clearly articulate the legal and ethical bases for achieving certain objectives involving animal welfare and public health, including humane handling and knowledge of animals' anatomy and physiology and temperament of the species being handled or terminated.

2. Collaboration and Effective Disaster Communication: Animal exponents (local farmers, veterinarians, civil defense and protection servants, animal welfare experts, epidemiologists, field workers, IACUC (Institutional Animal Care and Use Committee) representatives, concerned citizens) should have the opportu-nity to participate actively and directly in advanced planning and communication vis-à-vis emergency preparedness strategies. Disaster planning will be bolstered by these sources of local and specialized knowledge who are familiar with the day-to-day activities and patterns of behavior of various community members who care for, depend on, and/or use animals, as well as with knowledge of the animals themselves. Collaboration is particularly important to create spaces for

dialogue about positions that are ethically defensible, well-informed by science and local knowledge, and empirically relevant. Effective disaster communication is needed to prevent unnecessary abandonment of animals and minimize public panic and to reduce viral spread through effective physical distancing of animals. Public engagement can lead to clear communication of the risks aimed at minimizing disruptions to companion animals, wildlife, laboratory animals and livestock. It should also clearly highlight the interventions being deployed, how practicable they are for laypersons to follow and delineate how animal health and welfare will be advanced. Effective disaster communication will also highlight how priorities for resources and community services are allocated by related officials and disaster management teams and their ethical bases.

3. Strengthening Systems of Information Sharing, Surveillance, Scientific Research, Management and Training: Disaster managers should have up-to-date information concerning the numbers of animals and their whereabouts (e.g., a vulnerabilities database) and be responsive to generate the best available scientific evidence that contextualizes animal welfare in the referred disaster. Reference animal health and well-being research centers are essential for providing reliable information rapidly for the disaster management team, without only prioritizing the welfare of humans. Also, they should have the contact information of the responding/local veterinarians and related services and/or have access to the same real-time data as veterinarians and other collaborators who have jurisdiction to act regarding animals and their welfare (e.g., strike team or depopulation leaders) on the ground. Data management—that is data collection, organization, interpretation and dissemination about disasters—is an increasingly important asset. How data informs the practices and procedures adopted by the official epidemiological services should be reviewed carefully to enhance systematic ethics assessment and judicious priority setting, ideally by an interdisciplinary team. Technologies that mitigate and prevent animal disasters should also be included in any disaster management plan.

4. Community Outreach and Proactive Contact: Appropriate public involvement or civic engagement on animal issues can promote understanding and acceptance of necessary public health measures. Disaster managers and response teams should identify and map community assets and be in contact with communities so that in turn they can help identify and reach vulnerable populations and isolated groups, especially since disruptions to transportation and telecommunications are likely. Core procedures should consider how risk communication with the public and stakeholder involvement should be coordinated and how best to stockpile and ensure equitable and effective use of equipment and supplies.

5. Cultural Sensitivity and Attitudes Check: Disaster managers and responders should not over-generalize beliefs and attitudes or base emergency preparedness on untutored or unexamined assumptions concerning how animals might be vulnerable during the disaster or how they are valued (e.g., as largely moral subjects or commodities). Treatment of animals during a crisis should occur in a manner that minimizes animals' pain and distress as much as is practicable under the circumstances. Public perceptions of the humaneness of the procedures

used to handle or terminate animals are important for the success of a disaster management campaign and/or to mitigate the emotional and psychological toll of depopulating animals *en masse* by field personnel such as strike team members and veterinarians, when doing so becomes necessary, and efforts should be made to educate and gain public support.

6. Reflection, Review and Reform: Upon resolution of the emergency situation, it is important to review the humaneness and effectiveness of procedures involving the treatment of animals during the disaster in order to enhance future procedures and processes and minimize negative outcomes to animals. Doing so will reveal unintended biases regarding outcomes to animals, people and the environment and strengthen future policies and strategies, including improving on crisis standards of care. It will likely enrich understandings of the institutional expressions of social, moral and species inequities shaped by structures of power and politics that drive discourses of animal issues in the Anthropocene.

With the onset of the Anthropocene, humans have unwittingly created the conditions for an increasing amount and severity of disasters. In thinking further about the seemingly irreversible nature of how human beings have changed the planet, we owe it not only to our fellow human beings, but also to our fellow non-human animals to be prepared to deal with these calamities. Humane and respectful treatment of animals during disasters requires, amongst other things, the coordinated action of different professionals, informed by animal welfare science, and a reconsideration of the attitudes of a diverse set of stakeholders towards the moral status of animals.

Acknowledgements The authors would like to thank our colleagues who participated in the *Animals in Our Midst Expert Meeting* for their feedback in April 2019. In particular, we owe a debt of gratitude to Clemens Driessen for his guiding insights and thoroughness in reviewing our draft manuscript in conjunction with the Conference.

References

Allen, H., and A. Taylor. 2014. Evolution of US foot-and-mouth disease response strategy. *Disaster Prevention and Management* 23: 19–39.

American Veterinary Medical Association (AVMA). 2012. *Emergency Preparedness and Response Guide.* Available at: https://ebusiness.avma.org/files/productdownloads/emerg_prep_resp_guide.pdf. Accessed 1 March 2019.

American Veterinary Medical Association (AVMA). 2013. *Guidelines for the Euthanasia of Animals: 2013 Edition.* Available at: www.avma.org/KB/Policies/Documents/euthanasia.pdf. Accessed 1 March 2019.

American Veterinary Medical Association (AVMA). 2019. *Guidelines for the Depopulation of Animals* 2019 Edition. Available at: https://www.avma.org/KB/Policies/documents/AVMA-Guidelines-for-the-Depopulation-of-Animals.pdf. Accessed 3 June 2019.

American Veterinary Medical Association (AVMA). 2020a. Evaluating emergency euthanasia or depopulation of livestock and poultry. Available at: https://www.avma.org/sites/default/files/2020-04/Humane-Endings-flowchart-2020.pdf. Accessed 29 April 2020.

American Veterinary Medical Association (AVMA). 2020b. COVID-19 impacts on food production medicine. Available at: https://www.avma.org/resources-tools/animal-health-and-welfare/covid-19/covid-19-impacts-food-production-medicine. Accessed 29 April 2020.

Anthony, R. 2004. Risk communication, value judgments, and the public-policy maker relationship in a climate of public sensitivity toward animals: Revisiting Britain's foot and mouth crisis. *Journal of Agricultural and Environmental Ethics Special Supplement on Agricultural Crises: Epizootics and Zoonoses on Farm Animals* 17 (4–5): 363–383.

Arluke, A., and C. Sanders. 1996. *Regarding Animals*. Philadelphia, PA: Temple University Press.

Campbell, R., and T. Knowles. 2011. *The Economic Impacts of Losing Livestock in a Disaster: A Report for the World Society for the Protection of Animals (WSPA)*. Melbourne, Australia: Economists at Large.

Dalla Villa, P., S. Kahn, N. Ferri, P. Migliaccio, L. Possenti, and G. Vroegindewey. 2017. The role of the OIE in disaster management and risk reduction. *OIE Bulletin* 1: 20–28. https://doi.org/10.20506/bull.2017.1.2591.

De Paula Vieira, A., and R. Anthony. 2020. Recalibrating veterinary medicine through animal welfare science and ethics for the 2020s. *Animals (Special Issue on Veterinary Ethics)* 10: 654. https://doi.org/10.3390/ani10040654.

De Paula Vieira, A., V. Guesdon, A.M. de Passillé, M.A.G. von Keyserlingk, and D.M. Weary. 2008. Behavioural indicators of hunger in dairy calves. *Applied Animal Behaviour Science* 109: 180–189.

Delgado, C., M. Rosegrant, H. Steinfeld, S. Ehui, and C. Courbois. 1999. Livestock to 2020: The next food revolution. Discussion Paper for the International Food Policy Institute.

ECDC (European Centre for Disease Prevention and Control. Towards One Health preparedness). 2018. Stockholm: ECDC.

EM-DAT The International Disaster Database. 2020. Center for Research on the Epidemiology of Disasters—CRED. Available at: https://www.emdat.bc/classification. Accessed on 1 September 2020.

FAO. 2017. *The impact of disasters and crises on agriculture and food security*. Rome, Italy: FAO, 2018. Available online: http://www.fao.org/3/I8656EN/i8656en.pdf. Accessed 20 August 2020.

FAO. 2020. *Guidelines to Mitigate the Impact of the COVID-19 Pandemic on Livestock Production and Animal Health*. Rome. https://doi.org/10.4060/ca9177en.

Fraser, D. 2008. Understanding animal welfare. *Acta Veterinaria Scandinavica* 50 (Suppl 1): S1. https://doi.org/10.1186/1751-0147-50-S1-S1.

Fraser, D., and D.M. Weary. 2003. Quality of life for farm animals: Linking science, ethics and animal welfare. In *The well-being of farm animals: Challenges and solutions*, ed. G.J. Benson and B.E. Rollin, 39–60. Oxford: Blackwell.

Fraser, D., D.M. Weary, E.A. Pajor, and B.N. Milligan. 1997. A scientific conception of animal welfare that reflects ethical concerns. *Animal Welfare* 6: 187–205.

Glassey, S. 2020. Animal Welfare and Disasters. Available at: https://oxfordre.com/politics/view/10.1093/acrefore/9780190228637.001.0001/acrefore-9780190228637-e-1528. Accessed 18 August 2020.

Haynes, R. 2008. *Animal welfare competing conceptions and their ethical implications*. Netherlands: Springer.

Heath, S.E. 1999. *Animal management in disasters*. St Louis, MO: Mosby Year Book.

Heath, S.E., and R.D. Linnabary. 2015. Challenges managing animals in disasters in the U.S. *Animals* 5 (2): 173–192.

Hiko, A., and G. Malicha. 2016. Climate change and animal health risk. *Climate Change and the 2030 Corporate Agenda for Sustainable Development—Advances in Sustainability and Environmental Justice* 19: 77–111.

Institute of Medicine. 2003. *The Future of the Public Health in the 21st century*. Washington, DC: National Academies Press.

Institute of Medicine (US) Forum on Microbial Threats. 2007. Ethical and Legal Considerations in Mitigating Pandemic Disease: Workshop Summary. Washington, DC: National Academies Press (US).

Irvine, L. 2009. *Filling the ark: Animal welfare in disasters*. Philadelphia, PA: Temple University Press.

Itoh, M. 2018. *Animals and the Fukushima Nuclear Disaster*. Switzerland: Palgrave McMillan.

Jennings, B., and J.D. Arras. 2016. Ethical aspects of public health emergency preparedness and response. In *Emergency ethics: Public health preparedness and response*, ed. Bruce Jennings, John D. Arras, Drue H. Barrett, and Barbara A. Ellis, 1–103. New York: Oxford University Press.

Johnson, C.K., P.L. Hitchens, P.S. Pandit, J. Rushmore, T.S. Evans, C.C.W. Young, and M.M. Doyle. 2020. Global shifts in mammalian population trends reveal key predictors of virus spillover risk. *Proceedings of the Royal Society B* 287: 20192736. https://doi.org/10.1098/rspb.2019.2736.

Kevany, S. 2020. Millions of farm animals culled as US food supply chain chokes up. *The Guardian*. Available at: https://www.theguardian.com/environment/2020/apr/29/millions-offarm-animals-culled-as-us-food-supply-chain-chokes-up-coronavirus. Accessed 30 April 2020.

Knowles, T., and R. Campbell. 2014. *A Benefit-Cost Analysis of WSPA's 2012 Intervention in the Dhemaji District of Assam, India*. Melbourne: Economists at Large.

(LEGS) Livestock Emergency Guidelines and Standards. 2015. *Livestock Emergency Guidelines and Standards*, 2nd ed. Rugby, UK: Rugby Practical Action.

Martin, N.D., J.L. Pascual, J. Hirsch, D.N. Holena, and L.J. Kaplan. 2020. Excluded but not forgotten: Veterinary emergency care during emergencies and disasters. *American Journal of Disaster Medicine* 15 (1): 25–31. https://doi.org/10.5055/ajdm.2020.0352.

Meijboom, F.L., and E.N. Stassen (eds.). 2016. *The end of animal life: A start for ethical debate: Ethical and societal considerations on killing animals*. Wageningen, Netherlands: Wageningen Academic Publishers.

Mepham, B. 2016. Morality, morbidity and mortality: An ethical analysis of culling nonhuman animals. In *The end of animal life: A start for ethical debate. Ethical and societal considerations on killing animals*, ed. F.L. Meijboom and E.N. Stassen, 341–362. Wageningen, Netherlands: Wageningen Academic Publishers.

Midgley, M. 1983. *Animals and why they matter*. Athens: University of Georgia Press.

Murray, G., and S. McCutcheon. 1999. Model framework and principles of emergency management. *Revue Scientifique et Technique* 18: 15–18.

Nelson, C., N. Lurie, J. Wasserman, and S. Zakowski. 2007. Conceptualizing and defining public health emergency preparedness. *American Journal of Public Health* 97(S1). Available at: https://www.ncbi.nlm.nih.gov/pmc/articles/PMC1854988/. Accessed 31 January 2019.

OIE. 2016. *Guidelines on disaster management and risk reduction in relation to animal health and welfare and veterinary public health*, 1–8. Paris. Available at https://www.oie.int/filead min/Home/eng/Animal_Welfare/docs/pdf/Others/Disastermanagement-ANG.pdf. Accessed on 19 August 2020.

OIE. 2018. *Terrestrial animal health code*, 27th ed. Paris: OIE. Available at: www.oie.int/standard-setting/terrestrialcode/. Accessed 4 February 2019.

O'Mathuna, D.P., B. Gordijn, and M. Clarke (eds.). 2014. *Disaster bioethics: Normative issues when nothing is normal*. The Netherlands: Springer.

O'Toole, T., M. Mair, and T. Inglesby. 2002. Shining light on "dark winter". *Clinical Infectious Diseases* 34 (7): 972–983.

Pinillos, R.G. 2018. *One welfare: A framework to improve animal welfare and human well-being*. CABI International.

Possenti, L., L. Savine, A. Conte, N. D'Alterio, M.L. Danzetta, A. Di Lorenzo, M. Nardoia, P. Migliaccio., S. Tora, and P. Dalla Villa. 2020. *A New Information System for the Management of Non-Epidemic Veterinary Emergencies* 10 (6): 983. https://doi.org/10.3390/ani10060983.

Powers, M. 2016. Vulnerable populations in the context of public health emergency preparedness planning and response. In *Emergency ethics: Public health preparedness and response*, ed. B. Jennings, J.D. Arras, D.H. Barrett, and A.E. Barbara, 135–154. Oxford: Oxford University Press.

Rist, C.L., C.S. Arriola, and C. Rubin. 2014. Prioritizing zoonoses: A proposed one health tool for collaborative decision-making. *PLoS ONE* 9 (10): e109986. https://doi.org/10.1371/journal.pone.0109986.

Rollin, B.E. 2009. Ethics and euthanasia. *Canadian Veterinary Journal* 50: 1081–1086.

Rollin, B.E. 2011. Euthanasia, moral stress and chronic illness in veterinary medicine. *Veterinary Clinics of North America Small Animal Practice* 41: 651–659.

Salinsky, E. 2002. Public Health Emergency Preparedness: Fundamentals of the 'System'". National Health Policy Forum. Paper 82. Available at: https://hsrc.himmelfarb.gwu.edu/sphhs_centers_nhpf/8. Accessed 7 February 2019.

Sawyer, J., and G. Huertas. 2018. *Animal management and welfare in natural disasters*, 1st ed. New York: Routledge.

Schwartz, M.E. (ed.). 2020. *The ethic of pandemics*. Peterborough, ON, Canada: Broadview Press.

Splitter, J. 2020. Farmers face their worst-case scenario: 'Depopulating' chickens, euthanizing pigs and dumping milk. *Forbes*. Available at: https://www.forbes.com/sites/jennysplitter/2020/04/28/farmers-face-their-worst-case-scenarios-depopulating-chickens-euthanizing-pigs-and-dumping-milk/#571aee873003. Accessed 28 April 2020.

Stauffer, K.E., and L. Conti. 2014. One health and emergency preparedness. *Veterinary Record* 175 (17): 422–425.

United Nations Office for Disaster Risk Reduction (UNISDR). 2015. Sendai Framework for Disaster Risk Reduction 2015–2030. Available at: www.unisdr.org/files/43291_sendaiframeworkfordrren.pdf.

Van Herten, J., B. Bovenkerk, and M. Verweij. 2019. One Health as a moral dilemma: Towards a socially responsible zoonotic disease control. *Zoonoses and Public Health* 66 (1): 26–34. https://doi.org/10.1111/zph.12536.

Van Herten, J., S. Buikstra, B. Bovenkerk, and E. Stassen. 2020. Ethical decision-making in zoonotic disease control: How do one health strategies function in the Netherlands? *Journal of Agricultural and Environmental Ethics* 33: 239–259. https://doi.org/10.1007/s10806-020-09828-x.

Varner, G. 2012. *Personhood, ethics, and animal cognition: Situating animals in hare's two level utilitarianism*. Oxford and New York: Oxford University Press.

Vinícius de Souza, M. 2018. Medicina Veterinária de Megacatástrofes: Um Panorama do Maior Desastre Antropogênico na História do Brasil. *Revista CFMV Brasília DF Ano XXIV* 79: 55–61.

Vroegindewey, G. 2012. Animal welfare in disaster management. Proceedings of the Third OIE Global Conference on Animal Welfare, Implementing the OIE standards—Addressing regional expectations, 35–37. Kuala Lumpur, Malaysia, 6–8 November 2012.

Weary, D., and J.A. Robbins. 2019. Understanding the multiple conceptions of animal welfare. *Animal Welfare* 28 (1): 33–40.

Winders, D. 2020. Opinion: As Facilities Close for Covid-19, Stranded Animals Could Suffer. Available at: https://undark.org/2020/04/09/animal-facilities-covid-19/#. Accessed on 24 August 2020.

WSAVA. 2020. Covid-19—An Update for WSAVA Members. Available at: https://wsava.org/wp-content/uploads/2020/05/COVID-19-An-Update-for-WSAVA-Members-May-29.pdf. Accessed on 8 August 2020.

Zack N. 2009. The ethics of disaster planning: Preparation vs response. *Philosophy of Management* 8: 55–66.

Zinsstag, J., E. Schelling, D. Waltner-Toews, and M. Tanner. 2011. From 'One medicine' to 'One Health' and systemic approaches to health and well-being. *Preventive Veterinary Medicine* 101: 148–156.

Andreia De Paula Vieira (DVM, MS, Ph.D.) is an independent veterinarian, animal welfare scientist and One Health/One Welfare researcher, from Curitiba, Brazil. Between 2015–2020 was Professor of Animal Welfare, Public Health and Epidemiology at Universidade Positivo, Curitiba, Brazil. During the same period, she served as Research Manager at Centro de Pesquisa da Universidade Positivo (CPUP), a multidisciplinary Research Center. Prof. Dr. De Paula Vieira's specializations include Animal Behavior, Animal Welfare Science, Sustainable Animal Systems, Public Health, Disaster Management, Epidemiology, Knowledge Transfer Methods, Instructional and Educational Design and Public-Private Partnerships for the development of policy and best practices. Between 2012–2015, she served as project leader for the Animal Welfare Indicators Project (AWIN - WP4). Her team of IT developers and designers at Universidade Positivo worked together with international collaborators to develop the information architecture for the Animal Welfare Science Hub. Her publications have appeared in Biosystems Engineering, Journal of Dairy Science, Animals and Applied Animal Behaviour Science. She is currently the ISAE (International Society for Applied Ethology) Country Liaison for Brazil. Her current research projects include: wildlife conservation, facial recognition and neural networks and AI for animal systems, public health and human-animal relations in policies and management, values-aware research regarding the sustainability of animal systems and disaster management for animals. Prof. Dr. De Paula Vieira obtained her Ph.D. in Animal Science at the University of British Columbia, Canada.

Raymond Anthony is Professor of Philosophy at the University of Alaska Anchorage. His publications are at the intersection of environmental-animal-climate-food ethics and the philosophy of technology. He serves on the American Veterinary Medical Association's (AVMA) Animal Welfare Committee and is a co-author on the AVMA's Euthanasia, Humane Slaughter and Depopulation Guidelines, respectively. He has published peer-reviewed articles and book chapters with scientists, veterinarians and ethicists on ethical assessment, bioethical analysis, governance and decision-making, stakeholder engagement, sustainability, climate ethics and food and agricultural ethics, including a Council on Agricultural Science and Technology (CAST) report, Well-being of Agricultural Animals: Scientific, Ethical, and Economic Aspects of Farm Animal Welfare (2018). He was co-PI on an USDA research project focused on developing curriculum for agricultural animal bioethics and conducted a CAPES sponsored values-aware research that explores challenges to the sustainability of the dairy chain in southern Brazil. Prof. Anthony co-chaired the food and water systems working group for the Anchorage Climate Action Plan. Currently, he is PI for a National Institute of Food and Agriculture sponsored research project, WELLANIMAL, a project seeking to both map the ethical dimensions and epistemological challenges that affect human-animal relationships and develop strategies to promote farm animal welfare and care, during a novel pandemic.

Chapter 14
The Decisions of Wannabe Dog Keepers in the Netherlands

Susan Ophorst and Bernice Bovenkerk

Abstract Dogs have for long been humans' best friend, but the human–dog relationship can be problematic. A mismatch between dogs and their keepers can lead to welfare problems for both; for example: breeding for a specific look can result in health and welfare problems and importing dogs from other countries can lead to zoonoses. In our view, many of these problems could be avoided if wannabe dog keepers reflected better before deciding to obtain a specific dog. Attempting to influence this decision, however, assumes that we know what the right choice is. In this chapter, we discuss three cases: pups with pedigrees, pups without pedigrees, and adult dogs from (foreign) shelters. We show that, in each case, certain moral assumptions are made whose legitimacy can be problematised. We conclude that the decision about what dog to obtain is not a straightforward one and that it is often difficult to establish what is actually the right choice. However, we also pinpoint certain improvements that can be made to the current system and make a number of suggestions that make the right choice the easier choice. As Anthropocene conditions may lead to the domestication of an increasing number of wild species in the future, this analysis may support reflection on the ethical implications of domestication.

Terms like keeper and owner of animals are controversial with respect to the autonomy of animals. We deem keeper to be more neutral than owner. As we discuss the role of people in their desire to keep a dog in their life, we choose to use this term in this chapter. Owner is used only in the term ownership for lack of a suitable replacement. Although human companion might be a better term, it might give rise to confusion as we are focusing on the moment at which a human decides to obtain a certain dog.

S. Ophorst (✉)
University of Applied Sciences Van Hall Larenstein, Leeuwarden, The Netherlands
e-mail: susan.ophorst@hvhl.nl

B. Bovenkerk
Wageningen University and Research, Wageningen, The Netherlands
e-mail: bernice.bovenkerk@wur.nl

S. Ophorst
Radboud University, Institute for Science in Society, Nijmegen, The Netherlands

© The Author(s) 2021
B. Bovenkerk and J. Keulartz (eds.), *Animals in Our Midst: The Challenges of Co-existing with Animals in the Anthropocene*, The International Library of Environmental, Agricultural and Food Ethics 33,
https://doi.org/10.1007/978-3-030-63523-7_14

14.1 Introduction

As dogs are one of the most successfully domesticated species over the longest period of time, they are the quintessential example of the Anthropocenic animal. A long period of mutual influence has provided us with a lot of experience and knowledge regarding dogs; we could even say that humans and dogs have been domesticated alongside each other. From living side by side for mutual benefit in hunting and protection, to modern-day circumstances where the dog in Western countries seems totally dependent on humans, the human–dog relationship gives rise to discussion on domestication and its boundaries. Because of their long history of domestication, dogs provide an interesting illustration of the human–animal relationship and its multiple ethical challenges in the Anthropocene. The domestication of dogs could even be regarded as a paradigm case for the challenges that human–companion animal relationships face in the Anthropocene. As argued in the introduction to this volume, changing habitats and climatic circumstances lead to formerly wild animals increasingly becoming liminal. The next—perhaps inevitable—step may be the future domestication of currently wild species (see also the chapter by Palmer in this volume). Following Swart and Keulartz (2011), we view domestication as a gradual process; most animals lie somewhere on the continuum between wild and domesticated. As criteria for domestication, we take firstly the degree to which animals have adapted to their human environment and secondly the degree to which they are dependent on it. The more animals have adapted and the more dependent they are on humans, the more domesticated they are. Their level of dependence and adaptation has consequences for their agency, although these consequences are not clear-cut. For example, animals that are very dependent but not well adapted—such as a zoo animal—are frustrated in their agency, as they have little influence over their own life but likely do feel the need to express wild behaviour. Many dogs, on the other hand, are both very dependent and very well adapted. In contrast to wild dogs and stray dogs, companion dogs usually do not have the possibility to shape their own life, as their decisions are limited about, for example, where to live with what companions and with what conspecifics to mate. On the other hand, the fact that dogs are so adaptable means that they have learnt to express their agency in different ways, and they are very capable of conveying their wishes to their keepers. It is no coincidence that recent literature on animal philosophy often focuses on the human–dog relationship (Hearne 2016; Haraway 2007); dogs' adaptability and long history of domestication have enabled communication and collaboration between humans and dogs. For these reasons, we want to reflect on the lessons that can be learned from this 'successful' domestication story.

The human–dog relationship has undergone several changes: from living side by side for mutual benefit in hunting and protection, to the breeding of dogs specifically as workers, to modern-day circumstances where in Western countries dogs are primarily kept for their company and in order to confer status. All these situations have given rise to their own moral problems, leading to discussions on the justification of domestication and its boundaries. Problems faced today, for example, include dog health and welfare problems due to breeding for exaggerated looks, biting incidents, zoonotic diseases as a result of the importation of dogs, and health and welfare problems as well as a lack of socialisation of dogs in illegal breeding operations. In our view, many of these problems stem from the way in which people decide what dog to obtain and could be avoided if this decision moment was influenced. Of course, more ethical problems follow once one obtains the dog. Is one allowed to spay the dog for example? What should be done when the dog becomes ill and has to undergo a costly operation? Under what circumstances is one allowed to euthanise a dog? However, these problems are not the topic of this chapter, as they do not stem from the decision moment and can therefore not be influenced in the same way.

There are 1.5 million dogs in the Netherlands. As this number has been more or less consistent over the last decade, to maintain this number of dogs in society, with an average life expectancy of 10 years, each year a decision is made 150,000 times by humans to bring a new dog into their home. Of this number of dogs, Neijenhuis and Hopster (2017) established that 65% were born in the Netherlands, of which 26% had a pedigree. Of the remaining 35%, 16% were imported and 19% not registered—the latter being illegal in the Netherlands, which has had a mandatory identification and registration system for dogs since 2014. Different considerations by humans lead to different choices about dogs, each giving rise to its own ethical problems.

In the following sections, we discuss these choices and relate them to views expressed by animal ethicists on dog ownership in particular or pet ownership in general. Pups with pedigrees, pups without pedigrees, and adult dogs from (foreign) shelters form the cases that feature in these sections. At the end of this chapter, we discuss what can be learned from the perspective of different choices by humans and the steps forward that can be made in the interest of both dog and human. We show that, in each case, certain moral assumptions are made whose legitimacy can be problematised. We conclude that the decision about what dog to obtain is not a straightforward one and that it is often difficult to establish what is actually the right choice. This is important to realise, because if many of the problems that we encounter with dogs originate from the moment of the decision to obtain a specific dog, and if we want to somehow steer this decision moment in the right direction, we need some perspective on what that right direction is. By discussing different motivations for obtaining a specific type of dog and problematising these, we aim to make wannabe dog keepers reflect more on the implications of such a decision.

14.2   Animal Ethicists' Views on Dog Ownership

Numerous animal ethicists have engaged in the discussion regarding problematic aspects of dogs as companion animals. The ideas of a number of influential scholars are used in this chapter to show the variety of ideas on this topic. What they all share is that they have a view on animal welfare. In nearly all animal ethics theories, the central idea is that we should not harm animal welfare and that we should promote positive animal welfare. However, ethicists from different theoretical backgrounds may hold different views on what welfare entails. For example, some define welfare as having pleasant affective states, others focus on the ability of an animal to carry out species-specific or natural behaviour, and yet others hold a broad view of welfare as well-being over the course of the animal's whole life (see Bovenkerk and Meijboom 2013). Moreover, the significance that is attached to welfare differs between different theoretical frameworks. For example—generally speaking—for welfarists, the only criterion to determine correct treatment of animals is the effects on their welfare, whereas Kantians also take considerations beyond welfare into account, and ecocentrists contend that a certain amount of suffering is simply part of an animal's life. This implies that, if a good decision about obtaining a dog is dependent at least partly on dog welfare, determining the right decision will be dependent on what moral theory one holds and how one defines welfare.

Utilitarian philosopher Peter Singer (1973) sees no problem about keeping dogs as companion animals as such, as long as suffering is avoided. He does not differentiate between domesticated and wild animals; all animals have an equal interest in experiencing enjoyment and avoiding suffering. This obligates the keepers of dogs to treat them well and prevent them from being harmed. In contrast, legal scholar Gary Francione (and Garner 2010) thinks that pet keeping in general is problematic, as it depends on the idea that pets are human property. The assertion that pets are property suggests that they are things and condemns them to being mistreated, but animals are clearly not things. As animals have moral and legal rights in his view, and beings with rights should, most fundamentally, not be treated as human property, we should not keep pets. In this view, the purposeful breeding of puppies should be abolished, and with that the practice of keeping dogs as companion animals will die out eventually. In the meantime, we should only adopt dogs from shelters and treat them as equal companions rather than 'slaves'. Francione is opposed to domestication as this violates animal rights and makes animals thoroughly dependent on humans. Domesticated animals 'are dependent on us for everything that is important in their lives: when and whether they eat or drink, when and where they sleep or relieve themselves, whether they get any affection or exercise' (Francione 2012). This view contrasts with the idea of scholars such as Stephen Budiansky (1992) and Baird Callicott (1992) that domesticated animals have hypothetically signed an unspoken 'domestication contract'. In Budiansky's view, dogs initiated their own domestication; by choosing to associate with humans, they have gained many benefits and this has given them an enormous evolutionary advantage. However, both Budiansky and Callicott argue that, when animals in a specific situation are made worse off than they would have been in the wild, or when the relationship between humans and animals has been undermined by maltreatment, the contract has been broken by humans and a domestication contract can no longer serve to justify these practices. They think that this is the case mainly in relation to the way in which livestock are raised and do not appear to have a problem with the domestication of dogs. Clare Palmer (1997), on the other hand, rejects the notion of a domestication contract, as a contract presupposes informed consent, which is something that animals cannot give. Moreover, even if the ancestors of currently living animals had voluntarily entered into a domestication contract, this cannot be assumed to still hold to this day.

In Palmer's view (2010; Sandøe et al. 2015) then, we cannot justify our treatment of domesticated animals by assuming a domestication contract. The fact that we have domesticated animals gives rise to special obligations towards them. According to her contextual-relational ethics, there is a difference in humans' duties towards domesticated as compared to wild animals. Whereas a laissez-faire attitude towards wild animals is warranted, we have special obligations towards companion animals, livestock, or laboratory animals, because we have brought them into the situation in which they find themselves and they are dependent on us for their well-being. In fact, by domesticating them, we determine not only their freedom of movement and possibilities to make decisions for themselves, but even their genetic make-up. These actions of ours give rise to a moral commitment to take special care of their well-being.

This commitment is further elaborated regarding dogs by Kristien Hens (2009), who also argues from a relational-ethical perspective. She views the human–dog relationship as a reciprocal one; a relationship that is enabled by the fact that both human and dog are social animals. She argues, in contrast to Francione, that dogs are not treated as things or tools: 'It is questionable whether the relationship dog-human would have been so successful if they were merely man's tools.... it is more than just one of master versus slave' (Hens 2009, 6). Therefore, 'if we want to think of a proper ethic towards dogs, we must do so in the context of the dog and its specific niche, which is the human world, not using some vision of the dog as a wild animal' (Hens 2009, 5). Because dogs are 'natureculture' animals and we have decided to have a relationship with them and take them into our homes and families and in effect make them part of our communities, we have additional responsibilities towards them over and above our general responsibilities towards all sentient animals. Not only does she deem caring for the emotional and physical welfare of dogs as the responsibility of humans, but also sees special responsibilities, such as 'ensuring a bond of trust, which should not easily be broken' as part of the commitment (Hens 2009, 3). Dogs in her view, then, have an interest in maintaining a good relationship with humans and in being part of the human community. Special obligations to which this gives rise include, for example, the creation of dog parks, teaching schoolchildren how to handle dogs, and having strict government regulations on breeding.[1] The special obligations also extend to the moment of choice to obtain a particular dog; in Hens' view, this should be done only after thorough reflection. Moreover, simply taking a dog to a shelter before the holiday season would violate the relation of trust between dog and human.

Sue Donaldson and Will Kymlicka (2011) give a political turn to this relational and rights-based ethic: domestic dogs should be seen as individual agents with basic rights and citizenship rights, as they are already part of our society. As we brought them into our world by domesticating them, we owe them full inclusion as it is just as much their world as ours. Contrary to Francione, these authors do not see

---

[1]Despite the focus on special responsibilities towards dogs on the basis of their place in the family and mixed community, she argues that there are limits to these responsibilities: we do not have the same responsibilities towards dogs as towards our children for example.

the dependency of domesticated animals as necessarily undignified; for them, what matters is how we respond to dependency. Dependency should not necessarily be regarded as a weakness, but as a basis for a good relationship: 'If we don't view dependency as intrinsically undignified, we will see the dog as a capable individual who knows what he wants and how to communicate in order to get it – as someone who has the potential for agency, preferences, and choice' (Donaldson and Kymlicka 2011, 84). Different individuals in the human–animal relationship should be able to realise their own versions of the good life, and this means that dogs should be given the opportunity to make important choices for themselves. Such a choice could also mean that a dog no longer wants to live with a human family. The good life for a specific dog could be to become a stray dog and find a new pack outside a human family.

Finally, the good life is also central in Martha Nussbaum's capabilities approach (2006). According to this approach, justice demands that each individual is treated with respect for his/her dignity and flourishing. Each individual has certain innate capabilities and the good resides in the opportunity that the individual has to utilise those capabilities, as this is what makes the individual flourish. In order to find out what constitutes flourishing for an individual, one must create the opportunities for the individual to live up to his/her species norm. As dogs have co-evolved with humans for millennia, their species norm is different from that of wild dogs or wolves; a flourishing life for them entails their having possibilities for choice and for cultivating their capabilities, and this means that they need to be trained and disciplined to a certain extent by human guardians, just like human children. Just like Donaldson and Kymlicka, Nussbaum extends her theory of justice to the political realm. Although not giving a definite list, she suggests a number of capabilities that must be respected for animals to be able to flourish and that lead to political principles. Life and bodily health are obvious capabilities that must be met. Furthermore, she argues that, in the capabilities approach, animals are entitled to bodily integrity. For example, their bodies should not be 'mutilated' out of aesthetic motivations (Nussbaum 2006, 395). To the extent that they are capable of it, animals are entitled to make their own choices, receive suitable education or training, play, have room to move around, and have access to a variety of activities. They should also be able to form attachments to, and express love and care for, human or non-human others. Legally, animals should be granted political rights and the legal status of beings with dignity (Nussbaum 2006, 399). Finally, they have the right to the integrity of their habitat, either wild or domestic.

Against the background of animal ethics theories regarding domesticated animals, and dog keeping in particular, we next look at different practices in which dogs are obtained and point out a number of challenges in each case. We first discuss pups with pedigrees, then pups without pedigrees, and finally adult dogs from (foreign) shelters.

14.3   Pedigree Pups

As people who choose to bring a pup with a pedigree into their home often use the pedigree to legitimise the consciousness and the deliberation of their decision (Bovenkerk and Nijland 2017), we start with an explanation of what a dog pedigree entails. The Fédération Cynologique Internationale (FCI) is an international organisation consisting of members and contract partners in 94 countries, with one organisation per country being allowed to join. It recognises 346 breeds, which are all assigned to countries, mostly based on heritage, which are responsible for drawing up a standard for the characteristics that the breed should have, in terms of appearance, movement, and behaviour. A pedigree is a document that proves that the breed to which the dog belongs has a certain heritage, consisting of pedigree dogs from the same breed. In the Netherlands, this entails DNA testing for all pedigree dogs to verify that the pedigree is correct. Members (kennel clubs) can attach conditions to a pedigree, but this is not obligatory. In the Netherlands, requirements for a pedigree are based on the general welfare of the brood bitch, e.g. the age of her first and last litter and the time needed between consecutive litters. For just a few breeds, there are specific requirements in terms of health, without which a dog cannot receive a pedigree. The majority of breeds do not have such requirements however.

What ethical considerations play a role in the case of pedigree dogs? As Francione wishes to abolish dogs as property, a pedigree dog is certainly not an option for him, as the pedigree is proof of heritage, but also a registration of ownership. One of the rights of dogs that can be seen as violated in pedigree dog breeding is the right to choose their own mate for propagation. In most instances, the dog and the bitch are put together by humans, often without at least the bitch having much say in the matter and being held during the act. An analogy can be made with the purposeful breeding for predictability in offspring and genetic screening in humans, which are subject to ethical and legal discussion. When it comes to embryo selection, Dutch society struggles with the topic (De Haan et al. 2010). Dutch legislation condones embryo selection for the benefit of the embryo itself. This makes it possible to screen for invasive genetic diseases and to abort the embryo if it carries such a disease. It is prohibited, however, if anyone else is the beneficiary, such as a sibling who could benefit from stem cells of the embryo to fight his/her own disease, or parents with certain wishes with regard to their progeny. Grey areas arise where multiple parties benefit. If we transfer this stance to dog breeding, the rule should be that selection in dogs should only be done if the dog is the direct beneficiary of the practice. As the dog stands to gain nothing by being the best hunter, the fastest, or the best example of the breed standard, and this is for the most part just beneficial to humans, the practice of dog breeding certainly is not in accordance with rights that are similar to those of humans. This is where Donaldson and Kymlicka (2011) also have problems with the practice of pedigree dog breeding. For them, it is very important that animals can exhibit their own agency, and this includes being able to decide themselves with whom to mate. For Nussbaum, presumably such breeding would be allowed only if it respected dogs' capabilities and led to their flourishing. This may be the case if the selective breeding is carried out to enhance a dog's capabilities—for example when negative side-effects of prior breeding decisions are reversed—but not for harmful aesthetic reasons.

As Singer, Hens, and Palmer emphasise humans' responsibilities towards dogs, albeit to different degrees and for different reasons, another subject for discussion is the necessity to keep dogs safe and prevent them from harm. As the issue of pedigrees in the Netherlands depends on certain rules regarding the well-being of the brood bitch, pedigrees can be said to contribute to the safekeeping of dogs.

To obtain a pedigree, a dog must stem from parents from the same breed. This limits the possibilities of propagation. Not only is the dog's freedom to choose a mate for propagation limited by this practice, but pedigree breeding also increases the risk of hereditary diseases, which can be caused by breeding in a limited gene pool.

As some of these diseases, such as heart conditions, cancer, hip dysplasia, or epilepsy, cause suffering throughout the dog's life or shorten its life expectancy, dogs are not protected from harm in these instances.

On the other hand, pedigrees can be used to prevent hereditary diseases. A pedigree is a registration, which can also be used in health testing of dogs. The test results are connected to the pedigree of the tested dog and can then be traced back in a database. As some diseases are breed specific, DNA tests for certain conditions can vary for different breeds and therefore the ancestry of the dog is relevant. By using the information from these databases and checking for genetic closeness, choices in breeding can be made that diminish the risk of hereditary diseases. The pedigree in itself is not a guarantee of this, but the use that the breeder makes of possibilities to prevent the pups being harmed determines whether this goal is met. In other words, depending on how breeders and kennel clubs use their pedigrees, the registration system can be used either to guarantee healthy dogs or have harmful effects.

The standards for each breed are the guidelines for dog-show judges to assess the dogs. Also, breeders' interpretations of these standards play a part in how these guidelines work out in actual dogs. This has led to alterations in the appearance of many breeds over the years (McGreevy and Nicholas 1999), with the exaggeration of certain traits causing health and welfare problems to the dogs. Examples of this include an emphasis on broad chests, which cause problems with movement and with natural delivery, or problems in breathing for brachycephalic dogs, such as French Bulldogs. In these instances, surely harm is done to dogs, and the dogs' interests are not protected by dog-show judges, breeders, and buyers.[2] These excesses in dog breeding have given rise to an ongoing moral debate about where we should draw the line when changing the genetic make-up of companion animals, analogous to the debate that has been going on about selective breeding and genetic modification in livestock. This debate has focused not only on resulting health and welfare problems,

---

[2]However, it is not so easy to use ethical theory to explain why harm is done to these dogs, as the changes to the dogs have been made before the dogs were born and therefore one cannot say that a particular individual has been harmed. For a discussion on this application of the so-called non-identity problem to dogs, see Palmer (2012) and Bovenkerk and Nijland (2017).

but also on other ethical concerns. For example, some argue that extreme breeding violates dogs' integrity, that it objectifies or commodifies dogs, or that it is unnatural (Bovenkerk and Nijland 2017). On the other hand, a recent Danish study (Sandøe et al. 2017) shows that keepers of some breeds find that the problems that their dogs encounter are actually a reason for their strong bond, probably because the dog needs so much care and attention. Hens, in particular, emphasises the importance of the bond between dogs and their keepers. She probably assumes that this bond will primarily have advantages for the dog. The tension between dog welfare and the strong bond—as experienced by humans—in this situation establishes that this is not necessarily the case.

Historically, dogs were selected for function (e.g. hunting, guarding) and the dog's appearance merely needed to support this function. With large parts of these functions being taken over by newer developments (such as the meat industry and security cameras), the focus has shifted from function to appearance (Lindblad-Toh et al. 2005). This does not mean that all behavioural traits have vanished, as they are still present in the dogs today. It is a shift from an emphasis on function, with the best hunting dog being the most wanted, to appearance, where the dog that best looks the part is most popular in breeding. This entails a risk of humans choosing a certain breed, based mainly on appearance, without realising that the original function of the dogs requires them to roam around freely for hours on end, hunt other animals, be aware of others (including humans) entering their domain—needs that cannot necessarily be met in a dog's everyday life in the Netherlands.

As the information on the dog's heritage is guaranteed by the pedigree, this also contributes to the predictability of the dog's behaviour. It is sometimes argued that people have a better chance of finding a dog that is compatible with their circumstances when they choose a pedigree pup than it is when they choose a pup without a pedigree.[3] However, as the information in breed standards on behaviour is very limited—e.g. 15 of the 434 words in the Golden Retriever standard (FCI 2009)—one could question whether people can rely on this information. Obviously, more information is needed to evaluate compatibility. Breeders may be able to provide this information, as they are knowledgeable about the breed, at least as an experiential expert, but are not required to do so in order to be able to sell puppies with pedigrees.

As dogs with pedigrees are more expensive than dogs without pedigrees, there is a chance that the motive to make money will override informing prospective buyers on subjects that would possibly prevent them from buying a puppy of a certain breed. Singer would argue on this subject that humans' differing interests must be weighed against each other (earning a living versus living with a suitable dog) and this then must also be weighed against the interest of the dog (living in conditions where not all the dog's needs are met). The outcome of this weighing is not clear-cut. For Hens, the possible endangerment of the formation of a bond of trust between human and dog by withholding information weighs heavily, and she therefore condemns practices that put financial benefits before the human–dog bond. Palmer draws a line between commercial and non-commercial dog breeding, with pedigree breeders falling into

---

[3]Based on interviews carried out by Bernice Bovenkerk and Hanneke Nijland in 2015 and 2016.

the category of non-commercial breeding most of the time, because of the limited number of litters that they are allowed to produce in order to obtain a pedigree for the dogs. The commercial aspect in itself does not seem to cause her concern, but the emphasis on profit over animal welfare does.

If we look at the ethical acquirement of a dog and deem it most important to prevent harm to the dog, obtaining a pedigree dog does not have to be ruled out. On the other hand, pedigree dog breeding does not necessarily support the ethical acquirement of dogs either. The full potential of pedigrees to contribute to the well-being of dogs is currently not met. If the pedigree served more than is currently the case as a quality mark for adhering to dog welfare, testing on hereditary diseases, and informing prospective buyers, more problems in the breeding and keeping of pedigree dogs could be overcome. It is, however, important for wannabe dog keepers to realise that the pedigree at the moment is not in all cases a mark of good dog welfare.

14.4 Pups Without Pedigree

As only a relatively small proportion of the dogs acquired in the Netherlands have a pedigree, it is also interesting to look at pups without a pedigree. Maybe they will turn out to be the more ethical choice when people want to share their life with a dog.

Pups without pedigrees can be dogs from a breed that is not registered with the FCI and therefore not entitled to a pedigree. These dogs can have the advantages of a pedigree dog with regard to the predictability of, for example, size and personality as an adult, but registration does not even have the basic requirements that a pedigree has. One is therefore dependent on the reliability of the breeder. Mostly, these are breeds created by selection not so long ago. Examples of these new breeds or designer breeds are the Labradoodle and the Miniature Australian Shepherd, but the Pitbull also has no FCI registration. The problems encountered with pedigree dogs may also exist with this type of dog and maybe even more so as the trustworthiness of the registration papers is more questionable. Although the latter is debatable, as the Dutch Kennel Club is not monitored by the Dutch government either, at least procedures have for long been in place regarding self-control and upholding the rules.

Pups without pedigrees can also be pups from pedigree dogs, where the breeder did not apply for a pedigree. In most instances, a pedigree being too expensive and people not wanting to 'pay for a piece of paper' is given as the reason for not applying for the pedigree. It cannot be ruled out, however, that an important reason may also be to avoid the minimal requirements of a pedigree in terms of dog welfare. As in this case dogs' heritage is not confirmed by a pedigree, one can also question whether other breeds or untraceable dogs are in the ancestry of the pups. Dogs that look very similar to pedigree dogs but do not actually have a pedigree are called look-alikes. As these dogs are bred from the same gene pool as the pedigree dogs, most genetics-related problems in pedigree dogs will also be found in look-alikes. One could argue that the lack of testing and information on problems in certain lines of a breed might even enhance the problem. A study by Van Zeeland and Beerda (2015) showed that it is impossible to determine whether problems are more severe in dogs with pedigrees or in look-alikes, because of the lack of registration of pedigrees by veterinarians and the lack of control on the reports by dog keepers, who consider the dog to be a pedigree dog even if the dog does not actually have a pedigree. From the perspective of people acquiring these puppies, the money argument is mostly used to warrant this choice. There are people that find it ridiculous to spend a lot of money on a dog. Although the money transaction is proof of the dog being considered property and therefore not regarded as an ethical choice according to Francione, there is not much difference between paying a large and a small amount of money. As caring properly for a dog also involves at least feeding and veterinary costs, one can argue that not being willing to spend money on the 'purchase' of a puppy does not bode well for the intentions towards proper care of the dog after the puppy has entered the household, which can also be costly. As people tend to be more involved with a purchase when it involves a larger amount of money (Bauer et al. 2006), one could argue that a lower price for a dog can have a negative effect on the considerations made before getting a dog. A large number of the puppies in this category are distributed through illegal dog traffickers and so-called puppy mills, which are notorious for the deplorable state in which bitches and puppies are kept, with diseases and even premature death as a result (Radstake 2016). Here, certainly, people's responsibility towards dogs is not honoured and dogs suffer; this makes these practices reprehensible in the eyes of all animal ethicists.

Then there are the pups that look like originals, whose background can only be guessed. These are the dogs that are commonly referred to as mutts or mongrels. Despite these not very flattering names, positive traits are attributed to these dogs as they are said to be strong and healthy, often in comparison to pedigree dogs (Patronek et al. 1997). As a longer life expectancy in combination with fewer health problems is in the interest of the dog, this might be a preferred choice when a person is obtaining a puppy. As dogs nowadays are not normally free-ranging animals in the Netherlands, with rules for keeping dogs on the leash in most areas, dogs have little chance of meeting a mate without their keepers' interference. In the old days, dogs would just roam to nearby farms when there was a bitch in heat, but these 'accidental litters'— from a human perspective—are rare these days. In Kymlicka and Donaldson's work, the fact that humans make the procreative choices for dogs is already problematic. Keeping dogs on leashes can, however, also be seen as an instrument to keep dogs safe. In densely populated countries like the Netherlands, many more dogs would fall victim to traffic accidents if they had more freedom. Also, other animals would be at risk from dogs hunting.

With lots of dogs being castrated, especially when dogs and bitches are kept in the same household, there is an increasingly small chance of dogs in the same household procreating. The number of puppies born out of free encounters between dogs is not registered, but it is surely nowhere near the number needed to fulfil the demand for companion dogs. Then there are the encounters between dogs that are put together by their keepers. There are few such deliberate non-pedigree litters, although they might solve some problems that arise with pedigree dogs. In 2017, a Dutch foundation, Dier & Recht (Animals & the Law), known for lawsuits against pedigree dog breeders, launched an initiative called Cupidog. Cupidog aims to bring healthy dogs and bitches of different breeds or unknown descent together to create healthy puppies that are brought up under good conditions. Amongst the candidate dogs on the Cupidog website are dogs that are obviously of breeds that carry major problems. Cupidog aims for healthy puppies by ensuring that a dog will never be mated to a dog comprised of 50% or more of the same breed and by having every combination approved by a committee of two veterinarians. Interestingly, the majority of dogs enrolled in this service are male. The similarity with dating services for humans is remarkable, as male clients also predominate on those dating websites. In the case of Cupidog, obviously the dog does not choose to be put on the website, and no information is provided on the sex of the dog's keeper. It might be interesting to investigate the extent to which idealised online identities are portrayed of the dogs on Cupidog, as is found to be the case with human dating sites (Hancock et al. 2007). In the Cupidog case, a reason for the overrepresentation of males might be that the female dogs' keepers are expected to care for the puppies and might also take into account the risk for their bitch in delivering a litter, whereas the male dogs' keepers might see their dog as a great candidate to produce offspring or want to cater to their dog's sexual urges. For the keepers of male dogs, Cupidog involves no risk or work for the human and therefore might be an easier choice.

Although the health of mongrel dogs is used as a reason to favour these dogs when a puppy is being chosen, it cannot be ruled out that the low purchase cost of these dogs is also relevant to dog buyers. Cupidog sets the price for pups from litters that they mediate at between €500 and €700. This is lower than for most pedigree dogs, but higher than for dogs from a shelter or what people are used to paying for a non-pedigree dog. This might be another reason—besides the lack of potential mother dogs—why this initiative is not taking off with a flying start. As in the case of look-alikes, it is not very promising if humans are not willing to pay these prices when the costs of keeping a dog are much higher than the purchase cost and it may mean people pay less attention to the decision of getting a dog (Bauer et al. 2006).

Another aspect of mongrel pups is the uncertainty about what they will grow into. Pedigrees provide an estimation of the character and size of the dog, as also is largely the case with non-official breeds and look-alikes, but a mongrel dog can easily surprise one. This might lead to an adult dog that does not suit one's situation, and this can be detrimental to the welfare of both dog and human. From Hens' perspective, in these instances the dog should remain in the situation, as the bond between dog and keeper should be respected and preserved at all times. It can, however, be argued that it is not in the dog's best interest to be kept in inappropriate circumstances. This predicament is caused by the unpredictability of how a puppy will turn out as a mature dog. Although mongrel dogs, then, give people less control over the outcome, it could be argued in their favour that the need of humans to be in control over and 'manufacture' nature is in itself wrong and should not be encouraged (Bovenkerk and Nijland 2017). Dogs could, however, be the victim of mismatches and not in a position to alter their circumstances as their keepers might be. Responsibility in

Palmer's terms is a leading concept here. The dog's best interest should prevail in these situations and guide decisions on the dog's future.

In sum, acquiring a dog without a pedigree is, again, not necessarily a good or a bad choice. On the other hand, there is no ethically sound argument for purchasing a dog that stems from illegal dog trafficking or a puppy mill. This pleads for stronger regulations and enforcement of these regulations to ensure (breeding) dogs' welfare and discourage current practices. The option of obtaining a dog this way is simply a bad choice. This is not necessarily the case for non-pedigree look-alikes or new breeds. Here, the same arguments hold as previously put forward for the pedigree dogs: a registration system could be used to enhance dog welfare and eliminate some of the problems encountered in dog breeding. On the basis of ethical arguments, the original dogs are a promising choice, as long as they can be distinguished from pups from puppy mills or illegal dog trafficking. Favouring this option is not, however, carefree either. Considerations about possibilities to match the demand with the offer while still guarding the boundaries of welfare and dog rights are challenging. Creating awareness of the costs of dog keeping and the expectations regarding the dog will prove to be just as challenging.

14.5 Shelter Dogs

Most dogs mediated by animal shelters are adults. Sometimes, a pregnant bitch is brought into a shelter and delivers her pups there or in a foster home. Sometimes, puppies are found as strays and brought to animal shelters. Sometimes, puppies are confiscated from puppy traffickers and taken in by shelters. The majority of confiscated, stray, and relinquished dogs, however, are adult.

As taking in a shelter dog does not require deliberately bringing new dogs into the world, it is a practice that Francione condones as the proper solution for fading out the practice of keeping dogs as property. Most shelters do not offer the opportunity to buy a dog, just to adopt. This may be seen as an option to avoid the 'dogs as property issue'. Unfortunately, there still remains an 'owner', which in the case of shelter dogs is the shelter itself, even if the dogs are adopted out. One could argue that the possibility of putting dogs into shelters makes it easier for people to carelessly buy new dogs and therefore perpetuates this situation. In Francione's vision, it should be prohibited to keep dogs, and getting a dog from a shelter would be just a temporary solution for dogs that have already been brought into this world. Even though in Hens' view it is morally problematic to take a dog to a shelter, she also prefers people obtaining their dogs from shelters over breeding, in particular purebred dogs. She suggests that shelters should employ dog behaviourists in order to match the right dog to the right person (Hens 2009).

People who want a dog from a shelter can be motivated by a number of reasons: wanting to save a dog in need, to save money as shelter dogs are less costly to obtain than most puppies, wanting an older dog that is already housebroken, obedient, or calmer, or wanting an adult dog to avoid a mismatch. Dogs' welfare is an argument for people wanting to save dogs from the shelter and could also be an argument for people wanting to avoid a mismatch, as mismatches can be the source of welfare problems.

The current situation in Dutch shelters shows that the 'save the dog argument' is not particularly strong as an exclusive argument. This would entail people choosing to adopt the dogs that are most unhappy in the shelter situation or have been there for the longest time. Instead, people leave the long-stay dogs in the shelter, as the smaller and younger dogs or dogs of popular breeds get out of the shelter more quickly (Dierenbescherming 2018).

The 'saving argument', then, does not seem to be the decisive argument for a lot of people.

It could also be that people whose priority is to save dogs choose to get a dog from a foreign, mostly South or Eastern European, shelter. In the Netherlands, this has become a common option that has been growing in recent years (Radstake 2016). A reason given for choosing a foreign versus a Dutch shelter is the situation within these shelters. As Dutch shelters are mostly governed by the Dutch Society for the Protection of Animals (Dierenbescherming), basic welfare for dogs in the shelters is guaranteed, and euthanasia of shelter dogs is not common practice and is carried out only in very specific circumstances. In countries like Spain and Greece, shelters are mostly private initiatives run on tight budgets and have to provide shelter for too many dogs, leading to deplorable situations. Originally, tourists brought back stray dogs and shelter dogs from their holidays, but nowadays 130 organisations mediate between foreign shelters and Dutch people looking for a dog (Radstake 2016). As the willingness to save dogs is given as an important motivation for adopting a dog from abroad, it is important to take a closer look at this situation.

The main problems with foreign shelter dogs seem to be their lack of sociali-sation and potential health risks (Buckley 2020). As the Mediterranean countries have a different climate, different diseases exist there, for example because they are hosted by parasites that thrive on those weather conditions. Brucellosis and rabies are examples of zoonotic diseases that can be transferred from dogs to other dogs and to humans. Blood testing before adoption can help to prevent problems, but not all diseases can be tested for definitively, and some have long incubation times, which may lead to false-negative results (Fox et al. 1986). So far, no large outbreaks of diseases brought into the country by foreign shelter dogs have been reported, but, with the increasing number of this type of dog entering the country, this could quite easily happen. Besides the health risks for Dutch dogs, therefore also risks for Dutch people must be taken into account; dogs from foreign shelters form a potential public health risk.

It is not always easy to distinguish the origin of the dog. Cases are known of Dutch shelters bringing in dogs from foreign shelters to meet the demand for smaller and younger dogs (de Joode, n.d.). Instances have also been reported of organisations with just a commercial motive pretending to save foreign shelter dogs (Van Niekerk et al. 2014). These organisations play into the positive dog-saving image of the legitimate rescue organisations, when their practice is actually plain dog trafficking, often also without proper procedures followed and health precautions taken. People are therefore advised to check dogs' vaccinations, microchips, and the organisation's legitimacy. This requires extra action on the part of wannabe dog keepers, and it is uncertain to what extent they will actually follow this advice when the easy option is to fall in love with a sweet dog or give into the urge to save an innocent animal from life on the street or from a terrible shelter.

Many of the dogs in foreign shelters have led stray dog lives before being sheltered. This means that they are not used to living in a house, are used to lots of freedom to roam around, are used to being on the constant lookout for food, are wary of humans, and so on (Pal et al. 1998; Udell et al. 2010). Looking at this situation from a human point of view, the dogs might be suffering in these circumstances, with hunger, danger, and illnesses always lurking. This would require action on welfare grounds according to Singer and, in the case of abandoned dogs, also to Palmer. From Donaldson and Kymlicka's perspective, domesticated dogs have the right to food, shelter, and medical care. But what should be done about dogs' right to make their own decisions, to freedom, and to execute their agency? One could argue that the freedom that these free-ranging dogs enjoy is of great value (Majumbder et al. 2014; Paul et al. 2016) and is at risk when they are adopted out to the Netherlands. Dutch society is totally different than what the dogs are used to, and dogs that are adopted in the Netherlands out of such different circumstances certainly lose freedom and can have behavioural problems (Dietz et al. 2018). Moreover, the capabilities and interests of dogs that have formed attachments to other stray dogs and perhaps formed packs with them are not respected when they are suddenly taken out of their environment. In Nussbaum's view, this could be problematic, because capabilities are at least partly formed by one's relationships. This situation can cause a lot of anxiety in dogs that is by no means beneficial to their welfare or flourishing. On the other hand, in contrast to Donaldson and Kymlicka, Nussbaum does not distinguish clearly between domesticated dogs, who are part of our communities, and wild or stray dogs, who for Donaldson and Kymlicka have the right to have their sovereign communities respected. Nussbaum would be less hesitant to take in stray or wild dogs, as long as this does not interfere with their flourishing. Despite good intentions and professional help, however, there is often little that can be done to correct for the dog's bad or different start, as dogs' socialisation period ends at around four months of age (Freedman et al. 1961).

One might argue that there are similarities between adopting dogs and adopting children from other countries. Problems with attachment and adjustment have also been said to be issues in the adoption of children from other countries (Post 2008). However, there have also been reports to the contrary, as most children show a lot of resilience and the effect of growing up in poor circumstances may be even more negative (Juffer 2008). As the socialisation of dogs and humans differs and the 'window' for dogs to be socialised closes at an early age (Serpell and Jagoe 1995), the analogy between human and dog fails in this respect. Still, in the last decades, the view on adopting children from foreign countries has changed. Whereas saving children from detrimental circumstances was considered a noble action not so long ago, the current vision is that problems with human trafficking are prevalent in adoption procedures (Post 2008). Because of this, there is a tendency to leave children in the area from which they originate and provide care there. Obviously, this does not solve the desire for children of people who are unable to conceive children, but the children's welfare should take priority, and this has led to a decreasing number of children from other parts of the world being adopted in the Netherlands (Slot 2008). This can be compared to dogs, because, in the case of importing foreign shelter dogs, there is also mention of dog traffickers profiting from the transaction (Stray Animal Foundation Platform 2018). It would be beneficial to the dogs to remain in their country of origin and receive help there to ensure better living conditions for them. If anyone wanting to adopt a foreign dog did so without actually bringing the dog to the Netherlands and just supported it financially throughout its life, it would certainly make a contribution to the dog's welfare.

Adopting a dog from either a Dutch or a foreign shelter can be a better choice if prospective adopters look carefully into the organisation providing these dogs. Clear rules and quality characteristics are not easily found by the wannabe dog keeper, so improvements are necessary. Determining the suitability of the dog for the adopter's situation is another topic that can be easily trivialised and should receive more attention.

14.6 Discussion

As we have seen, ethical challenges exist in every choice when a dog is being obtained. As we define a good choice as a choice where at least the welfare of the dog is served and the welfare of other dogs and animals (including humans) is not harmed, one can wonder whether there is such a thing as a good choice of dogs. In Francione's view, abolishing altogether the practice of keeping dogs is the only option. The other animal ethicists that we discussed are less dismissive of the domestication process as such and the opportunities for dogs to experience good welfare or to flourish, although different ethicists use different definitions of welfare, and in practice these opportunities often fail to materialise.

However, all three channels/scenarios discussed provide options to better protect the welfare of all concerned. In the current situation, there is ample room for improvement in the dog-breeding system, with its forced mating, harmful breed characteristics, and restricted gene pools. On the part of the people wanting a dog, this entails careful consideration to determine what is in the best interest of all involved. This requires self-control at the moment of decision making about a specific dog. As Berkman et al. (2017) show, self-control is a value-based decision-making process in which people weigh up different aspects. In this process, easy choices are given more weight than difficult ones. Factors like the time it takes to acquire information, the effort it takes to process the information, or the financial costs can be barriers to making right decisions. Currently, agencies that want to improve decision making regarding dogs put a lot of emphasis on information that people should acquire before making their decision about a dog. With such an overload of information, it is not strange that people fall victim too easily to processing arguments such as nice memories of the dogs they used to have, the example of the dog next door, or a cute appearance.

Solutions can be found in multiple directions. One direction is the possibility of governmental control, possibly delegated to neutral controllers, on for example pedigrees and shelter licences (including more stringent breeding standards towards better health and welfare), an obligatory waiting period to enable people to do their research before bringing a dog home (which is currently standard practice in many Dutch shelters), or even the obligation to obtain a licence before being allowed to obtain and keep a dog. In a study by Packer et al. (2017) on the purchase of pedigree dogs, over a third of the respondents testified that they would do more pre-purchase research the next time they wanted to purchase a dog. Another direction is the possibility of different sectoral organisations implementing these measures. This has proved difficult, with sectoral organisations being dependent on support from their members, who may have different interests, resulting in slow change processes. The third direction lies in influencing wannabe dog keepers. If people are made aware of the consequences of bad decisions and are facilitated in making the right choices, this could be an essential step towards better dog welfare. As we have seen in the work of Berkman et al. (2017), the key is to make the good choice the easy choice. This requires the information to be presented in such a way that it can be easily accessed and processed by the wannabe dog keeper, and it may also entail a better infrastructure for dog acquisition practices. One could think here, for example, of making it more difficult to obtain a dog through less trusted channels, such as internet marketplaces. This is not possible for an individual to achieve without the assistance of all other parties. Moreover, it helps when wannabe dog keepers have positive role models or a social network that enables them to reflect on their decision. After all, the dog that someone has often becomes part of that person's identity, and a positive role model will help to shape an identity that matches well with the dog's welfare. An integrated approach towards sensible dog keeping is therefore the most promising route.

Humans' special responsibility towards dogs, in Palmer's and Hens' views, warrants the investment in these types of integral solutions. As dogs' welfare is served with this approach, an integral solution also complies with Singer's view. Moreover, in an integral approach, potential mismatches are avoided and guided choices are beneficial to the relationship between dog and human, as emphasised by Hens. The political solutions set out above would be supported by Donaldson and Kymlicka as well as Nussbaum.

What remains is the discussion on ownership and property. This is an issue that cannot simply be overcome by changing names or constructions such as adoption rather than ownership. Recognising that dogs cannot be seen or treated as a tool or an ornament, and therefore need advocates on their behalf, is a step on the route towards solutions that constitute a good choice, as already sketched. This seems to be the closest we can get other than abolishing companion animals altogether. After all, we can wonder how realistic the abolishment of animal domestication is. The destiny of many currently wild animals may be to become domesticated as a result of Anthropocene conditions. Human and animal habitats are becoming more and more intertwined, and animals are facing challenges to their survival consequent to changing climatic and environmental conditions. In order to help them survive, we may have to resort to technical and other interventions that may cause them to lose a measure of wildness (see the chapter by Palmer in this volume) and become more liminal or even domesticated. If an increasing number of animals become domesticated, we shall be facing challenges similar to the ones sketched in this chapter. Reflection on the pitfalls of dog keeping, and in particular the question of what dog to obtain, may shed some light on the challenges faced in the Anthropocene.

References

Bauer, H.H., N.E. Sauer, and C. Becker. 2006. Investigating the relationship between product involvement and consumer decision-making styles. *Journal of Consumer Behaviour* 5 (4): 342–354.

Berkman, E.T., C.A. Hutcherson, J.L. Livingston, L.E. Kahn, and M. Inzlicht. 2017. Self-control as value-based choice. *Current Directions in Psychological Science* 26 (5): 422–428.

Bovenkerk, B., and F.L.B. Meijboom. 2013. Fish welfare in aquaculture: Explicating the chain of interactions between science and ethics. *Journal of Agricultural and Environmental Ethics* 26 (1): 41–61.

Bovenkerk, B., and H.J. Nijland. 2017. The pedigree dog breeding debate in ethics and practice: Beyond welfare arguments. *Journal of Agricultural and Environmental Ethics* 30 (3): 387–412.

Buckley, L.A. 2020. Imported rescue dogs: Lack of research impedes evidence-based advice to ensure the welfare of individual dogs. *Veterinary Record* 186 (8): 245–247.

Budiansky, S. 1992. *The Covenant of the wild: why animals chose domestication: With a new preface.* London: Yale University Press.

Callicott, J.B. 1992. Animal rights and environmental ethics: Back together again. In *The Animal Rights/environmental Ethics Debate: The Environmental Perspective*, ed. E. Hargrove, 249–261. Albany, NY: SUNY Press.

De Haan, G., R. Benedictus, R. van Graafeiland, and M. Wissenburg. 2010. Gen-ethische grensverkenningen. Een liberale benadering van ethische kwesties in de medische biotechnologie. Report. The Hague: Teldersstichting.

de Joode, W. n.d. https://www.animalinneed.com/asiel-buitenland. Retrieved 14 March 2019.

Dierenbescherming. 2018. Jaarverslag Dierenbescherming 2017. https://www.dierenbescherming. nl/userfiles/pdf/Jaarverslagen/Jaarverslag_dierenbescherming_2017.pdf. Retrieved 8 July 2020.

Dietz, L., A.M.K. Arnold, V.C. Goerlich-Jansson, and C.M. Vinke. 2018. The importance of early life experiences for the development of behavioural disorders in domestic dogs. *Behaviour* 155 (2–3): 83–114.

Donaldson, S., and W. Kymlicka. 2011. *Zoopolis: A political theory of animal rights.* Oxford: Oxford University Press.

Federation Cynologique International (FCI). 2009. Breed standard Golden Retriever. http://www. fci.be/Nomenclature/Standards/111g08-en.pdf. Retrieved 13 March 2019.

Fox, J.C., H.E. Jordan, K.M. Kocan, T.J. George, S.T. Mullins, C.E. Barnett, and R.L. Cowell. 1986. An overview of serological tests currently available for laboratory diagnosis of parasitic infections. *Veterinary Parasitology* 20 (1–3): 13–29.

Francione, G. 2012. Blogpost. http://www.abolitionistapproach.com/pets-the-inherent-problems-of-domestication/#.VkS-L-s45p.

Francione, G., and R. Garner. 2010. *The animal rights debate: Abolition or regulation?* New York: Columbia University Press.

Freedman, D.G., J.A. King, and O. Elliot. 1961. Critical period in the social development of dogs. *Science* 133 (3457): 1016–1017.

Hancock, J.T., C. Toma, and N. Ellison. 2007, April. The truth about lying in online dating profiles. In *Proceedings of the SIGCHI Conference on Human Factors in Computing Systems*, 449–452.

Haraway, D. 2007. *When species meet*. Minneapolis, MN: University of Minnesota Press.

Hearne, V. 2016. *Adam's task: Calling animals by name*. New York: Skyhorse.

Hens, K. 2009. Ethical responsibilities towards dogs: An inquiry into the dog–human relationship. *Journal of Agricultural and Environmental Ethics* 22 (1): 3–14.

Juffer, F. 2008. De ontwikkeling van interlandelijk geadopteerden; een overzicht van onderzoek. *Justitiële Verkenningen* 34 (7): 38–53.

Lindblad-Toh, K., C.M. Wade, T.S. Mikkelsen, E.K. Karlsson, D.B. Jaffe, M. Kamal, et al. 2005. Genome sequence, comparative analysis and haplotype structure of the domestic dog. *Nature* 438 (7069): 803.

Majumder, S.S., A. Bhadra, A. Ghosh, S. Mitra, D. Bhattacharjee, J. Chatterjee, et al. 2014. To be or not to be social: Foraging associations of free-ranging dogs in an urban ecosystem. *Acta Ethologica* 17 (1): 1–8.

McGreevy, P.D., and F.W. Nicholas. 1999. Some practical solutions to welfare problems in dog breeding. *Animal Welfare-Potters Bar* 8: 329–342.

Neijenhuis, F., and H. Hopster. 2017. Reductie van gezondheidsrisico's bij import van honden; actieve communicatie als beleidsinstrument. *Wageningen Livestock Research, Internal Unpublished Livestock Research Report*.

Nussbaum, M. 2006. *Frontier of justice: Disability, nationality, species membership*. Cambridge, MA: The Belknap Press.

Packer, R.M.A., D. Murphy, and M.J. Farnworth. 2017. Purchasing popular purebreds: Investigating the influence of breed-type on the pre-purchase motivations and behaviour of dog owners. *Animal Welfare* 26 (2): 191–201.

Pal, S.K., B. Ghosh, and S. Roy. 1998. Agonistic behaviour of free-ranging dogs (Canis familiaris) in relation to season, sex and age. *Applied Animal Behaviour Science* 59 (4): 331–348.

Palmer, C. 1997. The idea of the domesticated animal contract. *Environmental Values* 6 (4): 411–425.

Palmer, C. 2010. *Animal ethics in context*. New York: Columbia University Press.

Palmer, C. 2011. The moral relevance of the distinction between domesticated and wild animals. In *Oxford handbook of animal ethics*, ed. T.L. Beauchamp and R.G. Frey, 701–723. Oxford University Press.

Palmer, C. 2012. Does breeding a bulldog harm it? Breeding, ethics, and harm to animals. *Animal Welfare* 21 (2): 157–166.

Patronek, G.J., D.J. Waters, and L.T. Glickman. 1997. Comparative longevity of pet dogs and humans: implications for gerontology research. *The Journals of Gerontology Series A: Biological Sciences and Medical Sciences* 52 (3): B171–B178.

Paul, M., S.S. Majumder, S. Sau, A.K. Nandi, and A. Bhadra. 2016. High early life mortality in free-ranging dogs is largely influenced by humans. *Scientific Reports* 6: 19641.

Post, R. 2008. De perverse effecten van het Haags Adoptieverdrag. *Justitiële Verkenningen* 34 (7): 25–27.

Radstake, C. 2016. *Immigratie van buitenlandse (zwerf)honden & (zwerf)katten naar Nederland: de cijfers 2016*. Stray Animal Foundation Platform.

Sandøe, P., S. Corr, and C. Palmer. 2015. *Companion animal ethics*. Chichester, UK: Wiley.

Sandøe, P., S.V. Kondrup, P.C. Bennett, B. Forkman, I. Meyer, H.F. Proschowsky, et al. 2017. Why do people buy dogs with potential welfare problems related to extreme conformation and inherited disease? A representative study of Danish owners of four small dog breeds. *PLoS ONE* 12 (2): e0172091.

Serpell, J., and J.A. Jagoe. 1995. Early experience and the development of behaviour. In *The Domestic Dog: Its evolution, behaviour and interactions with people*, ed. J. Serpell, 79–102. New York: Cambridge University Press.

Singer, P. 1973. Animal liberation. In *Animal rights*, ed. R. Garner, 7–18. London: Palgrave Macmillan.

Slot, B.M.J. 2008. Adoptie en welvaart; een analyse van vraag en aanbod van adoptiekinderen. *Justitiële Verkenningen* 34 (7): 11–24.

Stray Animal Foundation Platform. 2018. Wildgroei adopties buitenlandse honden aan banden – Stichtingen slaan handen in één, September 13. Retrieved from https://www.stray-afp.org/nl/nie uws/wildgroei-adopties-buitenlandse-honden-aan-banden-stichtingen-slaan-handen-in-een.

Swart, J.A.A., and J. Keulartz. 2011. Wild animals in our backyard: A contextual approach to the intrinsic value of animals. *Acta Biotheoretica* 59 (2): 185–200.

Udell, M.A., N.R. Dorey, and C.D. Wynne. 2010. The performance of stray dogs (Canis familiaris) living in a shelter on human-guided object-choice tasks. *Animal Behaviour* 79 (3): 717–725.

Van Niekerk, T.G.C.M., P. Potters, and M. Meeusen. 2014. *Rapportage kwalitatief onderzoek 'Identificatie en Registratie van honden'* (no. 818). Wageningen UR Livestock Research.

Van Zeeland, C.W.M., and B. Beerda. 2015. *Hereditary disorders in pedigree dogs and look-a-likes* (no. 317). Wageningen UR, Science Shop.

Susan Ophorst is senior lecturer at the bachelor Animal Management of the University of Applied Sciences Van Hall Larenstein in Leeuwarden, The Netherlands, and researcher (Ph.D.) in human-animal relationships at the Institute of Science in Society of Radboud University and University of Applied Sciences Van Hall Larenstein. Her main focus is the relationship between humans and dogs in Dutch society. This chapter would not have been possible without the financial support of the Dutch Research Council (NWO). This work is part of the research programme 2017-I BOO with project number 023.010.030, which is (partly) financed by the Dutch Research Council (NWO).

Bernice Bovenkerk is associate professor of philosophy at Wageningen University, the Netherlands. Her research and teaching deals with issues in animal and environmental ethics, the ethics of climate change, and political philosophy. Current topics are animal agency, the moral status of animals and other natural entities, with a particular focus on fish and insects, the ethics of animal domestication, animal (dis)enhancement, and deliberative democracy. In 2016 she co-edited (together with Jozef Keulartz) *Animal Ethics in the Age of Humans. Blurring boundaries in human-animal relationships* (Springer) and in 2012 she published *The Biotechnology Debate. Democracy in the face of intractable disagreement* (Springer). She received her Ph.D. title in political science at Melbourne University and her Master's title in environmental philosophy at the University of Amsterdam. Her homepage is bernicebovenkerk.com.

Chapter 15
Comment: Animals in 'Non-Ideal Ethics' and 'No-Deal Ethics'

Erno Eskens

Up until the seventies of the last century the idea prevailed that we should gradually improve animal welfare in husbandry systems and animal testing facilities, by focusing on a humane treatment of animals. But already in 1892 the idea emerged that more efforts were necessary. Henry Salt published *Animals' rights considered in relation to social progress*, a book in which he stated that animals needed to be seen as legal persons. In the nineteen seventies this idea caught on. Activists and ethicists embraced a more radical discourse on animal rights. They started to argue that animals should be given fundamental rights and that the exploitation of animals should be declared illegal. Abolitionist ethicists (like Francione 2000) demanded a complete stop of animal use on these grounds. The animal movement and animal ethicists have ever since been divided on this matter. This chapter is about this division, and more particularly on the dilemmas that the shift in thinking from 'humane treatment' to 'animal rights' brought about for the so called 'non-ideal animal ethicists'; i.e. those who stuck to the idea that we can improve animal welfare gradually by appealing to standards of human decency.

15.1 Non-ideal Animal Ethics and the Meat Industry

In this part of the book we have come across multiple chapters written by philosophers who favour an incremental approach to animal welfare. Most prominent is the American philosopher Paul Thompson. He regards himself a 'non-ideal ethicist'. The phrase is catchy. A non-ideal ethicist according to Thompson is someone who does

E. Eskens (✉)
Noordboek, Gorredijk, The Netherlands
e-mail: ernoeskens@xs4all.nl

© The Author(s) 2021
B. Bovenkerk and J. Keulartz (eds.), *Animals in Our Midst: The Challenges of Co-existing with Animals in the Anthropocene*, The International Library of Environmental, Agricultural and Food Ethics 33,
https://doi.org/10.1007/978-3-030-63523-7_15

not reach for the moon. He deplores the effort of animal rights activists and—ethicists who keep on trying to abandon the use of animals in the meat industry. It is basically a waste of time, since the strategy of the animal rights advocates is simply not working. An appeal to justice or fairness is simply too weak to actually protect animals in the ever growing animal industry. Nobody is able to beat the system that abuses animals, so, Thompson states, the best way is to accept the situation and to protect the animals as well as we can, within the abusive husbandry system. Thompson favors 'non-ideal ethics', that is, ethics that strives for what is attainable, not for what can be regarded as an ultimate outcome of fairness.

Non-ideal animal ethics, Thompson states, is more effective than any radical rejection of husbandry will ever be. "Many arguments for veganism, for example, hold that no form of animal agriculture is morally acceptable, but this does not logically vitiate the question of how the lives of animals living in these systems could be made better." Thompson has a point here. Principles do harm if they lead to a neglect of the actual animals in their actual situations and conditions. If we reach for the moon we easily lose sight on the possibility of earthly progress. The risk of abolitionism is that it sets welfare standards in such a manner that no farmer, politician or consumer can ever meet them, which takes away the inclination to move in the desired direction altogether. And let's be honest, most people will in fact ignore the call of justice. How many of them will ever become vegan?

Thompson has a point. Focusing on an incremental improvement of animal welfare will enable farmers, politicians and civilians to make small steps in the right direction. And finally, step by step we may or may not reach the bigger goal: the end of all animal abuse. Thompson adds that we do not have to sanctify the system, while making small improvements: "Inquiry into the welfare of animals aims to make morally compelling improvements in quality of life. It does not presume that improvements in welfare justify the continuation of these systems, on either animal welfare or environmental grounds." So non-ideal ethics can reject exploitation of animals in theory, while at the same time using every opportunity for practical improvement of the lives of animals.

I question Thompsons assumption on this point. I doubt whether we can gradually improve the welfare of animals within the system without at least implicitly justifying it. For can we actually reject the system as a whole and still ask for slightly better living conditions? Let me explain this dilemma by recalling the case of the Dierenbescherming, the Dutch equivalent of the British RSPCA. This moderate animal advocacy movement—the biggest in the Netherlands—introduced a three-star rating system for meat in supermarkets. Stars printed on the packaging ever since indicate the animal welfare level under which the meat is produced. The introduction of the stars was advertised as a 'major breakthrough', since starless meat was since seen as 'bad'. Supermarkets became hesitant to sell it, and most of them switched to one star meat, in order not be accused of animal abuse. The Dierenbescherming thereby succeeded in setting a new minimum requirement for meat quality in most supermarkets. Indeed a major improvement for the lives of many animals. But once introduced, the stars became a hindrance for further improvement. The Dierenbescherming now started saying one star meat wasn't all that good. It advised consumers to buy meat with at least two stars. But most consumers were satisfied with just one. They felt legitimized to buy poor quality meat, since it was obviously approved by the Dierenbescherming—it even had a star!—so why should they buy better meat? The supermarkets did not feel inclined to change their policy either. The Dierenbescherming now has a warning on its website that says the star system is meant for meat eaters who usually buy the cheapest meat. Others apparently are to ignore the stars.

Another example of stagnating progress due to non-ideal solutions, occurred when the Sophia Foundation for the Protection of Animals suggested to improve the cages of chimps that were held for animal testing in the Biomedical Primate Research Center (BPRC) facility in the Netherlands. The facility was under fire of animal rights activists who pleaded for a complete shutdown of the animal testing lab. Politicians were considering doing so, since the activists had a large following. But the small improvements the Sophia Foundation proposed to the cages, and funded, sanctioned the continuation of the testing facility.[1] It send the message to politicians that the situation just needed improvements and basically was under control. The BPRC, on the verge of bankruptcy and of being closed down, remained open (Meershoek 2005). These examples of non-ideal ethics in practice—and there are many more to be given—show the predicament we are in. On the one hand we like to applaud even the smallest welfare improvements—of course we want to better the life of chimps—yet, on the other hand in the eye of politicians and the broader public we do justify the system if are to go in that direction.

And there is a second problem. If we publicly welcome welfare improvements, yet at the same time more privately take an abolitionist approach to the system as a whole (something Thompson proposes), we can justly be called 'opportunists with a hidden agenda'. Many farmers see animal activists and non-ideal ethicists this way. Why should they consider adjusting their stables, cages, barns and machines, if they can predict that the minute they do so, animal activists and animal philosophers like Thompson will be back at their doorsteps?

The life of European chickens is a showcase example of this dilemma. Activists and ethicists have often successfully advocated better living conditions for the millions of chickens in European countries. But they kept on coming back with more demands. The minute the battery cages were improved, they asked for even bigger cages. The minute the cages where renewed, the activists demanded perches for chickens to sleep on and more room so they could spread their wings. Again the farmers were forced to make new cages. And then our non-ideal activists wanted to alter the cages again. They now needed to change into more open spaces with options for the chickens to go outside in the open air. How frustrating for all involved.

As we speak, the Dutch farmers block streets in the Netherlands with their tractors to force politicians to no longer continually change the playing field. They are sick and tired of incrementalism, which forces them to constantly reinvest in their business. Meanwhile the non-ideal ethicists are tired too. They are constantly

---

[1]On the 17th of October 2000, one day before a joint meeting on strategy by several animal welfare organisations, the Sophia Foundation donated money to the BPRC to make alterations in the cages, thus preventing bankruptcy for the BPRC.

more or less rightfully accused of a lack of integrity, and they struggle with their conscience, since they actually do have ideals, which they constantly, for the better good, suppress during their negotiations with farmers, politicians and supermarkets. Non-ideal philosophers do the same as they write their carefully crafted papers, in which they give hints, but mainly avoid to say what they actually think. Adhering to ideals while not putting them on the foreground, turns out to be hard.

15.2   Non-ideal Animal Ethics and Disaster Management

A third problem that follows from non-ideal ethics can be found in the chapter by Andreia De Paula Vieira and Raymond Anthony. They show how to improve the situation of animals during natural and man-made disasters. "Disasters are emergencies endured by people and animals and can be induced by natural or anthropogenic agents", they state. So it concerns a wide spectrum of events, ranging from floods and firestorms to toxicological crises, barn fires and even terrorist attacks. Next the authors delineate "six ethical stewardship caretaking aims for emergency preparedness and response." Their recommendations fall under the following categories: "Respect and Humane Treatment, Collaboration, Information, Community Outreach and Proactive Contact, Cultural Sensitivity and Attitudes Check, and Reflection, Review and Reform." The capitals underline the importance of these virtues, I guess, but as we shall see they are somewhat misleading.

In most cases the best way to protect animals during these emergencies, the authors state, is to focus on humans: "Framing a disaster in terms of public health emergency preparedness and response helps to highlight the adequacy of the infrastructure involved in advancing equity, inclusion, community relationships and galvanizing necessary political will." Putting the emphasis on human decency, and appealing to people of good will, is an essential characteristic of 'non-ideal ethics'. It often works out fine, but it has its limitations. This becomes clear if we look at a current disaster. As I am writing this, most of Europe is in a semi-lockdown due to the coronavirus. On the news we see hundreds of trucks stranded at the Polish border. Many of them transport livestock. The expectation is that most of the animals will perish in the next couple of days. A disaster, that draws a lot of attention. Non-ideal disaster ethicists encourage people and authorities to help the animals. And anyone with the slightest bit of decency hopes, of course, that their appeal will be successful. But there is a downside to this strategy of hope. It ultimately is an appeal to decency and charity. These events—animals suffering during transportation, will occur regularly if we do not fight the system as a whole. If we continue on this path we will need a lot of charity, and of course this is a scarce thing.

In ethical texts the appeal to charity is often hidden in a language of apparent deontology. Things *must* change, non-ideal authors state. Andreia De Paula Vieira and Raymond Anthony write for example: "In the event that depopulation is necessary (such as during the 2014-2015 Highly Pathogenic Avian Influenza Outbreak in the US) adherence to strong ethical standards and procedures, and state and federal laws should take precedence as a way to ensure that as much consideration as is practicable is given to respect the welfare of the affected animals." But notice how the deontological terminology of 'should' and 'ethical standards' is framed in a wider perspective of softening phrases like 'as much consideration as possible'. That is, as much as is possible within the general idea of a 'humane treatment' of animals. By humane treatment of animals we usually mean that the way animals live should match up to human decency standards (and therefore not necessarily to the standards of the animals themselves). If they do not, man is required to help animals in distress, by helping out and showing some mercy.

Non-ideal animal ethicists regard human decency a first requirement. The idea that animals should be able to live up to their own standards and deserve that their interests are being taken seriously, comes only second. And this is worrisome, since the primary focus on human dignity and decency, ultimately steers our attention away from the main principles of justice. In the chapter by De Paula Vieira and Anthony this results in a definition of a disaster which is human, all too human. Disasters are seen as 'emergencies'. Apparently they regard disasters as states of exception, as deviation of daily routine. What humans generally regard as decent and acceptable can therefore not be seen as a disaster. Those things belong to normality. So epidemics amongst chickens and pigs in barns are marked as disasters, while husbandry as such is not. Husbandry is normal. This normality—loads of neglected chickens and pigs in barns—is more or less taken for granted, since this is business as usual. De Paula Vieira and Anthony seem to be struggling with this point, as they note about disaster ethicists: "They are challenged to extend the humanitarian impulse directly to animals while doing so within the constraints of the human-centric world." So they would like to question normality, and in fact they do in a sense, but at the same time one gets the impression they first accept it as a given—as fate. What are you going to do about it? And by accepting this fate, the authors ultimately divert us from the underlying question: isn't livestock farming itself the real disaster that needs a disaster plan first?

All of the so-called animal disasters—zoonoses, problems with cattle during trans-portation, endangered wild life—are in the end results of a gross injustice: the discrim-ination of animals on irrelevant grounds and the enslavement of animals in husbandry systems, zoos and other facilities. Especially animals in husbandry are prone to lead a disastrous life: short, nasty and brutish. They spend their lives in darkened, foul-smelling, unhealthy barns and end up prematurely in the slaughterhouse. In these circumstances zoonoses, accidents, neglect and other 'disasters' are bound to occur. By framing these conditions as exceptions—as if they normally do not occur—and by declaring only the worst situations an emergency, these regular occurring events become framed as mere irregularities. And of course, they are not, since they are to be expected.

Why do the authors not mention the real disaster? Why do they try to manage a big disaster by focusing on its side effects? Probably, because they left philosophy and switched to a more practical and political mode of thinking. They are trying to convince farmers, consumers, politicians and everybody else involved to show some mercy, and they can only do so by pointing at the side-effects, while ignoring the fundamental rot in the system as a whole. Pointing at these facts would make negotiations and appeals difficult. It is somewhat understandable. It is difficult to kindly request politicians, farmers and others to change their behavior, while calling them fundamentally unjust at the same time.

Thompson and the other non-idealists will of course not agree with me that husbandry (including the so called CAFO's) is the real disaster. "Unlike David DeGrazia," Thompson writes, "I believe that the lives of animals in CAFOs *are* worth living, but like the majority of animal welfare scientists contributing empirical findings for husbandry ethics, I believe that their lives could be significantly better than they currently are." It reminds me of a fur breeder I once met in the Dutch parliament building at a hearing. He told me the skins of his animals look fine, so it must be clear to anyone that they lead a more than decent life. So his fur trade was not immoral. 'They wouldn't look like this, if they weren't happy.' Of course to make animals happy, you have to do more than keeping their skins healthy. The rhetorics of Thompson's chapter is somewhat similar. He points at the fact that the animals lead a live worth living. Well, yes, but this is just like pointing at the skins in the fur trade. It simply is besides the question. Perhaps every life is worth living. The question is, however, whether it is fair to treat them this way. In non-ideal ethics we are constantly being diverted away from this question.

I recall the seventeenth-century Dutch slave trader Willem Bosman. He truly was the non-ideal ethicists of his time. Bosman (1703) wrote a short instruction guide for improving the welfare of the enslaved during their shipment to the Americas. Bosman advises us to abstain from unnecessary violence and particularly to be kind to women while branding them, 'since they usually tend to be so tender'. Reading this, we feel uncomfortable. Yet, this is non-ideal ethics in practice. It is dealing with side-effects while turning a blind eye to the real disaster. Willem Bosman tells us his slaves look healthy too. And we should acknowledge, he might say, that these slaves *do* have a worthwhile life, while we have a moral obligation to improve their situation gradually. Of course Bosman is playing a villainous rhetorical trick on us, dodging the real question whether the situation is acceptable altogether. All non-ideal ethics have this flaw.

15.3   Non-ideal Ethics and Ethnographic Animal Studies

I suspect we can also find it, somewhat hidden, in the chapter by Leonie Cornips and Louis van den Hengel. They describe research on communication by young cows in husbandry systems. Cornips and Van den Hengel basically observed young cows for several months, a method they call 'ethnographic observation'. The phrase shows they apparently used observatory techniques common in the study of indigenous people, by noting all behavior, utterances and other sounds. "We have elaborated a posthumanist conception of language as a distributed effect of multiple interacting bodies in order to foreground the fluidity through which a cow, a calf, calves, a wheelbarrow, a farmer, an iron feed fence, a lock, the clattering of bars, sounds of chewing, sounds of puffing, sounds of urinating, the smell of food, urine, feces, other bodies in proximity or distance, movements up and down, become relationally entangled with one another and, crucially, with the anthropological machine of industrial animal production." They describe how cows bang against the bars of their cages in certain manners. This behavior belongs to a refined communication system. The cows certainly succeeded in developing a meaningful language, Cornips and Van den Hengel conclude. They call this 'place making'. By this they mean that animals are able to make the world *their* world by communicating with others.

Cornips and Van den Hengel state that cows turn out to be "intelligent, social, speaking beings, linguistic agents who even under poor conditions form rich and complex relationships with the world to make it a meaningful place." I like studies like these; since many people still see animals as dumb creatures, research like this can be used to improve the situation of animals. Yet, it has something troubling too. Let us go back for a while to our comparison with slaves. Suppose Cornips and Van den Hengel were to embark on a slave ship in order to do their ethnographic study on the use of language there. What would we think when they would report back that they found the slaves to have developed a rich language in 'poor conditions' and that they thus succeeded in place making? We would condemn the fact that they did not pass a harsh verdict on the system as such. Calling it 'poor conditions' is an understatement. It suggests that the conditions can be made richer, better, by making improvements. And of course this is not the case, since animals in cages will never be able to communicate as they would like to.

Again we are diverted from the real questions. The authors should have started by saying that animals ought to live a different, *more* meaningful, *more* worthwhile life. They should have condemned husbandry in a clear manner. Only after having done this, they could report their research without the loss of moral integrity. My point is not that we should refuse research in 'poor conditions', nor that we shouldn't give advice to improve the poor conditions, or to prevent further disasters; the point is we can only justify research in husbandry systems and that can only justify advising farmers to make small changes, if we undoubtedly distance ourselves from the gross injustice first. So yes, we can favor incremental progress. A small step in the right direction is a small step in the right direction. But it is simply wrong, to favor incremental progress, if we dodge the fundamental questions and forget who we are—philosophers. Justice first, politics second, if you ask me. It is much less tiring than taking thousands small steps and debating and regulating each of them while having a bad conscience.

15.4   Towards a No-Deal Animal Ethics

So what is our way out here? Well, there are no easy solutions to our problem. Either we deal with the devil, or we call the devil by his name and perhaps don't deal at all. If this is the case, we are caught up between a really not ideal 'non-ideal ethics' and an idealistic 'no deal ethics'. Having said this, we may start looking for a stance that bridges the gap between these two approaches. The first thing to do, however, is to accept that this society brings about people who do harm to animals, since speciesism (i.e. discrimination on the ground that one belongs to a certain species) is common and stands in a long tradition. The devil here is not some evil genius, he is just a type of person that is produced by this society and we should alter society if we want to prevent devilish things. The second is to acknowledge that no-deal ethics (animal rights ethics) has its problems too. It too leads to tiring situations. Think of all the activists and ethicists who fundamentally oppose animal testing. They convene every once in a while in front of the facilities. Outside, in the cold, not achieving much, except expressing their ideas and emotions and often fruitlessly demanding the shutdown of these facilities.

And let's face this too: no-one has a clear vision of the ideal situation. All animal rights activists talk about rights, but most of them cannot explain what they entail in detail. A growing number of thinkers, including myself, plea for political representation of animals, yet no one seems to know how to organize it. How do you represent ants, snakes and lions in politics for example? Of course, some attempts have been made to clarify things. The Australian philosopher Peter Singer (1975) started in the seventies by pointed out that animals are discriminated on the mere fact that they are animals, while some animals are actually quite similar to us in many respects. The American thinker Tom Regan (2004) pointed out that animals are 'subjects of a life', and therefore deserve consideration on their own grounds. They should be not be treated as means but as goals in themselves. And from the late ninety's, philosophers started thinking about political representation of animals. (My book *Democratie voor dieren* [2009], in which I suggested we should regard animals as citizens and give them full citizenship rights, only to take away those which are not useful, is part of the search for a new animal friendly politics). The Dutch philosopher Eva Meijer (2016) later pointed out that animals do actually have a (political) voice. Most animals use symbols and are quite eloquent. They indicate what they like or dislike. Donaldson and Kymlicka (2011) meanwhile developed a social philosophy and a political system for animal groups. They ascribed different social rights to different types of animals.

So the search for the clearer picture of the ideal way to treat animals has led to the rise of multiple approaches, which often are competing with one another. The ultimate image still is a bit blurry, yet all these developments in philosophy and society (the rise of veganism and animal rights advocacy) show quite clearly the general direction we are moving in. There is a general aspiration for justice, in which interests of animals are to be taken into account. The idea that this search for justice will lead to an ideal outcome, has never left some of the philosophers. The British philosopher Robert Garner (2013) for example tries to get a clear vision of the ideal situation, by starting with the fact that all animals, human and nonhuman, have a 'sentient position' in this world. Taking this as a starting point, Garner evokes 'the veil of ignorance', a thought experiment first proposed by John Rawls (who refused to apply it to animals, unfortunately) in his 1971 book *Theory of Justice*. Imagine you will be born sentient but you do not know under which conditions you will be born. Perhaps you will be born as a baby in Washington, or, as a stray dog in Istanbul or a piglet in a barn somewhere in the Netherlands. How would you like the world, into which you will be born, to be organized? If you place yourself in the position of others, you will be forced to take the interests of that sentient position into account, and will reach a fair judgement on how to weigh all these different interests. The veil of ignorance is of course a highly debated thought experiment. I won't go into the details here, but despite all skepticism, I think it is clear the thought experiment at least invites us to be fair and more empathic with other beings.

Empathy is something vegans usually have in abundance. Most of them share the ideal of a world without pain or discomfort for animals and humans alike. This is a strong ideal—so strong it is hard to live up to. It is practically impossible not to hurt animals. Even in your coleslaw there are small animals that won't survive dinner. Considering this, some vegans drop out and say: 'Well you have to draw a line somewhere. This is where I'll draw it.' It usually means they are not willing to proceed any further in the direction of the ideal.

The case of the vegans shows us that strong ideals can wear us down. Therefore I tend to advice new vegans to take it slowly. Take it one big step at the time, but always keep in mind the direction you are going. Keep in mind that you can and will never again accept the normality of meat eating or dairy consumption. You will not even accept that there is a tiny bug in your coleslaw. But it is ok to fail in your attempt to live up to your ideals. If you are 'a sinner' once in a while, fine, we all are in a sense, but never accept it as a normality. Never settle permanently for any non-ideal way of living. It may seem hard to do. But ignoring the real disasters in ethics and in our daily lives—meat eating, speciesism and animal enslavement—is even harder in the end, since it messes with our logic, our conscience and the moral foundations of our politics.

The answers I tend to give to vegans (and myself for that matter) is basically the answer I would suggest to ethicists. Try not to run from the real questions or from the ideal of justice. Stay on track by moving in the right direction. Be critical and alter your own behavior as much as possible, while being forgiving to those who fail once in a while. The whole aim is to keep up the spirit, to be as clear as possible that, despite the manifold roads ahead of us, we are moving in one direction, and to live up to one's conscience as much as possible. Deal with the demands of fairness. It is a doable strategy in daily live, and it may be too in ethics in general.

Perhaps this 'direction approach' can bridge the gap between incrementalistic non-ideal ethics and the no-deal ethics of the abolitionists. In contrast to non-ideal ethics it does not accept any normality, since it embraces a vision of another world, yet in contrast to animal rights idealists, it is pragmatic too, since it accepts failure as long as we make 'the biggest step forward at this point'. I guess moving forward ourselves, and demanding of others to do the same, is the best way to proceed. And, since we are philosophers, we shall never stop moving, since this is simply not what we as philosophers and ethicists do. Ethics is part of philosophy; part of a restless discourse, as it is favors continuous wonder and reaches constantly for better arguments.

As thinkers, we cannot hide behind any law, any tradition, any given situation, or any accepted standard of 'humane treatment'. Judges in our legal system may perhaps do so, since they have to deliver their verdicts within the boundaries of the law and the legal tradition. As philosophers, our task is to dig deeper and to define what criteria have to be met in laws and traditions. We can state, for example, that the containment of animals in cages, is under most circumstances a *malum in se*, i.e. as something that is accepted by the law, but can still be seen as a crime, since it violates basic moral standards and entails a gross disregard of those involved. It may not be forbidden, but it certainly should be. And we are moving in this direction.

References

Bosman, W. 1703. *Nauwkeurige beschrijving van de Guinese Goud-Tand en Slavenkust*. https://arc hive.org/details/nauwkeurigebesch00bosm/page/n10/mode/2up.
Donalson, S., and W. Kymlicka. 2011. *Zoopolis: A political theory of animal rights*. Oxford: Oxford University Press.
Eskens, E. 2009. *Democratie voor dieren*. Amsterdam: Contact.
Francione, G.L. 2000. *Introduction to animal rights*. Philadelphia, PA: Temple University Press.
Garner, R. 2013. *A theory of justice for animals: Animal rights in a nonideal world*. Oxford: Oxford University Press.
Meershoek, P. 2005. *De slag om de chimpansees*. Amsterdam: L.J. Veen.
Meijer, E. (2016). *Dierentalen*. Leusden: ISVW Uitgevers.
Raws, J. 1971. *A theory of justice*. Cambridge, MA: Harvard University Press.
Regan, T. 2004. *Empty cages: Facing the challenge of animal rights*. New York: Rowman & Littlefield.
Salt, H.S. 1892. *Animals' rights considered in relation to social progress*. New York: Macmillan.
Singer, P. 1975. *Animal liberation*. New York: HarperCollins.

Erno Eskens is publisher at Noordboek. He published several books on philosophy and on human-animal relations, including Democratie voor dieren (Democracy for animals) and Beestachtige geschiedenis van de filosofie (A Beastly History of Philosophy). Previously he worked as editor in chief of Filosofie Magazine, as a science publisher at Veen Magazines and as head of the philosophy department of the International School of Philosophy in the Netherlands.

Part III
Urban Animals

Chapter 16
Stray Agency and Interspecies Care: The Amsterdam Stray Cats and Their Humans

Eva Meijer

Abstract This chapter discusses the Stichting Amsterdamse Zwerfkatten (Amsterdam Stray Cat Foundation, afterwards SAZ), who work with and for stray cats. In their practices and views they challenge common assumptions about cat subjectivity and agency, the cats' right to a habitat and social relations, as well as the idea that there is a strict difference between cats and humans. Their approach offers an alternative way of thinking about cat agency and human agency, networks of cat-human relations, sharing the city with cats, and working towards more freedom for cats and humans. In the chapter I examine these relations from the perspective of agency, care, and politics, and investigate whether or not this can function as a model for building new communities with other animals, which centres their agency.

16.1 Introduction

Invisible to most, Amsterdam is home to many stray cats. These cats are neither wild nor domesticated: they live close to humans but usually do not depend on them.[1] Depending on their location and the humans in that area, their living circumstances can be anywhere on a spectrum between fairly precarious and fairly comfortable. Their lives may be more dangerous than those of companion cats, yet they also experience a larger degree of freedom, understood not only as the capacity to roam freely but also to make choices about where to live and with whom, and more generally how to lead one's life. Still, it is often assumed that companion cats are better off than stray or feral cats, and that cats living on the streets should be rescued, neutered and adopted by humans (Srinivasan 2013). Because they challenge dualisms between

---

[1] I use the word stray instead of feral or semi-feral to translate the Dutch 'zwerfkatten', which can refer to abandoned house cats, semi-feral and feral cats.

E. Meijer (✉)
Wageningen University and Research, Wageningen, The Netherlands
e-mail: eva1.meijer@wur.nl

© The Author(s) 2021
B. Bovenkerk and J. Keulartz (eds.), *Animals in Our Midst: The Challenges of Co-existing with Animals in the Anthropocene*, The International Library of Environmental, Agricultural and Food Ethics 33,
https://doi.org/10.1007/978-3-030-63523-7_16

nature and culture, domesticated and wild, many humans see them as out of place, or as not really belonging in the city (Van Patter and Hovorka 2018). They do not belong to a human, as domesticated animals do, but are also not understood as native to a certain area, such as for example deer or songbirds. Most of the Amsterdam stray cats are descendants of house cats from Amsterdam, who carved out their own lives in the city, have learned to negotiate its risks, and built relationships with one another. Assumptions about these cats as either belonging or not belonging in the city, and about human duties towards them, are usually based on the view that cats, like other animals, are categorically different from humans and that humans are hierarchically above them. The city belongs to the humans who built it, and other animals are seen as companions, guests or pests, but, perhaps with the exception of songbirds, not as rightful inhabitants. Interconnected with this is the fact that the cats, as other nonhuman animals, have no formal legal or political rights.

In this chapter I discuss the work of the *Stichting Amsterdamse Zwerfkatten* (Amsterdam Stray Cat Foundation, afterwards SAZ), who work with and for stray cats. In their practices and underlying ideals they challenge common assumptions about cat subjectivity and agency, their right to a habitat and social relations, as well as the idea that there is a strict difference between cats and humans. They emphasize cat agency, and have a non-anthropocentric view to cat-human relations, which shows us new ways of sharing the city with cats, and working towards more freedom for cats. I volunteered at the SAZ for a year and his paper is based on informal conversations with other volunteers and members of staff, written statements from the website of the SAZ (www.saz.amsterdam) and their biannual magazine Swieber, and my observations of cat-human interactions.

My aim in this chapter is not to provide a theory of just cat-human relations, or to develop a cat ethics for the Anthropocene. The chapter should be understood as a case study that draws on insights from political philosophy, ecofeminism and animal studies more broadly, to investigate what happens at the SAZ, and how their practices and views relate to mainstream ideas and practices. I focus specifically on agency and interspecies relations. In the first section I discuss the work of the SAZ in more detail. Section two zooms in on agency, discussing stray cat and human volunteer agency, and interconnections between these. The third section builds on these ideas about agency and relationality and turns the focus to interspecies care at the SAZ, and in the city of Amsterdam. In section four I move to the politics of the SAZ and the Amsterdam stray cats. The conclusion investigates what lessons we can learn from cat-human relations at the SAZ, in the context of establishing better relations with the other animals with whom we share our cities and households.

16.2   The Amsterdam Stray Cat Foundation

The SAZ have been working for and with the stray cats in and around Amsterdam since 1994. They 'trap, neuter and return' (TNR) the cats who are too wild to adjust to a life in a human home, and they socialise those who show interest in interacting with humans. Once these cats are used to humans, they are taken to shelters in the area from where they can be adopted. The SAZ also assist stray cat colonies in the city, making sure they are fed and watched, and receive healthcare if needed. They make sure all TNR cats are looked after when they are released back into their habitats, and they provide help for humans who take care of stray cats in their gardens or in parks near their homes by supplying food and winter homes, so called iso-boxes, for the cats.

They focus not only on stray cats: a large part of their work is offering financial support to humans who cannot afford veterinary care for their cats. In this context they also offer more general help to humans and cats in need, and are a last resort especially for those who need immediate help. They for example provide shelter for cats in cases of domestic abuse, or take over their medical treatment if their humans cannot provide the appropriate care. They also started a food bank for companion animal food. Those practices are documented in a biannual magazine and on social media, aimed at promoting better cat-human relations, and often drawing attention to the cats' perspectives on matters. They influence city policies with regard to dealing with stray cats, and developing neutering programs.

When the SAZ started in 1994, there were many genuinely feral cats in Amsterdam. Currently, most of the cats on the streets are either lost or abandoned, which is largely because of their TNR work and neutering programs. The SAZ see neutering companion cats as the most efficient way of reducing stray cat suffering, and run campaigns to convince humans to neuter their cats. In these and other efforts, they treat humans similar to the cats, sometimes adopting a slightly paternalistic attitude: humans need to be educated and/or nudged into having their cats neutered and taking good care of them. This attitude is also adopted with regard to financial help with medical issues. In summer, cats often fall from balconies, because they are playing or chasing birds or insects, or because they are old, or simply make a mistake. The SAZ will take care of these cats when necessary and pay for medical bills, but they will also visit the house and check if the balcony is fenced before humans get their companions back.

The SAZ receive a small amount of money from the municipality for their TNR program and neutering campaigns; the remainder of their activities are paid for by private donations and funding from animal welfare organisations. Only two of the humans who work at the SAZ are paid, both receiving the minimum wage. One of these humans cleans the cages, and the other is in charge of daily management. Most of the work however, including cleaning, socializing cats, administration, driving on the ambulance, and catching cats, is done by volunteers. Many of these volunteers have 'a backpack' as the Dutch say, usually referring to psychiatric problems, though some have physical disabilities as well.

I applied for a cleaning job at the SAZ after my cat companion Putih died in the summer of 2016, and was soon promoted to the office, where I answered phone calls, did administrative work, and applied for funding. In the year I worked at the SAZ, I got to know the other humans who worked on Tuesdays quite well, as well as some of the cats who stayed in the socialization kennels for a longer period of time. I was struck by the attention for cat agency I witnessed, and found out there are many invisible networks of care in the city of Amsterdam, formed by humans who look after stray cats, and often vice versa.

16.3   Degrees of Agency

Cat agency plays a major role in how the SAZ functions, both with regard to human behaviour and with regard to how different procedures are set up.[2] Catching cats, taking care of them afterwards, socialising the ones who show they want be socialised, releasing the others; all of this is done in ways that the cats co-determine, even though the power relations are not equal.

With feral and stray cats this works as follows. Before cats are caught, they are usually fed for a while at a certain time, in a certain place—both of which the cats co-determine. The cats are then caught, either because they allow humans to touch them, or, usually, by placing a cage at the feeding place. After they are caught they are taken to the SAZ headquarters, where they receive a health check and are neutered, and then are monitored for a while to see if they show interest in socialization. Sometimes it is immediately clear that they do not—experienced volunteers have developed an eye for it—at other times this needs more time. Cats often determine working hours of SAZ' volunteers, and of humans feeding cats, as well as cleaning techniques, and other SAZ procedures.

Cat agency also influences the lives of humans caring for them in the city. Many of the humans who call for assistance have adapted to having cats living in their gardens, and are willing to put great effort into caring for them or helping them. The cats brought them in this position: they normally take the first step in the relation by choosing a place to live, and picking specific humans to interact with.

While cat agency is taken seriously by the SAZ—paying attention to and respecting cat agency is one of the first things new volunteers learn—, cats have no official rights, and their political voice is not recognized by larger society. There is a gap between recognition of their agency at the micro-level, and official institutions and procedures. This is not a matter of all or nothing: cat agency, for example, does influence interaction with the city council from a distance, because of the habitats that the cats choose, and through the humans who advocate for them. They are informed and influenced by their relations, and sometimes negotiations, with the cats.

Nevertheless, the cats have no real voice in many of the decisions that concern their lives. Interestingly, many of the humans volunteering at the SAZ experience the same. They often complain that they do not speak the language of official institutions, which may lead to problems in their personal lives, and can also affect the effectiveness of their advocacy for the cats. In fact, many of them are more fluent in 'cat language' than in formal Dutch. This is partly due to a class difference: most volunteers at the SAZ have received no higher education and come from poor backgrounds, in contrast

---

[2]For similar observations, see Alger and Alger (1997, 2003).

to volunteers who work for private funds and government employees. It is also due to their neurodiversity. Many volunteers identify as autistic, having personality disorder, suffering from depression, or simply as 'crazy'. While their different perspectives sometimes lead to conflicts between volunteers, or between SAZ volunteers and outsiders, it also provides the organisation with valuable knowledge about groups of humans that are seen and treated as difficult by other organisations and official institutions. In fact, one of the great strengths of the SAZ is that they assist humans from marginalized groups, such as for example animal hoarders, psychiatric patients, incarcerated people, victims of domestic violence, and homeless people, if they need temporary or permanent help with their companions, offering shelter or medical care. These humans are often not too eager to interact with official institutions such as governmental organisations or even larger animal welfare organisations, but they trust the humans at the SAZ, or come to trust them after a while.

Humans have certain advantages in anthropocentric societies, but not all humans have them to the same degree.[3] At the SAZ there are many different relations between humans and cats, in which humans and cats exercise agency, and have their specific strengths and weaknesses, some of which are connected to being part of a specific social group, while others are individual. With agency I do not refer to the Kantian rational interpretation of this concept, but neither to object-oriented positions (Bennett Benett 2010; Latour 1993): it should be understood as relational and situated. Cats and humans enable each other to act in certain ways, and influence one another in a variety of ways (see also Bovenkerk and Meijer in this volume).

16.4 Networks of Care

Human care plays an important role in the lives of many of the stray cats. Outside cats are provided with food and healthcare, which is especially important when they get older. Cats in the socialization program receive quite a lot of care when they are at the SAZ, which continues after they are adopted. Kittens sometimes require intensive medical care, and they are usually placed in foster families.

The cats also take care of humans. Stray cats can provide their chosen humans with responsibility, a goal in life, mutual habits, and often also company, as many do at some point connect with the humans (often elderly females, sometimes males) who care for them. The volunteers who work at the SAZ have similar experiences. Cats interact with them, which leads to special relations with certain humans.[4] Caring for cats can for certain humans be helpful in itself. Being a volunteer creates a responsibility, and asks for them to be physically present, which means getting out of bed and going to the SAZ. The daily manager of the SAZ plays an important role

---

[3] See Taylor (2017) for a broader discussion of the interconnections between constructions of disabilities and animality.

[4] These cats are usually not cute at all, they often do not like humans, cannot be touched or approached, which can make the care difficult and demanding, but rewarding as well.

in this. If someone is for example depressed and in bed all day, she will call them and ask to bring in fish for a cat that has trouble eating. Humans will often feel obliged to come in for a specific cat, or because the cats generally need them. This can make a large difference in the lives of individuals. In casual conversations about the cats and their role in their lives, they often express feeling a special connection to the cats because they are outsiders too.

The SAZ also creates a culture of care, which transcends individual interactions. This is perhaps most visible at the SAZ headquarters, where volunteers are taught how to care for cats, through education and written rules, and cats learn to care for humans and one another. Care is not just feeding cats or keeping them safe. It can also involve play, for example. Younger cats especially often begin their contact with humans in the form of play—touching comes later. Similarly, black humour is for the humans an important tool in dealing with hardships in their own lives and the cats' lives. This culture is also found outside of the SAZ headquarters, in the network of humans who care for stray cats in the city. Not only do individuals take care of the cats, they know one another and work together to provide this care. They keep an eye on neighbourhood cats who have homes and make sure they are treated well, and they keep an eye on each other. The cats of course also form communities, in which they sometimes care for one another. Females in colonies for example sometimes share the care for kittens, including nursing.

Building on care ethical views Lori Gruen (2015) argues that there is a normative component to our entanglements with other animals: in order to do justice to those that we are connected to, we have to develop a kind of caring perception—an entangled empathy—that is focused on attending to the wellbeing of others. Developing empathy in this manner is a process in which both cognition and emotion have a role to play. With stray and feral cats one needs to pay attention, because otherwise you can get scratched or worse, and at the SAZ their perspective is brought to the front. Humans who start working there learn to be perceptive from the other humans, and from the cats; new cats learn from other cats and humans. Humans are often seen as givers of care, and other animals as recipients of care. The SAZ show that this is not necessarily the case, and that there are many different types of relationships possible. Recognizing that humans and cats at certain times depend on the other challenges a binary opposition between humans on the one side and cats on the other, with regard to agency and care. At the SAZ cats and humans are treated similarly: they are taken seriously as individuals, and attended to in special ways if necessary.

16.5   Cat Politics

The political position of the Amsterdam street cats is currently ambiguous. They are treated better than certain other inhabitants of the city, such as rats or pigeons, and perhaps even certain humans groups, such as illegalized refugees. Cats have a better reputation than pigeons or rats, who have traditionally been negatively stereotyped and are often unjustly associated with disease (see Meijboom and Nieuwland in this volume). Many humans more or less accept the cats' presence, and often do not even notice them. When there are problems, education can play a role: once humans learn that cat colonies are looked after and the cats are neutered, they tend to accept them. At the same time however the cats do not have political or legal rights, similar to other nonhuman animals, which makes their situation precarious and means they depend on the goodwill of humans for fair treatment. While current legislation in the province of Noord-Holland favours TNR over killing stray cats, this is up for debate in other provinces and could change.

The SAZ' attitude towards the cats reflects this ambiguity. Cat subjectivity is taken seriously—all cats are treated equally and seen as worthy of care and consideration—, and cats are treated as social beings who should have a place to live, opportunity to form relations with others, and lead a life that is good for them. They however do not argue for cat rights, and the cats' rights to life, bodily integrity, and autonomy are not always recognized, which seemingly contradicts their commitment to cat agency and subjectivity. This is most visible in their approach to euthanasia and neutering.

Older TNR cats who have lived in a certain habitat for a long time often depend on one human to take care of them. When this human dies and there is no one to take over the care, euthanizing, or, less euphemistically, killing, the cats is sometimes considered to be the best option. Old cats usually cannot settle in a new territory because they are too fragile, or be adopted by humans, because they are too wild. Living at the SAZ causes them great distress, and while this can be a temporary solution, it is not a permanent one. Similar problems arise with cats who are ill and need long-term treatments, who are permanently injured, or who have long hair, and who are too afraid of humans to live with them. Ideally, cats from these groups should have the chance to move into cat villages, where they can live at a distance from people but are regularly checked on by humans who have experience with wild cats. There are however not enough spaces in these cat villages for all cats, and when the waiting lists are long, older cats are sometimes killed. In the case of humans we would find this unacceptable, and create spaces where they could stay. This is of course part of a larger framework in which nonhuman animals' lives are not valued in the same way as human lives, but the SAZ are an actor in this system.

Another example of violating cats' bodily integrity and autonomy is neutering. The TNR program is based on neutering, and the SAZ run a program that allows humans with low income to neuter their companion cats for free. Their main reason for promoting neutering is reducing cat suffering. They rescue hundreds of stray kittens every year, many of whom are ill, and argue that shelters are full of cats who deserve to be adopted. Furthermore, intact males often fight—in cities, their territories often overlap—and it could be argued that females are better off if they are not constantly pregnant or with kittens.

Under current circumstances the SAZ's neutering programs can indeed probably be defended for the simple reason that they do reduce cat suffering in cities. There is however a more principal question of whether or not humans are ethically allowed to intervene in cats' lives and bodies in this way. This more principal question of whether humans have a right, or even a duty, to neuter cats deserves more space than I have here (see also Donaldson and Kymlicka 2011; Palmer 2013; compare

Narayanan 2017; Srinivasan 2013). It is however interesting to note that in the case of euthanasia and neutering the SAZ never mention what the cats want, or discuss whether or not this is problematic with regard to their bodily integrity, but instead simply assume that humans know best and are allowed to intervene.

16.5.1 Stray Cat Rights

These tensions bring us to the question of stray cat rights. The SAZ do not advocate for universal cat rights, but they do argue for specific kinds of legislation in relation to stray cats, for example in discussions with the city council and the province, and they promote what they see as good cat-human relations through their (social) media channels. Even though they do not use the language of rights or justice—they will speak of belonging, care, or being kind instead—, their underlying ideas correspond with Sue Donaldson and Will Kymlicka's (2011) views about denizenship rights for liminal animals.

Donaldson and Kymlicka use a citizenship approach to conceptualize the rights of different nonhuman animal communities, based on their relations to human political communities. Domesticated animals are conceptualized as citizens, who form inter-species communities with humans. Wild animals are sovereign communities, who prefer to stay away from humans and who are capable of taking care of themselves. Not all animals are wild or domesticated, and for those in between, Donaldson and Kymlicka introduce a third category. Liminal animals do not desire close relations with humans, but they do seek out human settlements for safety, shelter or food, and can have different types of interactions with humans. While most SAZ cats can indeed be seen as liminal, their relations with humans vary. Some street cats are domestic cats, who after a few days or weeks, or even hours, remember how nice it was to live inside a house. Others are feral and show a strong desire to get away from humans; they will try to escape and fight for their freedom if necessary. Between these poles we find shy cats, cats who after months of attention suddenly decide to try to trust humans, cats who usually like humans but remain unpredictable in their behaviour and therefore cannot be adopted, and many others.

According to Donaldson and Kymlicka (2011, Chapter 3), the different relations between groups of human and nonhuman animals are connected to different sets of rights. Liminal animals should, in addition to the basic negative rights all nonhuman animals should have, be awarded denizenship rights. These are the right not to be stereotyped negatively, to participate in reciprocal relations, and the right to a habitat. These three rights capture what the SAZ advocate for. The SAZ have influenced public opinion of stray cats and continues to do so; in the 1990s, they were seen as pests and now the image humans have of them is much more positive. Promoting better cat-human relations, based on reciprocity and mutual consent, in cat-human households and in cities, is also one of the key aims of the SAZ. This is interconnected with arguing for the cats' right to a habitat: the SAZ argue that the city is their home too.

16.5.2   Democratic Agency

One of the key features of Donaldson and Kymlicka's theory is their focus on political animal agency. Wild and liminal animals express political agency in different ways, for example through resisting oppression or leaving a certain territory. Domesticated nonhuman animals exercise democratic agency in relation to humans. They should not only be the recipients of rights or citizenship, but should be seen as political agents who co-shape communities, and whose voices should be taken into account in taking political decisions.

More specifically, they develop an account of 'dependent agency' in the case of domesticated animal citizenship. They distinguish three necessary features of exercising democratic agency, which in their opinion also apply to domesticated animals: the possibility of having and expressing a subjective good, the capacity to comply with social norms through relationships, and the capacity to participate in shaping the terms of interaction (2011, 104). To further conceptualize this, they turn to recent work in disability theory, and investigate parallels with theories that focus on how humans with severe mental disabilities can exercise agency through relationships that are based on trust. Exercising this dependent agency would, in the case of domesticated animals, mean that they communicate their standpoints to humans they know well and trust, and who know them well, who then communicate these to other humans. Domesticated animals have a right to be represented socially and politically through this form of interaction.

At the SAZ cats and humans exercise agency in different ways and to different degrees, and can depend on other cats and humans in a multitude of ways. Some cats and some humans live their lives without much help from others and prefer distance from them; others need more care, or prefer close relations. Politically, relations also do not follow the species line neatly. In the model that Donaldson and Kymlicka propose, the nonhuman companions depend on humans for social and political representation. At the SAZ humans speak up for the cats in official contexts, yet many humans at the SAZ also cannot or do not want to express themselves politically in this way. Furthermore, the cats act politically in other ways: they claim and defend their territories, resist captivity, vote with their feet, or cooperate with humans. While these acts differ from human political acts, they are meaningful and may have further political effects (see Meijer 2019 for a longer discussion on animal resistance).

16.6   Cat-Human Relations at the SAZ as a Model for Future Interactions

The SAZ's approach to stray cats can perhaps best be described as pragmatic. While they take seriously cat agency and subjectivity in many contexts, they restrict it at other points, as I discussed in relation to euthanasia and neutering. Furthermore, in ethical judgments they focus solely on cats and not on the wider environment. They feed the cats animal products, such as meat and fish,[5] and do not consider the impact of cat colonies on other liminal animals, such as rodents or songbirds.

While the first issue can be solved by feeding the cats vegan diets, the second question is more difficult to answer. Thinking about sharing communities and cities with cats raises the question of intervention in their predation behaviour (Donaldson and Kymlicka 2011; Palmer 2013), since individual animals and populations may suffer from their presence. In some places stray cats are seen and welcomed as rural working animals, providing 'rodent control' (Van Patter and Hovorka 2018), while in other places there is growing public concern over their impacts on songbirds (ibid.). This latter concern is often interconnected with a discourse and world view that relies on a distinction between nature and culture, and sees feral cats as intruders who are not native to certain areas and do not belong there. Taking issue with cats who kill songbirds but not minding them killing mice and rats is furthermore speciesist. While the question of predation is an important ethical question to consider in living with companion cats, and in supporting stray or feral cats, simply blaming them for the loss of wildlife is too easy. More empirical research is needed, into the relation between feeding cats and their hunting behaviours, human influence on wildlife—for example with regard to the effect of pesticides on insects and birds—, and differences in hunting and killing behaviours of companion cats and feral cats. More research into the moral and political standing of liminal animals is also needed, and this does not just apply to cats, but also rats, mice and birds.

While there are ethical and political challenges with regard to the SAZ's policies and practices, certain aspects of their work can serve as an inspiration for developing new relations with other animals. As a conclusion to this paper I will discuss these in three areas: forming ecologies of care, sharing the city, and interspecies resistance practices.

16.6.1   Ecologies of Care

Relations between humans and groups of liminal animals are often framed in terms of conflict. Humans usually see themselves as the dominant party, both in terms of rationality and of power. As we saw, the SAZ problematizes assumptions about power relations between humans and cats, and humans' right to space. While humans make certain decisions, cats know what is best—for them or the human opposite them—in other cases, and this wisdom needs to be recognized. Instead of focusing on eradicating city cats, or establishing stronger borders between cat colonies and humans, they accept that interactions and relations are inevitable, and do not aim to domesticate all cats. They work towards creating better relations, offer care where needed, and recognize that care goes two ways: helping cats can help humans. In

---

[5]Many volunteers also eat animals, though this is slowly changing because of some vegan volunteers who are creating awareness.

this way, the SAZ not only challenges the epistemic dichotomy between humans and other animals, but also offer another framework, one of care and not conflict.

Throughout the city of Amsterdam we find different networks of care in relation to stray cats: between cats, cats and humans, colonies and caretakers, socialized cats and volunteers, between different volunteers, between passers-by and cats in need. This is not always in perfect harmony: conflicts often occur. But care is not simply soft and sweet: it means taking responsibility for others also when that is difficult, and it can even involve making difficult decisions about life and death for others. Humans share cities and rural areas not only with cats, and relations and encounters with songbirds, mice, rats, seagulls, foxes and other animals are similarly often unavoidable. Focusing on care instead of conflict can be the first step for starting something new and perhaps even solving problems with co-existence (see Meijboom and Nieuwland on rats in this volume).

16.6.2   Sharing the City

Recent work in political philosophy (Cooke 2017; Hadley 2005; Milburn 2017) problematizes the assumption that humans automatically have all rights to the land they share with other animals, by arguing for nonhuman animal land rights, habitat rights, or territorial rights. Humans gave themselves a right to the land, but that is not the same as having a right to that land (Meijer 2019). The idea is not just that other animals belong in their habitats, have a right to their territory and should have access to the means they need to sustain their existence: it also imposes limits on human expansion, and even raises the question of giving land back to the animals who lived there in the first place (Donaldson and Kymlicka 2011).

Cities are seen as a product of human activity, and therefore as belonging to the humans who built them and not to the other animals who lived there before, or who came to live there when humans did. Jennifer Wolch (2010) sees a lack of nonhuman animals in urban theories and practices. This is related to the fact that cities are seen as human spaces, and the fact that urbanisation is based on a view of progress that favours culture over nature, which leads to exploitation of nature. To address this, she argues, we need not only to acknowledge and foster the presence of those other animals in cities, but also to build new relations with them.

Stray cats are not wild, and do not fall on the side of nature or wilderness, but are also not domesticated and living within human culture. They contest physical boundaries, by forming their own communities within city borders, often trespassing on land that belongs to individual humans and companies—they like courtyards, parks, big gardens, and industrial areas—, and symbolic borders, by invading a culturally human space (Van Patter and Hovorka 2018). Colonies have their own habits and norms, and the cats have their own maps of the city (compare Barua 2014). The cats know their habitat is theirs, even when humans do not recognize this.

Understanding cities as places that humans share with many others may mean humans need to make more space and take a step back. This however does not have to be a sacrifice: building new relations with other animals, attending to them and engaging with them differently can instead enrich one's life and lead to a deeper sense of connection with the world around you.

16.6.3 Interspecies Resistance as the Foundation for New Relations

New relations are not a matter of all or nothing, and are not formed in a utopian setting far away from daily realities. They can and do begin here. The SAZ shows us alternative ways of interacting with cats, in which power relations are flexible and cat agency is recognized. Some of the SAZ practices, such as caring for a group that no one cares about, can be seen as forms of resistance in a world that disregards nonhuman animal lives.

Michel Foucault (1998, see also Wadiwel 2016) conceptualizes everyday practices that challenge power structures as forms of resistance, which is a helpful way of looking at these cat-human relations. Similar to the ecologies of care I mentioned earlier, these patterns could perhaps best be described as ecologies of resistance, in which cats and humans resist oppression. The humans working at the SAZ sometimes describe their work in terms of resistance, to indifference and hostile attitudes to cats, and they take pride in being disobedient. They also often call themselves crazy and describe themselves as being literally from the streets, and it seems as if they take pride in being like the cats.

Their struggle and resistance is not aimed at large societal reform, and their acts are at times morally inconsistent. There are also often conflicts between this organisation and other animal shelters and organisations; they fight over resources and principles. But the SAZ does manage to bring about small, and sometimes big, changes in the lives of many cats and humans, they changed how cats are viewed and treated in Amsterdam, and they are a last resort for humans and cats who have nowhere else to go. With this they offer a glimpse of new interspecies relations, co-shaped by the cats.

References

Alger, J., and S. Alger. 1997. Beyond Mead: Symbolic interaction between humans and felines. *Society & Animals* 5 (1): 65–81.

Alger, J., and S. Alger. 2003. *Cat culture: The social world of a cat shelter*. Philadelphia, PA: Temple University Press.

Barua, M. 2014. Bio-geo-graphy: Landscape, dwelling, and the political ecology of human–elephant relations. *Environment and Planning D: Society and Space* 32: 915–934.

Bennett, J. 2010. *Vibrant matter: A political ecology of things*. Durham, UK: Duke University Press.

Cooke, S. 2017. Animal kingdoms: On habitat rights for wild animals. *Environmental Values* 26 (1): 53–72.

Donaldson, S., and W. Kymlicka. 2011. *Zoopolis: A political theory of animal rights*. Oxford: Oxford University Press.

Foucault, M. 1998. *The will to knowledge: The history of sexuality*, vol. 1. London: Penguin Books.

Gruen, L. 2015. *Entangled empathy: An alternative ethic for our relationships with animals*. New York: Lantern Books.

Hadley, J. 2005. Non-human animal property: Reconciling environmentalism and animal rights. *Journal of social philosophy* 36 (3): 305–315.

Latour, B. 1993. *We have never been modern*. Cambridge, MA: Harvard University Press.

Meijer, E. 2019. *When animals speak: Toward an interspecies democracy*. New York: New York University Press.

Milburn, J. 2017. Nonhuman animals as property holders: An exploration of the Lockean Labour-mixing account. *Environmental Values* 26 (5): 629–648.

Narayanan, Y. 2017. Street dogs at the intersection of colonialism and informality: 'Subaltern animism' as a posthuman critique of Indian cities. *Environment and Planning D: Society and Space* 35 (3): 475–494.

Palmer, C. 2013. Companion cats as co-citizens? Comments on Sue Donaldson's and Will Kymlicka's Zoopolis. *Dialogue* 52 (4): 759–767.

Srinivasan, K. 2013. The biopolitics of animal being and welfare: Dog control and care in the UK and India. *Transactions of the Institute of British Geographers* (1): 106–119.

Taylor, S. 2017. *Beasts of burden: Animal and disability liberation*. New York: The New Press.

Van Patter, L.E., and A.J. Hovorka. 2018. 'Of place' or 'of people': Exploring the animal spaces and beastly places of feral cats in southern Ontario. *Social and Cultural Geography* 19 (2): 275–295.

Wadiwel, D.J. 2016. Do fish resist? *Cultural Studies Review* 22 (1): 196.

Wolch, J. 2010. Zoöpolis. In *Metamorphoses of the zoo: Animal encounter after Noah*, ed. R. Acampora, 221–245. Plymouth, UK: Lexington Books.

Eva Meijer works as a postdoctoral researcher at Wageningen University (NL) in the project Anthropocene Ethics: Taking Animal Agency Seriously. She taught (animal) philosophy at the University of Amsterdam and is the chair of the Dutch study group for Animal Ethics, as well as a founding member of Minding Animals The Netherlands. Recent publications include Animal Languages (John Murray 2019) and When animals speak. Toward an Interspecies Democracy (New York University Press 2019). Meijer wrote nine books, fiction and non-fiction, that have been translated into sixteen languages.

Chapter 17
"Eek! A Rat!"

Joachim Nieuwland and Franck L. B. Meijboom

Abstract Rats are often despised. In what way does such aversion affect moral deliberation, and if so, how should we accommodate any distorting effects on our normative judgements? These questions are explored in this chapter with regard to recent proposals in (1) the ethics of pest management and (2) animal political theory. While ethical frameworks and tools used in the context of animal research can improve moral deliberation with regard to pest management, we argue based on psychological factors regarding the perception of rats that before implementing these methods in either animal research or pest management, one needs to ascertain that rats are owed genuine moral consideration. With regard to animal political theory, we identify three issues: truth-aptness, perception, and moral motivation. To complement as well as address some of the issues found in both animal research ethics and animal political theory, we explore compassion. Starting from compassion, we develop a pragmatist and interspecies understanding of morality, including a shift from an anthropocentric to a multispecies epistemology, and a distributed rather than an individual notion of moral agency. We need to engage with the experience of others, including rats and those who perceive these animals as pests, as well as pay attention to the specific way individual agents are embedded in particular socio-ecological settings so as to promote compassionate action.

Limits

> Who knows this or that?
> Hark in the wall to the rat

J. Nieuwland (✉)
Faculty of Veterinary Medicine, Utrecht University, Utrecht, The Netherlands
e-mail: j.nieuwland@uu.nl

F. L. B. Meijboom
Ethics Institute, Utrecht University, Utrecht, The Netherlands
e-mail: f.l.b.meijboom@uu.nl

B. Bovenkerk and J. Keulartz (eds.), *Animals in Our Midst: The Challenges of Co-existing with Animals in the Anthropocene*, The International Library of Environmental, Agricultural and Food Ethics 33,
https://doi.org/10.1007/978-3-030-63523-7_17

Since the world was, he has gnawed;
Of his wisdom, of his fraud
What dost thou know
In the wretched little beast
Is life & heart
Child & parent
Not without relation
To fruitful field & sun & moon
What art thou? His wicked cruelty
Is cruel to thy cruelty

Ralph Waldo Emerson

17.1 Introduction

Amidst human-dominated landscapes rats, both the brown (*Rattus norvegicus*) and black rat (*Rattus rattus*) are found in abundance. This particular co-existence of species can be found across history. Neither fully wild nor domesticated, in-between nature and culture, they are considered *liminal* (Donaldson and Kymlicka 2011). When they do venture away from human presence, the impact of rats can reverberate throughout landscapes, affecting whole ecosystems and species—though, as we should add, this pales in comparison to anthropogenic impact. Moreover, rats and humans alike display tremendous capacity to adjust and adapt to prevailing circumstances: modern urban space is home to both of them.

Any cross-species similarities have not kindled affection, as rats are one of the most despised creatures by humankind. Disease, destruction, disturbance, and death have become closely associated with their species:

> the Norway Rat is undoubtedly hated and feared by more people and in more countries in the world than is any other animal. These people see in it a filthy animal, destroyer of property, spoiler of food, carrier of bubonic plague and many other terrible diseases, attacker of human beings, particularly defenseless babies. (Richter 1968, 403)

We could offer a litany of frightful depictions of both brown and black rats to make our case, but we believe that the sentiment expressed just now suffices, and still holds. Aversion to rats is furthermore reflected in the way humans treat them. Slow-acting poison, neck-breaking spring traps, and body-fixating glue traps are some of the methods used in pest-management.

Such strong aversion in perception and the use of harsh measures appears to starkly contrast with the idea that human beings owe moral consideration to these non-human animals (henceforth "animals"). Indeed, rats are paid scant attention in moral and political philosophy, with some notable recent exceptions that we will discuss this chapter. Nonetheless, if moral concern is premised on being sentient—which we assume to impose a compelling sufficient condition for moral concern—rats are clearly among those who are owed direct moral consideration. So, despite the aversion that colors the way in which humans generally perceive rats, we need to take them seriously in our moral deliberations and actions.

Here, we explore this tension between moral consideration and adverse perception by looking at recent work in moral and political theory, starting with (1) proposals to extrapolate the methodology of ethical assessment of animal research to the field of pest management, to (2) then look at the political turn in animals ethics. Both hold great potential to improve human interactions with liminal rats but are not without issues themselves. One of the main problems for both accounts is the way in which moral agency is vulnerable to adverse perception of the rats subject of moral consideration. Moreover, the political turn in animal ethics (resulting in a field which we will call *animal politics*) is, in addition to the issue of adverse perception, confronted with a tremendous gap between theory and current treatment of animals, lack of moral motivation, and the skeptical question which political account of interspecies co-existence is true. These issues are probably more challenging when it comes to rats. In that sense, if philosophers would succeed in safeguarding ethical treatment of liminal rats, there appears not much in the way of accomplishing genuine interspecies justice. We believe that compassion provides the most promising route for doing so.

17.2 From the Lab to the Liminal

An historical account of medical development in the twentieth century would be radically incomplete without rodents such as the brown rat, *Rattus norvegicus*. Still, today, much of preclinical research continues in its reliance on rats as models for studying human disease.

Parallel to the rise of the "lab rat", the second half of the twentieth century witnessed an emergence of moral concern for animals, spurred in part by the burgeoning field of animal ethics that has thrown out its nets across the sundry and often separated human-animal interactions that characterize modern societies. Within the context of animal research, following up on discussions on the subject between, amongst others, Peter Singer and Tom Regan, a variety of ethical frameworks have been developed to engage in ethical deliberation regarding the exploitation of animals for the sake of research goals. As such, it provides a resource for other practices of animal treatment. Some suggest applying these frameworks to the context of pest management based on the principle of consistency (Meerburg et al. 2008; Yeates 2010). The implementation of the principles of refinement, reduction, and replacement paired with a structured way of weighing harms and benefits could greatly improve the management of animals that are regarded as pests (Meerburg et al. 2008).

Apart from the obvious benefits of implementing ethical scrutiny and general principles where reflection is found lacking, perhaps we need to ascertain that this type of ethical deliberation is apposite for dealing with liminal rats. To do so, let's unpack (a) the ethics of animal research in itself as well as (b) applying it to liminal rats. To start with the former, some would argue that the harm-benefit analysis between protecting human health and harming animals resembles something akin to a life-boat situation, where the only chance of survival is to throw one individual overboard, boiling down to the choice of saving a human or an animal life. But perhaps such a metaphor misses the point, as it is not clear that animals are in such a relation to human patients.[1] As Anders Schinkel (2008, 56) puts it,

> The animals are not in the lifeboat at all, until we take them on board. In reality, there is no proper analogue for the benefit the animals in the lifeboat would receive if human beings sacrificed themselves. Sick people that die 'because' no research has been done on animals do not sacrifice themselves for the animals. What happens is that human beings, like all creatures, fall ill sometimes. If it is serious, they may die. All this does not affect the lives of animals, so long as we do not make it so. To begin with, there is no lifeboat. Research on animals constitutes the lifeboat itself; only when we see research on animals as part of the initial situation— when we accept this as given—can we say that there is a lifeboat, that it is us or them.

In that sense, the harm-benefit analysis of animal research ethics is already biased towards humans as animals have often nothing to gain from the exchange. More accurately formulated then, it generally involves an "animal harm – human benefit analysis". The central moral issue is whether the use of animals in research for a human benefit can be justified, not who to sacrifice of those with similar interests at stake.

What justifies the combination of harm-benefit analysis and the three principles? The three principles (reduction of numbers of animals, refinement to decrease suffering, replacement in terms of alternatives) impose a constraint that can be supported by many, without touching the central moral question of whether it is justified to harm animals for the benefit of humans. Rather, they assume that this is the case, pointing towards opportunities to ameliorate the harms involved. So, the question most pertinent is: how does one justify harming animals for the benefit of humans?[2] For now, let's assume that animal research ethics in terms of harm-benefit analysis and the three R's proves sound. How does this translate to the context of liminal animals that are perceived as a threat to human interests?

A first thing to flag is the variety within animal research and management of pest animals. The need for, respectively, knowledge or control measures differs in strength. Some research is more important than other research. Similarly, some conflicts with liminal rats are more serious than others. In both cases an assessment of the weight of interests and potential harms is warranted, so a systematic harm-benefit analysis appears valuable for the management of conflict with liminal rats. The two also differ from each other. While research animals are wholly dependent and vulnerable

---

[1] And we could question the added value of such lifeboat-deliberation as thought experiment in the first place (Donovan 2006).

[2] Note that the ethical assessment does leave room to reject any invasive research involving rats.

in relation to humans, liminal animals are under the influence of human beings and sometimes even rely on human activity for their survival but remain undomesticated.[3]

As long as we restrict ourselves to situations where the interests of liminal animals and humans conflict, the model of ethical assessment used in animals research could work; assuming that the aforementioned concerns about animal research ethics itself have been successfully addressed. It does, however, not tell us anything about the way we should configure our relations with liminal rats apart from the conflicts that could arise. Moreover, some conflicts only arise precisely because a broader outlook is lacking. Do humans have any obligations towards liminal rats beyond the negative duties of not harming them unnecessarily (inflicting suffering, depriving of life)? And if so, how should we implement such imperatives in our lives? So, while the model of animal research does provide certain opportunities to improve our treatment of liminal animals in situations of conflicting interests, its scope is rather narrow in the light of the inextricable nature of human-liminal animal co-existence. Moreover, if we would restrict ourselves to conflicts with liminal animals and develop ethical frameworks for this purpose specifically and only, then we may inadvertently bolster negative associations like fear and disgust already attached to these animals.[4] Before looking at a model for co-existence, we need to address the ways in which rats are perceived and how this affects moral judgments regarding their fate.

17.3 How Fear and Disgust Impair Moral Judgment

Associations like fear and disgust can run deep. When Albert Camus wrote his book *La Peste*, published in 1947 and generally interpreted as an allegory about the emergence of fascism, he did so by reference to rats and their supposed role in spreading diseases like bubonic plague.[5] In the book, both in literal—as harbingers of disease, suffering, and death[6]—as well as in metaphorical sense—infesting society with

---

[3]Part of the ethical assessment of animal research could pertain to the conditions in the lab that shape rat's lives, whereas this does not directly or only in part applies to liminal rats; their lives are heavily shaped by human action but not determined to the same extent as their lab counterparts.

[4]A similar mechanism is at work in the increase of attention for wild animals in the context of infectious disease emergence (Stephen 2014). The sort of attention matters. As wild animals become associated with disease reservoirs, the dependencies and vulnerabilities of their own health stay in the background. Only a partial representation of them, as participants within disease ecology of zoonotic disease (pathogens that jump species boundaries from animals to humans), make their appearance on stage. So not only are they reduced to their role in the emergence of disease, they only capture the interests of researchers if they play a role in the emergence of diseases that affect humans (zoonoses) or their interests (for example, economic interests related to diseases that affect livestock production).

[5]https://www.theguardian.com/books/booksblog/2015/jan/05/albert-camus-the-plague-fascist-death-ed-vulliamy. Accessed 23 March 2020.

[6]See Dean et al. (2018) for a challenge of the hypothesis that rats were the primary vectors during the Second Plague.

authoritarianism and exclusion/extermination of "the other"—rats signify misery; a mingling of fear and disgust.

Aversion appears to cut even deeper than mere fear for infectious disease. The aversion for certain liminal animals, especially rodents, could have evolutionary roots. Jonathan Haidt and Jesse Graham (2007, 116) argue from an evolutionary perspective on human psychology that "(d)isgust appears to function as a guardian of the body in all cultures, responding to elicitors that are biologically or culturally linked to disease transmission (feces, vomit, rotting corpses, and animals whose habits associate them with such vectors)". Aversion to rodents could prove valuable to protect human health if these animals are indeed associated with the threat of infectious disease. Perhaps humans are hardwired to dislike certain animals.

Nevertheless, there is some plasticity to disgust. Disgust is, like taste, relative to one's cultural setting and background. We could learn to overcome primal disgust for certain foods, such as durian fruit,[7] that would almost without question instill disgust if we encountered it at a buffet being previously unaware of its existence. Perhaps we would have liked durian if we were acquainted with its taste from early age on. The other way around, people can learn disgust. In addition to primary objects of disgust, where the emotion of disgust functions as a "guardian of the body", "in most human societies disgust has become a social emotion as well, attached at a minimum to those whose appearance (deformity, obesity, or diseased state), or occupation (the lowest castes in caste-based societies are usually involved in disposing of excrement or corpses) makes people feel queasy" (Haidt and Graham 2007, 106). Does disgust as social emotion affect our beliefs about rats? It is not much of a stretch to take seriously the possibility that in addition to the evolutionary hard-wired disgust of rats, disparaging social representation—in part most likely driven by this evolutionary backstory—further fans the flames of aversion.[8] This is all the more relevant considering the possibility that disgust works as a "moral magnifier" (Ivan 2015). When we are confronted by a moral problem, requiring ethical judgment, there is some indication that the emotion of disgust could throw us off guard, doubling down on our negative dispositions regarding the situation at hand.

Whereas liminal rats are generally associated with filth and disease, those in the lab are perhaps freed from the adverse associations of their con-specifics "out there" but at the same time objectified as epistemological resources, as models for human disease.[9] Or perhaps there are different perceptions in play alongside each other. While some point out that "advantages of rodents include their small size, ease of maintenance, short life cycle, and abundant genetic resources" (Bryda 2013, 207) make for a great

---

[7]Durian fruit can be found in various countries in Southeast Asia and generally people either hate or love its scent. https://en.wikipedia.org/wiki/Durian. Accessed 23 March 2020.

[8]Moreover, "having a rat problem" is associated with stigma itself (Van Gerwen and Meijboom 2018).

[9]Are the sorts of ethical reflection with regard to animal research structured in a way to arrive at trustworthy judgments, and are they truly attuned to the suffering inflicted on individuals? Or do the number of animals and the way they are described and brought to the attention of those making the ethical assessment foreclose or otherwise heavily sway intuitions?

"animal model", aversion could very well explain why, since the beginning of the twentieth century, rodents, and not dogs, have become a primary "animal model" for invasive research. Narratives about animals have significant power to structure human action (see also Robin et al. 2017). The way perception affects our moral psychology often remains hidden in the background, and because these "are not always introspectively accessible, even a moral action can take place against the background of unconscious, non-virtuous tendencies of cognitive response" (Westerhoff 2017, 299). Given explicit aversion regarding rats, there is at least work to be done in unraveling human perception and its effect on moral judgment, so as to ascertain that moral psychology and its dynamics do not hinder fair assessment of the interests of rats involved, liminal or lab. As we will discuss now, these concerns also plague models for co-existence.

17.4   Rat Politics

While animal research ethics provides a source for an ethics of conflict resolution between liminal rats and humans, animal politics builds upon the work in human political theory. The most prominent and recent example is found in the work of Sue Donaldson and Will Kymlicka (2011), who develop a theory of denizenship for liminal animals.[10] Such political accounts go beyond mere conflict as they aim to establish the ground-rules for human-liminal animal co-existence. Whereas animal research ethics substantially but modestly sets new standards, animal politics ambitiously ups the ante by speaking in terms of moral rights and justice. Such justice will not come easy, considering (1) the way in which humans have subjugated non-humans throughout history, something apparently ingrained in culture at large, and (2) the way in which economic and self-centered interests have become institutionalized and vested (Donaldson and Kymlicka 2011, 252). Political philosophy can help to engender novel perspectives on animals, but we need to be careful not to get carried away, as

> (d)eveloping new and expanded theories of animal rights may be intellectually stimulating and challenging, but can it make any difference to real-world campaigns and debates? We are not optimistic about the prospects for dramatic change in the short term, and we certainly have no delusions that one can somehow change the world simply by articulating better moral argument ... [which] ... are notoriously ineffective when they run so fully against the grain of self-interest and inherited expectations. (ibid.)

Compounding these concerns, the language of justice may prove insufficiently persuasive with regard to rats due to their liminal status and association with disease,

---

[10]This is part of a three-way distinction of groups of animals, a delineation that determines the extent and existence of positive obligations to animals in addition to universal basic negative animal rights. Next to understanding liminal animals as denizens, domesticated animals become citizens, and animals in the wild are best understood as members of sovereign wildlife communities. Group-membership of individual animals differentiates specific rights and obligations, resulting in three distinct and political understandings of human-animal interaction.

fear, and disgust. If animal politics has to work against the grain of institutionalized anthropocentrism and egoism, any theoretical defense for a species so often despised, however compelling in itself, will likely fail to capture the public imagination. Up for the challenge, several political theorists (including Donaldson and Kymlicka) have begun to look into the factors that motivate individuals to act upon philosophical theory. Perhaps this search also provides a way to take liminal rats seriously in moral deliberation.

17.5   Failure of Imagination

Before looking at liminal rats specifically, why does the vast majority of people not live up to the tenets of animal ethics and politics? Perhaps it is due to "failings and limits of the imagination" (Cooke 2017, e4). Imagination provides a portal to approximate the experience of others. A failure of imagination renders one ignorant of the needs and interests of others, paving the way to moral inertia. Arguments emerging out of animal ethics and politics require fertile soil, which is why, according to Steven Cooke, we need to identify and promote the social conditions that shape moral imagination "as a precondition for non-human animals to be properly recognised as beings owed justice for their own sakes" (ibid.).

Not any sort of imagination will do. Conservative works of art and fiction, Cooke argues, probably often hamper the kind of imagination needed to recognize the entitlements of animals. More specific, we assume that the work of, say, J.M. Coetzee serves as a progressive example of cultivating imagination attuned to ethical interaction with animals. Distinguishing conservative from progressive works of art is not sufficient, however, as the latter could in reference to individual freedom unhinge individuals from the collectives that form the sometimes-unbeknownst fabric of their lives. Amitav Ghosh, for example, painstakingly makes the case in his book *The Great Derangement*, published in 2016, that those anointed with the task of imagination (novelists in particular) too often overlook a collective more-than-human imagination due to their preoccupation with autonomous individuals, all the more unsettling given the immense ramifications of climate change.

Will imagination indeed motivate moral agents to act justified or just towards liminal rats? Granting some level of epistemic access to the experience of others— walking a mile in their shoes—could prompt consideration. While imagination plays a role, we should perhaps not overstate is ability to spur agents into morally praiseworthy action, nor expect it to dispel all negative associations that cloud our judgements. The sort of imagination matters.

17.6 Sympathy for the Rat

If imagination is not sufficient in and of itself, what is? Let's distinguish between different ways of relating to the experience of others, including (1) sympathy, (2) empathy, and (3) compassion. Although none of these can be articulated in any undisputed sense, we understand sympathy as the ability to relate to the suffering of others. While it attunes one to the other, this engagement could remain rather cognitive, predominantly based and geared toward one's own suffering—"I would suffer if that would happen to me!"—and lacking in motivational impulse (Bloom 2017, 59). To improve upon imagination, narrowing the motivational gap, we need something more. Empathy goes further as it involves the ability to feel (approximately) what others feel, so more affective, making "it possible to resonate with others' positive and negative feelings alike — we can thus feel happy when we vicariously share the joy of others and we can share the experience of suffering when we empathize with someone in pain" (Singer and Klimecki 2014, r875).

Empathic concern can weigh heavily on one's shoulders, possibly resulting in empathetic distress, a state of suffering because others suffer. Such distress can lead one to turn away from the suffering of others in order to relieve suffering or result in overall fatigue when one continues to remain in the grip of empathy. So while empathy can spur individuals into action, it could do so primarily out of one's wish to stop the experience of empathic distress rather than an other-regarding act out of beneficence (Halifax 2011). There are further reasons why various theorists see in empathy a shaky foundation for fostering moral action. Whereas empathy links one to the other, it does so in ways that could actually hamper genuine moral concern. Empathy falters in the face of collective suffering, as "a single individual can evoke feelings in a way that a multitude cannot" (Bloom 2017, 127). Moreover, empathy can strongly bias moral concern to the suffering of those near and dear (e.g. ibid.; Gruen 2013; Kasperbauer 2015).

Avoiding the pitfalls of empathy, some put their trust in compassion instead. Whereas empathy involves feeling (approximately) what the other is feeling, compassion interweaves the awareness of suffering of others with the motivation to alleviate it: "In contrast to empathy, compassion does not mean sharing the suffering of the other: rather, it is characterized by feelings of warmth, concern and care for the other, as well as a strong motivation to improve the other's wellbeing. Compassion is feeling for and not feeling with the other" (Singer and Klimecki 2014, r875).

Don't we need the empathetic ability to feel what the other is feeling in order to jump in action? No. If "I see a child crying because she's afraid of a barking dog. I might rush over to pick her up and calm her, and I might really care for her, but there's no empathy there. I don't feel her fear, not in the slightest" (Bloom 2017, 64). Feeling for rather than feeling with allows compassion to become rational, as Bloom puts it, informing our reasoning and deliberation without too much of a danger of falling victim to the perils of empathy. Compassion makes one susceptible to the suffering of others in a different way altogether compared to empathetic distress. Whereas the latter is preoccupied with oneself, inciting negative feelings with possible detrimental consequences for one's health and wellbeing, compassion turns towards others (both in terms of affect and motivation) with feelings of kindness or even love, giving rise to wholesome benefits for those who act compassionately (Singer and Klimecki 2014). Compassion attunes one to the suffering of others, prevents empathetic distress, and could perhaps help to bridge the motivational gap bothering animal theorists.

17.7   Compassion: A Stepping Stone?

The potential of compassion to motivate people to care about and alleviate the suffering of others cannot but entice moral and political theorists. Cheryl Abbate (2018, 45; emphasis original) argues that we should view compassion as a *"prerequisite* to being just. And if we have a duty to act justly towards animals (human and nonhuman), as the philosophy of animal rights holds, it follows that we have a duty to fulfill the prerequisites of being just".

One needs to foster compassion in order to act in accordance with justice. Compassion becomes a stepping stone for political theory in making sure that the moral agents are able to abide by its principles. Here we could wonder whether there is more to compassion beyond being instrumentally valuable in facilitating just action. Abbate herself asks whether we should reserve a key epistemic role for compassion in our moral inquiries given its tremendous moral potential. A reason for rejecting this suggestion, as she argues, is to point out that compassion does not help the "lifeboat deliberator" we encountered earlier in this chapter. If one finds oneself in a lifeboat, pondering the question who to throw overboard, the following question appears difficult to answer: "Who should I show compassion to: the laboratory mice or the sick children?" (ibid., 41). This is why, while hesitant to attribute a more substantial role to compassion in moral inquiry, Abbate instead claims that: *"If moral agents have a duty to treat animals justly, and if being compassionate is necessary for moral agents to act justly, then moral agents also have a duty to cultivate compassion"* (ibid., 45–46; emphasis original). Animal rights theory sets the standard, and compassion helps individuals to make it happen. If liminal rats have rights, we should foster compassion in order to act in accordance with the demands of their entitlements. This could help to overcome the strongly negative and biased outlook many humans hold with regard to rats.

Abbate's imperative raises two questions: (1) how does one cultivate compassion and (2) is it true that moral agents have a duty to treat animals justly? To start with the former, Buddhist philosophy offers a rich source of contemplative practices, some of the specifically geared towards cultivating compassion. *"Metta bhavana*, or loving-kindness practice", as Abbate explains, "typically involves meditation-related techniques that foster feelings of benevolence and kindness for all beings, human and nonhuman" (ibid. 44; emphasis original). Several research studies support the claim that fostering compassion results in more prosocial and altruistic action (e.g. Singer and Klimecki 2014). Abbate also highlights humane education, referring to the initiatives of both Jane Goodall and Marc Bekoff to foster compassion and respect for life trough training and education in various sorts of settings. Meditative practice, together with humane education, can help to cultivate compassion, creating the conditions for moral agents to ascertain animal justice.

So it appears there is a way to foster compassion, so as to make individuals attuned to the suffering of liminal rats and inclined to help. But to return to our second question: What if people remain unpersuaded by animal rights theory? Abbate starts from the assumption that animal rights are theoretically convincing enough to jump into the facilitating potential of compassion. However, if indeed the problem of truth-aptness raises its head, the motivation to foster compassion could turn out question-begging. What if one is not convinced by the imperatives derived from animal rights theory, or on the fence about which theory proves sound? The instrumental reason to engage in compassion then loses much of its grip. Do we indeed first have to see the truth of animal rights theory to then develop our compassionate capacity in order to act in accordance with it?

Fleshing out the relation between compassion and rights-theory is helpful at this point. Should we indeed regard compassion instrumental to the demands of justice? Are there other ways of drawing the links, and how do these stack up to this claim? One could distinguish between at least three perspectives on this relation. As Abbate has it, (1) animal rights theory delivers the moral imperatives putting compassion in place as a psychological condition to get moral agents to do the right thing.[11] Others (2) understand the relation more in terms of opposition, endorsing either compassion or justice. The rift between care-ethics (Luke 1992) and rationalistic ethics (Singer 2011) comes to mind. At the far side of the spectrum, we find those who put compassion at the center, not at the disregard of rights-theory per se but providing the possible ground for such arguments to arise (Garfield 2001).

Is compassion the handmaiden of animal rights? We argue that putting compassion front and center comes with particular virtues, especially with regard to human interactions with liminal rodents. What are these virtues? Let's first identify several challenged faced by animal rights theory:

(1) any specific animal rights theory is vulnerable to the challenge of truth-aptness. While this is not the place to discuss the issues raised by error-theorists, anti-realists, relativists and the like, as long as there is widespread disagreement on whether one particular political theory of animal rights has correctly identified the true way to guide human-animal interaction, a gap between theory and practice appears difficult to bridge by reference to truth.[12]

(2) the language of rights, entitlements and impartial requirements appears especially ineffective with regard to animals who have become almost inseparable from adverse associations shaping human perception.

(3) finally, in line with the previous concern, and apparent in the attempts to make moral agents inclined via imagination (see Cooke 2017) or compassion (see Abbate 2018) to act in accordance to the precepts of animal rights theory, there is a problem of moral motivation. Moreover, rights theory builds (implicitly or explicitly) upon the idea of autonomous agency and the ability to transcend one's culturally and in other ways shaped behavior based on arguments primarily.

---

[11] Like Cooke who takes moral imagination as a social condition for endorsing animal rights theory.

[12] See DeGrazia (1996) for an approach to animal ethics that tries to accommodate this concern.

How does compassion navigate these issues? Below we will outline a notion of morality that requires endorsement of claims by individuals across species divides. Understanding morality in terms of interspecies engagement is less vulnerable to sceptic concerns, requires humans to develop clear and unbiased perception of liminal rats, directs us to compassion, and likely motivates humans in the process.

17.8   Compassion: Cornerstone of Interspecies Morality

For any sentient being, suffering is an inevitable part of one's life, and undeniably undesirable in a primordial way. In a sense, moral systems and morality in general appear fundamentally dependent on the existence of suffering, attested for example by the guiding strength of non-maleficence and beneficence within Western moral philosophy, and the often central role of compassion in for example Buddhist philosophy.[13]

We have already encountered a way to infuse animal politics with ideas originating from Buddhist philosophy; i.e. Abbate's rendering of loving-kindness meditation as a way to foster compassion, which in turn functions as an imperative for moral agents so as to be able to act in accordance with animal rights theory. This approach, however, is vulnerable in terms of its starting point. What if one, on a theoretical level, doubts the validity of right claims of liminal rats? The reason to foster one's compassion evaporates if not for the sake of motivating moral agents to acknowledge animal entitlements not only in theory but also in practice. Are there any reasons to take compassion more seriously apart from its instrumental value?

Why should we start from compassion? Here, we limit ourselves to two particular ways in which compassion can emerge. First, compassion can arise out of a way of seeing reality as fundamentally interdependent. Rather than seeing oneself as an individual, unhinged from the interspecies fabric of life, we could lessen the grip on ourselves, cultivating compassion in the process (Garfield 2001). Here, the moral imperative of compassion follows from a metaphysical realization of reality as radically interdependent, including oneself. In this sense, compassion is intimately tied up with one's perception of oneself. We need to ask ourselves whether there is "something very special, very independent about the self, something that could justify the distinction between my suffering or well-being and that of others as a motive for action" (Garfield 2015, 90). The less one attaches to a strong independent notion of self, the more one opens up to suffering of others, including other sentient beings such as rats.

But we need not go metaphysical. Some see the imperative to alleviate suffering not to arise primarily from a metaphysical realization, or derived from ethical theory, but as a task right in front of us. Rather than offering a solution of how to deal with

---

[13]Of course, compassion is found in many other traditions and religions, including Christianity, Taoism, etc. Here we follow up on the thread that connects to Buddhist philosophy so as to complement Abbate's angle.

the moral complexity of everyday live, the quest for truth in metaphysics and ethics could cloud our moral perception, numbing us so that we do not recognize the moral salience of suffering and the compassionate presence it calls for (Batchelor 2012).[14] Instead of falling into metaphysical dispute, or fervent theorizing, we should face suffering wherever we find it and simply address it as best as we can (Glassman 2003). Of course, theorizing can be useful in fostering moral action, but when it fails to do so, it becomes superflouos or even detrimental to achieving moral goals. Indeed, this aligns with pragmatists, who largely opt out of protracted ethical discursive dispute as well as the opposition between absolute truth and full-blown skepticism. Rather than continuing to ponder the question of justification, they usher each and every one towards the moral imperatives right in front of us.[15]

If pragmatism indeed dodges the charge of truth-aptness, haven't we then lost our moral bearings, let alone compassion as a guiding force? Whereas some pragmatists, perhaps most famously Richard Rorty, dismiss the notion truth altogether, we do not think there is a need for them to do so, if we hold that

> engaging in genuine moral inquiry – searching for principles and for particular judgements which will not be susceptible to recalcitrant experience and argument – requires that we take our beliefs to be responsive to new arguments and sensibilities about what is good, cruel, kind, oppressive, worthwhile, or just. Those who neglect or denigrate the experiences of others because of their gender, skin colour, or sexual orientation are adopting a very bad means for arriving at true and rational beliefs. They can be criticised as failing to aim at truth properly. (Misak 2002, 104)

We can salvage legitimacy of our moral claims, as Cheryl Misak proposes, by putting them to the test across a wide range of individual experiences, so as to aim at truth. Such a pragmatist account of morality aims at truth by means of its methodology, buttressing certain basic moral claims on the condition that they have been endorsed across a diversity of individual human experience. This will not get us absolute truth, nor sway all sceptics, but neither is required from a pragmatist perspective.

Of course, among other contingent characteristics such as age, we should add species to Misak's list of "gender, skin colour, or sexual orientation" as well. Note that this goes beyond the recognition of individuals as recipients of mere moral concern. The methodology sets down the conditions for an *interspecies morality*. We uncover what we should do in interspecies engagement, consulting and approximating the

---

[14] Secular Buddhists, including Stephen Batchelor, have interpreted Buddha's teachings as tasks rather than metaphysical truths to avoid the pitfalls of ongoing theoretical dispute. Buddhism, on this reading, involves a rather pragmatist attempt to address the inevitable and unmistakable suffering that permeates all sentient life. Of course, this is not the place to engage in exegesis, nor provide a comprehensive overview of the many different strands within Buddhist philosophy, or the various commitments to compassion across cultures throughout history. The reference to Buddhist philosophy provides a distinct perspective on compassion, again, further exploring the connection that Abbate draws between current animal politics and Buddhist philosophy going back all the way to the teachings of Buddha himself.

[15] Several authors, including William James himself, have recognized affinity between Buddhist philosophy and pragmatism (e.g. Scott 2000).

experience of other sentient beings as best as we can.[16] It will certainly not always be clear what should be done, though deliberation is not always necessary as there will be many situations where a compassionate outlook readily informs us what to do. As issues become more complex and diverse, however, insufficient endorsement is expected, which requires us to tread carefully, and/or allow for pluralism. Such an approach builds morality from the ground up, again and again putting normative claims against the test, requiring robustness and awareness of a diverse range of perspectives before letting such claims guide our moral interactions (Misak 2002).

Buttressing moral claims in line with the methodology, remarkably, gets us on track towards compassion, as it requires individuals to consult the experience of others (human and non-human) before endorsing a moral claim and act on it. Perhaps it is easier to determine which claims would be rejected by individuals across species, rather than to determine what they genuine would want. We anticipate that at least significant suffering will be vetoed across sentient species. It is hard to fathom the possibility that others endorse one's claim to inflict suffering upon them or reject your assistance to alleviate suffering if feasible. Of course, this goes both ways. We would want others to not inflict suffering upon us nor turn away from our suffering if they could easily help us. Not being made to suffer would appear to garner endorsement from sentient beings if we would (be able to) consult them. The requirement of *interspecies experiential endorsement* invigorates any moral agent to not only imagine but also engage with the other's perspective. In requiring engagement with the experience of others, this perspective on morality fosters compassion in the process. Moreover, as far as we can trust interspecies inquiry only when pursued as Misak (2002: 155) advocates "as far as it could fruitfully go", engagement with the experience of others avoids the empathetic pitfalls identified earlier.

17.9   From Anthropocentric to Multispecies Epistemologies

Seeking endorsement requires an effort to approximate the experience of others as best as we can. Many epistemic issues arise in trying to bridge species boundaries—or even intra-species ones. We cannot get to a subjective, first-person perspective from an objective stance.[17] What do we know (e.g. Donaldson and Kymlicka 2016) about the capacities of animals to make choices in a certain socio-ecological setting? These are concerns at a rather cognitive level, susceptible to empirical informed

---

[16]See also Clemens Driessen (2014) for the idea of animal deliberation. Josephine Donovan (2006) argues for a dialogical development of care ethics. The work of Donaldson and Kymlicka is rife with examples of acknowledging and promoting animal agency, so that humans can reasonably infer, from carefully reading their behavior, what animals truly want. Eva Meijer (2019) goes beyond behavior by investigating language from a multispecies and interspecies perspective. See Kate Manne (2017) for the idea of bodily imperatives, where the normative content and force is believed to reside in the vulnerability of being embodied. All these examples could very well fit within the idea of building an interspecies morality based on interspecies experiential endorsement.

[17]See Nagel (1974).

reasoning and research. We also need to address the more affective challenges that could hinder our ability to cross the species barrier. Some animals elicit disgust, fear, or any other sort of outright aversion that, as we have already indicated, could bias moral judgment. On the view of morality outlined just yet, these aversions not only distort our moral judgments but prevent genuine morality from coming into existence in the first place. Aversion for rats could (a) negatively affect our deliberations about them as moral subjects, and (b) make us rather unwilling to explore what they really want as participants in developing an interspecies morality. So, we not only have a moral but also an epistemic reason to uncover the aversions that color the way we perceive rats. It is why we need to go beyond anthropocentric epistemology and turn towards a multispecies version instead.

A clear view paves the way for genuine moral concern and to open up our minds to rat agency.[18] Strong feelings of aversion are not the only thing that distorts our perception. Less obviously, scientific modeling of rats also shapes the way we see rats. As epistemological models, they are viewed in terms of how they shed light on the intricacies of human health and disease.[19] Despite swaths of data, knowledge about rats often drifts atop underlying currents of human interests and perspectives. We fail to genuinely know rats, as we investigate them within the self-imposed limits of their preordained usage (Despret 2015).

In the field of pest control research, a similar epistemology is present albeit with a notable difference. Where in medical research the rat functions as a model for humans, in pest control research there is a genuine interest in the rats for the animals that they are. Nonetheless, such knowledge about rats is readily turned against them. In response to poison- and bait-shyness in rats—respectively the avoidance of food containing poison and avoidance of food free from poison of the sort previously encountered as containing poison—researchers developed baits containing slow-acting anticoagulants (substances that prevent blood from clotting) in order to sever the association between eating something and developing illness due to internal bleeding (Naheed and Khan 1989). As a result, rats die a slow and excruciating death. Another example: rats apparently map their environment in terms of predatory risks, which is why "(r)odent management could be more efficient and effective by concentrating on those areas where rodents perceive the least levels of predation risk" (Krijger et al. 2017, 2396), ironically, there where the rats feels most safe. As a third example, the inescapable sexual attraction of pheromones excreted by the male brown rat proves fatal for those female rats who cannot withstand the allure of the synthetic counterpart of these pheromones used to lure them into traps (Takács et al.

---

[18]While evolutionary history appears to underpin human aversion for rats, and these associations could very well prove difficult to overcome, the phenomenon of black rats considered as holy creatures at the *Karni Mata Temple* in Deshnoke, India, gives us a reason to be hopeful about the possibilities of replacing overtly negative associations with at least more positive ones. https://en.wikipedia.org/wiki/Karni_Mata_Temple. Accessed 23 March 2020.

[19]Whether pre-clinical animal research indeed provides a sound methodology for developing human health interventions is highly questionable due to the myriad biological differences between species (e.g. Greek and Menache 2013; Pound and Ritskes-Hoitinga 2018).

2016). Again, knowledge about rats cannot be separated from the human interests to control them.

All the more striking, considering the above, is the lacuna of knowledge about liminal rats in the urban context. While one of the species studied most extensively in medical research, rats remains elusive in their urban setting. Precisely there where rats are perceived as a problem, ecological knowledge is scarce. Why is that? One reason could be that

> (u)rban rat ecology [...] remains vastly unexplored because these animals are cryptic, crepuscular, difficult to identify, and hazardous to handle. Additionally, the high-rise buildings that block satellite link-ups, underground sewers and subway tunnels, and rebar enforced concrete covered landscape make it difficult—if not impossible—to track urban animals using traditional radio telemetry. Consequently, there are few ecological studies with free-ranging urban rats. (Parsons et al. 2015, 1)

In addition to the fact that obtaining knowledge about rats in an urban environment is difficult, perhaps the lack of research interest is in part due to the lack of perceived urgency and relevance. As long as the perceived problem is addressed in terms of pest management and conflict there is no need for further scientific research to improve upon the practice.[20] The lack of knowledge about liminal rats hinders development of policy to address human-animal co-existence. Whereas rats are generally associated with threats to health, it is difficult to quantify an acceptable level of risk. As common in policy making, one has to decide in the face of uncertainties and incomplete knowledge. Still, the predicament of such decision-making does require one to ascertain whether enough has been done to gather information—and, getting back to the point of interspecies morality, whether rats themselves would endorse the measures taken.

What about research that shows that rats giggle when tickled (Panksepp and Burgdorf 2010), release another rat from confinement to share a piece of chocolate (Bartal et al. 2011), or come to the rescue of a distressed conspecific (Sato et al. 2015). All of this research strives to highlight its relevance to understanding human social behavior, while incidentally illuminating rat empathy in the process. Of course, this type of research raises questions of moral legitimacy. Even if we would like to know more about rats unperturbed by any other interests, developing a multispecies rather than anthropocentric epistemology, moral concerns restrict the possible ways of getting to know each other. Still, we can learn from what we already got while at the same time thinking about new ways for interspecies engagement. Rats rescuing each other to share chocolate could tell us something about the evolutionary shared characteristics across species, shedding light on human social behavior. However, rather than seeing the ingenuity and empathetic concern displayed by rats primarily as proxies for human counterparts, or turn these capacities against them, these findings can help us to establish a clear view of rats, and, furthermore, seeing how rats themselves would endorse compassionate action.

---

[20]Of course, ecological approaches to liminal rodent management do demand more research into the urban ecology of liminal rats.

Whereas an anthropocentric epistemology is geared towards using or eradicating rats, a multispecies epistemology helps to envision new perspectives on co-existence. It escapes the boundaries of how we ordinarily see our shared landscapes by trying to approximate the way in which rats and other animals see reality. Whereas interspecies morality disavows invasive animal experimentation as a way to know animals, multi-species epistemology requires that we go beyond the human perspective in order to acknowledge the manifold other ways in which other individuals, including those from other species, experience the world. It requires that we shift perspective and see the urban environment as an animal collective embedded in myriad socio-ecological interconnections. Rats are remarkable inhabitants of urban and human-dominated landscapes, and as apparent from the discussion above, humans need to do much more to really get to know the rat's perspective, especially considering that morality demands endorsement across the species divides.[21]

17.10   From Philosophical Deliberation to Compassionate Engagement

What do metaphysical disputes, skewed perception of animals, and wavering moral motivation tell us philosophers? These concerns are even more pressing with regard to (liminal) rats and taking them seriously in moral consideration: theories remain provisional in terms of their specific moral and political status, anthropocentric epis-temologies cloud human perception, and rats are unlikely candidates to benefit from uncertain moral motivation. Fostering compassion could remedy these concerns. By means of contemplative practices and humane education, we might develop a less biased view of rats, getting familiar with their experience of the world, and extend our compassion based on the recognition that their suffering is not categor-ically different from those of other sentient beings, ourselves included. However, the moral imperative for autonomous agents to foster compassion through loving-kindness meditation perhaps assumes some sort of *moral privilege*. Whereas being a competent moral agent already sets a substantial standard, a requirement to engage in such contemplative practices even ups the ante.

In addition to personal growth by means of contemplation, compassion ushers us to look at the situation at hand in all its complexity. We need to acknowledge individual and situational differences regarding capacity for compassionate action. Some people

---

[21] What about the gap between "is" and "ought"? Are animals not locked within the domain of "is"? Perhaps 'the problem' as Steven Shaviro aptly puts it 'is not to derive an "ought" from an "is," but to see how innumerable "oughts" already *are* ... nonhuman animals *do* continually ascribe value to things, and make decisions about them—even if they do not offer the sorts of cognitive justifications for their value-laden actions that human beings occasionally do ... (o)ur own value activities arose out of, and still remain in continuity with, nonhuman ones—as we have known at least since Darwin. We perpetuate anthropocentrism in an inverted form when we take it for granted that a world without us, a world from which *our own* values have been subtracted, is therefore a world devoid of values altogether' (2015, 24; emphasis original).

are naturally endowed with a caring demeanor, whereas for others compassion does not come easy. Some are nudged by their social environment, whereas others have to swim upstream in waters inimical to compassionate action. If we put compassion front and center, we need to carefully consider where someone is coming from so as to foster compassionate action as best as we can. Here a difference between a justice and compassion approach becomes apparent. As the former will speak to the will of individuals to live up to the imperatives of theory, or else they are blameworthy, the latter will primarily look at the situation at hand in order to spot opportunities to promote compassionate action.

One of the ways in which to foster compassion is, as we have argued, to genuinely engage with the other. In engagement with the experience of others and taking this inquiry "as far as it could fruitfully go" (Misak 2002, 155), we develop an awareness of suffering across the boundaries of species. An interspecies view of morality in terms of engagement is not merely cognitive, but also socially interactive. Reflective deliberation, imagination, and individually cultivated compassion (for example by means of meditation) indeed provide important building blocks for moral engagement but remain insufficiently imbedded in social interaction. Considering that our perception of the world and our moral agency emerges out of social interaction, it is perhaps too much to expect individually-oriented practices to dramatically improve moral action. Notwithstanding the relevance of philosophical reflection, the arts, and contemplative practices, we furthermore need forms of engagement like humane education, as Abbate emphasizes, and other ways of engendering genuine interest in the experience of others, so as to know what we should do.

Moreover, shifting our emphasis in morality from deliberation to engagement, we can inform our method by various insights from moral psychology. What does it take for individuals to act with compassion? Rather than relying on individual willpower alone, we should explore supporting socio-ecological conditions of compassionate action. Such a perspective can help us track *compassion determinants*: what socio-ecological factors affect the ability of individuals to perceive and act with compassion when confronted with liminal rats? And how do we bring such an approach to bear on the management of human-rat relations?

Preventive approaches to human-liminal rat conflicts emphasize an ecological awareness so as to take measures that avoid future conflict as much as possible. Knowledge about the ecology and behavior of liminal rats here can function as a way to open up the minds of people to understand the inevitability of conflict when ecological processes are ignored in the management of the human-liminal rat interface. In addition, we suggest, professionals in the field of "pest management" could play a role as *multispecies epistemologists* in promoting compassionate human-liminal rat interaction. They could do so by, in addition to (a) communicating knowledge that fosters ecological awareness with the aim to prevent future conflict, also via (b) certain narratives and other communicative approaches to make others engage with the lived experience of individual rats. In doing so, they could clear up any overtly negative associations attached to the rats in question and motivate clients to take an interspecies approach to their moral decision-making. Insights from psychology could help at this level to promote a compassionate outlook. While we should be careful not to put all responsibility on the shoulders of these individuals at the danger of forgetting the various forces shaping human-rat interactions (legislation, availability of animal welfare compromising measures, dynamics between service-providers and clients in a market-economy, etc.), both in terms of conflict-negotiation and co-existence mediation, these professionals do play a key role in shaping interactions between humans and liminal rats.

17.11 Conclusion

As James Albrecht (2004, 25) interprets Ralph W. Emerson's poem, "the rat is 'cruel to thy cruelty' – that is, 'wicked' and cruel only when defined as such from a narrow anthropocentric perspective". We have argued that moral perception is vital in ascertaining genuine moral consideration, which puts the onus on those making moral judgments about the fate of (liminal) rats to critically evaluate the way in which negative associations color the way in which they perceive these animals. Moreover, being one of the most despised animal species across human cultures, rats can be considered a litmus test for moral philosophers and animal politics. If people take the interests of rats seriously in moral consideration, then nothing would appear in the way of a genuine just interspecies society. In that way, thinking about and with rats could engender a novel perspective on morality; as something that emerges out of engagement across species-divides, with an emphasis on fostering compassion. We need to compassionately engage with the experience of others, including rats and those who perceive these animals as pests, as well as pay attention to the way we are all embedded in particular socio-ecological settings so as to promote compassionate action. Rather than viewing compassion as instrumental to attaining animal rights, instead we should view elaborate theoretical accounts of interspecies co-existence are ways of expanding moral imagination and fostering compassion.

It is up to philosophers, among others, to find ways to promote and facilitate moral actions that are robustly endorsed across species. Indeed, Buddhist philosophy provides a valuable insight for its general devotion to compassion, and we have indicated a way forward based on this insight in line with pragmatist inclinations. Perhaps compassion lacks the epistemic role in guiding specific moral action but we have argued that in the specificity of our lives and moral problems, we should tread carefully anyway. In other words, the "lifeboat-deliberation" is not necessary better off in the possession of theoretical knowledge, considering earlier discussed concerns about the truth-aptness such moral claims. The buck stops somewhere, perhaps such that "(w)hen we are faced with the unprecedented and unrepeatable complexities of this moment, the question is not "What is the right thing to do?" but "What is the compassionate thing to do?" (Batchelor 1998, 48).

References

Abbate, C. 2018. Compassion and animals: How to foster respect for other animals in a world without justice. In *The moral psychology of compassion*, ed. J. Caouette and C. Price, 33–48. Lanham, MD: Rowman & Littlefield.

Albrecht, J.M. 2004. "What does Rome know of rat and lizard?": Pragmatic mandates for considering animals in Emerson, James and Dewey. In *Animal pragmatism: Rethinking human-nonhuman relationships*, ed. E. McKenna and A. Light, 19–42. Bloomington, IN: Indiana University Press.

Bartal, I.B.-A., J. Decety, and P. Mason. 2011. Empathy and pro-social behavior in rats. *Science* 334 (6061): 1427–1430.

Batchelor, S. 1998. *Buddhism without beliefs: A contemporary guide to awakening*. New York: Penguin.

Batchelor, S. 2012. A secular Buddhism. *Journal of Global Buddhism* 13: 87–107.

Bloom, P. 2017. *Against empathy: The case for rational compassion*. New York: HarperCollins Publishers.

Bryda, E.C. 2013. The Mighty Mouse: The impact of rodents on advances in biomedical research. *Missouri Medicine* 110 (3): 207–211.

Camus, A. 1947. *La peste*. Paris: Éditions Gallimard.

Cooke, S. 2017. Imagined Utopias: Animals rights and the moral imagination. *Journal of Political Philosophy* 25: e1–e18.

Dean, K.R., F. Krauer, L. Walløe, O.C. Lingjærde, B. Bramanti, N.C. Stenseth, and B.V. Schmid. 2018. Human ectoparasites and the spread of plague in Europe during the Second Pandemic. *Proceedings of the National Academy of Sciences of the United States of America* 115 (6): 1304–1309.

DeGrazia, D. 1996. *Taking animals seriously: Mental life and moral status*. New York: Cambridge University Press.

Despret, V. 2015. Thinking like a rat. *Angelaki: Journal of the Theoretical Humanities* 20: 121–134.

Donaldson, S., and W. Kymlicka. 2011. *Zoopolis: A political theory of animal rights*. New York: Oxford University Press.

Donaldson, S., and W. Kymlicka. 2016. Rethinking membership and participation in an inclusive democracy: Cognitive disability, children, animals. In *Disability and political theory*, ed. B. Arneil and N. Hirschmann, 168–197. Cambridge: Cambridge University Press.

Donovan, J. 2006. Feminism and the treatment of animals: From care to dialogue. *Signs: Journal of Women in Culture and Society* 31: 305–329.

Driessen, C.P.G. 2014. Animal deliberation. In *Political animals and animal politics*, ed. M.L.J. Wissenburg and D. Schlosberg, 90–104. London: Palgrave Macmillan.

Emerson, R.W. 1994. *Collected poems and translations*. New York, NY: Library of America.

Garfield, J.L. 2001. *Empty words: Buddhist philosophy and cross-cultural interpretation*. Oxford: Oxford University Press.

Garfield, J.L. 2015. Buddhist ethics in the context of conventional truth: Path and transformation. In *Moonpaths: Ethics and emptiness*, ed. The Cowherds, 77–95. New York: Oxford University Press.

Glassman, B. 2003. *Infinite circle: Teachings in zen*. Boston: Shambhala Publications.

Ghosh, A. 2016. *The great derangement: Climate change and the unthinkable*. Chicago: University of Chicago Press.

Greek, R., and A. Menache. 2013. Systematic reviews of animal models: methodology versus epistemology. *International Journal of Medical Sciences* 10 (3): 206–221.

Gruen, L. 2013. Entangled empathy: An alternative approach to animal ethics. In *The politics of species: Reshaping our relationships with other animals*, ed. R. Corbey and A. Lanjouw, 223–231. New York: Cambridge University Press.

Haidt, J., and J. Graham. 2007. When morality opposes justice: Conservatives have moral intuitions that liberals may not recognize. *Social Justice Research* 20: 98–116.

Halifax, J. 2011. The precious necessity of compassion. *Journal of Pain and Symptom Management* 41 (1): 146–153.

Ivan, C.E. 2015. On disgust and moral judgments: A Review. *Journal of European Psychology Students* 6: 25–36.

Kasperbauer, T.J. 2015. Rejecting empathy for animal ethics. *Ethical Theory and Moral Practice* 18 (4): 817–833.

Krijger, I.M., S.R. Belmain, G.R. Singleton, P.W.G. Groot Koerkamp, and B.G. Meerburg. 2017. The need to implement the landscape of fear within rodent pest management strategies. *Pest Management Science* 73: 2397–2402.

Luke, B. 1992. Justice, caring, and animal liberation. *Between the Species* 8 (2): 100–108.

Manne, K. 2017. Locating morality: Moral imperatives as bodily imperatives. *Oxford Studies in Metaethics* 12: 1–26.

Meerburg, B.G., F.W.A. Brom, and A. Kijlstra. 2008. The ethics of rodent control. *Pest Management Science* 64: 1205–1211.

Meijer, E. 2019. *When animals speak: Toward an interspcies democracy*. New York: New York University Press.

Misak, C. 2002. *Truth, politics, morality: Pragmatism and deliberation*. London: Routledge.

Nagel, T. 1974. What is it like to be a bat. *Philosophical Review* 83: 435–450.

Naheed, G., and J.A. Khan. 1989. "Poison-shyness" and "bait-shyness" developed by wild rats (*Rattus rattus* L.). I. Methods for eliminating "shyness" caused by barium carbonate poisoning. *Applied animal behaviour science* 24 (2): 89–99.

Panksepp, J., and J. Burgdorf. 2010. Laughing rats? Playful tickling arouses high-frequency ultrasonic chirping in young rodents. *American Journal of Play* 2 (3): 357–372.

Parsons, M.H., R.J. Sarno, and M.A. Deutsch. 2015. Jump-starting urban rat research: conspecific pheromones recruit wild rats into a behavioral and pathogen-monitoring assay. *Frontiers in Ecology and Evolution* 3. https://doi.org/10.3389/fevo.2015.00146.

Pound, P., and M. Ritskes-Hoitinga. 2018. Is it possible to overcome issues of external validity in preclinical animal research? Why most animal models are bound to fail. *Journal of Translational Medicine* 16 (1). https://doi.org/10.1186/s12967-018-1678-1.

Richter, C.P. 1968. Experiences of a reluctant rat-catcher the common norway rat-friend or enemy? *Proceedings of the American Philosophical Society* 112: 403–415.

Robin, C., E. Perkins, F. Watkins, and R. Christley. 2017. Pets, purity and pollution: Why conventional models of disease transmission do not work for pet rat owners. *International Journal of Environmental Research and Public Health* 14 (12): 1526.

Sato, N., L. Tan, K. Tate, and M. Okada. 2015. Rats demonstrate helping behavior toward a soaked conspecific. *Animal Cognition* 18 (5): 1039–1047.

Schinkel, A. 2008. Martha Nussbaum on animal rights. *Ethics & the Environment* 13 (1): 41–69.

Scott, D. 2000. Walliam James and Buddhism: American pragmatism and the Orient. *Religion* 30 (4): 333–352. https://doi.org/10.1006/reli.2000.0292.

Shaviro, S. 2015. Consequences of panpsychism. In *The nonhuman turn*, ed. R. Grusin, 19–44. Minneapolis, MN: University of Minnesota Press.

Singer, P. 2011. *Practical ethics*. Cambridge, UK: Cambridge University Press.

Singer, T., and O.M. Klimecki. 2014. Empathy and compassion. *Current Biology* 24 (18): 875–878.

Stephen, C. 2014. Towards a modernized definition of wildlife health. *Journal of Wildlife Diseases* 50: 427–430.

Takács, S., R. Gries, H. Zhai, and G. Gries. 2016. The sex attractant pheromone of male brown rats: Identification and field experiment. *Angewandte Chemie International Edition* 55: 6062–6066.

Van Gerwen, M.A.A.M., and F.L.B. Meijboom. 2018. The Black Box of rodents perceived as pests: On inconsistencies, lack of knowledge and a moral mirror. In *Professionals in food chains*, ed. S. Springer and H. Grimm, 392–397. Wageningen: Wageningen Academic Publishers.

Westerhoff, J.C. 2017. Madhyamaka and modern Western philosophy: A report. *Buddhist Studies Review* 33 (1–2): 281–302.

Yeates, J. 2010. What can pest management learn from laboratory animal ethics. *Pest Management Science: Formerly Pesticide Science* 66 (3): 231–237.

Joachim Nieuwland is a junior assistant professor of veterinary ethics at the Faculty of Veterinary Medicine, Utrecht University, and post-doc researcher at the Adaptation Physiology group (ADP) of Wageningen University. At the Centre for Sustainable Animal Stewardship (a collaboration between the Faculty of Veterinary Medicine and the Animal Sciences Group of Wageningen University & Research), he works on the value of knowledge in animal ethics.

Franck L. B. Meijboom studied theology and ethics at the Universities of Utrecht (NL) and Aberdeen (UK). As Associate Professor he is affiliated to the Faculty of Veterinary Medicine of Utrecht University, the Ethics Institute of Utrecht University (Faculty of Humanities) and the Adaptation Physiology group (ADP) of Wageningen University. Additionally he is Head of the Centre for Sustainable Animal Stewardship (CenSAS). His fields of interest are in ethics of animal use and veterinary ethics, in agricultural and food ethics, and the role of public trust and debate in these domains. His research covers ethics of animal welfare, ethics of livestock farming, fisheries and aquaculture, ethics of innovative and translational animal research and ethical aspects of the use of technologies for sustainable animal breeding. Furthermore, he is member of the national Council of Animal Affairs (RDA), the Supervisory Board of the Animal Protection Society (Dierenbescherming), Editor in Chief Journal of Agricultural and Environmental Ethics and Vice President of the European Society for Agricultural and Food Ethics (EurSafe).

Chapter 18
Interpreting the YouTube Zoo: Ethical Potential of Captive Encounters

Yulia Kisora and Clemens Driessen

Abstract YouTube hosts a vast number of videos featuring zoo animals and humans actively reacting to each other. These videos can be seen as a popular genre of online entertainment, but also as a significant visual artefact of our relations with animals in the age of humans. In this chapter we focus on two viral videos featuring captive orangutans interacting with zoo visitors. The interpretations of ape-human interactions arising from the extensive number of comments posted to the videos are ambivalent in how they see the animals and their assumed capabilities. We argue that the YouTube Zoo could figure as a snapshot of human-animal relations in late modern times: mediating artificial conditions of animals suspended between the wild and the domestic, while offering a screened account of a deeply surprising interaction. The chapter shows the potential of close interactions between humans and animals to destabilise or reinforce the neat divisions between the human and the animal. It also shows the ethical potential of these interactions to either reinforce or question common practices of dealing with wild animals.

18.1 Introduction

The very first video posted on the video sharing platform YouTube was uploaded on the 23rd of April 2005, at 8:27 pm (CET) by Jawed Karim. He named it: 'Me at the zoo'. In this video, Jawed was filmed standing in front of an elephant enclosure in San Diego zoo, saying: "All right, so here we are in front of the, uh, elephants, and the cool thing about these guys is that, is that they have really, really, really long, um, trunks, and that's, that's cool, and that's pretty much all there is to say". The elephants in the background remain seemingly unimpressed with Jawed or his comment, as they casually consume hay scattered next to the fence separating them from the visitor area. Nothing else happens in this video, similar to many other online mementos of

Y. Kisora (✉) · C. Driessen
Wageningen University & Research, Wageningen, The Netherlands
e-mail: Yulia.kisora@wur.nl

© The Author(s) 2021
B. Bovenkerk and J. Keulartz (eds.), *Animals in Our Midst: The Challenges of Co-existing with Animals in the Anthropocene*, The International Library of Environmental, Agricultural and Food Ethics 33,
https://doi.org/10.1007/978-3-030-63523-7_18

zoo visits that can be found on the internet. Nonhuman animals (hereafter "animals") are lying down or pacing, serving as an exotic background to a family and friends' day out.

As an alternative to this somewhat dull representation of captive animal lives, YouTube hosts innumerable videos of 'funny zoo animals', featuring lions and otters, polar bears and lamas, chimps and giraffes and many others. Different as these animals are, the videos follow a similar plot: animals and humans actively react to each other in an apparently unexpected and highly entertaining manner. It triggers viewers' curiosity and entices them to come forward, comment and vigorously defend 'the right' interpretation of what is happening in the videos. Some videos go 'viral', gain millions of views, shares and comments, reaching viewers across continents and over the years. With the global reach of YouTube of 1.9 billion monthly active viewers in 91 countries (in September 2019) these videos have become, without a doubt, a popular genre of online entertainment.

Yet they are more than that. In line with literature (Driscoll and Hoffmann 2018), films (Bousé 2000; Burt 2002) and videogames (Driessen et al. 2014) YouTube videos can be seen as a curious artefact of our relations with animals, an emerging cultural genre imbued with power to (re)configure the ways we see and relate to animals. Or, as philosopher Vinciane Despret argued, "the proliferation of these videos attests not only to new habits but to the creation of a new interspecific ethos, of new relational modalities, that at the same time construct knowledge" (Despret 2016, 195).

The fact that some of these videos take place in a zoo deserves a closer examination due to a heavily laden history of zoos as places of human-animal engagement. Defining human against animal and culture against nature has been one of the most powerful ordering practices in Western culture. Metropolitan zoos have played a crucial role in this process of self-definition. As Anderson (1995) argued, "zoos are spaces where humans engage in cultural self-definition against a variably constructed and opposed nature" (Anderson 1995, 276). In the interpretation of critical scholars, zoos are hosting disaffected visitors staring at pale representations of animals, thereby reinforcing essentially violent, unjust and unequal relationships between humans and animals (Acampora 2010; Malamud 1998).

To (ethically) analyse human-animal relationships (in zoos and beyond), we commonly rely on binaries such as human/non-human, wild/domestic, natural/unnatural etc. (Bovenkerk and Keulartz 2016). However, these notions can be contested by drawing on what are in practice messy engagements, calling for more contextual ethical analysis (Palmer 2010). In this case, zoos can be investigated as spaces productive of peculiar types of interactions, resulting from the proximity between humans and what are generally considered to be 'wild' animals (Kiiroja 2016; Park et al. 2016).

If we believe that YouTube as a medium has something to offer in terms of ethics and production of knowledge about (zoo) animals we need to look closely at the character of the documented interaction between human and animal. If human-animal relationships are produced and re-configured in the process of visual representation (Burt 2002), what can YouTube videos tell us about the ways we relate to animals, especially in relation to traditionally meaningful categories of animal/human and wild/domestic? And what does the phenomenon of these videos' immense online popularity mean for early twenty-first century mediated wildlife and for the figure of the animal in contemporary society?

In an attempt to answer these questions, this chapter discusses two YouTube videos together with multiple comments of their viewers. The videos feature zoo orangutans and have jointly gained more than 70 million views and 22k comments, indicative of their global reach and capacity to trigger reactions of the online public. They document one interaction (not compilations thereof) so we were able to relate the comments to the interaction, and they have inspired multiple interpretations. We have paid special attention to the conversations in the branches of comments, treating them as a 'focus group' of a kind. As Wemelsfelder et al. (2000) have shown, it can be a suitable approach to interpret animal behaviour qualitatively by arranging for their interpretations to be shared and discussed. Here we tentatively extend this approach to the YouTube comments section, not to establish the true interpretation of the meanings of certain animal behaviours or to assess their welfare, but primarily to show how processes of interpreting can be thoughtful and deliberative, drawing on a range of considerations, and having particular implications for discussions on human-animal encounters and zoos.

We have attempted to present an overview of YouTube viewers' narratives related to human-animal relations, focusing on the comments that explicitly attempt to interpret orangutans' behaviour. Below, we will highlight the details of the filmed interaction that grasped attention of the viewers and prompted them to discuss the featured beings as animals, either as representatives of species or as particular individuals. Based on that, we will discuss the ambivalence of zoo orangutans' status in relation to categories of human/animal, culture/nature, domestic/wild, and the ethical implications this ambivalence seems to carry.

As social scientists we believe there is no such thing as a neutral, 'merely descriptive' position with regard to contested phenomena and institutions such as the zoo. Besides concerns about malpractice and the welfare of zoo animals, there is a broader debate about the legitimacy of holding wild animals captive, involving a range of arguments that have been extensively put forward elsewhere (Acampora 2010; Bovenkerk 2016; Keulartz 2015). Being aware of the critiques, in our research we try to look at the practice without condoning or promoting it, or trying to offer 'recipes for justifying moral positions and producing logically straightforward moral arguments' (Driessen and Heutinck 2014, 6). Like Irus Braverman in her in-depth investigation of workings of the institution of captivity 'Zooland', we would like to position ourselves 'on the fence': being driven by curiosity and ambivalence regarding what happens in zoos, without seeking to revert to judgments or position ourselves as inherently pro or against the institution itself (Braverman 2012). With Braverman we are aware that this is not only a shaky but also not a neutral position, as it—quite literally—is a position afforded by the institution of captivity and the fences it erects. Nevertheless, we feel that this vantage point can be productive to get a sense of the intricacies and ambivalences in the relationships between animals and humans that emerge under captive conditions.

The videos we focus on are available on Youtube under the respective titles 'Clever orangutan makes a fair-trade with human' and 'Monkey sees a magic trick'. We strongly advise you to watch the videos now and come back to reading once you have done so.

18.2   Interpreting the YouTube Zoo

The video 'Clever orangutan makes a fair-trade with human' was posted on YouTube by Vitaly R. in August 2016. Within less than a year it gained 6,982,528 views and almost 4000 comments. Later the video was probably sold and re-posted from the account of Rumble viral, a 'live viral video tv show' on YouTube, and racked up as many as 25,882,723 views and 7,208k comments (as of December 2019). In the beginning of the video, a man throws something to an orangutan, who is sitting across the moat with his hand open in a begging gesture. The man then stretches his arm towards the orangutan, asking him (in Russian) to give something in return. The orangutan looks around, reaches for a melon rind on the ground and throws it directly into the hands of the human. As the man thanks the orangutan in English, the cameraman suggests the man to throw the item back to the orangutan. After the man does so, the orangutan catches it, but immediately tosses it into the bushes. The description goes: "During a trip to Bali, Vitaly R. decided to throw a few treats towards an orangutan. To his surprise, his newfound friend decided to repay him with a treat of his own!"

The first detail that has drawn attention of the viewers is the fact that the orangutan

glances sideways before throwing the rind. Visitors feeding animals is considered by the zoo community as a 'perplexing' problem (Bitgood et al. 1988). No wonder, that in the very beginning of the video, we see and hear a passing zookeeper asking Vitaly not to feed the orangutan. The camera tilts, and we see Vitaly disappointedly walking away from the moat, but the cameraman encourages him to proceed 'Let's do it, they are gone'. Vitaly quickly looks around before throwing his item to Jacky. Although there is no sign saying 'don't feed the tourists', Jacky also looks to the left and to the right before launching the item. The commenters pick up on these 'furtive' glances:

- LOL I love how the orangutan also looks around to see if the guards are watching before he throws the banana'. (User Adolf)
- Watching if there's no snitches or zookeepers around. (User rondle berik)
- Exactly, they would have ruined the fun (User Anthony O'Brian)

The discussion of these glances speculates on a broad variety of assumptions about orangutans' cognitive and emotional processes. Is it merely mirroring of Vitaly's behaviour? Is he aware of the illegitimate character of their interaction? Or is he mocking Vitaly's behaviour, in an act of meta-communication about the meaning of throwing the rind? Soon enough via several commenters 'an orangutan' becomes a personality: commenters identify him as Jacky, a male orangutan who is 30-something years old and resides in Zoo Bali. We also learn that throwing items at visitors is a habit that Jacky might have adopted from keepers who would 'chuck' (in words of one of the viewers) food at him. He learned to use it for his own means, namely, to communicate his discontent with visitors passing by or taking photos while not giving anything in return. The story we are looking at turns from 'a zoo visitor feeding a zoo animal' to 'Jacky throws stuff at visitors to tell them they are not welcome' and even 'Jacky throws stuff at visitors when they are not looking to entertain himself'.

> i had an encounter with that same monkey when i was in bali…… so he was throwing poo (with excellent accuracy) but he would ONLY throw when we weren't looking. so funny. one point i turned around as he was throwing it, he had his arm in the air, stretched above his head, filled with poo…. but because i was looking he relaxed it behind his head as if nothing was happening….. very funny. (User matt ward)

Throwing objects at visitors is an issue more widespread among zoo chimpanzees and is generally considered to be a sign of stress (Martin 2008). We can speculate that zookeepers may have tried to prevent Jacky from throwing things at the visitors (at least a few viewers recall being warned about Jacky's habit by the zoo staff). Yet, as we've seen, some commenters have instead interpreted the habit as a specific communication medium between Jacky and the visitors. Understood in a context of a complex interaction with layered meanings, it actualises anecdotes that distinguish Jacky from other animals in zoos and give him a biography and an individual character.

One of the dominant interpretations of Jacky's behaviour in the video, as well as the description of the video itself, emphasizes the economic nature of the exchange between the human and the orangutan. Viewers assume that what Vitaly threw was a treat, as Jacky immediately ate it. The item he throws back looks to many as a banana, so the situation reminds of a treat for a treat fair deal. Yet when Vitaly returns 'the banana', Jacky tosses it into the bushes (which viewers ascribe to a deliberate intention, not a failure to throw properly), effectively stopping the interaction. Viewers wonder if it was Vitaly who failed to understand the meaning of the situation and respond accordingly and was therefore 'punished' by Jacky with stopping the exchange:

> He [Vitaly] should have just kept the fruit and pretended to eat it throwing it back was kinda of an insult he was honestly paying for the treat he was giving (User MrSnapy)
>
> Who is more fair in this case? (User La Nomia)

The second video with a (taxonomically misleading) title 'Monkey sees a magic trick' was published on December 2015 by Dan Zaleski (US). The video gained more than 3,000,000 views within three days after the original publication date, and became a headliner on news outlets such as the Daily mail, Time, Metro etc. As of December 2019, it has 56,480,311 views and 14,926 comments, making it one of the most popular zoo animal videos in YouTube history. The video starts with a close-up of a young orangutan behind glass. We see a man sitting down on a concrete step next to the glass. The orangutan (female orangutan from Barcelona zoo named Jinga, as we later find out from the comments) makes herself comfortable, as she sits down and puts one of her hands underneath her chin. The man places a cup on the step and demonstratively lowers a lychee into the cup, as the orangutan follows it with her eyes. The man then closes the cup and shakes it behind the step, removing the lychee. He then demonstrates the empty cup to the orangutan. She looks at it intently for 3 s and bursts into open-mouth laughter, falling on her back (ROFL-type, in YouTube vernacular). The camerawoman and the man are laughing out loud. After 15 s the orangutan sits up straight again, while the man puts the lychee back in the cup and the woman behind the camera comments: "You're crazy guy". The video ends as the man lifts the lychee again and the orangutan follows it with her eyes, seemingly eager to watch the trick anew.

Jinga's response to the 'magic trick' has proven to be puzzling for the viewers. Her visceral reaction seems to suggest, for many, that she understands the trick and finds it amusing, which leads them to praise her 'almost human' intellectual capabilities. For others, the primitivity of the trick makes the one who falls for it 'stupid'. Alternatively, other viewers observe kind-hearted and polite nature of Jinga in the fact that she appreciates the trick despite its basic level. In other words, the interpretation of the situation goes on to represent Jinga not only as an individual of a certain species with certain capabilities, but as a relational being joyfully engaged in a social situation.

– The best part is the obliging, polite attitude of the orang to watch the trick, like a loving grannie with a 7-year old grandson. (User anthro2)
– My thoughts exactly. The orangutan was thinking "I'm not stupid, but I'll laugh at this to make you happy" (User Mind Speaker)

And again, in a fierce debate on possible explanations of the orangutans' behaviour, personal knowledge of Jinga as an individual comes to play an important role. A commenter who introduces herself as a Barcelona zoo keeper describes her as an extraordinary intelligent and fun-loving personality.

> She was born in the zoo six years ago. Her name is Jinga and she definitely understands each situation and game you show her. When she plays with her little brothers she always makes the same funny face and she starts rolling on the floor. she is legit having fun here… Normally they aren't that intelligent but this one right here is by far the most intelligent member of the group, smarter than her mother even. She can do leggo and a lot of things and she is always the first to understand and solve problems or games that we show her! (User Maria Castilla)

As we have seen, the two videos feature different interactions and evoke various interpretations speculating on mental and emotional capabilities of orangutans. This raises the question of how we can make sense of the videos and viewers' reactions to them if we see them as a distinctly telling type of engagement with the animals. We suggest that the YouTube Zoo could figure as a deeply ambivalent phenomenon indicative of a new phase in human-animal relations in late modern times: mediating artificial conditions of animals suspended between the traditional categories of human/animal, nature/culture and wild/domestic. This screened account of a deeply surprising interaction may not merely contribute to the commodification of an encounter but also has an ethical potential to transform our view of animals and our ways of relating to them. We will discuss this in detail below.

18.3   YouTube Orangutans Unsettling Binary Concepts

Apes are among the most charismatic animals, appreciated by zoos for their ability to attract visitor crowds (Carr 2016). This is of course not accidental, considering that, as Corbey puts it, "our fascination with apes is only rivalled by our rebuff of apes: <humans tend to feel somewhat baffled by the paradoxical experience of recognizing something human in them, while at the same time tending to deny any identification with these beastly creatures>" (Corbey 2005, 7). Whenever an act of interpretation of ape behaviour occurs, it is not only scientific accuracy that is at stake. YouTube videos trigger comments that expose a constant and unsettling negotiation of categories or, if you wish, social orderings, in which we try to fit orangutans. Interpretation of animal behaviour is indeed an exercise in negotiating the human-animal divide, with both scientific and ethical implications.

Although none of the commenters interpret the actions of the orang-utans in the video as explicitly aggressive or dangerous, some of them point towards the alleged physical strength of the animals and graphically describe vicious ways in which they would have used it if given a chance. Some of these comments might have a note of irony to it, especially in the case of Jinga, whose peaceful outlook contradicts the alleged bloodthirsty intentions. Jacky's carefully calibrated strength does not appear to indicate aggression either. Seemingly puzzled by observing the somewhat weird, not easily explainable, but apparently peaceful interactions that the wild animals engage in, viewers warn each other that the glass wall and the moat separating animals and zoo visitors are the only boundaries that keep the former from ripping various body parts (the arm/neck/head/ears etc.) off the latter. "Be careful though, these ones can bite your face off" warns user Robin.

It is not surprising that some viewers see in Jacky and Jinga nothing more than 'just animals', given the long-standing tradition of Cartesian understandings of animals as mechanical and subsequent reductions of animal behaviour as instinct-driven. Reproducing the rather low opinion of animal emotional and mental capacities, which can be considered widespread in our species (De Waal 2019), they focus on significant absences and deficiencies in capabilities, which makes orangutans less than human. Along these lines, interpretations that put forward explanations more complicated than those based on instincts and aggression are dismissed as amateurish and fallacious. Jinga's laughter is considered to be an anthropomorphic projection; the orangutan is not laughing, some viewers insist, but gapes 'its' mouth in terror, triggered by something 'it' cannot explain. It only looks like laughing to humans not trained in interpreting orangutan behaviour, they conclude. Probably unknowingly, viewers comply with the canon of animal sciences that demands a strict separation of vocabularies for referring to human and nonhuman animal behaviour. Those who dare to speak about animal behaviour in terms of intentions and feelings risk being accused of anthropomorphism (in the past this even happened to Darwin himself) (De Waal 2019).

YouTube comments suggest that seeing captive animals might normalise the captivity for some viewers. While academic literature discusses various justifications for animals being held in captivity (Bovenkerk 2016), the one that seems to play out most in the comments is based on the assumed superiority over animals that gives humans the right of doing what they see fit. In this moral stance, the difference in cognitive capabilities justifies the difference in the treatment of human and non-human animals. Comments justifying the captivity for human entertainment tend to interpret the behaviour in a rather simplistic way, presenting an animal as an aggressive beast operating on instincts. Seeing Jacky and Jinga as, first and foremost, animals, triggers discussions of many ways in which animals are less than humans and how that is exactly why they are kept in zoos. Rationality, consciousness, technology and language skills are mentioned as uniquely human capabilities that orangutans (in this context repeatedly referred by the commenters as 'monkeys') wouldn't even dream of having. Thousands of comments attest animals (the species is often not mentioned) as aggressive and 'stupid', which explains the strict division between 'us' on this side of the cage and 'them' behind the bars. Importantly, in line with this reasoning it is human entertainment as the goal of the captivity that gets picked up and justified, as the viewers seem to be oblivious to the conservation claims of contemporary zoos. The animals' inferiority not only justifies the captivity, but also attests to the inability of animals to fully comprehend the lack of freedom and thus suffer from it.

> An animal does not have the same level of intelligence as us. They (and that includes monkeys) are incapable of forming complex thoughts an consider variables. They act solely on instinct. Saying humans do that too is cliché and just not true. Some act on instinct more than others, but everyone will consider the variables at some point again. These animals don't mind, as long as they have food, space, toys and the chance to socialize. (User MetroVerse)

Yet contrary to the mid-twentieth century way of looking at animals as displaying primitive behavioural patterns as units of evolution, there has been a recognition of a need for a broader vocabulary (Despret 2016). Unease about mechanistic descriptions of apes' behaviour has been expressed by, among many others, famous field primatologists, such as Jane Goodall, Diane Fossey and Barbara Smuts, as well as more traditional primatologists with a preference for experiments and observation in controlled environments, such as Frans De Waal. It has become apparent that sometimes, especially in the case of apes, denying any similarities may be more unscientific than acknowledging them (Weiss et al. 2012). In fact, the fear of anthropomorphism itself has been charged as a construct culturally specific to Western science (Allen et al. 1994). Mechanomorphism (or anthropodenial) (De Waal 2019) is beginning to be recognised as a fallacy no less serious than uncritical anthropo-morphism. Similarly to this development in primatology, and in contrast to some commenters describing orang-utans as primitive and less than human, others high-light their intellectual, emotional and moral capabilities. Although some viewers might attribute fairness, sense of humour and self-awareness to orang-utans lightly, or even as a joke, in many cases these hypothesised explanations cannot be easily dismissed as anthropomorphist projections, since many of them seem to be in line with research on orangutan minds. Jacky's 'furtive glances' can actually be explained by his awareness of himself and others (Shillito et al. 2005). His fair trade—by calcu-lated reciprocity, readiness to exchange goods and services "based on weighing costs and benefits when giving or returning favours and keeping track of them over time" (Dufour et al. 2009, 172) or, indeed, perception of fairness (Bekoff 2004). Jinga's reaction to the magic trick might have something to do with her understanding of object permanence (Rooijakkers et al. 2009), while laughing-out-loud can be a play reaction (Davila Ross et al. 2008), indicative of what humans would call 'having a good time'. Indeed, "the knowledge of animals survives in places where academics would never want to tamper" (Hearne 1994, 176)—even on YouTube.

'Almost' human behaviour and a looming realisation that they are "smarter than we think" (as User Dario Pavlovic puts it) seem to destabilise some viewers' views on what orang-utans are. They recall our common evolutionary background, wonder whether the distance between species is actually as big as usually thought and contem-plate whether supposedly 'unique' human qualities are actually unique. That leads to a discussion of orang-utans in terms of categories such as 'a beast', 'a thing' or 'a person', exemplified in the following branch of comments.

– I wouldn't hug that thing, pffft LMAO, I'm dying. (User Truth)
– It's not a "thing" it's a person (User Golvan)
– It's certainly an intelligent beast, but I wouldn't call it a person. (User Truth)
– Person definition: a human being, adult or child (User Robin)
– Fine, a non-human person then (User Golvan)

As the renowned primatologist Jan van Hooff once admitted, "studying apes creates an 'empathic unrest' because they evoke 'the subjective appreciation of animals as experiencing, judging, and striving beings', begging interpretations of their behaviours in terms of subjective valuations and calculated intentions" (Van Hooff 2000, 126, cited in Corbey 2005, 7). Starting to wonder about subjective experiences of orang-utans and meanings of their behaviours, some YouTube viewers seem to fall prey to the same uncomfortable condition as primatologists, closely observing their subjects. Arguably, they may be just about to display 'the nobler instincts of inquiry', with the absence of which Malamud so poignantly charged zoo visitors (Malamud 1998, 225). In this process, the human-animal divide starts to shift: while for some the category of an animal pre-determines their explanation of orang-utans' behaviour as instinctive and primitive; for others the observed behaviour and exposed capabilities serve as a springboard for seeing beyond the human-animal divide and recognising Jacky and Jinga as non-human persons. The status of orang-utans as non-human persons seems to make the spectator position less comfortable. As the intellectual, emotional and social capabilities of the animals come to the fore, the viewers also start to question the legitimacy of their captivity: 'animals this smart' or 'animals who can laugh' cannot be held captive.

> They're unnervingly smart, and seem to understand us better than we'd like. Makes me feel like a creep watching them, because just a short time with them and you realise they're pretty much another person. (User Tails Clock)

> If they can understand and emote like that wth are they locked up for our pleasure? (user Marie Watson)

While zookeepers may routinely treat their animals as persons and individuals (Park et al. 2016), zoos in general seem to have a complicated relationship with portraying their animals as persons to the public. For example, Artis Zoo in the Netherlands made headlines in 2016 when its director announced that their animals are not going to be given public names anymore so as to escape humanising them: "Giving animals a name blocks our educational message. (If) the public focuses on the name, we will not be able to tell other stories. Also: they are not domestic animals, they are wild animals".[1] In general, zoos have become cautious of any practices that might make their animals seem a 'human in an animal skin' (Mullan and Marvin 1999), trying to avoid charges of anthropomorphism and Disneyfication that can undermine the scientific image of contemporary zoos (Carr 2018) as well as their credibility as conservation and education actors. Contemporary zoos have been fighting for distancing themselves from expositions of curiosities, aiming to gain the status of a scientific institution. It has become important for them to use the scientific discourse in talking about their animals, presented as wild ambassadors of species (Anderson 1995; Braverman 2012). Thus, close interactions with animals have become a thing of the past with many zoos (Carr 2018). Similarly, seeing Jacky and Jinga as wild animals, YouTube viewers disapprove of such intimate encounters. They argue that it would be right for the orang-utans to interact in the nature with their conspecifics rather than with humans.

At the same time, in contrast to the dominant zoo discourse that frames animals as species ambassadors and genetic material to conserve (Anderson 1995; Braverman 2012), animals in the YouTube videos are boisterously described by commenters as subjects with biographies and personalities. While the focus on species aspires to nurture conservation awareness, talking about zoo animals as individuals seems to generate an interest in the animals as subjects, with intentions towards and experiences of their situation. These interpretations de-centre the experience of the human ('we are in front of, uh, elephants'). Instead, they foreground the animal's experience (what is it like for the animal to live in this situation?), their perspective (what is he or she thinking or feeling about this interaction?) and their capabilities (what do they know, and what are they capable of?).

This raises the question, whether recognising orangutans by name disturbs their wild image, and whether it can indeed be disruptive for conservation narrative that orangutans as species are part of. The 'wild' imagery of zoos has been subjected

---

[1]The translation from Dutch is ours, source: https://www.metronieuws.nl/binnenland/amsterdam/2015/11/dieren-in-artis-nu-zelf-een-naam-geven.

to criticism, as they have been 'exposed' as a purely cultural institution (Anderson 1995; Grazian 2012). Based on this, critics have argued, the educational promise of the zoo to teach their visitors about nature does not hold true. While it is hard to contest that the zoo world has everything to do with our culturally established ideas about nature, we might argue that abovementioned critique of zoos might result from a rather rigid understanding of culture and nature as two distinct poles, a view that Sarah Whatmore has famously criticised in her *Hybrid Geographies* (2002). It can be argued that in the age of Anthropocene the infamous zoo fence in practice may not be so strict and impermeable. With the ever more intensive active biopolitical management of wildlife in conservation practices (Biermann and Mansfield 2014; Srinivasan 2014), in situ and ex situ conservation can be seen as a gradual rather than absolute distinction (Bovenkerk and Keulartz 2016). Especially so for orangutans, who increasingly live in institutionalized conditions also in what used to be their home range (Parrenas 2018).

The peacefulness of the interactions and the positive experiences (in the case of Jinga—even fun) that the orangutans seem to enjoy while being captive, together with a belief that apes are 'intelligent enough' to enjoy interactions with humans, justifies their captivity as a better way of living than outside of captivity under the conditions of poaching and habitat loss. These interactions are not deadly for them, as interactions in the wild would be, viewers claim, demonstrating familiarity with orangutan conservation discourse, with a hint of western superiority (the locals don't know how to deal with the local wildlife) and critique of capitalism (big corporations are ruining the habitats).

– Yes but still better than having to survive their homes being felled and then being stoned by villagers scared of them. At least here they are safe and looked after. (User ladyjbriritsh)
– Sad to see that adorable creature in a cage but glad to see it laughing, that was really funny. I would rather like to see something like above than the animal being poached (User Melina)

In line with the conservation discourse, the commenters urge each other to check the household products for the presence of palm oil, a leading cause of orangutan extinction. The imagery of Jinga, conducive to this type of responses, bears similarity to imagery of orangutans used in Greenpeace promotional videos. For example, in the animated video "Rang-tan", part of the campaign #DropDirtyPalmOil, a child finds a young orangutan in her own bedroom. The story of disturbance caused by the Rang-tan in the girl's bedroom ("She throws away my chocolate and howls at my shampoo") is mirrored by Rang-tan's story of humans destroying her home ("He destroyed all of our trees for your chocolate and shampoo"). The video ends with Rang-tan and the girl embracing, as the girl promises to 'spread her story far and wide'. What seems to be implied here is that we, as urban dwellers, are ignorant of our involvement in suffering of wild animals and responsibilities to change that. The context of Jinga's video is more complicated and multi-faceted than that of the animated video, and we can wonder if it conceals the uneven power relationships of having wild animals in captivity, separated from their authentic nature and environment, replacing it with joy from observing a de-contextualised 'funny' animal. Or can it still present an opportunity to relate and recognise the responsibility we as urban dwellers have for orangutans as species?

To sum up, YouTube frames orangutans as beasts and non-human persons, wild animals and individuals. These orderings expose the shifting understanding of orangutans in the context of opportunities of wider publics to be exposed to intimate lives of wild animals via online resources. Apart from a site of mindless entertainment, YouTube can function as a platform for discussing the latest developments in ethology, and extend understanding of our responsibilities towards wild animals and their (captive and wild) environments. In the last section before drawing conclusions, we would like to tease out the entertainment value of the videos and their potential to enable a more explicitly moral gaze.

18.4   The YouTube Zoo: Increasing Encounter Value or Enabling a Moral Gaze?

Zoo-based animal celebrities have played a double role for the image of zoos in the last few years. On the one hand, they undoubtedly increase the 'encounter value' of captive charismatic animals, thereby contributing to the appeal and economic viability of the zoo and animals as 'lively commodities' (Barua 2015; Collard and Dempsey 2013; Haraway 2010). Knut the polar bear, famous enough to share a Vanity Fair cover with Leonardo di Caprio, made millions of euros for Berlin Zoo (Giles 2013). On the other hand, zoo celebrities have opened a gateway for many uncomfortable discussions about certain practices of zoos and their justifiability in general. The case of Marius in Copenhagen Zoo can be seen as a spectacular demonstration of how conservation logic and popular culture logic collided and caused a public shock. While for the zoo the giraffe seemed to have been a representative of a gene pool and hence killable as not having enough value, for the public an animal, with a name and personality, could not be killed. YouTube videos of Jacky and Jinga, while functioning as a viral online entertainment phenomena, also seem to trigger a moral gaze—that is, questioning the conditions of captivity that made them possible in the first place.

While, as we've mentioned earlier, ethologists are fighting over the correct interpretations of apes' behaviour that would put them not too close to the humans, but not too far either, popular culture has always been quick to capitalise on tensions between closeness and distance for comic effect. Morphological and psychological proximity to humans have accounted for apes' popularity as entertainment and especially comedy performers: "The intelligence, enthusiasm, and attractiveness of young pongids (chimpanzees, gorillas, and orang-utans) has for decades made them popular performers in a variety of entertainment fields (Morris and Morris 1966, cited in Allen et al. 1994). Thousands of laughing emojis posted in response to the videos represent the same-old human reaction to apes, "a sign of recognition but also of unease with the uncomfortable closeness" (De Waal 2019, 17).

Jacky and Jinga do not disappoint their viewers. "My favourite video ever", they exclaim, "this video made my day!" The videos seem to therefore promote zoos as places where animals are having fun, and animals are fun themselves. It would probably not be wrong to assume that for some viewers the videos might become a reason to visit a local zoo in search for interaction. As we read the comments in which commentators lament that they didn't get to see orangutans being so fun, we can't help but recall the infamous regret of Berger that animals in zoos are generally disappointing (Berger 2013). Unlike the disinterested, mechanically scanning animals described by Berger, the YouTube orangutans are readily responding to human actions, thereby satisfying what the visitors are truly looking for—an opportunity to be reacted to by the animals (Rosenfeld 1981, cited in Woods 2015). Given that, the videos present a dream-coming-true visit—not only seeing, but also being seen by the other. As Cahill argued in his YouTube bestiary, "this desire for animal attraction frequently serves as a narcissistic affirmation, that each of us merits the rapt attention of animals. It is not enough to reduce animal beings to their "to-be-looked-at-ness" as displays in a zoo, we must also force them to take interest in us" (Cahill 2016, 277). Yet there seems to be more going on than tickling human narcissism.

In the YouTube videos it is the orangutans who seem to have the upper hand: Jacky chucks whatever he is being 'fed' in the bushes, and Jinga does not perform for the public, the public performs for her. This switch of roles seems to suggest that the joke is on us, humans: we are the ones, who are less fair and have rather primitive ideas of what other animals, and orang-utans specifically, might be capable of. Similar to animal trainer and philosopher Vicky Hearne's account of orang-utan comedians, we can suspect that what is mocked here is the 'importance and state' of humans rather than behaviour of the animals. Or, as Vinciane Despret noted earlier in relation to YouTube animal videos in general, we are laughing not at animals but at ourselves, as the animals do something surprising, something we thought they were unable to do (Despret 2016). In addition, the frame of captivity creates a bittersweet impression for some viewers: they are entertained by the animals, and amazed by their very being, and find it disturbing to notice their captive condition. As some commenters reluctantly note, humans involved proceed to see the next exhibit, while the animals are left behind. The lack of symmetry in these relationships (Driessen et al. 2014) puts the viewers at unease, contributing to the empathic unrest.

18.5   Conclusion

The videos function as an online artefact of human-animal relations in the age of humans, highlighting the ambiguous nature of the encounter and its potential to expose existing assumptions about animals inherent in human-animal relationships. To conclude, we shall elaborate on how the videos trigger the shifts of established categories that frame our relationships with animals and what role zoos play therein.

Firstly, the tensions that arise in the comments in regard to the attribution of cognitive and emotional capabilities to Jacky and Jinga reflect ongoing negotiations of how people understand the human-animal divide. In line with the long-standing tradition of anti-anthropomorphism (in turn critically labelled as mechanomorphism) and overall sobriety in sciences interpreting animal behavior, some viewers interpret orangutans as lacking complicated capabilities and therefore merely less than human. Others, however, employ a broader vocabulary to tease out the meanings and emotions behind the actions and reactions of Jacky and Jinga. This emerging genre of Youtube animal videos, powered by ubiquitous access to cameras and endless sharing possibilities, seems to have remarkable potential to flesh out emotional and cognitive capabilities of animals. Jacky, fully aware of zoo rules and demonstrating accurate aiming skills and grasp of a social situation, and laughing Jinga who seems to love being entertained, provide a powerful demonstration of mental prowess and individual character. With the arising empathic unrest exemplified by many online commenters, the videos help establish practices of interpreting animals as intelligent individual characters who can have meaningful encounters with us. At the same time, these entertaining videos can be seen to justify the existing uneven power relationships and normalise them. The vicarious encounter between the orangutans and the viewers thus appear as deeply ambivalent, as it both reinforces and destabilises a human-animal divide (Oakley et al. 2010). Secondly, the comments reflect a tension between the tendency to comprehend orangutans as wild animals and the proximity of the encounter that seem to push them more in the category of familiar individuals. We've argued that while it is a tendency of zoos to portray their animals as wild, seeing orangutans as individuals might be indicative of a recognition of muddled boundaries between nature and culture and therefore raising a necessity for rethinking our ways of relating to animals. It can also result in stripping the context and flattening the conditions that made captivity possible. Yet at the same time it may position orangutans more firmly in the web of responsibilities experienced by humans and therefore contribute to their conservation.

Finally, we would like to reflect on the role of videos for our view of zoos. Zoos have undergone a significant historical transformation, from private animal collections of curiosities and exotic beasts, to centres of family entertainment with far-reaching conservational and educational aspirations. This transformation is reflected in the way zoos represent their animals—from captive beasts or humanised chimps drinking tea (Allen et al. 1994) to naturalised ambassadors of endangered species (Braverman 2012). Can these YouTube videos be seen as an online remake of the chimp tea party, a throwback to the past, when zoos were more overtly about entertainment? Or can the celebrification of animals, or appreciating them as particular individuals, be a way of entering a discussion on how captivity and its conditions need serious revision/justification? The videos seem to play along increasing the encounter value, while triggering a moral gaze. This emphasises how the morality of keeping animals captive in zoos is something that is not self-evident and needs to be discussed. It also emphasises that looking closely at captive animals and what their interactions with us may mean may not just be entertaining, but required, on moral and political grounds.

Looking closely at human-wildlife encounters of the kind found and recorded in the zoo could be also relevant in relation to semi-domesticated or acculturated groups of wild animals, that are not confined in an institution, but increasingly folded into practices of encounter and value (Barua 2015). More and more research on animal capabilities and the increasing emergence of difficult cases in conservation practice due to climate change and other human-induced planetary changes (Palmer 2010; Parrenas 2018), shows that the ethics of relating to non-human animals in the age of humans can not be seen as a self-evident set of rules. Instead it requires an empirical investigation, in which encounters can play a role, when we make an effort to engage in and interpret them.

References

Acampora, R.R. (ed.). 2010. *Metamorphoses of the zoo: Animal encounter after Noah*. New York: Lexington Books.

Allen, J.S., J. Park, and S.L. Watt. 1994. The Chimpanzee tea party: Anthropomorphism, orientalism, and colonialism. *Visual Anthropology Review* 10 (2): 45–54. https://doi.org/10.1525/var.1994.10.2.45.

Anderson, K. 1995. Culture and nature at the Adelaide Zoo: At the frontiers of "human" Geography. *Transactions of the Institute of British Geographers* 20 (3): 275–294. http://www.jstor.org/stable/622652.

Barua, M. 2015. Encounter. *Environmental Humanities* 7 (1): 265–270. https://doi.org/10.1215/22011919-3616479.

Bekoff, M. 2004. Wild justice and fair play: Cooperation, forgiveness, and morality in animals. *Biology and Philosophy* 19 (4): 489–520. https://doi.org/10.1007/sBIPH-004-0539-x.

Berger, J. 2013. Why look at animals. In *The animals reader: The essential classic and contemporary writings*, ed. L. Kalof and A. Fitzgerald, 251–261. London: Penguin.

Biermann, C., and B. Mansfield. 2014. Biodiversity, purity, and death: Conservation biology as biopolitics. *Environment and Planning D: Society and Space* 32 (2): 257–273. https://doi.org/10.1068/d13047p.

Bitgood, S., J. Carnes, A. Nabors, and D. Patterson. 1988. Controlling public feeding of zoo animals. *Visitor Behaviour* 2 (4): 6.

Bousé, D. 2000. *Wildlife films*. Philadelphia, PA: University of Pennsylvania Press.

Bovenkerk, B. 2016. Animal captivity: Justifications for animal captivity in the context of domestication. In *Animal ethics in the age of humans*, ed. B. Bovenkerk and J. Keulartz, 151–173. Cham, Switzerland: Springer.

Bovenkerk, B., and J. Keulartz (eds.). 2016. *Animal ethics in the age of humans*. Cham, Switzerland: Springer.

Braverman, I. 2012. *Zooland: The institution of captivity*. Stanford, CA: Stanford University Press.

Burt, J. 2002. *Animals in film*. London: Reaktion Books Ltd.

Cahill, J.L. 2016. A YouTube bestiary: Twenty-six theses on a post-cinema of animal attractions. In *New Silent Cinema*, ed. P. Flaig and K. Groo, 263–293. Taylor & Francis. https://doi.org/10.4324/9781315819297.

Carr, N. 2016. Ideal animals and animal traits for zoos: General public perspectives. *Tourism Management* 57 (C): 37–44. https://doi.org/10.1016/j.tourman.2016.05.013.

Carr, N. 2018. Zoos and animal encounters: To touch or not to touch, that is the question. In *Wild animals and leisure: Rights and welfare*, ed. N. Carr and J. Young. London: Routledge.

Collard, R.C., and J. Dempsey. 2013. Life for sale? The politics of lively commodities. *Environment and Planning A: Economy and Space* 45 (11): 2682–2699. https://doi.org/10.1068/a45692.

Corbey, R. 2005. *The metaphysics of apes. Negotiating the animal-human boundary*. Cambridge: Cambridge University Press.

Davila Ross, M., S. Menzler, and E. Zimmermann. 2008. Rapid facial mimicry in orangutan play. *Biology Letters* 4 (1): 27–30. https://doi.org/10.1098/rsbl.2007.0535.

De Waal, F. 2019. *Mama's last hug: Animal emotions and what they tell us about ourselves*. New York: W.W. Norton.

Despret, V. 2016. *What would animals say if we asked the right questions?*. Minneapolis, MN: University of Minnesota Press.

Driessen, C., K. Alfrink, M. Copier, and H. Lagerweij. 2014. What could playing with pigs do to us? Game design as multispecies philosophy. *Antennae: The Journal of Nature in Visual Culture* 9 (30): 81–104.

Driessen, C., and L.F.M. Heutinck. 2014. Cows desiring to be milked? Milking robots and the co-evolution of ethics and technology on Dutch dairy farms. *Agriculture and Human Values* 32 (1): 3–20. https://doi.org/10.1007/s10460-014-9515-5.

Driscoll, K., and E. Hoffmann (eds.). 2018. *What is zoopoetics? Texts, bodies, entanglement*. Basingstoke: Palgrave Macmillan.

Dufour, V., M. Pelé, M. Neumann, B. Thierry, and J. Call. 2009. Calculated reciprocity after all: Computation behind token transfers in orang-utans. *Biology Letters* 5 (2): 172–175. https://doi.org/10.1098/rsbl.2008.0644.

Giles, D.C. 2013. Animal celebrities. *Celebrity Studies* 4 (2): 115–128. https://doi.org/10.1080/19392397.2013.791040.

Grazian, D. 2012. Where the wild things aren't: Exhibiting nature in American zoos. *The Sociological Quarterly* 53 (4): 546–565.

Haraway, D. 2010. When species meet: Staying with the trouble. *Environment and and Planning D: Society and Space* 28 (1): 53–55. https://doi.org/10.1068/d2706wsh.

Hearne, V. 1994. *Animal happiness. A moving exploration of animals and their emotions*. New York: Skyhorse.

Keulartz, J. 2015. Captivity for conservation? Zoos at a crossroads. *Journal of Agricultural and Environmental Ethics* 28 (2): 335–351. https://doi.org/10.1007/s10806-015-9537-z.

Kiiroja, L. 2016. Semiotics in animal socialisation with humans. In *Animal umwelten in a changing world: Zoosemiotic perspectives*, ed. T. Maran, M. Tonnessen, and S. Rattasepp, 182–204. Tartu, Estonia: University of Tartu Press.

Malamud, R. 1998. *Reading zoos: Representations of animals and captivity*. New York: New York University Press.

Martin, A.L. 2008. Functional analysis and treatment of human-directed undesirable behaviors in captive chimpanzees. Thesis Georgia Institute of Technology.

Mullan, B., and G. Marvin. 1999. *Zoo culture*. Champaign: University of Illinois Press.

Oakley, J., G.P.L. Watson, C.L. Russell, A. Cutter-Mackenzi, L. Fawcett, G. Kuhl, J. Russell, M. van der Waal, and T. Warkentin. 2010. Animal Encounters in Environmental Education Research: Responding to the "Question of the Animal". *Canadian Journal of Environmental Education* 15: 86–102. http://cjee.lakeheadu.ca/index.php/cjee/article/view/826.

Palmer, C. 2010. *Animal ethics in context*. New York: Columbia University Press.

Park, J., N. Malone, and A. Palmer. 2016. Caregiver/Orangutan relationships at Auckland Zoo. *Society & Animals* 24 (3): 230–249.

Parrenas, J.S. 2018. *Decolonizing extinction. The work of care in Orangutan rehabilitation* Durham, NC: Duke University Press.

Rooijakkers, E.F., J. Kaminski, and J. Call. 2009. Comparing dogs and great apes in their ability to visually track object transpositions. *Animal Cognition* 12 (6): 789–796. https://doi.org/10.1007/s10071-009-0238-8.

Shillito, D.J., R.W. Shumaker, G.G. Gallup, and B.B. Beck. 2005. Understanding visual barriers: Evidence for level 1 perspective taking in an orang-utan. *Pongo Pygmaeus. Animal Behaviour* 69 (3): 679–687. https://doi.org/10.1016/j.anbehav.2004.04.022.

Srinivasan, K. 2014. Caring for the collective: Biopower and agential subjectification in wildlife conservation. *Environment and Planning D: Society and Space* 32 (3): 501–517. https://doi.org/10.1068/d13101p.

Weisberg, Z. 2019. The problem with the personhood Argument Zipporah. *ASEBL Journal* 14 (1): 33–36. https://www.sfc.edu/uploaded/documents/publications/ASEBLv14n1Jan19.pdf.

Weiss, A., M. Inoue-Murayama, J.E. King, M.J. Adams, and T. Matsuzawa. 2012. All too human? Chimpanzee and orang-utan personalities are not anthropomorphic projections. *Animal Behaviour* 83 (6): 1355–1365. https://doi.org/10.1016/j.anbehav.2012.02.024.

Wemelsfelder, F., E.A. Hunter, M.T. Mendl, and A.B. Lawrence. 2000. The spontaneous qualitative assessment of behavioural expressions in pigs: First explorations of a novel methodology for integrative animal welfare measurement. *Applied Animal Behaviour Science* 67 (3): 193–215. https://doi.org/10.1016/S0168-1591(99)00093-3.

Woods, B. (2015). Good zoo/bad zoo: visitor experiences in captive settings. *Anthrozoos* 15 (4): 343–360.

Yulia Kisora studied humanities, animal sciences and cultural geography. Her multifaceted educational background was applied in her work as an education officer in zoos in Russia and Finland, as well as a junior lecturer in cultural geography at Wageningen University. Her current research interests are located in the productive intersections between natural and social sciences: animal geographies and animal ethics in the context of urban wildlife.

Clemens Driessen studies the ways in which the lives of animals, plants and humans are intertwined in our technological culture. Often in collaboration with designers, artists, and sometimes with reluctant nonhumans, he seeks to reimagine meaningful relations to build a more-than-human world. He is based in the Cultural Geography group at Wageningen University, the Netherlands.

Chapter 19
Wild Animals in the City: Considering and Connecting with Animals in Zoos and Aquariums

Sabrina Brando and Elizabeth S. Herrelko

Abstract Connecting people with nature is a powerful concept that opens doors for relationship building and conservation messaging. The roles of wild animals in the city (e.g., in zoos and aquariums) and how we interact with them—and vice versa—must evolve along with our theoretical discussions and animal management practices in order to advance the field. While taking into consideration the long history of animals in captivity, where we are today, and were we should go in the future, this chapter reviews animal welfare and its ethical frameworks, human-animal interactions and its effect on both animals and people, wildness in zoos and how we perceive different states of origin, compassionate education programs and their efforts to instil empathy and empower people to become agents of change, and the power of modern technology in providing real connections with artificial means. In this ever-changing world, living responsibly together has never been more important.

19.1 Introduction

Professional zoos and aquariums (henceforth zoos) can function as a powerful connection between humans and non-human animals (henceforth animals) and the natural world, as well as playing critical roles in conservation, education, and research programs. These human-animal connections and relationships introduce opportunities to explore the different interactions that can occur and evolve, with the animals in the zoo as well as urban wildlife. Wild animals housed in zoos in cities around the world are the objects of delight, fascination and of criticism, and a topic of

S. Brando (✉)
AnimalConcepts, Teulada, Spain
e-mail: sbrando@animalconcepts.eu

E. S. Herrelko
National Zoological Park, Smithsonian Institution, Washington, DC, USA

S. Brando · E. S. Herrelko
University of Stirling, Stirling, Scotland, UK

© The Author(s) 2021
B. Bovenkerk and J. Keulartz (eds.), *Animals in Our Midst: The Challenges of Co-existing with Animals in the Anthropocene*, The International Library of Environmental, Agricultural and Food Ethics 33,
https://doi.org/10.1007/978-3-030-63523-7_19

ongoing debates. Through our combined work of 44 years in zoological facilities, we aim to provide the reader with a behind-the-scenes view of professional zoos and acquaint them with some of the theoretical concepts zoos consider when assessing the opportunities and challenges of connecting people to animals, and we reflect on whether what animals like to do and what we think they 'should' be doing might be very different things. This chapter proposes that professional zoos are places where: Animals can experience optimal wellbeing and provide opportunities for a variety of connections with humans and other animals; people, including staff and visitors, can exhibit empathy by considering animals from an individual perspective, including the conflicting or contradicting situations these scenarios might create; and language and procedures used to describe and manage animals should reflect a commitment to the 24/7 across the lifespan approach (Brando and Buchanan-Smith 2018) and consider the dynamic social environments in zoos, including the concept that additional approaches going beyond today's standard may be needed.

Animal welfare[1] is defined as "an animal's collective physical, mental, and emotional states over a period of time, and is measured on a continuum from good to poor" (AZA 2020a). When considering the primary components of welfare, the Association of Zoos and Aquariums' (AZA) Animal Welfare Committee further explains this as "an animal typically experiences good welfare when healthy comfortable, well-nourished, safe, able to develop and express species-typical relationships, behaviours, and cognitive abilities, and not suffering from unpleasant states such as pain, fear, or distress. Because physical, mental, and emotional states may be dependent on one another and can vary from day to day, it is important to consider these states in combination with one another over time to provide an assessment of an animal's overall welfare status."

Humans are animals too, but for the purpose of this chapter we will apply the terminology used in animal welfare science when referring to interactions, bond and relationships of humans with other animals as human-animal interactions (HAI) and human-animal relationships (HAR). HAI in zoos can be diverse, including the expectations people have of these interactions, from personal and friendship perceived bonds, to the idea that we should hardly interact with animals we deem 'wild'. The ways in which people interact and the expectations they have of free-ranging wildlife on zoo grounds sometimes differs from that towards the animals housed within the zoo. These differences highlight opportunities to ask new research questions within the field of HAI. Although HAI research in zoos is relatively new, it has been present in the companion animal and agricultural fields since the 1980s, each with a different focus: companion animal research focuses on its impact on humans whereas agriculturally-based research focuses on the impact to animals (Hosey and Melfi 2014). The need for additional zoo-based research is essential for a better understanding of experiences on both sides of the interaction.

---

[1]The terms "animal welfare" and "animal wellbeing" have both been used over the years (Moberg 2000, 1), to describe the state of the animal, and are used in this chapter interchangeably.

AZA. (2020a). Animal Welfare Committee. https://www.aza.org/animal_welfare_committee. Accessed 29 March 2020.

Zoos also play a crucial role in being a new home and safe haven for confiscated wildlife from the illegal trade and rescue. The roles of zoos today are many, with the increased pressure on the natural world and wild animals and the continuing trend of people migrating to cities and urban areas. The role of zoos can be to provide a place where people can connect with animals and nature, be education and conservation focussed, as well as to consider the lives of other animals and our relationship we have with them.

19.2 Animal Welfare

Zoo professionals and laypersons alike have questions about animal welfare: Can zoos promote optimal welfare for the animals in their care? Are zoos providing opportunities for choice and control over the environment and activities? Those who oppose captivity speculate that animals cannot experience optimal welfare no matter what is done for them. What do we know and what evidence can be provided to address animal welfare concerns? How do we navigate the different ethical frameworks on what animal welfare entails?

We have an ethical responsibility to provide animals in zoos with environments that allow them to experience good welfare (Brando and Buchanan-Smith 2018). Zoos have seen a significant evolution, today promoting optimal welfare from "cradle-to-grave", birth to death (Seidensticker and Doherty 1996) and 24/7 across lifespan (Brando and Buchanan-Smith 2018). This approach is in stark contrast to how zoos started, most as menageries approaching the keeping of animals like a stamp collection, with little understanding and regard for animal welfare and lacking a science, education and conservation approach. The ability to challenge the status quo through the review of emerging scientific evidence and ethical considerations and frameworks (De Mori et al. 2019) remains necessary and forms an important aspect in the process of changing long-held beliefs or practices. Being professional is to continue to ask if this is the best that we can be, if this is what is in the best interest of the animal 24 h a day, seven days a week. Professionalism entails an 'animal-first' approach, and striving to achieve the goals of education, research, conservation and recreation goals (Brando and Coe 2020).

Contemporary animal welfare thinking is approached from a holistic perspective and encompasses physical, behavioural and psychological aspects, and is increasingly emphasizing the promotion of positive states and the centrality of the animal feelings (Wemelsfelder et al. 2001; Wemelsfelder 2007; Mellor and Beausoleil 2015; Mellor 2016; Veasey 2017; Brando and Buchanan-Smith 2018). Welfare pertains to the individual; positive or negative welfare is not something we can give to animals, but something they experience based on the circumstances in which they live. Through a holistic approach, professional zoos are responsible for providing care and environments which promote optimal welfare for all animals.

Many accredited facilities such as the Smithsonian's National Zoo, Lincoln Park Zoo, and Chester Zoo, as well as a few unaccredited but contemporarily operating zoos,[2] are highly functioning conservation and education organisations, with many *in-* and *ex situ* programs and projects. Change is a dynamic and necessary aspect to the running of a modern zoo, and there is always room to improve, hence the importance of staying up to date with the latest developments in education, conservation and animal welfare. It is important to note that most zoos in the world are not accredited by a one of the major regional accrediting bodies, and therefore many malfunctioning and bad facilities exist.

Different yet collaborative tools for animal welfare assessments, combining natural history and species needs, as well as tailoring programs to individuals advances optimal welfare. Some facilities have also stopped housing certain species, not necessarily because they think it cannot be done well, but also because species needs may outweigh the location parameters or facility resources. A universal animal welfare framework for zoos (Kagan et al. 2015), and an animal welfare risk assessment process (Sherwen et al. 2018) can be used to identify risks and determine priorities.

Professional zoos endeavour to provide animals with meaningful choice and control to meet their own needs and preferences. The potential benefit of choice and control have been well described long ago by Chamove and Anderson (1989) and Snowdon and Savage (1989), later by Brando (2009), and recently by Allard and Bashaw (2019). There is considerable empirical evidence that not having control of one's environment leads to behavioural and physiological problems (e.g., Mineka and Hendersen 1985; Perdue et al. 2014). Thus, a common goal of these approaches is to allow animals far greater control (or agency) over their own lives with less dependence on caregivers (Brando and Buchanan-Smith 2018; Coe 2018; Allard and Bashaw 2019). Several studies show that simply having meaningful choices, whether or not they are acted upon, is rewarding to animals (e.g., Owen et al. 2005; Leotti et al. 2010; Kurtycz et al. 2014).

We can provide many different environmental enrichments, which are planned and designed according to certain goals (e.g., increasing space use, species-specific behaviours such as climbing and jumping) however, we must be OK with the choices the animals make. Some activities such as playing with plastic toys or old telephone books or being trained to sit on a scale for weighing are not natural but can be used to reach behavioural and animal care goals that are still very enjoyable and engaging for animals. Artificially presented habitats and objects are regularly prohibited, based on the argument that these environments and enrichment negatively affect conservation and education efforts. However, considering the data proving this are lacking (e.g., Perdue et al. 2011; Jacobson et al. 2017) and artificial items are often successful in enriching animals' lives (e.g., review of touchscreen activities with apes by Egelkamp and Ross 2019), we believe zoos should focus on the potential benefits for animals and opportunities to help visitors understand the link between behavioural goals reflecting animals' natural histories and their interactions with artificial items. Educators, signs,

---

[2]Zoos may opt out of accreditation processes for various reasons, yet still show contemporary practices based on latest science and best practices.

talks by care staff and other means could be used to convey to the public that non-naturalistic activities can promote positive wellbeing, that animals enjoy them and while they do not naturally play with phonebooks in the wild, the message that we should protect and conserve animals should not be jeopardized by a rejection of artificial items.

The natural versus the unnatural is a debate which has been ongoing for decades, and we want to make the argument that all items and activities which are enriching an animals life are suitable for the animal, natural or not, and should not conflict with connecting people to animals and nature. In these circumstances, if we consider animals to be agents (i.e., have agency), we should let the animals decide if something is enriching or not. This is the challenge for professional zoo staff: identify activities that are meaningful for animals that can simultaneously help make connections to other zoo goals. It is our moral obligation to promote optimal welfare for animals in our care for many reasons, but as it relates to HAI, the most important reason might be because we find active and engaging animals the most interesting to watch and easiest to learn about.

This is not to say that choice and control is an all or nothing concept that could be used to rationalize omitting tasks that optimal well-being is reliant upon (e.g., vaccinations and medications). The concept should be used proactively, yet judiciously. When a choice is provided, it should be meaningful to the animal, and the animal care staff must be ready to accept whatever choice the animal selects. If we provide an animal with a choice of two non-preferred social partners, does it really count as a meaningful choice to improve their welfare experience? Yes, it would mean they have an opportunity to control this segment in time, but if both choices potentially lead to negative outcomes, does that feel like control? If an animal is not feeling well and the option is to take medication or not, how long would not choosing to take the medication serve an animal's welfare before it becomes a significant problem?

Professional animal welfare programs today strive to operate on an evidence-based approach as set out by Maple and Lindburg (2008) for the "empirical zoo" and include evidence-based animal welfare programs (Melfi 2009). These programs reflect best practice processes of care and conducting research, including the importance of good human-animal relationships and interactions (Hosey and Melfi 2012). Choice and control should also revolve around human-animal interactions and relationships, including staff, the general public and others who interact or are in the direct or indirect space of the animals.

19.3   Human-Animal Interactions

Zoo professionals and laypersons alike have questions about HAI: What types of interactions and relationships do we find at the zoo? What do animals think about staff and/or visitors? How well do visitors deal with choices animals make, such as not interacting with or looking at them? What is the effect of the human gaze or demand on the animals? What type of relationships with animals do zoo staff want? What do animals want from the people around them?

There are many formal and informal ways to discuss HAI and HAR, from the more formal evidence-based to the idiosyncratic, from the animals', staff members' or individual visitor's perspective. HAI and HAR can be viewed from exchanges that people directly have with animals, such as during a feeding session in the children's farm between children and goats, the play bout between a young chimpanzee and the person on the other side of the window, or the relationship between the care staff and the animals, engaging in hide and seek, and voluntary care behaviours like weighing and shifting from one area to another.

In the book '*Animals in the Age of Humans*' (2016), Brando discussed the concept described by Gruen as 'wild dignity', and that our making animals look ridiculous or our portraying them as something different than what they are, violates this dignity. According to Gruen animal dignity needs to be dynamic, i.e., adapted to the individual animal, and is only valuable when it is expressed and recognized as contributing to the well-being or flourishing of that individual. Giving animals more choice and control through environmental complexity, enrichment and animal training, as well as interactions with people might not be 'natural' but might serve a meaningful function in restoring some of the individual's agency and therefore 'wild dignity'.

'Wild dignity' only comes into play when animals are taken out of their wild context and put into the human context. It appears, then, that for Gruen, the basis of the attribution of wild animal dignity rests in our own attitude: When we change animals' species-specific behaviours and take away the control of their own lives, we dominate them and thereby we violate their dignity. The challenge here is that this is assumed without regard to what the animal in question experiences. Therefore, when housing captive wild animals, we need to consider and respect their needs and preferences, and what contributes to and interferes the least with their 'wild dignity'. This can concern behaviours that we might find off-putting and consider indecent from a human perspective, for example, masturbation or aggression that animals engage in (killing prey) or are important to maintaining social relationships. It is important to note that an animal can have good welfare, but still his or her 'wild dignity' can be violated, to acknowledge that meeting both welfare needs of animals and respect their wild dignity creates tension and might not be possible.

This concept also has relevance for how wild animals are portrayed to the public, housed, or used in the various activities offered by zoos, such as shows, presentations, and interactions. Gruen writes, In contrast we dignify the wildness of other animals when we respect their behaviours as meaningful to them and recognize that their lives are theirs to live. Although there will be restrictions to life in captivity, the concept of 'wild dignity' can serve to identify conflicts between human expectations versus animal needs and preferences and to propose ways to mitigate them (Brando 2016). Should contemporary zoos focus more efforts on facilitating agency and less on how we think animals should behave, to enquire what it means to be, or to be 'truly wild'? If animals can never be returned to the wild, to what extent should captive wild animals have agency over their life?

Professional zoos consider the wide variety of HAR from the animal's perspective. Braverman (2012) notes that at the zoo, direct physical contact is prevented between the animals and visiting public using moats and fences. Many of the barriers are not only in the interest of the visitor, but many actually exist to protect animals from inappropriate feeding, touching, and other harmful behaviour. Physical barriers should not only be seen as protecting staff and the visiting public but can also as a form of privacy and barrier for the animals to use to their liking.

Caregivers who are sensitive to the possible impacts of insufficient flight distance, or few or no opportunities to view someone approaching around a corner or in the surrounding area, will interact in ways that try to offer animal choice and control. Developing positive HAI and clear ways of communication are needed for new and well-designed environments, and urgently needed in circumstances where animals have limited agency, such as in smaller spaces and/or social groups, when they do not have access to complex environments, choice or control. While these circumstances should be avoided where possible, sometimes for reasons of law, weather, health and safety reasons, smaller areas may be part of an animal's environment. Well-designed habitats provide different micro-environments with complexity, with choices for animals to decide where they want to be, what they want to do, and who they want to do it with (e.g., Herrelko et al. 2015). This includes different fixed and flexible structures, vistas, different climatic zones to sun, bathe or be in the shade, species-appropriate social groups as well as opportunities to hide and find shelter or quiet areas. Designing environments in ways that animals have choice and control and allow for visitors to observe the animals without disturbance and disruption are key factors in professional habitat design (Bonnie et al. 2016; Brando and Buchanan-Smith 2018). Well-designed habitats can also allow animals to watch humans, to interact with them in a safe manner, and for the animals to disappear out of sight.

While some animals actively avoid eye contact with visitors, others may seek it out. Some zoos provide information on how to behave in these different scenarios, e.g., when a gorilla looks your way, with content such as "nod and lower your head, glance away [and] don't stare! Crouch or kneel down, so the gorilla is above or across from you, this posture puts them at ease" (Braverman 2012, 79). Animals who do not look for visitor or staff visual interaction, might not feel comfortable doing so from a species perspective, like many primates avoiding the human gaze as they also avoid gaze between group members depending on hierarchy. Or even when these behaviours can be learned, they might just not be interested in connecting with you in this way. Some animals are very interested in visual contact and looking at you or others, positioning themselves so they can see better, or actively trying to get in your field of vision.

Even when animals cannot see visitors, but visitors can see them, people are interested and concerned about the implications for animals. This concept of 'big brother is watching you' is an interesting discussion to have with staff and public, with many people understanding that none of us want to be watched all the time. Even the discussion on whether if animals disappear out of sight, it is OK to then watch them on camera and screens is engaging as some members of the public mention that they feel like they are spying on the animals, while other think it is OK as the animals do not know that we are watching. In a sense this falls in the same category of animals performing in shows behaviours like dancing or being a pirate. Some think this is fine, as the animals do not know, while other say we should not ridicule animals (Brando 2016; Brando et al. 2018). These thoughts and feelings towards animals point to people being interested and concerned what we do to and ask of animals, regardless of the welfare impact it has on the animals.

From a welfare perspective it is fundamental to understand how other animals see us (Hosey 2013; Hosey and Melfi 2012). The ability to behave and interact appropriately in relation to a variety of species is crucial for navigating captivity and the field. For animal care professionals, their ability to provide safe and productive management practices relies on the relationships built with animals. If they are not seen as a benefit to the animals (e.g., as food provider, someone who appropriately responds to their behavioural cues), the ability to create opportunities for optimal welfare decreases. Creating positive relationships and engaging in positive interactions can also elicit inter-species play and communication (for a review and examples on animal play see Brando and Burghardt 2019). In the book *Smiling Bears*, Else Poulsen tells a story of the specific nesting behaviour of one of the polar bears she cared for; when preparing his sleeping place, he separated some materials out from others and left it next to his nest. She would only remove what he put aside, and it seemed these interactions created a form of mutual understanding and communication between her and the bear.

Of course, scientific evidence-based understanding of the effects of interactions on animals are important in order to behave appropriately and effectively towards the animals promoting choice, control, and positive wellbeing. However, relationships and interactions can be subtle and complex with regards to individual preferences of the animals and the people who care for them. To date little attention has been given to the qualitative aspect of these interactions, including the thoughts and feelings people have towards the animals, and perceived friendship or bond, a topic which is still controversial in zoos. While many zoos will acknowledge the bond that exists between staff and animals, some may consider this to be anthropomorphic and inappropriate between humans and wild animals.

HAI opportunities such as touch tanks and animal feeding demonstrations are other, low-tech immersion techniques where visitors physically share the same space as an animal, even if just with their hands. Immersion, and specifically hands-on activities, are wildly sought after by the public, e.g., Shedd Aquarium's Stingray Touch exhibit, an exhibit that is only open five months out of the year, reaches 400,000 visitors annually (Ruppenthal 2018; Shedd 2019), and has been linked to increased knowledge about aquatic life and perceived likelihood that participants would make an effort to protect aquatic species (Ogle 2016). While these activities are very popular with the general public, the challenges zoos face are safeguarding animal welfare for fragile animals such as sea stars and small fishes. As one of many of the activities offered in zoos, the touch tank is an example that revolves around something we do to animals (i.e., entering their space) and behavioural and health assessments should be made to help facilitate an optimal welfare experience for the animal. Research on human health parameters shows that the touch tank is similar to nature experiences; mental stress decreased, but similarly to an exciting event, heart rate increased (Sahrmann et al. 2016). Beetz et al. (2012) suggests that animal contact plays a role in reducing perceived stress in humans because it activates the oxytocin system, informally referred to as the "cuddle hormone" (see Shamay-Tsoory and Young 2016 for commentary on advances in the study of oxytocin).

While the purpose of outreach, education and interactive programs with hands-on experiences may be laudable and enjoyable for visitors, it has the potential to present a conflict (i.e., animals may not enjoy the interaction) or a contradiction (i.e., people touching and taking photos with animals in the zoo, which would be discouraged in nature). For these reasons, professional zoo staff are trained to select animals who exhibit traits that thrive in these environments (e.g., have a personality which is open and inquisitive, who tolerate being touched and handled, and stay calm with loud noises and sudden events). It is not always easy to manage how people interact and handle animals, and while most animals are purposefully desensitized to these types of interactions and people are informed on proper etiquette prior to the encounters, there is the potential for the unknown. This is why staff must be trained to identify signs of stress and empowered to end encounters when animals signal it is time through their behaviour.

Animal welfare scientists are exploring the program animal[3] world to identify best practices. An evidence-based approach will help clarify needs for standard operating procedures related to program animals that spend extended periods in environments that may limit their range of movement, provide them limited choice and control, and require them to live in species-atypical social groups. Animals typically selected as program animals tend to be small and easy to handle, such as guinea pigs, tarantulas, bearded dragons and chameleons, which does not necessarily make them the best candidates as they are often considered prey species, as they might need more habituation and safe havens, or less manual handling.

While there is much research showing that being in nature and interacting with animals is positive for human wellbeing, there is little empirical evidence on whether touch tanks and interactive programs (Brando and Buchanan-Smith 2018) also have a positive influence on animals; and existing publications largely focus on physical health. For example, a study of the physical health of cownose rays showed that health parameters were comparable for individuals in a visitor accessible touch tank or in an off-exhibit system with minimal human interaction (Johnson et al. 2017). Some complementary behavioural research exists for similar human-animal interaction experiences (for a detailed review, see Sherwen and Hemsworth 2019). Research suggests visitor-feeding programs with captive giraffes serve as enrichment for animals that could lessen the foraging needs linked to oral stereotypies (Orban et al. 2016) and more recently in a penguin-visitor experience, researchers found that encounters did not disrupt colony behaviour, particularly when it came to affiliative or aggressive behaviours (Saiyed et al. 2019). Of course, in these scenarios,

---

[3] A program animal (or ambassador animal) is defined by the Association of Zoos and Aquariums as "an animal whose role includes handling and/or training by staff or volunteers for interactions with the public and in support of institutional education and conservation goals" (AZA 2020b).

management plans surrounding the activities are likely to play a significant role; not all touch tanks and visitor-interaction programs are equal. Even with safety and welfare protocols in place, if visitors do not follow the instruction well-being could be compromised (a review of concerns and recommendations, specific to aquatic animal touch tanks, is available online: see Dogu et al. [2011]).

We believe this area of research greatly deserves attention so that we can have additional quantitative data to support policy development. Multiple zoos have introduced animal-first approaches to welfare; each with their own perspective: (1) NZP/SCBI (Moore et al. 2013) presented their Animal First approach as a proactive take on communication channels to provide a safe environment in which animal care staff may identify and address welfare concerns. (2) Lincoln Park Zoo (LPZ) (2019) introduced their "animal-first" philosophy with changes in programming for ambassador animals. Programs are only deemed to be an "animal-first" program if animals are deemed to have choice and control in participating in an activity, can remain in the comfort of their habitat, and the program is to the benefit of the animal.

Zoos want to avoid that visitors freely feed animals (unsupervised) because it is deemed not good for the animal's welfare. However, the animal might actually enjoy or encourage it." Kisora and Driessen (in this volume) describe the content of a video of an orangutan interacting with a member of the public, and while the person and the animal 'know' that this is not acceptable behaviour, the video clearly shows that both parties have an interest in the continued interaction.

Big cats in zoos can be seen to 'hunt' smaller children along the window, or ambushing them, a behaviour which is sometimes discouraged by zoo staff (as it could scare the public or potentially hurt the animal if they run into the glass window), sometimes it is encouraged, however, as both parties run and seem enjoy the activity. Chimpanzees have been seen to ask visitor or staff what is in a bag by pointing at it, others sit with their back to the window steeling glances while others sit up high and away. Orangutans clean the windows with a cloth and scoop nesting materials in a bucket like they see the care staff do. What animals like to do versus what we think they 'should' be doing might be very different things.

19.4   Wildness in Zoos

When considering how well people translate information from the screen to the world in which they live, we put forward another concept: Is it important to be truly wild? If an emotional connection to an animal in a zoo is real, does the location of their birth (e.g., in situ, ex situ) matter and or affect perceptions? Public perception can be split based on experience versus perceived notions. Reade and Waran (1996) discovered that zoo visitors had positive perceptions as well as a greater awareness of key topics within animal care science than the general public.

The fear of wild animals becoming 'less wild' in zoos is a concern raised by the zoo and conservation field, when considering a potential loss of behavioural diversity necessary for survival in the wild (Rabin 2003) and a potential decrease of reintroduction success. Many wild animals in zoos today do not encounter stimuli relevant to life in the wild, however, most zoos do not reintroduce their animals to the wild. As such, the debate on what stimuli these animals should be exposed to in order to maintain species-specific behaviours is complex and ongoing. One particularly relevant example that should be discussed is when stimuli could cause a temporary decrease of wellbeing, e.g., exposure to a predator to maintain social cohesion and hiding behaviour.

What does wildness mean in a zoo context? In general, zoos speak of wild animals as animals who are not domesticated. They probably make the distinction between wild and domesticated animals because many zoos also house domesticated animals such as horses, cows, goats, sheep, and rabbits, and feed domesticated farm animals to the wild animals. Some zoos will not formally train the animals to participate in their daily care as they advocate a 'hands off' approach as 'wild animals should stay wild and not have too close interactions with people'. Zoo professionals knowledgeable of animal learning and training know that animals learn all the time, and that there is no such thing as a 'hands off' approach. However, how the animals are habituated and desensitized to people, situations and objects, what type of behaviours they engage in, and how much room there is for the animals to choose what they want to do differs between zoos and their philosophies on what is acceptable and appropriate.

As described above, HAI and HAR might create conflicts between the goals of different programs within a zoo, such as animal welfare, education and conservation. Relationships might be experienced by the animal as desirable and positive, while from an education and conservation perspective it might be undesirable seeing a wild animal interact with a member of the public or staff member, including e.g., hand feeding, as the educational and conservation message likely revolved around wild animals being wild, to respect them in the wild, and to refrain from interacting and handfeeding them. Animals who are in conservation programs with the aim of reintroduction are in a different role than exhibit animals who will live their lives in the zoo. Our relationships with them can be fundamentally different than with animals who will not be exposed to direct human contact. Nevertheless, we must remember the importance of visual context for endangered species. As Ross et al. (2011) discovered, even showing pictures of chimpanzees alone versus near a human or in a typical human environment (e.g., office) negatively influences peoples' perceptions of their endangered status and their appeal as a pet.

When animals are not to be reintroduced but have an endangered status it may be important from the point of view of the conservation message to still portray them as wild which may create a tension between 'being or looking wild' versus the impression of 'being domesticated'.

Summing up, animals with different roles (e.g., program animals) have different relationships with humans and requirements we should consider, e.g., they should not live in fear or anxiety of humans and they should have choice and control over their environment and what happens to them. Animals who are truly going to be released need to maintain their fear for humans, and to be fit and independent for life in the wild and will have a different program and trajectory.

Whether for release or to serve as ambassadors for their species, it is clear that roles are assigned by people, it is not the animals who have chosen their role. This is a responsibility that contemporary zoos take seriously, outlining specific criteria for each role. In the past, an animal which did not have a suitable social group might have ended up in the ambassador animal or education area, but today staff make these decisions with careful consideration of each animal's personality and capacities.

Interactive programs such as ambassador and education programs, as well as shows and presentations have evolved to include more choice and control for the animal, with bird presentations having birds fly into a larger arena (or not if they do not want to), and goats and dolphins having areas to retreat to if they do not want to interact with the public. At any given moment in a program, or when animals do not 'show up' or 'have left the building', care staff and educators take the opportunity to highlight the relationship the care staff has with the animals. It is the animal's choice to participate or not, and they have some control over the activity as such that they can leave if they are bored, not interested, or tired. It is a moment to talk about how we build relationships through play, positive reinforcement, varied activities, and the intrinsic value of choice and control. Most people understand, find it funny, and pretty cool that animals are choosing what they want to do, that they have their own individual ways and decisions. It sheds another light on the animals than the standard information of weight, distribution range and scientific name, and also opens up opportunities for visitors to experience empathy to better understand animal perspectives.

Cultural effects may dictate how care staff interact with, train, and manage animals. For example, with a long history of Asian elephants employed as working animals, many may utilize the traditional range-country management tool of an ankus or bull hook. A different cultural approach is the movement in professional zoos to work in protected contact, where care staff are separated from elephants by barriers. In these scenarios their tools of the trade are different, relying mainly on relationship-building and positive reinforcement-based animals training techniques.

For a long time and in many environments, Asian elephants were chained overnight in outdoor yards due to a lack of funds and elephant-proof housing. Now with better materials, a deeper understanding of sleeping behaviour and social dynamics, and a change in housing conditions, one might propose that elephants who are under less pressure and control of humans and voluntarily participate in their own care (e.g., protective contact), and gain more agency and 'wild dignity'.

19.5   Compassionate Education Programs

Education programs which aim to encourage and instil compassion, empathy and kindness, and to empower younger (and older) people to become active agents for animals and the environment, can be called Compassionate Education Programs. An example of such programs in zoos is showing people how to help animals in their neighbourhoods by drawing comparisons on zoo grounds, such as building safe frog crossings in their community for similar frog species to those kept in the zoo, or for native frog species living around the zoo as (urban) wildlife. Education of caring for and connecting to animals does not have to be for exotic animals only, some zoos house rabbits, a common companion animal in households today and another animal in the city. They have housing and information which models good rabbit environments that allow the animals to perform species-typical behaviours, e.g., hop, stretch, and socialise. Compassion education programs thrive within and outside of zoos to connect people to nature. Roots and Shoots, for example, is a non-zoo-based program which envisions "a healthy planet where people live sustainably and in harmony with animals and our shared environment".[4]

Compassionate education programs can also teach children to break away from animal stereotypes. For example, snakes are not slimy, and spiders are not scary but should be admired and protected. Studies report that having animals in the classroom improves children's learning outcomes (Trainin et al. 2005) and participation in interactive programs and petting zoos produces positive cognitive and/or emotional impacts on the welfare of humans (e.g., DebRoy and Roberts 2006; Sahrmann et al. 2016).

The types of interactions experienced in a zoo can shape how people interact with animals outside of a zoo. For example, a structured program which informs people how to behave around a type of animal they are not familiar with may set the tone for animal interactions outside of that program. People who learn to observe and respect animals tend to interact differently and are more understanding of the choices animals make (e.g., approaching a stranger or not), than those unaware of behavioural cues.

Structured programs aiming to safeguard the safety of the animals, staff and public might miss out on the flexibility, free interpretation and space of unplanned or orchestrated activities. This tends to arise from a risk-aversive perspective and inhibits activities such as spontaneous play. An example of this is in creativity and play sessions with animals in training, where animals are asked to create something new, invent a game, or choose what to do, where to go, or identify food and item preferences. These activities can even surprise the care staff, as what they think animals want and prefer does not always align with what animals choose when given the opportunity.

It is important to note that animals are not the only ones with preferences, visitors have preferences, too. How visitors choose to experience environments or types of programs and activities differs from person to person. They can choose to go alone, with the family, or only have the opportunity to go in the weekend or school holidays. These choices can affect the type of experience they will have, either quiet or busy. The visitors of a zoo are many and each differs in interest, attitude, and conduct. Some are quiet and interested in watching animals, volunteer at the zoo for enrichment and behavioural observations, and help with education programs. Others will come for quick entertainment and picnics on the lawn. Some are zoo members who frequently visit and feel a personal connection to their favourite animals, while others come

---

[4]See https://rootsandshoots.global. Retrieved online February 23, 2020.

once or twice a year without much connection. We mention this not only to posit that people experience zoos as connection points with the natural world in different ways, but also as a parallel to help the reader identify and compare their own interests and needs with that of other species.

19.6 Real Connections with Artificial Means

The various ways in which people connect with animals prompt us to wonder if the type of connection matters. In the age of the Anthropocene, where many species are going extinct, will connecting people to nature help in the conservation and protection of animals? Could computer simulations and 'experiences' through video or other artificial means, replace our in-person connection to nature? If the emotion associated with a connection is real, does it matter if the content was real (i.e., living animal versus computer-generated imagery, CGI)? Digging deeper, does it matter if the connection with a real animal happened to be an individual not born in the wild? Based on what we know about the long-term impact of atypical life histories on animals such as performer chimpanzees (Freeman and Ross 2014), we are fortunate to live in a period when films rely on CGI animals instead of animal actors. Recent movies, as well as holograms used in circuses and some zoos created lifelike experiences which may provide another opportunity for people to view, learn, and enjoy animals and their stories (e.g. Guangzhou Zoo's virtual reality exhibit in China).

In films we become emotionally invested in non-human animals in a variety of forms, even when not real or remotely realistic. From fearing a devious octopus-humanesque sea witch in the Little Mermaid (cartoon), to admiring the wisdom and optimism of Falkor, the Luckdragon, from the Neverending Story (animatronic puppet), and cheering for Aslan, the talking lion, to come back to life in the Chronicles of Narnia (CGI), we easily experience a range of emotions. Our minds appear to cross the barrier of reality and still remain emotionally invested the concept of believing in the stories told by animals, represented by fictional characters, the mission of the film and storyline.

Given what we know about perceptions of conservation status for primates (Ross et al. 2011; Leighty et al. 2015), context is important. Do these characters, even when exhibiting limited species-typical behaviour, trigger misinformed judgements? Do the emotional connections we make with realistic characters transition into caring about animal welfare and conservation? What is the impact of digitally simulated wilderness? Natural history documentaries delve into the animal world in ways some of us can only dream: Following the sardine run along the South African coast (Nature's Most Amazing Events: The Great Tide, 2009), witnessing Darwin's beetles lock horns in competition for mates (Life: Insects, 2009), or experiencing an extremely close encounter with playful mountain gorillas (Life on Earth: Life in the Trees, 1979). Mike Bossley suggests that the 'Attenborough effect' (Bulbeck 2005), may influence viewers, where incredible films like these set a tone and expectation for tourists in terms of how close they may get to animals or how active animals will be when they see them. The effect reaches beyond the *in situ* environment; it affects expectations in zoos as well, ranging from realistic to unrealistic, e.g., how close one can get to an animal, or how many different behaviours can be seen in a short period of time. There is scant literature on the topic, but Silk et al. (2017) reviews the risks and benefits of harnessing the power of Hollywood to positively affect conservation challenges. They propose a framework to mitigate negative impacts by focusing on biodiversity conservation impact, behaviour change, engagement with film industry, and raising audience interest and awareness. We also recommend an impact assessment. Without incorporating metrics to identify the success or failure of these endeavours, the questions will still exist.

Zoos strategically collaborate with each other to maintain genetic diversity and population sustainability for at least one hundred years. These animals represent the same species as their wild conspecifics, but behavioural differences are present. No matter how much skill has gone into training zoo animals (e.g., to voluntarily participate in their own healthcare by sitting for radiographs or receiving voluntary blood draws), their connection with a human should not be confused with domestication. We consider these animals to fit somewhere in the middle of the continuum from wild to domesticated (e.g., Swart and Keulartz's specific care model, 2011). The ambiguity of where most zoo animals fall on the continuum raises questions about assessing welfare. We often rely on comparing zoo animal behaviour to that of wild conspecifics, but is a wild animal the appropriate measuring stick to identify "normal"? We are reminded again that context is important. Zoos provide carefully crafted programs for each individual to address medical, nutritional, environmental, and behavioural needs. For example, if food is provisioned, they do not need to hunt or forage, and if that means they do not need to patrol miles and miles of land in search of their next meal, they will likely walk less in zoos. If there is a difference in the behaviour between a zoo animal and a wild animal, it does not specifically mean there is a compromise in welfare (Veasey et al. 1996). Behavioural flexibility plays a significant role in an animal's ability to thrive in differing environments, including zoos. Their innate ability to adapt is strengthened when professional zoos can support their needs with care staff who exhibit the patience, knowledge and capability to build relationships, identify preferences and appropriately respond to each individual's needs.

19.7 Conclusion

In this chapter we provided examples and insights into how animals in zoos are cared for, not just focusing on how humans interact with animals, but also on how animals interact with humans. We highlighted historic changes in approaches to care and choice and control, as well as in the complexity of animal habitats. We discussed evidence-based animal welfare programs to understand how animals see us and how we promote, monitor and assess well-being, as well as elude to more elusive bonds and interactions animals and humans have with each other, and the conflicting or contradicting situations or views to which this can lead.

Insight from conservation psychologists indicates that visitors want to know two things when visiting a zoo: How are the animals taken care of in the zoo and what are we doing in the wild to save them (Fraser, personal communication)? More importantly, they identified that if visitors do not think we take good care of our animals, they are not willing to listen to our conservation stories. Animal welfare is the key determining factor to connecting with zoo visitors and potentially inspiring behaviour change. In order to make this happen, professional zoos today need to have advanced animal care and welfare programs, incorporating the latest best practice, rooted in science and evidence-based approaches.

An indicator of peoples' connections with animals is how they respond to news of the death of a beloved zoo animal. When Inuka, the only polar bear in Singapore Zoo was euthanised due to age-related ailments, the zoo communicated information about this difficult decision to their visitors ahead of time. This allowed an opportunity for visitors to say their goodbyes and hundreds of people came to see Inuka one last time. He died surrounded by the people who cared for him.[5] Terrible tragedies such as the recent fire in a zoo in Germany killing many primates highlighted the public's sadness for the animals who lost their lives, and concern for the staff who had to deal with this dreadful accident. Hundreds of candles, flowers, cards, and messages where left on social media, through the mail, and at the gates of the zoo. There is no shortage of stories zoo staff can tell of zebras running to them when they walk over, the chimpanzees who line up to be groomed with a soft brush through the fence, or the fishes who gather in front of the window when feeling the vibrations of the footsteps of their carer. There are many examples of care staff creating remembrance gardens for those who have passed, have photo collages in their lockers, and buy treats and toys with their own money to spoil the animals in their care. These examples above of joys and sorrows give us insights into the relationships care staff and the public feel they have with the animals in the zoo. In this ever-changing world where people and animals are living closer to one another, and interacting more and more, inspiring a world to live responsibly together has never been more important.

Acknowledgements We would like to thank Yulia Kisora, Bernice Bovenkerk, Jozef Keulartz, and Clemens Driessen for valuable comments and a critical review of earlier drafts of this chapter.

References

Allard, S., and M.J. Bashaw. 2019. Empowering zoo animals. In *Scientific foundations of zoos and aquariums*, ed. A.B. Kaufman, M.J. Bashaw, and T.L. Maple, 241–273. Cambridge: Cambridge University Press.

[5]See https://www.straitstimes.com/singapore/inuka-died-surrounded-by-his-keepers-singapore-zoo-releases-photos-of-polar-bears-last. Retrieved February 23, 2020.

AZA. 2020a. Animal welfare committee. https://www.aza.org/animal_welfare_committee. Accessed 29 March, 2020.

AZA. 2020b. AZA ambassador animal policy. https://www.aza.org/aza-ambassador-animal-policy. Accessed 29 March, 2020.

Beetz, A., K. Uvnäs-Moberg, H. Julius, and K. Kotrschal. 2012. Psychosocial and psychophysiological effects of human-animal interactions: the possible role of oxytocin. *Frontiers in Psychology* 3: 234.

Bonnie, K.E., M.Y.L. Ang, and S.R. Ross. 2016. Effects of crowd size on exhibit use by and behavior of chimpanzees (*Pan troglodytes*) and Western lowland gorillas (*Gorilla gorilla*) at the zoo. *Applied Animal Behaviour Science* 178: 102–110.

Brando, S. 2009. Exploring choice and control opportunities applied in enrichment and training (presentation). International Conference on Environmental Enrichment, Paignton, UK.

Brando, S. 2016. Wild animals in entertainment. In *Animal ethics in the age of Humans*, ed. B.B. Bovenkerk and J. Keulartz, 295–318. Cham, Switzerland: Springer.

Brando, S., and H.M. Buchanan-Smith. 2018. The 24/7 approach to promoting optimal welfare for captive wild animals. *Behavioural Processes* 156: 83–95.

Brando, S., D.M Broom, C. Acasuso-Rivero, and F. Clark. 2018. Optimal marine mammal welfare under human care: Current efforts and future directions. *Behavioural processes* 156: 16–36.

Brando, S., and G. Burghardt. 2019. Studying play in zoos and aquariums. In *Scientific foundations of zoos and aquariums*, ed. A.B. Kaufman, M.J. Bashaw, and T.L. Maple, 558–585. Cambridge: Cambridge University Press.

Brando, S., and J. Coe. 2020. Confronting back-of-house traditions: Primates as a case study. Submitted.

Braverman, I. 2012. *Zooland: The institution of captivity*. Standord, CA: Stanford University Press.

Bulbeck, C. 2005. *Facing the wild: Ecotourism, conservation and animal encounters*. Abingdon, UK: Earthscan.

Chamove, A.S., and J.R. Anderson. 1989. Examining environmental enrichment. In *Housing, care and psychological well-being of captive and laboratory primate*, ed. E.F. Segal, 183–202. Park Ridge, NJ: Noyes Publications.

Coe, J.C. 2018. Embedding environmental enrichment into zoo animal facility design. Retrieved from http://www.zoodesignconference.com/wp-content/uploads/2017/10/03.-Coe-Embedding-enrichment.pdf.

DebRoy, C., and E. Roberts. 2006. Screening petting zoo animals for the presence of potentially pathogenic Escherichia coli. *Journal of Veterinary Diagnostic Investigation* 18 (6): 597–600.

De Mori, B., L. Ferrante, D. Florio, E. Macchi, I. Pollastri, and S. Normando. 2019. A protocol for the ethical assessment of wild animal–Visitor interactions (AVIP) evaluating animal welfare, education, and conservation outcomes. *Animals* 9 (8): 487.

Dogu, H., S. Wehman, and J.M. Fagan. 2011. Touch exhibits for aquatic animals. Best management practices for touch exhibits of AZA accredited aquariums. https://rucore.libraries.rutgers.edu/rutgers-lib/38467/.

Egelkamp, C.L., and S.R. Ross. 2019. A review of zoo-based cognitive research using touchscreen interfaces. *Zoo Biology* 38: 220–235.

Freeman, H.D., and S.R. Ross. 2014. The impact of atypical early histories on pet or performer chimpanzees. *PeerJ* 2: e579.

Herrelko, E.S., H.M. Buchanan-Smith, and S.-J. Vick. 2015. Perception of available space during chimpanzee introductions: Number of accessible areas is more important than enclosure size. *Zoo Biology* 34: 397–405.

Hosey, G., and V. Melfi. 2012. Human–animal bonds between zoo professionals and the animals in their care. *Zoo Biology* 31 (1): 13–26.

Hosey, G. 2013. Hediger revisited: How do zoo animals see us? *Journal of Applied Animal Welfare Science* 16 (4): 338–359.

Hosey, G., and V. Melfi. 2014. Human-animal interactions, relationships and bonds: A review and analysis of the literature. *International Journal of Comparative Psychology* 27: 117–142.

Jacobson, S.L., L.M. Hopper, M.A. Shender, S.R. Ross, M. Leahy, and J. McNernie. 2017. Zoo visitors' perceptions of chimpanzee welfare are not affected by the provision of artificial environmental enrichment devices in a naturalistic exhibit. *Journal of Zoo and Aquarium Research* 5: 56–61.

Johnson, J.G., L.M. Naples, W.G. van Bonn, A.G. Kent, M.A. Mitchell, and M.C. Allender. 2017. Evaluation of health parameters in cownose rays (*Rhinoptera bonasus*) housed in a seasonal touch pool habitat compared with an off-exhibit habitat. *Journal of Zoo and Wildlife Medicine* 48: 954–960.

Kagan, R., S. Carter, and S. Allard. 2015. A universal animal welfare framework for zoos. *Journal of Applied Animal Welfare Science* 18 (sup1): S1–S10.

Kurtycz, L.M., K.E. Wagner, and S.R. Ross. 2014. The choice to access outdoor areas affects the behavior of great apes. *Journal of Applied Animal Welfare Science* 17 (3): 185–197.

Leighty, K.A., A.J. Valuska, A.P. Grand, T.L. Bettinger, J.D. Mellen, S.R. Ross, et al. 2015. Impact of visual context on public perceptions of non-human primate performers. *PLoS ONE* 10 (2): e0118487.

Leotti, L.A., S.S. Iyengar, and K.N. Ochsner. 2010. Born to choose: The origins and value of the need for control. *Trends in Cognitive Sciences* 14 (10): 457–463.

Lincoln Park Zoo. 2019. Changing the approach to caring for ambassador animals. https://www.lpzoo.org/blog/changing-approach-caring-ambassador-animals Accessed 29 March, 2020.

Maple, T.L., and D.G. Lindburg. 2008. Empirical zoo: Opportunities and challenges to research in zoos and aquariums. *Special Issue of Zoo Biology* 27 (6): 431–504.

Melfi, V.A. 2009. There are big gaps in our knowledge, and thus approach, to zoo animal welfare: A case for evidence-based zoo animal management. *Zoo Biology* 28 (6): 574–588.

Mellor, D.J., and N.J. Beausoleil. 2015. Extending the 'Five Domains' model for animal welfare assessment to incorporate positive welfare states. *Animal Welfare* 24 (3): 241–253.

Mellor, D.J. 2016. Updating animal welfare thinking: moving beyond the "Five Freedoms" towards "A Life Worth Living". *Animals* 6: 21.

Mineka, S., and R.W. Hendersen. 1985. Controllability and predictability in acquired motivation. *Annual Review of Psychology* 36 (1): 495–529.

Moberg, G.P., and J.A. Mench. 2000. *The biology of animal stress: Basic principles and implications for animal welfare*, 1st ed. Wallingford, UK: CABI Publishing.

Moore, D., E. Bauer, and B. Smith. 2013. "Animals First": Good communication may be the missing element in nonhuman animal welfare systems. *Journal of Applied Animal Welfare Science* 16 (4): 393.

Ogle, B. 2016. Value of guest interaction in touch pools at public aquariums. *Universal Journal of Management* 4 (2): 59–63.

Orban, D.A., J.M. Siegford, and R.J. Snider. 2016. Effects of guest feeding programs on captive giraffe behaviour. *Zoo Biology* 35: 157–166.

Owen, M.A., R.R. Swaisgood, N.M. Czekala, and D.G. Lindburg. 2005. Enclosure choice and well-being in giant pandas: Is it all about control? *Zoo Biology* 24 (5): 475–481.

Perdue, B.M., A.W. Clay, D.E. Gaalema, T.L. Maple, and T.S. Stoinski. 2011. Technology at the zoo: The influence of a touchscreen computer on orangutans and zoo visitors. *Zoo Biology* 31: 27–29.

Perdue, B.M., T.A. Evans, D.A. Washburn, D.M. Rumbaugh, and M.J. Beran. 2014. Do monkeys choose to choose? *Learning & behavior* 42 (2): 164–175.

Rabin, L.A. 2003. Maintaining behavioural diversity in captivity for conservation: Natural behaviour management. *Animal Welfare* 12 (1): 85–94.

Reade, L.S., and N.K. Waran. 1996. The modern zoo: How do people perceive zoo animals? *Applied Animal Behaviour Science* 47: 109–118.

Ross, S.R., V.M. Vreeman, and E.V. Lonsdorf. 2011. Specific image characteristics influence attitudes about chimpanzee conservation and use as pets. *PLoS ONE* 6: e22050.

Ruppenthal, A. 2018. Do stingrays like being touched? New research says 'maybe'. https://news.wttw.com/2018/02/21/do-stingrays-being-touched-new-research-says-maybe. Accessed 27 March 2019.

Sahrmann, J.M., A. Niedbalski, L. Bradshaw, R. Johnson, and S.L. Deem. 2016. Changes in human health parameters associated with a touch tank experience at a zoological institution. *Zoo Biology* 35: 4–13.

Saiyed, S.T., L.M. Hopper, and K.A. Cronin. 2019. Evaluating the behavior and temperament of African penguins in a non-contact animal encounter program. *Animals* 9 (6): 326.

Seidensticker, J., and J.G. Doherty. 1996. Integrating animal behavior and exhibit design. In *Wild mammals in captivity*, ed. D. Kleinman et al., 180–190. Chicago: University of Chicago Press.

Shedd. 2019. Animal encounters. https://www.sheddaquarium.org/plan-a-visit/visitor-guide/experiences/animal-encounters/. Accessed 27 March 2019.

Sherwen, S., L. Hemsworth, N. Beausoleil, A. Embury, and D. Mellor. 2018. An animal welfare risk assessment process for zoos. *Animals* 8 (8): 130.

Sherwen, S.L., and P.H. Hemsworth. 2019. The visitor effect on zoo animals: Implications and opportunities for zoo animal welfare. *Animals* 9 (6): 366.

Shamay-Tsoory, S., and L.J. Young. 2016. Understanding the oxytocin system and its relevance to psychiatry. *Biological Psychiatry* 79 (3): 150–152.

Silk, M.J., S.L. Crowley, A.J. Woodhead, and A. Nuno. 2017. Considering connections between Hollywood and biodiversity conservation. *Conservation Biology* 32 (3): 597–606.

Snowdon, C.T., and A. Savage. 1989. Psychological well-being of captive primates: General considerations and examples from callitrichids. In *Housing, care and psychological well-being of captive and laboratory primates*, 75–88. Park Ridge, NJ: Noyes Publications.

Swart, J.A., and J. Keulartz. 2011. Wild animals in our backyard. A contextual approach to the intrinsic value of animals. *Acta biotheoretica* 59 (2): 185–200.

Trainin, G., K. Wilson, M. Wickless, and D. Brooks. 2005. Extraordinary animals and expository writing: Zoo in the classroom. *Journal of Science Education and Technology* 14 (3): 299–304.

Veasey, J.S., N.K. Waran, and R.J. Young. 1996. On comparing the behaviour of zoo housed animals with wild conspecifics as a welfare indicator. *Animal Welfare* 5: 13–24.

Veasey, J.S. 2017. In pursuit of peak animal welfare; the need to prioritize the meaningful over the measurable. *Zoo Biology* 36 (6): 413–425.

Wemelsfelder, F., T.E. Hunter, M.T. Mendl, and A.B. Lawrence. 2001. Assessing the 'whole animal': A free choice profiling approach. *Animal Behaviour* 62 (2): 209–220.

Wemelsfelder, F. 2007. How animals communicate quality of life: The qualitative assessment of behaviour. *Animal Welfare* 16: 25–31.

Sabrina Brando is the director of AnimalConcepts and is pursuing a PhD at the University of Stirling in Scotland on the topic of animal and human wellbeing and its interconnectedness". She is a Research Associate with the Smithsonian Institution, manages 24/7 Animal Welfare and is the Primate Care Training Program Coordinator for the Pan African Sanctuary Alliance. She has worked in and with the global zoo and wildlife profession, including zoos, aquariums, wildlife centres and sanctuaries for 28 years. Sabrina has a BSc. in psychology and an MSc. in animal studies, and is trained as a Compassion Fatigue Educator.

Elizabeth (Betsy) Herrelko manages the animal welfare program at the Smithsonian's National Zoo and facilitates research within Animal Care Sciences. As a behavioral scientist, Herrelko's interests focus on the pursuit of advancing animal welfare science with an emphasis on animal management and how animals think. She created the Welfare Laboratory of Animal Behavior (WelfareLAB) to focus on research, applied practice, outreach, and compliance, and serves as a member of the Association of Zoos and Aquariums (AZA) Animal Welfare Committee. In

2012, she joined the Smithsonian's National Zoo as the David Bohnett Cognitive Research Fellow studying cognitive bias (a measure of emotional affect) in apes and animal welfare topics with various species around the zoo. Prior to the Smithsonian, Betsy earned her master's degree in animal behavior and conservation from Hunter College, CUNY (New York), was an animal keeper at the Gorilla Foundation (California), and earned her PhD at the University of Stirling (Scotland), where she continues to be affiliated as an honorary research fellow. While studying the lives of the chimpanzees at the Royal Zoological Society of Scotland's Edinburgh Zoo, she assessed the development of a cognitive research program and large-scale introductions of zoo-housed chimpanzees. The project's use of touch screen technology and on-exhibit research was the first of its kind for the UK and was the focus of the BBC Natural World documentary, The Chimpcam Project (2010). Her work on introductions has been highlighted in the BBC documentary Origins of Us (2011) and she has contributed to the Discovery Channel's World's Scariest Animal Attacks (2012) film as a primate expert presenting the behavioral perspective.

Chapter 20
Comment: Encountering Urban Animals: Towards the Zoöpolis

Lauren E. Van Patter

20.1 The Urban, the Animal

We are living in the age of the urban, with the majority of the planet's human population now inhabiting cities. Urbanisation creates novel challenges for many animals who are either driven out of these 'human-dominated landscapes' or forced to adapt to drastically different conditions to survive. There are pressing questions about what it means to coexist with other species amidst the precarity of life in the Anthropocene, in which the urban is both a central driver and outcome (Amin and Thrift 2017; Ruddick 2015).

Cities are predominantly constructed in our imaginaries, policies, and practices as uniquely human spaces—in opposition to 'nature' or 'wilderness'—and thus are distinctive contexts in which to discuss animals. Counter to modernist bifurcations of culture/nature, urban/rural, and domestic/wild, we need to recognize cities as a porous matrix of landcover types with transposed infrastructural networks and habitat corridors which create opportunities for the circulation of animals into and around the city (Amin and Thrift 2002).

Discussing urban animals becomes more challenging when we trouble understanding of the 'urban' alongside a recognition of the vast heterogeneity of 'The Animal'[1]? Urbanization and all it entails has drastically different import for the bear and the crow. Does the snail differentiate between the farm field and the city park? For many, the assumption remains that the urban is an 'unnatural' dwelling for wild animals—a space of danger. But the success of many species—even those as

---

[1] As Derrida (2008) famously delineates.

L. E. Van Patter (✉)
Queen's University, Kingston, ON, Canada
e-mail: lauren.vanpatter@queensu.ca

© The Author(s) 2021
B. Bovenkerk and J. Keulartz (eds.), *Animals in Our Midst: The Challenges of Co-existing with Animals in the Anthropocene*, The International Library of Environmental, Agricultural and Food Ethics 33,
https://doi.org/10.1007/978-3-030-63523-7_20

unlikely as leopards[2]—is a testament not only to animals' resilience and adaptability, but to the very real opportunities often afforded by this heterogenous space we call the urban. In landscapes otherwise dominated by agriculture, cities may provide oases of greenspace (parks designed for aesthetic or recreational purposes, riparian corridors, and renaturalized urban natures) amidst vast areas of insecticide-treated monocultures. Within such urban-rural matrices, where is the 'natural' landscape in which animals belong? We need to find new ways of conceptualizing and caring for 'recombinant ecologies' (Barker 2000), wherein our assumptions about the neat divides between the natural and unnatural, the feral and the wild, the native and the exotic/invasive are reconfigured in Anthropocene environments of globalization and urbanization.

This in part involves asking how animals make a living within cities, either with our help—as with intentionally provisioned species such as songbirds—or in spite of our best efforts to 'manage' them—as with 'nuisance' or 'pest' animals like rats. This is both a spatial question, stemming from our judgements about who does or does not belong in certain spaces and how particular animals should live, but it is also an ethical question, in terms of what rights other-than-humans might have to the city, and what responsibilities we might have to promote, or at least not impede, their flourishing. Despite the challenges posed by the dominant anthropocentric design and ethos in cities, animals survive, inhabiting their own rich 'storied' worlds (van Dooren and Rose 2012), which raises questions about how they experience and know the city (Barua and Sinha 2019), and what animal-friendly cities might look like. Moving towards multispecies spatial justice—towards the Zoöpolis—means recognizing the many nonhuman Others who live alongside us in our shared urban ecologies, and developing creative solutions aimed at flourishing in the more-than-human city.

In this chapter I briefly consider the politics of spatial access within three settings of animal encounters—the home, the zoo, and the street/park/margins—and reflect on three avenues that merit further engagement in thinking towards the Zoöpolis: 'articulating with' animals; making visible relationalities; and re-storying the city to imagine otherwise.

20.2 Urban Animal Encounters and the Politics of Spatial Access

Spatial inclusions and exclusions in the city have long been a concern of animal geographers, who have explored the expulsion of farmed animals from the urban sphere (Philo 1995), the border practices which keep wild 'intruders' from the space of the home (Power 2009), and the fostering of very particular forms of animality, human-animal relations, and nature, within cities (Griffiths et al. 2000). Although the city is usually thought of as a primarily human space, many animals are welcome within its bounds. The most obvious are domestic 'pets', which in modern cities are valued for

---

[2]See Braczkowski et al. (2018).

companionship and often made to fit within visions of the heteronuclear family and neoliberal individualistic and consumerist cultures (McKeithen 2017; Nast 2006). Certain species of wild animals are also embraced, enrolled in the production of biodiversity and valued for their aesthetic benefits or contribution to ecosystem services. These welcomed wild animals are expected to remain in 'natural' areas and are subject to biosecurity measures when they engage in practices viewed as disruptive or dangerous, enter spaces in which they are deemed not to belong,[3] or come to be viewed as 'pests' due to perceptions of overpopulation or association with filth and disease.[4] Overall, contentions around the 'place' of animals in modern cities raises questions about who belongs and where. Encounters within diverse spaces of the home, zoo, and streets/parks/urban margins are subject to their own complex affective and political dimensions, with animals disciplined in the production of particular modes of value and visions of nature and culture.

20.2.1 The Home

Within the home, we find intimate relations of companionate cohabitation. Urban 'pets' are increasingly seen as members of the family, a status at odds with socio-legal/capitalist operations wherein animals remain property—commodities able to be bought, sold, traded, used, and disposed of with little regard (Instone and Sweeney 2014; Pallotta 2019). But home is more than a physical space, it encompasses particular relations of companionship that stretch outwards into the broader realm of anthropogenic urban public space, with associated contention in the case of domestic animals' spatial access. For example, debates surround whether or not canine companions should be given their own space in the form of designated dog parks, which arguably represent urban planning's response to more-than-human agencies and corporealities (Urbanik and Morgan 2013). Or the contradiction wherein dominant understandings of environmental responsibility increasingly dictate that domestic cats be confined indoors, while at the same time we are increasingly finding it morally indefensible to bar animals such as chickens raised for food from having access to the outdoors and the opportunity to exercise natural behaviours. But these debates around spatial access are part of a larger landscape of negotiated borderlands which make visible the porosity of the domus, as the wild are invited in as exotic pets (Collard 2014), ingress against our will by burrowing under our porches or into our walls and ceilings (Power 2009), and the domestic leaks out from under our control, becoming stray or feral.

---

[3]For instance raccoons in daycares (Pacini-Ketchabaw and Nxumalo 2016) or cougars in suburbs (Collard 2012).

[4]Such as cormorants in Toronto (Sandilands 2017), or pigeons in New York (Jerolmack 2008).

20.2.2 The Zoo

Zoos are another space in which animals are made to live within the city, with contention surrounding the purpose of the Modern zoo: Do they exist primarily for entertainment, conservation, or education? For the benefits of individual animals, species, or ecologies, or ultimately to serve human interests, reinforcing our supremacy? As Brando and Herrelko highlight in their chapter *Wild Animals in the City: Considering and Connecting with Animals in Zoos and Aquariums*, debates surround: what constitutes 'natural' or 'wild' versus 'unnatural' behaviours in these situations of confinement; the implications of this for zoos' conservation goals which include reintroductions—for which animals are required to be maintained as sufficiently 'wild'; and the inherent tensions between providing for animal welfare versus purported conservation and educational aims.

Alongside these concerns are layered considerations surrounding public engagement, perceptions, and expectations. As Kisora and Driessen point out in their chapter *Interpreting the YouTube zoo: ethical potential of captive encounters*, what is most often sought by the zoo-goer is "a dream-coming-true visit—not only seeing, but also being seen by the other". Proximate encounters, even touching and handfeeding, may be afforded at zoos, but are discouraged in the case of animals in 'the wild', potentially sending mixed messages to the public about appropriate interactions and boundaries with wildlife (Brando and Herrelko).

Our relationships with wildlife take shape not only through proximate encounters, but through complex media ecologies, which enrol diverse actors, including zoos. The ways we come to know about, and our expectations for relating to, animals and 'nature' more broadly, are increasingly shaped through 'spectacular environmentalisms', highlighting the currency between the dramatic and the everyday (Goodman et al. 2016). Within these 'fabulous ecologies' (Howell and Taves 2019), the boundaries and purpose of the zoo are increasingly blurred, as 'electronic zoos' create virtual encounters through advances in surveillance technologies, within which animals live digital 'second lives' (Adams 2020) with limited opportunities to shape recounted narratives about Nature (Davies 2000). A more expansive understanding of the 'zoo' takes into account that various animal bodies are cultivated for particular encounters in the service of entertainment, and/or the (re)production of particular visions of Nature. Furthermore, it is important to consider the dispersed spatialities and temporalities of encounter, as the immediacy of the zoo travels through virtual networks, bringing animals into the spaces of homes, workplaces, theatres, and pockets around the world. As discussed by Kisora and Driessen, within such 'Youtube Zoos', the ambivalence surrounding boundaries between animals as familiar individuals versus autonomous wildlife, and the ways in which their virtual representations both challenge and reinforce assumptions about and unequal relations with animals, come to the fore. In user-generated venues of online videos and commentary, the meaning of animal encounters can be contested. Digital technologies and connectivity trouble the boundary between public and personal spaces of encounter, as the 'Youtube Zoo' opens up the private lives of animal celebrities to discussion and debate. Consequences of such negotiated meanings ripple out to shape lived encounters between people and animals—domestic, zoo, wild, liminal—throughout the city.

Materially, zoos represent unique spaces, bubbles designed for certain valued animal bodies within cities, with architecture that promotes wellbeing and/or encounterability, fostering particular affective atmospheres and encounter value (Barua 2017, 2019). Within these spaces charismatic species—cherished megafauna fetishized as the epitome of 'wild' nature—are accumulated, lively commodities in the currency of rare genes (Lorimer 2015). Alongside considerations of animal welfare, labour, and multispecies relations of power lie questions about the role, function, and implications of the zoo as simulacrum of nature (Braverman 2015): Which Natures do zoos conserve, and for whom? What is the value of ex situ conservation in the context of ever-disappearing 'natural' habitats? What is the relationship between captive animals in zoos and the wild animals which may share their genetic makeup, but which lead vastly different lives, embedded in vastly different experiences, memories, and relations which, arguably, are core to what it means to be that animal[5]? Such questions illuminate the biopolitical realities of these quintessential spaces of 'wild' encounters within the city, wherein animals are "not only confined and subdued…but also interpreted and classified" (Anderson 1995, 283).

20.2.3   The Streets/Parks/Margins

In the streets, parks, and 'marginal' (Gandy 2013) city spaces, we encounter 'liminal' stray, feral, or wild animals who negotiate their own existences, often counter to human intentions. These unintentional or spontaneous spaces and beings, and in particular the feral and synurbic, transgress expectations around who should be using natural versus anthropogenic spaces and resources, and who should be making a living autonomously versus through dependency on humans. Feral animals are domesticated animals who have 'gone wild', in that they are no longer under human care—or at least not in the traditional sense of 'belonging' to one particular owner or household. Synurbic species are wild animals that thrive in cities (Luniak 2004), like racoons, rats, crows, and seagulls. Both types of animals transgress nature/culture divides: the feral by escaping the realm of culture, becoming unruly, independent, and wild; the synurbic by crossing boundaries from nature into culture—into cities—disrupting our expectations of the safe, tame, ordered space of the human.

Urban animals are often both materially and discursively invisible. Materially, behavioural patterns are shifted to avoid attracting notice, for instance, through becoming increasingly nocturnal (Gaynor et al. 2018). Discursively, they are backgrounded, dismissed, or devalued as 'common', 'pest', or 'trash' animals (Nagy and Johnson 2013). Liminal animals are often pitied, with assumptions about their

---

[5] As discussed by Evernden (1985) in the case of gorillas, and Whatmore (2002) in the case of an African elephant.

inevitable poor quality of life as they eke out an existence in such an 'unnatural' manner. But the heterogeneity of livelihood opportunities experienced by synurbic and feral/stray/street animals defies simplistic assumptions about their lives and practices (Meijer, this volume; see also Van Patter and Hovorka 2018). They are also, at times, celebrated, their transgressions, resistances, and agencies in coshaping urban spaces and relations with humans indicating to some the resilience of nature.[6] But, as Meijer emphasizes in her chapter *Stray agency and interspecies care: The Amsterdam stray cats and their humans*, these relationships are inherently unequal, as despite animals having a degree of choice in the spaces they occupy, resources they access, and proximity to or avoidance of humans and other species, we ultimately have the power to discourage, remove, or destroy unwanted or 'nuisance' individuals and populations.

The city takes shape through myriad more-than-human relations within diverse spaces, from the porous home, to the ambivalent zoo, and the liminal urban interstices. Along with considering these urban 'animal spaces', it is important to consider the politics of knowledge through which belongings are negotiated, and the 'beastly places' of material animal lifeworlds.[7] In terms of the former, this involves attending to the ways in which power operates in the maintenance of expertise whereby 'nature' becomes intelligible in particular registers. For instance, in the chapters contained herein, who can legitimately interpret the actions and motivations of Jacky and Jinga? What happens when cat caretakers and more powerful agencies or institutions disagree about what is best for cats, or for the community? Or when practitioners and theorists are at odds about zoo animals' lives, needs, and welfare? In terms of the latter, our lack of knowledge about animals' ecologies and lived realities can present limitations for coexistence, and foregrounding their experiences, knowledges, and practices is key to engaging with challenging questions of shared life in the Zoöpolis.

20.3   Towards the Zoöpolis

Over two decades ago Wolch (1996; Wolch et al. 1995) advanced that realising the Zoöpolis[8]—the imagined city of multispecies cohabitation and belonging—requires that we take animals seriously as legitimate matters of concern within urban policies and practices. A number of interventions have advanced approaches which work towards this, for instance: a '*cosmopolitics*' in which space is made for diverse actors to participate in a politics that resists narrow nature/society binaries (Hinchliffe et al. 2005); an understanding of '*commoning*' as a more-than-human practice through

---

[6] See, for example, Montford and Taylor (2016).

[7] See Philo and Wilbert (2000), who formulate this distinction.

[8] The term is also mobilized by Donaldson and Kymlick (2011) to advance a political theory of animal inclusion within urban governance frameworks through a model of citizenship for domesticated animals and denizenship for 'liminal' animals who live around humans but not in direct relations of companionship.

which the needs and benefits of diverse urban inhabitants are negotiated (Cooke et al. 2019); and an 'ethics of *conviviality*' which demands that we "find multiple, life enhancing ways of sharing and co-producing meaningful and enduring multispecies cities" (van Dooren and Rose 2012, 17). But many questions remain to be addressed, and in the remainder of this chapter I reflect briefly on three avenues that merit further engagement in thinking towards the Zoöpolis: 'articulating with' animals; making visible relationalities; and re-storying the city to imagine otherwise.

20.3.1 *'Articulating With' Animals*

The Zoöpolis requires an approach to urban policies and practices in which animals "bring their own politics of recognition" (Narayanan 2017, 488). But attending to animals' 'political voices' (Meijer 2013) presents challenges in terms of how we typically interpret, represent, and engage with animals. For instance, Meijer highlights that there are often inherent tensions within practices of care enacted for urban animals, including paternalistic assumptions about what is 'good' for them (such as sterilization in the case of feral/stray animals[9]). In asking questions about what matters to animals and how they want to live, we need to shift away from *speaking for* more-than-human Others, and towards experimental and generous modes of *articulating with* them (Giraud 2019; Haraway 2003).

As Nieuwland and Meijboom point out, methodologically, we need to attend to "the urban environment as an animal collective" by engaging "multispecies epistemologies". For instance, Barua and Sinha's (2019) etho-geographical approach asks what animals' knowledges and practices can tell us about life in the city, and the material, ecological, and phenomenological dimensions of urbanization. By foregrounding animals' experiences and lifeworld, we can begin to take seriously more-than-human modes of inhabitation and claims to space. In so doing, we move towards seeing "urbanisation not as something merely going on in cities, but as a process where dense traffic in commodities and materials transforms lifeworlds of humans and animals, with asymmetric and often disturbing effects" (Barua and Sinha 2019, 1174).

20.3.2 *Making Visible Relationalities*

The Zoöpolis require a 'politics of sight' (Hunold 2019)[10] wherein we learn to see the city as legitimate habitat for many more-than-human Others embedded in rich social and ecological relations. As delineated by Nieuwland and Meijboom in their chapter "*Eek! A Rat!*", being sensitive to particular animals' circumstances and ways

---

[9]See also Srinivasan (2013).

[10]Drawing on Schlosberg (2016).

of life can be a meaningful starting point for compassionate action. But, as the authors note, compassion and care can be thwarted by the potency of affective dimensions—such as visceral responses of disgust or fear—which are central to the 'terrain of killability' (Gillespie and Collard 2015, 15) that constitutes our relations with many urban animals. In overcoming these barriers to coexistence, there is value in both pragmatic attention to suffering, and a metaphysics of interconnection. Only by realizing the complexity of ecological interconnectedness—that we are all in this together—can we hope to create futures of flourishing amidst the threats of global crises like climate change, zoonotic pandemics, and extinction.

Rather than focusing narrowly on conflict mitigation, we need to make visible the 'ecologies of care' and resistance which permeate the urban (Meijer). Rethinking care for the more-than-human city involves attending both to animals as individuals, and as relationally-embedded within complex socio-ecological networks. For instance, the intersections between individuals or groups of humans and animals are often ignored when we consider urban animal management policies and their implications. As Narayanan (2017) discusses, close relationships of the urban poor with street dogs, who provide security and companionship, means that programs to 'manage' these dogs often adversely impact the most marginalized human inhabitants of cities as well. Similarly, Meijer notes the shared precarity of particular animals (stray/feral cats) and the humans with whom they often associate (e.g. homeless, neurodiverse, and economically disadvantaged individuals). It is imperative to carefully consider the ways in which more-than-human identities, differences, and inequalities intersect and are (re)produced within multispecies relations of power (Hovorka 2019). Making visible these relationalities within complex colonial-capitalist realities requires that we resist oversimplifications and grasping for tidy answers. Though no easy task, we have a responsibility to 'stay with the trouble' (Haraway 2008) and work to make visible the violent histories and ongoing injustices and dispossessions in the post-/settler colonial city. The Zoöpolis requires that we unlearn the anthropo- and Eurocentric frames through which the city is typically understood, with their "erasure of existing kinship relations that have been nurtured for generations" (Porter et al. 2020, 10).

20.3.3   Re-Storying the City to Imagine Otherwise

Working towards the Zoöpolis requires that we 're-story' the city, imagining 'as well as possible' multispecies futures (Puig de la Bellacasa 2017) amidst the shared precarity of the Anthropocene. One way in which to do so is by attending to the lives of animals to weave a different narrative of urban life, one with the potential to envision "flourishing landscapes of coexistence, rather than battlescapes of violence" (Narayanan 2017, 488). Engaging in such exercises of 'imaging otherwise' recognizes the interconnected nature of our imaginings, understandings, thoughts, emotions, and practices, and their implications for material worldly becomings (Walker 2013).

How can we imagine the city otherwise, as a "co-emergent world based on inti-mate human-more-than-human relationships of responsibility and care" (Bawaka Country et al. 2016, 470)? We need to think carefully with 'care' in its asymmetrical reciprocities (as Meijer notes, "cats also take care of humans") and its messiness and noninnocence (Puig de la Bellacasa 2017). Chrulew (2011, 139) notes that the "caesura between the overloved and the unloved, between the politics of life and death, bios and thanatos, brings into stark relief one of the central ethical questions of our time: how should we love in a time of extinction?" Within the city—in the zoo, the home, and the streets/parks/margins—this 'caesura' plays out dramatically: with the hyper-visible overloved—doted upon 'fur babies' and carefully cultivated spaces of 'biodiversity'—and the made-invisible unloved—intensively eradicated 'pests' and surplus stray/feral bodies. Thus, central to re-storying the city is confronting these inequalities and the challenging questions surrounding what it means to love, to care, to 'live well' with more-than-human Others in a time of planetary urbanisation.

Imagining the Zoöpolis requires an openness to "risky worldings" (Haraway 2008, 27) in which outcomes and optimal approaches and configurations remain uncertain. It involves asking questions which include: Can we think of wild animals as compan-ions, as Haraway (2008) suggests, but in ways that are attentive to the tensions of space, boundaries, and remaining responsible for the futures that are created through situated relatings? Can we be open to the recombinant ecologies of the Anthropocene, and rather than gazing into the past and clinging dogmatically to divisions of 'native' versus 'exotic' or 'invasive', ask ourselves what opportunities arrivant species may offer, what we can learn from other animals about living together, and what our responsibilities to these new configurations might be; as Reo and Ogden (2018) suggest, drawing from Anishinaabe teachings and practices[11]? Can we think of feral or stray animals as legitimate and valued components of urban socio-natures, as Srini-vasan (2019) advances, moving towards non-dualistic understandings of belonging in the multispecies city? Imagining the city otherwise—as a place of more-than-human belonging—requires "a speculative commitment to think about how things could be different… attached to situated and positioned visions of what a livable and caring world could be" (Puig de la Bellacasa 2017, 60).

20.4   Conclusion

This chapter has briefly considered the spatial and politico-ethical dimensions of multispecies urban cohabitations within three settings: the home, the zoo, and the street/park/margins. Cities are heterogenous spaces composed of myriad actors, most

---

[11]Reo and Ogden (2018) discuss the concept of 'aki' within the Anishnaabe land ethic, which denotes the "cosmological sense of the sacredness of place", wherein teachings "hold land as sacred and as the embodiment of Creation, as are all the living beings such as plants and animals, as well as water, stones, and supernaturals" (1446). Within such an understanding, "the agency of plants and animals, as persons, relatives, nations and teachers, are all central to how [Anishnaabe] make sense of introduced species" (1445).

of whom are not human, and are routinely made invisible within urban planning and policies. Despite this, everyday practices reveal intimate interconnections of care and violence, discipline and transgression. Urban animals become emplaced conceptually and materially within dominant visions of what the city, the animal, and nature ought to be.

But urban animals also shape spaces and relations according to their own needs and lifeways. Addressing the pressing questions of coexisting with other species in a time of planetary urbanisation requires that we see the city in a new light: as a space of multispecies cohabitation and possibility. This chapter briefly advances three avenues that could help in thinking towards the Zoöpolis: 'articulating with' animals, making visible relationalities, and re-storying the city to imagine otherwise. It is crucial that we continue working towards new understandings of urbanization and animals in the Anthropocene which foreground multispecies justice and opportunities for co-flourishing in the more-than-human city.

References

Adams, W. 2020. Digital animals. *The Philosopher* 108 (1): 17–21.

Amin, A., and N. Thrift. 2002. *Cities: Reimagining the urban*. Cambridge: Polity Press.

Amin, A., and N. Thrift. 2017. *Seeing like a city*. Chichester: Wiley.

Anderson, K. 1995. Culture and nature at the Adelaide Zoo: At the frontiers of 'human' geography. *Transactions of the Institute of British Geographers* 20 (3): 275–294.

Barker, G. (ed.). 2000. *Ecological recombination in urban areas*. Peterborough: The Urban Forum/English Nature.

Barua, M. 2017. Nonhuman labour, encounter value, spectacular accumulation: The geographies of a lively commodity. *Transactions of the Institute of British Geographers* 42 (2): 274–288.

Barua, M. 2019. Affective economies, pandas, and the atmospheric politics of lively capital. *Transactions of the Institute of British Geographers*. https://doi.org/10.1111/tran.12361.

Barua, M., and A. Sinha. 2019. Animating the urban: An ethological and geographical conversation. *Social and Cultural Geography* 20 (8): 1160–1180.

Country, Bawaka, S. Wright, S. Suchet-Pearson, K. Lloyd, L. Burarrwanga, R. Ganambarr, and J. Sweeney. 2016. Co-becoming Bawaka: Towards a relational understanding of place/space. *Progress in Human Geography* 40 (4): 455–475.

Braczkowski, A.R., C.J. O'Bryan, M.J. Stringer, J.E. Watson, H.P. Possingham, and H.L. Beyer. 2018. Leopards provide public health benefits in Mumbai, India. *Frontiers in Ecology and the Environment* 16 (3): 176–182.

Braverman, I. 2015. *Wild life: The institution of nature*. Stanford: Stanford University Press.

Chrulew, M. 2011. Managing love and death at the zoo: The biopolitics of endangered species preservation. *Australian Humanities Review* 50 (1): 137–157.

Collard, R.C. 2012. Cougar—Human entanglements and the biopolitical un/making of safe space. *Environment and Planning D: Society and Space* 30 (1): 23–42.

Collard, R.C. 2014. Putting animals back together, taking commodities apart. *Annals of the Association of American Geographers* 104 (1): 151–165.

Cooke, B., A. Landau-Ward, and L. Rickards. 2019. Urban greening, property and more-than-human commoning. *Australian Geographer* 51: 1–20.

Davies, G. 2000. Virtual animals in electronic zoos: The changing geographies of animal capture and display. In *Animal spaces, beastly places: New geographies of human-animal relations*, ed. C. Philo and C. Wilbert, 243–267. London, UK: Routledge.

Derrida, J. 2008. *The animal that therefore I am.* New York: Fordham University Press.

Donaldson, S., and W. Kymlicka. 2011. *Zoopolis: A political theory of animal rights.* Oxford: Oxford University Press.

Evernden, L.L.N. 1985. *The natural alien: Humankind and environment.* Toronto: University of Toronto Press.

Gandy, M. 2013. Marginalia: Aesthetics, ecology, and urban wastelands. *Annals of the Association of American Geographers* 103 (6): 1301–1316.

Gaynor, K.M., C.E. Hojnowski, N.H. Carter, and J.S. Brashares. 2018. The influence of human disturbance on wildlife nocturnality. *Science* 360 (6394): 1232–1235.

Gillespie, K., and R.C. Collard (eds.). 2015. *Critical animal geographies: Politics, intersections and hierarchies in a multispecies world.* New York: Routledge.

Giraud, E.H. 2019. *What comes after entanglement? Activism, anthropocentrism, and an ethics of exclusion.* Durham: Duke University Press.

Goodman, M.K., J. Littler, D. Brockington, and M. Boykoff. 2016. Spectacular environmentalisms: Media, knowledge and the framing of ecological politics. *Environmental Communication* 10: 677–688.

Griffiths, H., I. Poulter, and D. Sibley. 2000. Feral cats in the city. In *Animal spaces, beastly places: New geographies of human-animal relations*, ed. C. Philo and C. Wilbert, 56–70. New York: Routledge.

Haraway, D.J. 2003. *The companion species manifesto: Dogs, people, and significant otherness.* Chicago: Prickly Paradigm Press.

Haraway, D.J. (2008). *When Species Meet.* University of Minnesota Press.

Hinchliffe, S., M.B. Kearnes, M. Degen, and S. Whatmore. 2005. Urban wild things: A cosmopolitical experiment. *Environment and planning D: Society and Space* 23 (5). 643–658.

Hovorka, A.J. 2019. Animal geographies III: Species relations of power. *Progress in Human Geography* 43 (4): 749–757.

Howell, P., and I. Taves, 2019. The curious case of the croydon cat-killer: Producing predators in the multi-species metropolis. *Social & Cultural Geography*: 1–20.

Hunold, C. 2019. Green infrastructure and urban wildlife: Toward a politics of sight. *Humanimalia* 11 (1): 150–169.

Instone, L., and J. Sweeney. 2014. Dog waste, wasted dogs: The contribution of human-dog relations to the political ecology of Australian urban space. *Geographical Research* 52 (4): 355–364.

Jerolmack, C. 2008. How pigeons became rats: The cultural-spatial logic of problem animals. *Social Problems* 55 (1): 72–94.

Lorimer, J. 2015. *Wildlife in the Anthropocene: conservation after nature.* Minneapolis: University of Minnesota Press.

Luniak, M. 2004. Synurbization–adaptation of animal wildlife to urban development. In *Proceedings of the 4th International Symposium on Urban Wildlife Conservation*, 50–55. Tucson, AZ.

McKeithen, W. 2017. Queer ecologies of home: Heteronormativity, speciesism, and the strange intimacies of crazy cat ladies. *Gender, Place & Culture* 24 (1): 122–134.

Meijer, E. 2013. Political communication with animals. *Humanimalia: A Journal of Human/Animal Interface Studies* 5 (1): 28–51.

Montford, K.S., and C. Taylor. 2016. Feral theory: Editors' introduction. *Feral Feminisms* 6: 5–17.

Nagy, K., and P.D. Johnson II. (eds.). 2013. *Trash animals: How we live with nature's filthy, feral, invasive, and unwanted species.* Minneapolis: University of Minnesota Press.

Narayanan, Y. 2017. Street dogs at the intersection of colonialism and informality: 'Subaltern animism' as a posthuman critique of Indian cities. *Environment and Planning D: Society and Space* 35 (3): 475–494.

Nast, H.J. 2006. Loving… whatever: Alienation, neoliberalism and pet-love in the twenty-first century. *ACME: An International E-Journal for Critical Geographies* 5 (2):300–327.

Pacini-Ketchabaw, V., and F. Nxumalo. 2016. Unruly raccoons and troubled educators: Nature/culture divides in a childcare centre. *Environmental Humanities* 7 (1): 151–168.

Pallotta, N.R. 2019. Chattel or child: The liminal status of companion animals in society and law. *Social Sciences* 8 (5): 158.

Philo, C. 1995. Animals, geography, and the city: Notes on inclusions and exclusions. *Environment and Planning D: Society and Space* 13 (6): 655–681.

Philo, C., and C. Wilbert. 2000. Animal spaces, beastly places: An introduction. In *Animal spaces, beastly places: New geographies of human-animal relations*, ed. C. Philo and C. Wilbert, 1–34. London, UK: Routledge.

Porter, L., J. Hurst, and T. Grandinetti. 2020. The politics of greening unceded lands in the settler city. *Australian Geographer*: 1–18. https://doi.org/10.1080/00049182.2020.1740388.

Power, E.R. 2009. Border-processes and homemaking: Encounters with possums in suburban Australian homes. *Cultural Geographies* 16 (1): 29–54.

Puig de La Bellacasa, M. 2017. *Matters of care: Speculative ethics in more than human worlds*. Minneapolis: University of Minnesota Press.

Reo, N.J., and L.A. Ogden. 2018. Anishnaabe Aki: An indigenous perspective on the global threat of invasive species. *Sustainability Science* 13 (5): 1443–1452.

Ruddick, S. 2015. Situating the Anthropocene: Planetary urbanization and the anthropological machine. *Urban Geography* 36 (8): 1113–1130.

Sandilands, C. 2017. Some "F" words for the environmental humanities: Feralities, feminisms, futurities. In *The Routledge companion to the environmental humanities*, 459–467. London: Routledge.

Schlosberg, D. 2016. Environmental management in the Anthropocene. In *The Oxford handbook of environmental political theory*, ed. T. Gabrielson, et al. Oxford: Oxford University Press.

Srinivasan, K. 2013. The biopolitics of animal being and welfare: Dog control and care in the UK and India. *Transactions of the Institute of British Geographers* 38 (1): 106–119.

Srinivasan, K. 2019. Remaking more-than-human society: Thought experiments on street dogs as "nature". *Transactions of the Institute of British Geographers* 44 (2): 376–391.

Urbanik, J., and M. Morgan. 2013. A tale of tails: The place of dog parks in the urban imaginary. *Geoforum* 44: 292–302.

Van Dooren, T., and D.B. Rose. 2012. Storied-places in a multispecies city. *Humanimalia* 3 (2): 1–27.

Van Patter, L.E., and A.J. Hovorka. 2018. 'Of place' or 'of people': Exploring the animal spaces and beastly places of feral cats in southern Ontario. *Social and Cultural Geography* 19 (2): 275–295.

Walker, R.L. 2013. Environment imagining otherwise. *Journal of Curriculum and Pedagogy* 10 (1): 34–37.

Whatmore, S. 2002. *Hybrid geographies: Natures cultures spaces*. London: Sage.

Wolch, J. 1996. Zoöpolis. *Capitalism, Nature, Socialism* 7 (2): 21–47.

Wolch, J., K. West, and T. Gaines. 1995. Transspecies urban theory. *Environment and Planning D: Society and Space* 13 (5): 735–760.

Lauren E. Van Patter is a critical human-environment geographer and doctoral candidate at Queen's University in Kingston, Canada, where she works with The Lives of Animals Research Group. She draws on posthumanist and feminist traditions to investigate ethical entanglements and spatial politics in more-than-human lifeworlds. Current projects explore how more-than-human actors negotiate, shape, and experience shared life in multispecies cities, including work on the feral ecologies of cat colonies, and the phenomenologies and mobilities of urban coyotes in Ontario, Canada. She is co-editor (with Alice Hovorka and Sandra McCubbin) of the forthcoming volume A Research Agenda for Animal Geographies (Elgar). You can read more about her research and teaching at https://levanpatter.wordpress.com/.

Part IV
Wild Animals

Chapter 21
Should We Provide the Bear Necessities? Climate Change, Polar Bears and the Ethics of Supplemental Feeding

Clare Palmer

Abstract This paper considers whether we have any moral responsibility to offer supplemental feeding to wild animals who have lost food access due to climate change. It takes as a particular case the situation of some individual polar bears who, over the next decade, are likely to be threatened with abrupt loss of food access due to changes in sea ice, potentially causing starvation. The paper argues that, as is implied by most positions in animal ethics, there are ethical reasons to assist individual polar bears by supplemental feeding. However, there are also good reasons to hesitate, and to consider potential harms both to bears and to other animals, as well the loss of wildness value that may be involved. From some ethical positions, the likely harms involved make euthanasia ethically preferable to supplemental feeding. But on other plausible ethical arguments, these likely harms are not decisive. We need to know more about the possible effects of supplemental feeding of polar bears. So, the paper concludes that when the first bears are threatened by abrupt loss of food access, a trial of supplementary feeding should be considered in consultation with relevant native peoples.

21.1 Introduction

Pied flycatchers arriving in the Netherlands from migration go hungry because their migration is not shifting early enough to match even more rapid shifts in spring insect peaks (Both et al. 2006). Rain falls on snow in the Arctic tundra in winter, freezing into an ice sheet and making it impossible for reindeer to break through the ice to

---

[1] See this news story in *The Guardian*, February 24, 2019: https://www.theguardian.com/enviro nment/2019/feb/25/decline-in-bogong-moth-numbers-leaves-pygmy-mountain-possums-starving. Accessed 23 March 2020.

C. Palmer (✉)
Texas A&M University, College Station, TX, USA
e-mail: c.palmer@tamu.edu

© The Author(s) 2021
B. Bovenkerk and J. Keulartz (eds.), *Animals in Our Midst: The Challenges of Co-existing with Animals in the Anthropocene*, The International Library of Environmental, Agricultural and Food Ethics 33,
https://doi.org/10.1007/978-3-030-63523-7_21

the browse below (Forbes et al. 2016). Massive declines in bogong moth numbers in Australia, a vital food for breeding mountain pygmy possums, seem to be leading possum litters to die from starvation in their mothers' pouches.[1] All these instances of wild animals losing food access seem to be related to climate change. They are ways in which human emissions of greenhouse gases are making life harder and causing suffering to members of at least some wild animal species.

Do we have any moral responsibility to respond to such suffering? If so, what form do these responsibilities take? Should we euthanize animals if a painful, drawn-out death seems inevitable? Or should we try to help at least some of these animals, where we can, by attempting to restore food access patterns, or by offering supplemental feeding?

Proposals to assist wild animals made hungry due to climate change, however, raise a host of ethical (not to mention practical) challenges. One inevitable effect of assistance would be to bring wild animals into new relationships with humans. This may in itself be regarded as a problem, since it could be perceived as making wildlife less "wild". And even ethical positions that, in principle, accept that there's reason to assist wild animals undergoing food access difficulties due to climate change are likely to diverge in terms of whether and how to assist—or whether to euthanize—in particular cases (see Palmer 2019).

In this paper, I will focus on one case of climate-induced food access problems: the situation of *some individual polar bears* over the next decade or two, who are likely to be threatened with abrupt loss of food access, potentially causing starvation. Should we help these bears? In particular, should we offer them supplemental feeding?

Some of the ethical issues raised by starving polar bears are unique. But this case also raises a number of more general ethical questions about supplemental feeding in the context of climate change, relating to multiple possible *harms* (to the animals receiving supplementary feeding, to other animals, and to humans) as well as *benefits*, to concerns about the loss of animals' agency, and about the loss of the wildness or naturalness of animals or places. I'll argue here, as is implied by most positions in animal ethics, that there is at least ethical *reason* to help individual polar bears by supplemental feeding, with the goal of avoiding their suffering and death from starvation and the possible extirpation of entire wild bear populations. But it's also clear that there are good reasons to hesitate, and to consider alternative actions—including euthanasia—in the case of polar bears. And alongside ethical concerns about individual animals, there are also worries about the creation of what Arctic scientists Derocher et al. (2013) call "semi-managed bear parks", which are likely to reduce the amount of wildness value placed both on bears and their Arctic environment.

Unfortunately, this situation presents a choice between only bad options. As I'll suggest, on some ethical views, euthanasia will normally look like a better choice than supplemental feeding; but on other ethical views, we need to know more than we currently do about what would actually happen if bears are fed, both in terms of short and long term harms, and wildness impacts. So, I'll end by tentatively suggesting a feeding trial, where the first bears facing this desperate situation are fed, and the results carefully assessed and monitored. But this suggestion also raises problems, since in many cases, polar bear habitat is shared with native peoples; feeding the bears without engaging the humans that live nearby presents potential justice issues. So, as I'll conclude, any attempt to put some form of this suggestion into practice also requires the meaningful participation of local native communities at an early stage in the consideration of such a proposal.

21.2 Some Basic Premises of This Paper

This paper rests on several assumptions. First, I'll assume that climate change is anthropogenic. Second, I'll also accept that humans are not just *causally* responsible, but also *morally* responsible for climate change. I'll base this conclusion (for simplicity) on Nolt's (2011) argument: We are morally responsible for a harm when (1) We can cause or prevent the harm, (2) We can recognize it as morally significant, (3) We can anticipate it with some reliability, and (4) We can act in less harmful or more beneficial ways. Nolt maintains that while not all of these four conditions held in the past with respect to climate change, this is no longer the case, at least in industrialized countries. We know we are causing climate change; we can see that its effects, not only on humans but also on non-humans, are morally significant; we have a reasonable idea of what is likely to happen, and we can do something about it. Of course, the details of how this moral responsibility is distributed over human nations, populations and individuals over time and space, are contested, as is determining who has responsibilities to do what as a result. I do not have space to try to untangle these important but complex matters here. For now, while recognizing this as an over-simplification, I will merely claim that "we" are morally responsible for climate change and for the harms it causes.

The particular harms I'll be discussing here are those caused by climate change to *individual* polar bears. The focus here is not on the possible extinction of the polar bear *species*; indeed, it seems likely that some populations will, in fact, persist into the foreseeable future.[2] I'll assume what's fairly uncontentious—that polar bears are sentient—and that the wellbeing of sentient animals matters morally. More controversially, I'll interpret the wellbeing or welfare of sentient animals in terms of their *subjective experience*—what happens to them matters to them in terms of how it feels; in particular, they have an interest in not suffering. I include as an aspect of subjective wellbeing the expression of *agency*, in the sense of animals being able to direct their own bodies and activities, pursuing goals and making choices about what they do, and having motivational states (such as wanting things) (see Steward 2009).

There are, I should note, other interpretations of animal welfare. On a *perfectionist* view, things can matter for animals' welfare even if they are not experienced by the animal itself—in particular, the performance of natural behaviors, such as hunting. On this view, if bears could not perform natural behaviors, this would matter for their welfare, even if the bears themselves experienced no negative feelings as a

---

[2] See Palmer (2009) for a discussion about whether the polar bear *species* can be harmed.

consequence. It might be argued, for instance, that they have lost something central to their identity (Hettinger, *pers. comm.*) However, this is not the interpretation of welfare I am adopting here. While it may be reasonable to argue that there's a loss of wildness value *to us*, in some sense, if bears can no longer perform natural behaviors owing to human-derived constraints, if this loss is not experienced negatively by the bears themselves, it does not (on my account) mean that their welfare has been impacted.

Climate change also raises what's known as the *non-identity problem* in the context of wild animals. The non-identity problem draws attention to the possibility that climate impacts could play such a significant role in wild animals' lives that any particular genetic individual would not have existed, but for the influence of a changing climate. For instance, suppose a male polar bear only tracked and mated with a particular female bear because melting ice forced him into an area he would not otherwise have visited. The particular bear cub that resulted would, in part at least, owe his or her existence to the melting ice brought about by a changing climate. The *problem* then would be in saying later on, if that cub was starving due to climate change, that he or she had been "harmed" or "unjustly treated". After all, had climate change not existed, this particular cub would not have existed either; the mother would have had some other cub with a different father, or no cub at all.

While this is an interesting and important problem, and it will pose challenges for thinking about climate justice both for future people and animals, I won't consider it in any detail here. Polar bears are relatively long lived (they can live for up to 20 years in the wild); their solitary nature and dispersed populations mean that who mates with who is much less chancy than among some other wild animal populations; and the changes I'm considering are abrupt. Purves and Hale (2016), in contrast, maintain that a polar bear doesn't live long enough "to perceive the effects of climatic change that are the consequence of actions performed during its lifetime". In terms of this paper, though, (a) most individual bears currently in existence and coming into existence would exist or have existed as particular genetic individuals without climate change, i.e. climate change has not yet been identity-affecting; and (b) the relevant changes in Arctic ice are likely to happen very swiftly, so that the individual bears will be made *abruptly* worse off by climate change—the likelihood is one of sudden emergency, rather than gradually unfavorable changes (though those are happening too). So, for now, I'll set the non-identity problem to one side.[3]

Finally: one likely objection to the premise of this paper is that responsive strategies such as supplemental feeding fail to target what's really important: that we reduce greenhouse gas emissions, thereby reducing the amount that the planet will warm, and so (among other things!) lessening impacts on wild animals. Of course, this *is* the ideal strategy. However, while a slowing of emissions increases may be within reach, significant overall global drops in emissions are not likely any time soon. And, even if such drops occurred, there would still be a time lag before any benefit was felt in terms of the preservation of sea ice. As Gardiner (2013) notes, the

---

[3] It's also the case that much of what's said here, relating to reducing animal suffering from climate change, would work cast in impersonal terms, not just individual-affecting terms.

effects of climate change are backloaded. It is possible that some form of geoengineering might change the situation, but this is not near at hand either. We are locked into a warming climate for the near future. And that's enough to make it very likely that the polar bear suffering I'll be considering here will be triggered. Decisions about assisting individual polar bears are likely to face us within the next decade or so, whatever happens to global greenhouse gas emissions in the near term.

21.3   The Situation of Polar Bears

There are around 26,000 polar bears in total across the Arctic, with sub-populations (scientists have designated 19 of these) across Norway, Alaska, Greenland, Russia and Canada. The bears live and hunt mostly on sea ice. It's because of this dependence on Arctic sea ice that they are threatened by climate change across their circumpolar range, but especially in the Canadian Arctic Archipelago (IUCN Red List 2015). One of the most striking effects of climate change so far has been the unprecedented melting of sea ice, and this has already had effects on some (although not all) polar bear populations.

Polar bears use sea ice as their base from which to hunt ringed and to a lesser extent bearded seals, their main prey and food source (Stirling et al. 1993). If there's no ice, bears find it very difficult to hunt. In the summer months, when sea ice has always melted out to some degree, some bears come onto land, while others move out onto the pack ice further from the shore; some bears currently swim between shore and pack ice (Pongracz and Derocher 2017). While bears are on land, if there's no sea ice, there's no access to seals. So bears mostly fast, living on their own stored fat, and what they can scavenge in terms of birds' eggs, whale carcasses left from native human communities' hunting, and other food from human communities (potentially leading to bear-human conflicts).

The looming problem for bears, however, is that sea ice is melting out earlier and re-forming later. Those bears that are on the land are already forced to live off body fat for longer periods. As the pack ice is further from the shore, it will become increasingly difficult, dangerous and exhausting to swim to and from it, and bears will effectively have to stay on land and fast until the sea ice forms again. Difficulty in finding sufficient nutrition over the lengthening summer has already resulted in a significant worsening in the body condition of bears in some populations (Stirling and Derocher 2012). But as the Arctic warms disproportionately to the rest of the Earth, in a year that is likely approaching, sea ice is likely to melt out too soon and re-form too late for some bears to survive the fasting period involved. Polar bear scientists Derocher et al. (2013), in a controversial paper I'll draw on in several places, conclude: "Malnutrition at previously unobserved scales may result in catastrophic population declines and numerous management challenges" (Derocher et al. 2013, 370). So, what—if anything—can and should be done to meet such "management challenges"?

21.4 Possible Responses to Abrupt Polar Bear Starvation

Derocher et al. (2013, 368) propose several possible "proactive conservation and management options" for polar bears in the event of "sudden negative population-level effects". It's worth noting that these scientists' concern is not primarily for the welfare of *individual* polar bears (though occasionally this worry is suggested in their paper), but for the persistence of polar bear populations and the avoidance of human/bear conflicts that may threaten either human safety or species conservation goals. However, while the scientists' emphasis may be on species conservation, in this case, the strategies to protect species also appear to be the only plausible ones available to assist individual bears as well. Since I have no special expertise in polar bear management, I'll confine this discussion to the suggestions outlined in the Derocher et al.'s (2013) paper:

(a) **Cessation of existing bear harvests** to help promote population persistence.
(b) **Diversionary feeding**: feeding bears to move them away from human communities. Risks to bears include disease and parasites from foodstuffs, and these risks are particularly problematic if wild food from other ecosystems is used.
(c) **Supplemental feeding**: providing bears with "sufficient short-term energy to help individuals survive periods of food deprivation" (Derocher et al. 2013, 372). This also poses risks of disease, and may just postpone bear extirpation, as well as being costly and logistically difficult. But the authors consider there will be cases where this is one of the most plausible strategies available: "we believe supplemental feeding will be a conservation option for some populations" (Derocher et al. 2013, 372).
(d) **Translocating animals** either into temporary or long-term captivity in local facilities or distant zoos, or into alternative wild habitats. Any kind of captivity would probably need to become permanent, the authors note, given the nature of the threat, while translocation into new habitat, according to the authors, is very risky to the bears, not least because bears show "strong geographical fidelity". The authors "do not advocate this as a viable conservation alternative" (Derocher et al. 2013, 373).
(e) **Intentionally euthanizing starving bears**: This "may be the most humane option for individual bears that are in very poor condition and unlikely to survive" – and, they argue, guidelines need to be developed to clearly identify such bears (Derocher et al. 2013, 373).
(f) **Do nothing**: The authors reject this as a *conservation* option, but it's clearly what they fear as the most likely scenario, unless the alternative options are widely discussed in advance of an abrupt starvation event occurring.

Looking at these five options: (a), reducing or ending bear harvest is not a way of responding to individual starving bears, so is not directly relevant here. Derocher et al. (2013) are largely negative about (d), captivity/relocation, as being risky, short term, difficult to scale up and unlikely to be successful, so I will not discuss relocation options here (although this does not mean that translocation/captivity could not be an effective strategy in *other* species where food access has been negatively impacted). This leaves (b), (c), (e) and (f). I'll focus on these here. There's no obvious reason not to merge (b) and (c) and consider supplemental feeding that's *also* diversionary; I will consider option (e) euthanizing bears (probably by shooting them) as a potentially "humane" alternative; and assume that (f) is the default situation. So, in a situation of abrupt starvation, should we offer supplemental, diversionary food to the bears?

21.5 Ethical Reasons for Supplemental Feeding of Starving Bears

From a number of different ethical positions, we have reason to feed bears starving due to climate change. Some of these positions are rights- or justice-based; others are based on beneficence; and yet others are based on environmental values.

Justice-based arguments, of the kind I have previously defended (Palmer 2010, 145) take something like this form: Polar bears are sentient animals, and their lives and wellbeing matter morally. Climate change—for which humans are morally responsible, following Nolt's argument above—threatens polar bears with severe suffering and death. What's more, polar bears have gained no benefits from the production of the fossil fuels that have led to climate change (indeed some individuals may have been additionally harmed by oil and gas extraction or transportation). So, the severe impacts of climate change on bears look like a distributive injustice, in terms of the infliction of morally significant harms on sentient beings that have nothing to gain from the process that produced the harms.

One particular version of this argument is based on animals' *rights*. So, it's frequently argued that climate change violates *human* rights. Caney (2010), for instance, maintains that climate change violates basic (negative) human rights not to be deprived of life, of food and water, or of health. If sentient animals have similar basic negative rights, then climate change can violate animals' rights too (an argument specifically made by Pepper [2018]). The polar bear case, after all, is exactly a case about deprivation of access to something very basic—food: a deprivation that leads to a decline in bear health and ultimately to death; exactly the concerns of negative rights theories.

One need not defend a rights argument, however, to see this as an injustice or, at least, a wrongful harm, as I have argued elsewhere (Palmer 2010). Either way, this injustice or rights violation cannot be halted any time soon, given the nature of climate change. So, the next best option is to try to mitigate the effects of the injustice or rights violation by offering rectificatory assistance to the animals worst affected. Although rectificatory justice has not, as yet, been widely discussed by animal ethicists, especially in the context of climate change, such discussions as there are have indeed argued that negative impacts of climate change may warrant rectificatory responses. I have argued, for instance, (Palmer 2010) that where sentient animals suffer from climate change, special obligations of assistance are created.

Milburn (2016) maintains that on Nozick's entitlement theory of justice, since climate change arises as a result of human appropriation, it violates animals' rights, creating a state of injustice that can only be remedied if every individual animal negatively affected by climate change were compensated in ways that leave them no worse off. Pepper (2018) argues that, in the context of climate change, moral agents have a "general duty to facilitate adaptation" where climate change threatens wild animals' basic rights. On all these arguments, we have reasons or duties to assist polar bears in this situation.

In a slightly different justice argument, Donaldson and Kymlicka (2011, 2017) propose that sentient wild animals should be recognized as members of self-organizing, sovereign wild animal communities. The purpose of designating these communities "sovereign" is to protect them from human incursion—for instance, by human colonization, displacement, and "spillover harms" such as environmental pollution. (Donaldson and Kymlicka 2011, 157) Climate change is, on this view, also an injustice; it's a human incursion into sovereign communities that undermines "the ecological fabric they [wild animals] depend on" (Kymlicka 2014). Applying this to the bears' case would suggest that in causing bears to starve, anthropogenic climate change both violates the rights of individual bears, and the boundaries of the sovereign wild community.

For Donaldson and Kymlicka, though, this does not *necessarily* mean that if we can't realistically stop an unjust incursion (as in this case) we should assist the wild animals concerned as a matter of rectificatory justice. These wild communities are, after all, supposed to be sovereign. Attempts to assist should not undermine the sovereignty of the wild community. So, for Donaldson and Kymlicka, in the bears' case, while we clearly have reason to assist, whether we should actually do so depends on the kind of intervention, and how sustained and continuous it would be.

A second group of views in animal ethics focus on duties of beneficence rather than justice. They maintain that what matters is just relieving and reducing wild animal suffering, whether the origin of the suffering is human or not. As Horta (2015) puts it, the moral relevance of animal suffering "gives us reason to conclude that we should intervene in those cases where it is feasible, in order to reduce the disvalue suffered by nonhuman animals", although only in "cases in which we can reduce it [suffering] as a whole, not in some isolated way that reduces disvalue for some in ways that trigger processes that result in more suffering elsewhere". This seems to have a fairly simple application to the bears case: Since bears are clearly suffering, and since feeding them would, at first glance anyway, seem likely to alleviate their suffering, we clearly have reason to feed them. The question, though, is whether doing so would "trigger processes that result in more suffering elsewhere." I'll return to this question below.

Lastly, there are ethical views that emphasize the value of wildness or naturalness. One particularly sophisticated version of this view has been proposed by Hettinger (2018), who develops a principle of 'Respect for an Independent Nature' (RIN). However, even if the primary principle here is RIN, animal suffering still matters. Indeed, Hettinger says: "wild-animal suffering too is intrinsically bad. And its badness does give us a moral reason to consider alleviating or preventing it." So, this also sounds like a beneficence-based reason for alleviating or preventing wild animal suffering. Certainly, relieving wild animal suffering could be in tension with the principle of respect for an independent Nature; in the case of the polar bears, we might expect supplemental feeding to be ethically problematic on the grounds that it compromises naturalness. Hettinger's position, however, is more nuanced. While "preserving naturalness value typically outweighs the importance of alleviating animal suffering," he notes that some interventions can relieve suffering while causing little loss in naturalness value, and also that "some intentional human influence on nature can lessen human impact overall, as when we remove the first few members of a human-introduced invasive species before it has time to spread" (2018, 69). So, on this view, we do at least have reason, based both on beneficent responses to the intrinsic badness of animal suffering, and the potential loss of wild populations in natural places, to offer supplemental feeding to starving bears. The ethical concern here is the degree to which doing so would cause loss of naturalness value overall.

From almost all positions in both animal and environmental ethics, then, there's at least *reason* to assist the bears by offering supplemental feeding, whether on the basis that their starvation would be an injustice (as I have argued elsewhere) or a rights violation, or because we generally have duties of beneficence to relieve suffering where we find it. Almost anyone who thinks that sentient animals are morally considerable, and that their suffering matters, will agree. However, this isn't to say that *all things considered* feeding polar bears is the right thing to do. Assistance may cause further suffering to bears or to other animals. It may undermine the autonomy of wild animal sovereign communities. Or it may reduce overall naturalness or wildness, and thereby fail to respect independent nature. So, what do we need to know to reach an all things considered view?

21.6 Ethical Reservations About Feeding Bears

First, we should consider what supplemental feeding would require in a sudden starvation event where, for at least one bear population, sea ice forms so late that bears won't make it through the summer. It's likely that, initially, this event will be short, it won't happen every year, and only some populations of bears will need feeding for a week or two. In this case, food could be air-dropped to the bears on a one-off basis. However, given continuing sea ice loss, after a few years, supplemental feeding is likely to involve more populations, every year, for longer periods, possibly amounting eventually to many or all bears in some populations requiring feeding for several months each year.

Clearly, feeding polar bears in this way would reduce or remove their suffering from hunger. And we've already seen that there are ethical reasons to prevent serious bear suffering from hunger due to sea ice loss. However, significant ethical reservations about assistance also exist. Here I'll look at two major ethical worries: Would feeding bears cause more suffering, or additional harm/injustices to bears or to other sentient animals? And: Does loss of the sovereignty of wild animal communities, or the loss of wildness/naturalness, involved in creating what Derocher et al. (2013) call "semi-managed bear parks" outweigh ethical reasons for assistance?

21.6.1   Would Feeding Bears Harm the Bears Themselves?

Studies of human intentional supplemental feeding across a range of wild species show mixed benefits to wild animals' welfare; the balance of benefits and risks depends on the nature and purpose of the supplemental feeding, and how consistent and well controlled it is. Dubois and Fraser (2013), in a wide-ranging review of publications on supplemental feeding, argue that supplemental feeding can be hazardous to the animals being fed, as well as leading to potential harms to other animals and to people. They argue that some managed conservation projects such as, for instance, a project that restored the Mauritius Kestrel, use supplemental feeding to good effect, in terms of both conservation and animal welfare. But many other instances of supplemental feeding, in particular where opportunistic and uncontrolled, and especially where associated with tourism, can have wide-ranging negative effects on the welfare of members of the assisted populations. Dubois and Fraser ultimately recommend that supplementary feeding should only be undertaken in cases where the feeding could be carefully controlled, is intended to benefit populations, and may improve animal welfare (Dubois and Fraser 2013, 984).

This description, though, matches pretty closely what we would expect supplemental feeding of polar bears to look like. It would need to be controlled; it would support the continuation of bear populations that may otherwise be extirpated; and the primary goal, as discussed here at least, would be to improve the welfare of individual bears. But still, there are risks to the bears from being fed. One obvious risk is of conflict with human beings, if feeding brought bears into humans' orbit, increasing threats to people. However, I've already suggested that feeding would need to be diversionary as well as supplemental—i.e., it would need to be located away from human communities to avoid generating human/bear conflict (DeRocher et al. [2013] suggest it should be placed by helicopter). In some Arctic communities, diversionary feeding of bears is already occasionally used,[4] and ways of preventing conflict between bears, and between bears and other species, have been devised; so this risk may be manageable. Another risk is that disease could spread between bears at feeding stations; and while this risk can be minimized, it can't altogether be eliminated (a feline leukemia outbreak in endangered Iberian lynx was connected back to feeding stations: Palomares et al. 2011). Another worry may be that while feeding bears has the effect, in the short term, of reducing their suffering and keeping them alive, in the long term it may change bears' behavior in ways that reduce their welfare by limiting agency or increasing suffering, if the fed bears were no longer able, or willing, to hunt for themselves. I'll discuss this worry in more detail later. But it's

---

[4]See   https://polarbearsinternational.org/research/research-qa/why-are-managers-not-feeding-polar-bears-in-the-wild/. Accessed 23 March 2020.

still worth noting that unlike most of the cases of supplementary feeding considered by Dubois and Fraser, without supplemental feeding, *these* bears will starve and die. The bears' suffering is acute, and their need is urgent, even though with feeding their futures are also uncertain.

21.6.2 Would Feeding Bears Harm Other Sentient Animals?

Polar bears are predators, "the most carnivorous of all bears" (IUCN 2015). They are adapted to eat extremely high levels of fat—much more so than most other carnivores—and they eat very little plant matter. Their primary food source is ringed and bearded seals; they will scavenge on whale carcasses, and sometimes eat fish, walruses or birds' eggs (in fact in Norway a new problem created by hungry bears unable to catch seals is that they are eating goose eggs, and so depleting goose populations by up to 90% [Hoffman 2017]).

This raises two kinds of concern about harm to members of sentient species *other* than polar bears. One direct concern is: Given their highly carnivorous natures, what should they be fed on? And another, more indirect, concern is: If we save the lives of bears, and they go back to preying on seals, aren't we allowing (or even responsible for causing) harm to seals?

Both these concerns are difficult to address, given that the motivation for the feeding here is to help sentient animals, either because reducing suffering is in itself a goal, or because humans are morally responsible for causing polar bear suffering. But if the worry is just about suffering, it isn't species specific; the suffering of members of other species also must be important, if we are to be consistent.

Let's begin, then, with the food itself. It needs to be nutritionally adequate, especially if used for longer periods. It must include very high quantities of fat; no nutritionally adequate alternative to animal products currently exists, or is likely to exist on the timescale required, i.e. in the next ten years. So, realistically, feeding bears now means feeding them with other animals (as zoo polar bears are currently fed). The two most plausible alternatives are feeding them with seals that are killed and air-dropped, or using polar bear chow developed for zoo polar bears (although chow is currently only recommended as 50% of a bear's diet in a zoo—it is normally supplemented by raw meat and fish [Lintzenich et al. 2006]).

So: first, should bears be fed with seals? This would require the killing of ringed seals, thus harming them or violating their rights, depending on the philosophical perspective taken; although it probably could be done without causing much suffering to the seals actually killed (though there may be stressful effects on the seals that remain). On some moral views, including most rights views, killing wild seals to feed bears would be absolutely unacceptable in principle. In addition, doing so could be interpreted as attempting to compensate for an ongoing injustice to one group of animals by harming a second group of animals to benefit that first group. However, Abbate (2016) tentatively proposes, in the context of a rights view (she draws on Regan [1984], but other rights views might work here too), that where animals are victims of injustice—as the polar bears are in this case—it may be justified to harm other animals in order to fulfil our duty to the victims of injustice. But even if this argument works in some cases, it seems problematic in the case of feeding polar bears ringed seals.

Recent research suggests that populations of ringed seals are also threatened by climate change. They give birth in lairs protected by snow roofs, so that bears can't easily see them. However, as snow and ice melt out earlier due to climate change, the snow roofs on ringed seal lairs are collapsing, exposing pups to bear attacks. Although bears kill these exposed seal pups, the pups are still too small to have nutritional value to the bears, and sometimes bears don't even eat them (Stirling 2017). Climate change, then, is making ringed seal pups more vulnerable to attack, and preventing many female seals from successfully raising pups. So, it seems reasonable to say that ringed seals are also being harmed by anthropogenic climate change, or that they too, like bears, are in a state of injustice. And it certainly seems problematic to try to compensate one set of victims of injustice by killing another set of victims of injustice (Derocher et al. [2013] may in fact be gesturing at this by commenting— somewhat elliptically—that "There are also ethical questions around killing one marine mammal species to supplement the diet of another"). Abbate (2016) argues that, from Regan's rights view at least, if it's absolutely necessary to feed an obligate carnivore who is a victim of injustice with meat from another animal, it's better to kill wild animals than to use the products of animal agriculture. Although wild animals are harmed by being killed, she claims, they can still be treated with respect; their entire lives are not instrumentalized, as the lives of agricultural animals are. However, she also maintains that animals who are victims of injustice—like the ringed seals in this case—should not be killed to feed other victims of injustice. So, there seem to be serious ethical obstacles to using ringed seals to feed bears. While harp seals would be an alternative, and would meet Abbate's criteria, since they don't yet seem to be negatively affected by climate change, they would need to be brought in from different ecosystems. Derocher et al. (2013) rule out this possibility on the grounds that there's a high risk of importing disease and parasites to the bears by feeding them harp seals from other ecosystems.

The alternative option here is polar bear chow. Current commercial chow, formulated especially for polar bears, includes fish meal and a high proportion of extruded pork fat and bone meal. The fish meal is made from menhaden, wild fish high in fat, normally used for fish oil and fishing bait. On Abbate's terms, as wild fish, the menhaden don't have instrumentalized lives, and they aren't obviously already victims of injustice like ringed seals. So, using menhaden would at least be somewhat ethically preferable to using seals (perhaps, also, on some views, menhaden may be regarded as having fewer complex psychological interests than polar bears, and so being of less moral significance; but then, on the other hand, more of them would need to be used). The pork fat and bone meal in the chow is almost certainly derived from parts of pigs that humans don't eat, rendered into animal by-product, and would otherwise be used for fertilizer or feeding other animals in pet food. So, pigs would not be directly killed to feed bears, although of course the production of bone meal and meat byproduct is part of the meat industry. So, while the production of bear chow may be somewhat ethically preferable to killing seals, obviously it still involves harm to other animals.

Let's move on to the second worry: Suppose we feed the bears for a period of time, achieve the goal of keeping them alive, and then they go back out hunting again— and killing ringed seals. Should we be worried about the predation of bears that have been kept alive by us? The answer here depends on the ethical approach taken. For those theorists primarily concerned with wild animal suffering in general, the bears' continued predation hooks into a more general deep unease with predation (see for instance, McMahan 2014). Cowen (2003), one such theorist, even maintains that "we should count negative impacts on carnivores as positive features of ... human policy" and rejects subsidizing "the propagation of carnivorous wild animals". From positions like Cowen's, then, helping predators is in principle unacceptable, since predators will inevitably go on and cause suffering. The best option here, then, of the options above is (e). We should euthanize the bears and end their suffering, and at the same time, prevent them from causing suffering to seals by future predation.

However, we should not be too hasty here, since we can't be sure that euthanizing polar bears would actually reduce seal suffering; the effects of removing members of one predatory species are unpredictable. Seals may suffer from other causes, or become too abundant for available resources; and in any case, seals are *themselves* predators causing suffering. Trying to estimate the broad-range effects of euthanizing polar bears rather than feeding them to reduce overall suffering thus runs into huge epistemological difficulties, as Delon and Purves (2018) point out. So, it may be that in this case, from the suffering-reduction view, it's just best to look at the immediate suffering with which one is faced—the starving bears—and deal with that particular problem, rather than trying to work out what the down-the-line impacts on wild animal suffering overall would be.

For justice theorists, a somewhat different conundrum is presented. Rectificatory justice to polar bears may, after all, look like an injustice, or a responsibility for harm, to the ringed seals that later become the surviving bears' prey. After all, if humans had not helped these bears, these seals would not *be* bear prey. This raises difficult questions for justice-based approaches (questions that are also, of course, found in human justice cases, for instance where reparations to one group that has suffered an injustice leaves others, who have not themselves caused harms, much worse off). How far one should be concerned about this is contested in the human case—for instance, there are significant disputes about how far indirect effects of rectificatory interventions should be counted as part of the intervention itself. In addition, in this case, it might be argued that the bears' continued predation, in effect, just restores the status quo for the seals. After all, supplemental feeding means that the same number of bears are out on the ice preying on seals as would have been there had the anthropogenic climate threat of sudden starvation not arisen. It's not as though we have been breeding *additional* polar bears with a view to having them attack ringed seals for some purposes of our own! Nonetheless, from some justice perspectives, feeding bears is unjust to the particular seals that are preyed upon by human-saved bears, and this may be a decisive ethical factor in decision-making.

So, some of those deeply concerned about justice or suffering reduction for sentient animals, may consider that there are very strong reasons not to offer supplemental feeding. The cost to other sentient animals of feeding bears is too high, either in terms of increasing overall animal suffering (despite reducing it for polar bears) or causing new injustices to other animals, even while rectifying injustice to polar bears. The consequence is, though, that some polar bears, in a reasonably near year, will very likely starve to death. All those who take this kind of ethical view would recommend shooting bears (or euthanizing them in some other way, if that's more painless) as preferable to letting them starve. Shooting them ends their suffering, and would then be the best we can do for them in terms of rectificatory justice.

From some ethical positions, this conclusion seems definitive. However, I want to continue to explore the possibilities here a bit further. There do *also* seem to be reasonable ethical positions focused on individual sentient animals that could still defend feeding bears as the best of a set of bad options. For instance, someone taking a rectificatory justice view could argue that feeding bears polar bear chow *can* be justified, despite harms to other animals, because we are dealing here with a non-ideal situation where there are only "imperfect solutions to horrific problems" (Emmerman 2014). With respect to what is actually fed to polar bears, intensive animal agriculture and fishing for wild menhaden is also not going away any time soon, and the vast majority of people will not be vegetarian or vegan. Nor will their dogs and cats. In this context, why pick out polar bears—obligate carnivores, unlike people and dogs, at least—whom we have already unjustly deprived of the ability to feed themselves, and insist that they should starve or be shot since they can't at present eat vegan? This potentially adds a further layer of injustice to the ones from which the bears are already suffering by, as it were, picking them out for conditions that aren't generally insisted upon for our companion animals. Add this to the view that feeding bears chow to enable their survival just returns to the status quo in terms of subsequent seal predation, and such a justice position is not firmly decided against supplemental feeding. Similarly, on a beneficence view, if bears were fed on fatty animal by-products, so not causing additional suffering to intensively farmed animals, *and* it was accepted that the outcomes for overall suffering that would result from feeding bears were incalculable, then there would at least be a possible ethical case for supporting feeding bears on polar bear chow for the times in the near future when sea ice has not yet formed.

21.7   The Problem of "Semi-Managed Bear Parks"

So far, I've only considered supplemental feeding of bears with respect to justice, harms and suffering to individual animals. I've suggested that from some ethical positions, concerns about these factors would support euthanasia over supplemental feeding, but that from other perspectives, the door is not entirely closed to supplemental feeding. But these were only some of the values with which I began the paper: I also mentioned naturalness/wildness values, bear agency, and what Donaldson and

Kymlicka (2011) call "sovereign wild animal communities". And with respect to these values, we're not yet out of the woods. One of the issues Derocher et al. (2013) raise is that a supplementary feeding program might result in commitment to a "semi-managed bear park model if habitat conditions continue to decline." Wouldn't the existence of such "bear parks" challenge naturalness values, the agency of bears, and the "sovereignty" of wild animal communities?

21.7.1   The Worry About Naturalness Value

Let's start with wildness or naturalness value, taking Hettinger's (2018) idea of "Respect for Independent Nature" as paradigmatic of this view. To recap, naturalness, Hettinger says, "involves an overall judgment of the degree of independence of an entity from humans, that is, of the extent to which a being is autonomous vis-à-vis humanity." Clearly, anthropogenic climate change has reduced naturalness in this sense. But it has not entirely eliminated it, so that "anything goes". We can still treat animals in ways that are problematically more "unnatural" than others. So, Hettinger (2018, 78) comments: "For example, that anthropogenic climate change has dramatically increased the rate of interbreeding between Grizzly Bears and Polar Bears does not mean there is no naturalness left to protect in our treatment of them or their ecosystems. This impact would not undermine the unnaturalness of relocating Polar Bears from the Arctic to Antarctica, even ignoring the negative consequences this would have on penguins and other southern species."

So, how would feeding the bears relate to this idea of naturalness value? First, it's important to note that Hettinger's concern is about what's likely to protect naturalness best *overall*. So, a temporary intervention that would wash out, with the purpose of protecting long-term naturalness, would overall be a good thing. However, feeding the polar bears is unlikely to be temporary—that's why there's a concern about semi-managed bear parks. Some bears may become dependent on people, potentially in the long term, for significant parts of the year. This does look like a major loss of naturalness value, and suggests that if naturalness value is a priority, we should not feed polar bears—independent of any concerns about what they are actually fed on.

However, Hettinger (2018) does mention one case that leaves the door ajar to justifying feeding starving polar bears on the basis of naturalness value. When discussing an (imagined) case of genetically modifying members of an animal species, the American pika, in order to prevent extinction from climate change impacts, Hettinger (2018, 69) notes that if the pika were not modified "it is arguable that it [the extinction] would give us a much greater impact on nature than we would have with our rescue attempt." So, the impact of anthropogenic species loss on naturalness value *could* be greater than the impact of a human intervention to save the species, even a long-lasting intervention like genetic modification.

Paralleling this case, then, we could see feeding polar bears as one way of keeping bears on the landscape. An Arctic landscape that retains polar bears may be regarded as more natural than an Arctic landscape that lacks them, even if human intervention is required to keep them there. (This may depend on whether landscape naturalness is understood more in terms of *composition*, that is what kinds of things are present, than *history*, that is how things got to be or to remain on a landscape.)

However, the bear case is potentially more naturalness-undermining than the pika one. First, the pika intervention aimed to avoid extinction; but here we are talking about feeding some individual bears for the *bears'* sake, not because they are the last bears (although certainly particular bear populations in particular places could be lost, and this would be a big human impact on *those* landscapes). And second, the idea of modifying the pika was to allow them to go on living their normal pika lives, not to live in semi-managed pika parks. I'm not sure what semi-managed bear parks would look like, but if they involved human structures (beyond feeding stations), human populations, tourists etc. then they are likely to undermine naturalness values more significantly. In this sense, while a short-term intervention for a couple of weeks, preventing bears from starving and helping them back out onto the ice, is likely to *enhance* naturalness value overall, if feeding becomes long term and institutionalized, and bears can no longer hunt from sea ice much at all, feeding bears is likely to *diminish* overall naturalness value.

21.7.2   The Worry About Bear Agency and Longer Term Vulnerability

Earlier in the paper, I defined agency in the context of sentient animals as about the subjective experiences involved in being able to direct their own bodies and activities, to pursue goals and make choices about what they do, and to have motivational states (such as wanting things). One worry about "semi-managed bear parks" is that feeding bears over time may constrain their agency, creating negative experiences, and in doing so also making them more vulnerable to other threats later on.

First, it's worth noting: Climate change itself creates a major restriction on bears' agency in this sense. Bears can no longer use sea ice to hunt seals for significant portions of the year; and sea ice is the main arena of bears' activity—where they make choices about when and where to hunt and so on. Bears' inability to pursue these characteristic behaviors in their own individual ways presumably causes them stress and distress.

Unlike climate change, supplemental feeding does not *intrinsically* restrict bears' agency. Indeed, in one sense it offers them an opportunity to satisfy their hunger that they won't refuse, so long as the food is palatable; they will choose to eat it. But there is a deeper worry about agency here: that supplementary food will eventually make bears dependent on human beings for provision, and will harm them by depriving them of the ability to make *future* choices. This is unlikely to happen initially, if bears are just fed for a week or two, and then, when ice forms, feeding is withdrawn, forcing the bears back to the shore and to hunting seals. However, if, as warming continues, feeding is extended to a month or several months, dependence is more likely to follow, possibly becoming year-round (although at the moment we just don't know what effects more extended feeding of wild bears would have on their behavior).

As Donaldson and Kymlicka (2011) rightly argue, being dependent is not intrinsically a bad state. However, in the context of wild polar bears, dependency raises questions both in terms of agency and further future vulnerability. In particular, it makes bears more vulnerable to human policy changes—for instance, if those sponsoring the feeding decide that feeding bears costs too much, or that preserving naturalness value in Arctic landscapes is more important than feeding bears in semi-managed bear parks. And so, feeding now might in principle lead to less expression of bear agency and more bear suffering in future.

I say "in principle" though, because we mustn't forget the context here. If there's no supplementary feeding, these particular hungry bears have no choices at all, either now, or in the future, because they'll starve. As Horta (2015, 118) notes, "remaining alive is a condition to enjoy autonomy". Seen from that perspective, feeding may be better in that, at least, bears retain their lives, and some ability to express agency (after all, not every choice that a bear makes is about what it's going to eat)! So, even though climate change and supplemental feeding likely do mean reduced bear agency, as their lives are more constrained than they would otherwise be, from the perspective of these particular bears, given the intensifying pervasiveness of climate change, we can reasonably imagine that supplemental feeding is what they would choose, were they able to understand the nature of the choices involved.

21.7.3   The Worry About Sovereign Communities

Alongside concerns about individual animals' agency, Donaldson and Kymlicka's political account of our relation to wild animals also includes a concern about wild animal *communities*. Wild animals"should be recognized as having the right to live autonomously on their own territories, and hence as exercising their own sovereignty" (Donaldson and Kymlicka 2017). Human interventions to relieve suffering are permissible if they are small scale or temporary, thus maintaining the ongoing sovereignty of the community. However, this is unlikely to remain the situation here. Feeding the bears might become permanent and ongoing—exactly the kind of intervention Donaldson and Kymlicka reject: "a kind of permanent paternalistic management in which we take over responsibility for feeding and sheltering them" where "we are basically turning wilderness into a zoo" (Kymlicka 2014).

However, Donaldson and Kymlicka's objections to sustained interventions focus on fundamentally "competent" wild animal communities, where members of the community can normally meet their own needs. But climate change in the Arctic is changing all this. Its pervasive effect on particular places is essentially rendering the existing wild animal community in those places "incompetent". Ideally, of course, on Donaldson and Kymlicka's theory, recognition of this injustice to wild animal sovereign communities should force humans sharply to reduce emissions, and prevent or at least reduce the unjust incursion at its origin. But in the non-ideal political circumstances that currently prevail, this isn't going to happen, and even if it did, as noted, there's enough warming already built into the earth's systems for some bears to starve in the next few years. So the question of intervention doesn't go away.

The key question here seems to be: Is there any point, when a sovereign wild animal community is so anthropogenically compromised that members of the community are starving, in continuing to protect the community's "sovereignty" by rejecting sustained intervention? If the primary purpose of sovereignty rights is, ultimately, to best protect wild animals, then *not* intervening here on the grounds of sovereignty would seem to undermine exactly what that sovereignty was established to protect. A 'semi-managed bear park' would clearly be far from ideal, and it's obviously far removed from the kind of independent sovereignty Donaldson and Kymlicka imagine would be best for wild animals. But it can be seen as reducing bear suffering, making a move towards reparation for injustice, and keeping bears alive in the face of a very specific, severe anthropogenic threat. It might be the best among bad options, even on their terms.

21.8   A Tentative Proposal: A Trial of Feeding Bears Without Injustice to People

In a nearby year, if we do nothing, some polar bears in some wild populations will starve to death because of anthropogenic climate change. We could just let them starve, but from most of the ethical perspectives discussed here, there is at least ethical reason to consider intervention. One intervention that can clearly be ethically justified is to relieve the suffering of bears that are going to starve by shooting them. The alternative is to feed them. As we've seen, while there are ethical reasons in favor of feeding them, there are also significant ethical concerns that count against doing so (which may be viewed as more or less important from different ethical positions):

- Risking harms to humans if bear feeding is not sufficiently diversionary
- Risking harms to bears through diseases/conflicts at feeding stations
- Keeping bears alive, meaning that they cause suffering to, and kill, seals
- Making bears' food from sentient animals we have killed, whether these are wild or part of industrial production
- Satisfying bears' immediate choices (to eat!), but risking restricting future options available to them, so also reducing bears' agency and increasing future vulnerability
- Loss of naturalness value through the establishment of semi-managed bear parks
- Further significant erosion of the sovereignty of Arctic wild animal communities

Two of these problems are inevitable, if bears are fed. First, bears will be fed on other animals. And second, if feeding is successful, bears will go back out and hunt seals. On some of the ethical views outlined above, that just rules feeding bears out. We should instead kill them as painlessly as possible.

But as I've also argued, there are views on which those two ethical concerns are not complete blockers. From these other perspectives, we need to know more about what would happen if bears are fed. Will semi-permanent bear parks be created in the long-term? Will there be disease at feeding stations? Will bears become habituated and consequently more vulnerable? There are still many things we don't know about this situation. Perhaps feeding won't have a significant effect on bears' long-term behavior; perhaps we won't need to create semi-managed bear parks. One way forward is to consider a trial of supplementary and diversionary feeding in the first polar bear population that becomes seriously at risk to see how bears respond, how quickly they return to the sea ice, and whether they come back to feeding stations once ice has formed, or go back out to the seal hunt (and also whether other hazards manifest themselves, such as disease outbreaks or conflicts at feeding stations that cannot be managed). At least trying this out with one bear population could assist in finding out more about the implications of feeding bears, and help in making better informed choices between feeding and shooting them, as sea ice shrinkage continues across the Arctic.

However, committing to do this raises an additional concern: justice with respect to *native communities* in the Arctic that are located in areas potentially affected. These communities have deep cultural and subsistence relations with polar bears, and also carry the burden of danger brought by polar bears coming into their communities. To avoid procedural injustice, any decisions about feeding bears should involve meaningful consultation with local Arctic native peoples. The importance of such consultation is already enshrined in the 2013 Declaration of the Responsible Ministers of the Polar Bear Range States,[5] which notes that polar bear conservation should: "Engage Arctic local people in management decision-making processes and promote the collection and maintenance of Traditional Ecological Knowledge (TEK) by acknowledging the important role polar bears play in the cultural heritage and subsistence of Arctic indigenous people, as well as the role that they play in the long-term conservation and survival of the polar bear." However, as Young (2016) notes, meaningful consultation and engagement with TEK is likely to be very difficult, given the ways in which native perspectives on polar bears have diverged from "techno-managerial solutions", especially in terms of bear hunting quotas. In principle, though, protecting threatened polar bear individuals and local populations by feeding, providing the feeding is diversionary, and follows from a consultative process, may be in the common interest of polar bear scientists, conservationists, and local native communities—as well as the bears themselves.

The hungry bears case is in some ways unique. It's urgent, dramatic, it's likely inescapable, and it's going to afflict individuals that are members of one of the most favored, charismatic species on earth. But it's also just one example of a case where

---

[5] Available at: https://polarbearagreement.org/resources/agreement/declaration. Accessed 23 March 2020.

climate change will require us to make difficult choices between unpalatable alternatives. Whatever we do, some sentient animals are going to suffer, and interventions, over time, will compromise wildness values. For this reason, it's particularly important that we think about these issues in advance. As Derocher et al. (2013, 370) plausibly argue: "it is critical to contemplate and discuss options ahead of the need to respond…Although some of the topics may seem radical… future conditions may be well outside the range of past circumstances and necessitate very different actions than today. The success of interventions will be partly determined by the degree of advanced planning."

References

Abbate, C. 2016. How to help when it hurts: The problem of assisting victims of injustice. *Journal of Social Philosophy* 47 (2): 142–170.

Both, C., et al. 2006. Climate change and population declines in a long-distance migratory bird. *Nature* 441: 81–83.

Caney, S. 2010. Climate change, human rights and moral thresholds. In *Human rights and climate change*, ed. S. Humphreys, 69–90. Cambridge: Cambridge University Press.

Cowen, T. 2003. Policing nature. *Environmental Ethics* 25 (2): 169–182.

Delon, N., and D. Purves. 2018. Wild animal suffering is intractable. *Journal of Agricultural and Environmental Ethics* 31 (2): 239–260.

Derocher, et al. 2013. Rapid ecosystem change and polar bear conservation. *Conservation Letters* 6 (5): 368–375.

Donaldson, S., and W. Kymlicka. 2011. *Zoopolis: A political theory of animal rights*. New York: Oxford University Press.

Donaldson, S., and W. Kymlicka. 2017. Animals in political theory. In *The Oxford handbook of animal studies*, ed. L. Kalof. New York: Oxford University Press. https://doi.org/10.1093/oxf ordhb/9780199927142.013.33.

Dubois, S., and D. Fraser. 2013. A framework to evaluate wildlife feeding in research, wildlife management, tourism and recreation. *Animals* 3: 978–994.

Emmerman, K. 2014. Sanctuary, not remedy: The problem of captivity and the need for moral repair. In *The ethics of captivity*, ed. L. Gruen, 213–230. New York: Oxford University Press.

Forbes, B.C., et al. 2016. Sea-ice, rain-on-snow and tundra reindeer nomadism. *Biology Letters* 12: 20160466. http://dx.doi.org/10.1098/rsbl.2016.0466. Accessed 23 March 2020.

Gardiner, S. 2013. *A perfect moral storm: The ethical tragedy of climate change*. Oxford: Oxford University Press.

Hettinger, N. 2018. Naturalness, wild Animal suffering and Palmer on laissez-faire". *Les ateliers de l'ethique* 13 (1): 65–84. https://doi.org/10.7202/1055118ar.

Hoffman, T. 2017. Polar bears shift from seals to goose eggs as arctic ice melts. *New Scientist, Daily News*, May 12. Available at: https://www.newscientist.com/article/2130821-polar-bears-shift-from-seals-to-bird-eggs-as-arctic-ice-melts/.

Horta, O. 2015. The problem of evil in nature: Evolutionary bases of the prevalence of Disvalue. *Relations Beyond Anthropocentrism* 3 (3.1): 17-32.

IUCN. 2015. Ursus maritimus. The IUCN Red List of Threatened Species. Retrieved from: http://dx.doi.org/10.2305/IUCN.UK.2015-4.RLTS.T22823A14871490.en.

Kymlicka, W. 2014. Will Kymlicka on animal denizens and in the wilderness. Interview with Arianno Mannino. Retrieved from: http://gbs-schweiz.org/blog/will-kymlicka-on-animal-den izens-and-foreigners-in-the-wilderness-interview-part-2/.

Lintzenich, B.A., et al. 2006. *Polar bear nutrition guidelines.* Retrieved from: wildpro.twycrosszoo. org/000ADOBES/Bears/D315FinalPolarBearSG2007.pdf.

McMahan, J. 2014. The moral problem of predation. Retrieved from: http://jeffersonmcmahan. com/wp-content/uploads/2012/11/The-Moral-Problem-of-Predation.pdf.

Milburn, J. 2016. The demandingness of Nozick's 'Lockean' Proviso. *European Journal of Political Theory* 15 (3): 276–292.

Nolt, J. 2011. Non-anthropocentric climate ethics. *WIRES Climate Change* 2 (5): 701–711.

Palmer, C. 2009. Harm to species—Species, ethics, and climate change: The case of the polar bear. *Notre Dame Journal of Law, Ethics & Public Policy* 23 (2): 587–683.

Palmer, C. 2010. *Animal ethics in context.* New York: Columbia University Press

Palmer, C. 2019. Assisting wild animals vulnerable to climate change: Why ethical strategies diverge. *Journal of Applied Philosophy* 36 (2). https://doi.org/10.1111/japp.12358.

Palomares, F., J.V. López-Bao, and A. Rodríguez. 2011. Feline leukaemia virus outbreak in the endangered Iberian and the role of feeding stations: A cautionary tale. *Animal Conservation* 14: 242–245.

Pepper, A. 2018. Adapting to Climate Change: What we owe to other animals. *Journal of Applied Philosophy.* https://doi.org/10.1111/japp.12337.

Pongracz, J.D., and A.E. Derocher. 2017. Summer refugia of polar bears (*Ursus maritimus*) in the southern Beaufort Sea. *Polar Biology* 40: 753–763.

Purves, D., and B. Hale. 2016. Non-identity for non-humans. *Ethical Theory and Moral Practice* 19 (5): 1165–1185.

Regan, T. 1984. *The case for animal rights.* Berkeley: University of California Press.

Steward, H. 2009. Animal agency. *Inquiry* 52 (3): 217–231.

Stirling, I., D. Andriashek, and W. Calvert. 1993. Habitat preferences of polar bears in the western Canadian Arctic in late winter and spring. *Polar Record* 29: 13–24.

Stirling, I., and A.E. Derocher. 2012. Effects of climate warming on polar bears: A review of the evidence. *Global Change Biology* 18 (9): 2694–2706.

Stirling, I. 2017. What about the ringed seals as the Arctic climate warms? Retrieved from: https:// polarbearsinternational.org/news/article-climate-change/what-about-the-ringed-seals-as-the-arc tic-climate-warms/.

Young, J. 2016. Polar bear management in a digital Arctic: Inuit perspectives across the web. *The Canadian Geographer* 60 (4): 466–478.

Clare Palmer is the George T. and Gladys H. Abell Professor of Liberal Arts and Professor of Philosophy at Texas A&M University in the USA. She studied both for her BA and her DPhil at Oxford University. She works in animal ethics, environmental ethics, and the ethics of new technologies, with a particular interest in ethical questions raised by wildlife conservation and management, and the use of emerging technologies for conservation goals. She is the author or co-author of four books, including Animal Ethics in Context (New York: Columbia University Press 2010). With four co-authors from several disciplines, she is currently working on a book project, Wildlife Ethics, contracted to Wiley-Blackwell. She is an Associate Editor for the British Ecological Society journal People and Nature and served as a member of a National Academies of Science, Engineering and Medicine committee that produced the report Forest Health and Biotechnology: Possibilities and Considerations in 2019.

Chapter 22
Understanding and Defending the Preference for Native Species

Ned Hettinger

Abstract The preference for native species, along with its concomitant antipathy toward non-natives, has been increasingly criticized as incoherent, obsolete, xenophobic, misanthropic, uncompassionate, and antithetical to conservation. This essay explores these criticisms. It articulates an ecological conception of nativeness that distinguishes non-native species both from human-introduced and from invasive species. It supports, for the most part, the criticisms that non-natives threaten biodiversity, homogenize ecological assemblages, and further humanize the planet. While prejudicial dislike of the foreign is a human failing that feeds the preference for natives, opposition to non-natives can be based on laudatory desires to protect natural dimensions of the biological world and to prevent biological impoverishment. Implications for our treatment of non-native, sentient animals are explored, as well as are questions about how to apply the native/non-native distinction to animals that share human habitats and to species affected by climate change.

Nature doesn't care about conservationists' artificial divide between…native and alien species…The concept of natural has outlived its usefulness in conservation…Aliens are rapidly changing from being part of the problem to part of the solution. (Pearce 2015b)

Their demonization says more about us and our fears of change than about them and their behavior…This hostility is generally justified by outdated and ill-founded ideas about how nature works…We need to lose our dread of the alien and the novel…Conservationists must stop spending all their time backing loser species–the endangered and the reclusive. They must start backing some winners. (Pearce 2015a, xii–xvi)

Biological invasions are fundamentally analogous to natural disasters…the annual combined economic cost of invasions worldwide exceeds that of natural disasters. (Ricciardi et al. 2011, 312)

N. Hettinger (✉)
College of Charleston, Charleston, SC, USA
e-mail: HettingerN@cofc.edu

© The Author(s) 2021
B. Bovenkerk and J. Keulartz (eds.), *Animals in Our Midst: The Challenges of Co-existing with Animals in the Anthropocene*, The International Library of Environmental, Agricultural and Food Ethics 33,
https://doi.org/10.1007/978-3-030-63523-7_22

22.1   Introduction

I once planted a *Mimosa* tree in my yard on a barrier island off the coast of the U.S. state of South Carolina. I'd seen the trees around the island and thought their gorgeous, showy pink flowers would make a good addition to our palms, live oaks, cedars and wax myrtles. When I learned that *Mimosa* was a species from east Asia, I was upset. I wanted our yard to fit in with the native ecosystems of barrier islands off the Southeast U.S. coast. I believed that planting "alien" or "exotic" species was incompatible with the integrity of our island native ecosystem. While it was true that may *Mimosas*, as well as other non-natives species, inhabited our island, I did not want to be a part of what I considered a degradation. It was like planting California redwoods in Northern Europe or transporting Camels into the U.S. southwestern desert: These species did not belong in those places.

Although the idea that non-natives should be kept out of native ecosystems is strongly endorsed by most environmentalists, over the last twenty years it has been subject to increasing criticism. "Nativists" (those with a preference for native species and an antipathy toward non-natives) are accused of xenophobia, not only for having an irrational fear of and hostility toward alien species, but also for being complicit with the recent and growing prejudice against human immigrants. Nativists are also accused of being in the grip of a scientifically uniformed conception of natural systems, mistakenly believing that they are unchanging, tightly knit communities whose species composition remains fixed. It is argued that in a time when human alteration and domination of the natural world is extensive and ever increasing ("the Anthropocene"), with consequent loss of endemic species and collapsing ecosystems, we need to embrace non-natives and the novel ecosystems they create. Such ecological novelty is a positive contribution to the only nature that we can hope of sustaining on earth. Rather than treating non-native species as a plague to be eradicated (as when the U.S. National Park Service shoots Mountain Goats that have strayed into the parks), non-native species should be seen as a part of nature and part of the biodiversity we should be preserving.

Some scientists have even suggested that the discipline that studies non-native species (invasion biology) constitutes a "pseudoscience" (Theodoropoulos 2003) and should be dissolved. Critics within the field have worried about its use of misleading war-like metaphors and claimed that it often misrepresents the behavior of colonizing species, which are far more benign than they are made out to be (Davis et al. 2011). They argue against "judging species by their origin" and against the presumption of guilt, insisting instead that non-natives should be treated as innocent until proven otherwise. In response, some invasion biologists accuse the supporters of this more open attitude toward non-natives of engaging in "science-denialism," analogous to those who deny the reality of climate change, arguing that the threat non-natives pose is undeniable (Russell and Blackburn 2017).

This essay defends the preference for natives over non-natives. While the critics raise a number of important issues that a defender of native species must accommodate, their criticisms can, by and large, be successfully addressed. A precising

definition of nativeness will be offered along with a defense of respect for independence nature, a nature threatened by the vast spread of human-introduced non-natives.
Further, while non-natives add to biodiversity in some ways, in more important ways,
and overall, they lessen it. And while a welcoming and cosmopolitan approach to the
foreign has its virtues, so too does promoting traditional communities and unique
ways of life. Non-natives and novel ecosystems have a place in this increasingly
humanized world, but, more importantly, so too do natives and traditional ecosystems.

22.2 The Distinction Between Native and Non-Native Species

There is no agreed upon understanding of what makes a species native or not. Some
think the distinction and related concepts are hopelessly muddled, deceptive, and
should be given up (Chew and Hamilton 2011). One analysis argues that, "Without
an explicit criterion, *exotic* and *native* are problematically imprecise concepts and are,
consequently, often used inconsistently by ecologists and conservation biologists"
(Justus 2009, 413). Frequently it is claimed that designating a species as non-native
("exotic" or "alien") is used to manipulate public support for its removal. In an era
of anthropogenic global change, some noted scientists think:

> It is time for scientists, land managers and policy-makers to ditch this preoccupation with
> the native–alien dichotomy and embrace more dynamic and pragmatic approaches to the
> conservation and management of species — approaches better suited to our fast-changing
> planet. (Davis et al. 2011, 153)

In one of the first careful philosophical treatments of the topic, Woods and Moriarty
(2001) distinguish five possible accounts of what it means to be a native species: The
human introduced criterion, the evolutionary origin criterion, the historical range
criterion, the degradation criterion, and the community membership criterion. They
argued that no one of these was correct and instead that we should think of the idea
of native species as a "cluster concept." In their view, none of these features is either
necessary or sufficient for being native, but the more of them a species possesses,
the more likely they should be considered native. Correlatively, they argue, the less
a species instantiates these features, the more likely it is non-native.

The most common idea is that what makes a species non-native is that it was
introduced to an ecosystem by humans. The Netherlands' Minister of Agriculture
has defined native species thus:

> I define an exotic species as a non-native plant, animal or micro-organism that is not able to
> enter the Netherlands by its own efforts, but through human activity (transport, infrastructure)
> and has entered nature in the Netherlands, or threatens to do so in the near future. Species that
> enter the Netherlands by their own efforts, due to climate change for instance, fall outside
> this definition and are not included in the policy. (Verburg 2007)

Such a definition will not do. On the one hand, this account ignores that species can
and do travel to new and radically different habitats on their own, as when ballooning

spiders arrive on remote oceanic islands or vagrant birds get blow from one continent to another. For example, when Cattle Egrets first arrived in South America having been carried by winds from Africa, it was not plausible to think of them as native there. On the other hand, the human-introduced criterion fails to allow for the possibility of human restoration of native species, as was done with the 1990s restoration of wolves to Yellowstone National Park in the U.S. Wolves had been in the Park until they were eradicated in the early 1920s. Seventy years later the Park Service transported wolves from Canada back into Yellowstone. On the human-introduced criterion, it would make no difference whether the park had trucked in wolves from Canada or flown in Siberian snow leopards from Russia, both were brought there by humans and would thus be non-native.

An almost equally popular (and problematic) idea of non-native species is that they are species that "invade" and wreak havoc in local ecosystems. In fact, "non-native species" and "invasive species" are often used interchangeably. Such species come in and take over an ecosystem causing massive amounts of damage, often extirpating local endemic species. An often-cited example are the introduction of cane toads to Australia. In 1935, hoping South American Cane Toads would control the beetles that were plaguing their sugar cane crop, farmers brought in about a hundred. The toads spread quickly, and today, they exist "from coast to coast," sometimes in densities of a thousand an acre. There are now an estimated 1.5 billion of these toxic critters in Australia poisoning native animals who eat them (Crawford 2018; Slezak 2015). Another often cited example is Kudzu (Pueraria lobata) in the U.S. A Japanese vine imported in the early 1900s and planted to reduce soil erosion, it has a growth rate of up to a foot a day, spreads over 200 square miles (518 sq. km) a year, and now covers millions of acres in the southeastern United States, killing trees and shrubs by heavy shading. Or consider that European zebra mussels, arriving into the U.S. Great Lakes in ship ballast water, clog water intake pipes at a cost of hundreds of millions of dollars annually.[1]

While it is undeniable that some non-native species have devastating consequences both for nature and for humans, it is a mistake to equate non-natives with damaging invasives. For the most part, non-natives do not cause problems, either because they quickly die out or, if they do establish a permanent presence, they do not aggressively spread and cause problems. While not scientifically rigorous, a ballpark estimate is the often cited "tens-rule" that claims one in ten introduced non-natives survives in the wild and that only one in ten of these goes on to spread problematically.

Not only are many non-natives ecologically benign, but native species also can irrupt and cause massive damage. While "invasive" suggests entering from outside, it also can mean encroaching or infringing, and aggressively spreading from within can also count as "invasive" in this sense. For example, the native pine beetle in the U.S. "is currently suspected to be killing more trees than any other [insect] in North

---

[1] Why are such non-natives so successful? The science behind this success is not settled, but some of the factors often mentioned are that the parasites, predators, and competitors that keep these species in check in their native habitats are not present in their new home and that their new neighbors (unlike their neighbors back home) have not developed defenses against them.

America" (Davis et al. 2011, 153). One study titled "The natives are restless, but not often and mostly when disturbed" (Simberloff et al. 2012) suggest that while native species do sometimes spread dramatically, they do so less frequently than non-native species and most often the irruption is the result of human-caused changes (such as grazing pressure or fire suppression). So, non-natives cannot be equated with damaging invasive species.

Nor should we can we think of natives as those that originally evolved in a location and non-natives as those that originally evolved someplace else. For species routinely move around and colonize new areas where they make their home for thousands of years and thus are clearly native. Such a criterion would implausibly entail that long-lived colonizers on volcanic islands are non-natives while species that just recently evolved in that local are native. Or consider a species that evolved in one locale and then migrated to other places, while dying out in its original location. The story of Camel migration is instructive. The ancestor of the modern camel species evolved in North America and then spread to South America and Asia. Camels went extinct in North America when humans migrated to the continent thousands of years ago. Depending on how strictly we use the term "species" (species per se or closely related taxon), the evolutionary origin criterion would suggest that returning Camels to North America would be restoration of a native species. The idea that a herd of Camels roaming the American Southwest should be considered native species to those ecosystems is on its face preposterous!

I suggest that we think of native species as species that have considerably inter-acted/adapted to the local biota and abiota. Natives are tied to other residents via some sort of interaction, whether competition (such as predation) or cooperation (such as mutualism). Natives have influenced other natives and adapted to local climate and landscape. Similarly, the locals have interacted with and adapted with them. Perhaps the local abiota have been affected by their presence as well. So native species are those that have *significantly* adapted to or interacted with the local biota and abiota (and vice versa). Non-natives are species that have not significantly adapted or interacted with the local inhabitants or abiota.[2]

This notion of having significantly adapted or interacted with locals is not the same as fitting in functionally (that is, performing certain ecological functions). A species that has never actually adapted to local species might fit in fine, but is not native. For example, Snow Leopards in Yellowstone might serve the same roles as wolves, but they are non-natives as they have not actually adapted with local species. Note too that this conception of native species does not assume some ideal balance of nature or harmony in ecosystems. Natives are those that have interacted and adapted to each other; whether that interaction is harmonious or not, in balance or not, is left open.

This conception of the native/non-native distinction is *ecological* and not *geographical*. For example, the idea of a species being "native to North America" makes no sense on my account. There are too many different ecological assemblages for a species to have interacted and adapted with all of them. Nor is it helpful to think

---

[2] I first proposed this idea in Hettinger (2001).

of species as native to a state or region, unless such geographical designations are uniform in terms of their ecology.[3] On this account of native species, "native to a place" makes no sense, unless place means ecosystem.

Note also that on this conception being native or not is a matter of degree. Obviously, the extent to which a species has significantly adapted/interacted can increase (or decrease) over time. This allows that a species that comes into a foreign habitat can "naturalize" over time as it interacts with the locals and they interact with it.[4] Camels that evolved in North American ecosystems 50 million years ago and then moved to Asia via the Bering land bridge presumably became native in the new desert ecosystems they inhabited for millions of years. Given uncertainty in biological taxonomy and the looseness in the use of the concept of species–including whether we are talking about members of the same sub-species, the same species, or perhaps even the same genus–whether or not organisms are "native species" would also seem to admit of degrees for that reason. For example, what if the wolves used to restore the Yellowstone population had been from a region of Alaska with somewhat different habitat than what existed just north of the Montana-Canadian border where the restored wolves were captured? Or what if the restored wolves had a greater percentage of red wolf, coyote, or even African Golden wolf (a jackal) genes that those that previously resided there? Presumably in these cases the restored wolves would be somewhat less native. Or consider that polar bears and grizzly bears have increased their interbreeding due to climate change. If such a hybrid was either transported to or found its way into Yellowstone, it seems reasonable to say it is "somewhat native" to the Park. Additionally, because the boundaries of ecological assemblages or ecosystems are also not always clear, whether a species is native or not, and to what extent, depends of how one understands the boundaries of the ecosystem it is in.

Recognizing that nativeness is a degree phenomenon helps us see that climate change, while problematic for the concept, need not undermine it, as many seem to think. It will, of course, put additional pressure on native species and increase the relative abundance of non-natives. How much nativeness will survive depends on how dramatically climates change and in what ways. In general, climate change will lessen the degree of nativity of stationary natives as they become less adapted to local abiotic conditions (such as temperature and moisture). Such species will also become less adapted to the local biota insofar as ecosystems "reshuffle":

> New climates are expected to cause ecosystem reshuffling as individual species, constrained by different environmental factors, respond differently. One tree may be limited by summer rains that hold back seedling recruitment, for instance, whereas another species may be limited by winter freezes that control insect pests. Some species may migrate up-latitude or

---

[3]Perhaps some generalist species are counterexamples. For example, coyotes now inhabit much of North America (though not the arctic regions). European starlings are another similar example.

[4]Note that I am not using "naturalized" identically to the usual biological understanding according to which any non-native that establishes a self-sustaining community of individuals in a new habitat has naturalized. Note further that the possibility of naturalization over time implies that novel ecosystems (human-caused but not maintained ecosystems with unique combinations of species or functions) might eventually be composed of thoroughly native species.

up-elevation, while others stay put. An ecosystem might see many species vanish—but also new arrivals. (Fox 2007, 823)

Some native animals and plants will no longer be able to survive in their historical habitats and so will either have to move or die out. In so far as ecological assemblages shift en masse, say migrating north as the temperature rises, their component species will remain (relatively) native. But if "reshuffling," rather than mass migration, is the norm, nativeness of species will dramatically decline and many native species will become non-natives, either in their original location as their neighbors and abiota have changed, or in their new location, as they join species to which they have not adapted/interacted. "As climate change pushes more species out of their home ranges and into new areas, the number of so-called invaders is likely to multiply exponentially" (Goode 2016). There remains the possibility of naturalization over time when and if the ecological disruptions climate change causes cease.

This account of the native/non-native distinction has some intriguing implications for conservation policies. For example, assisted migration of endangered species need not be exotic introduction. Species moving from an historical range can remain native if they have sufficiently adapted to species in the new location. Thus, assisted migration would count as native introduction when the species and abiota in the new site are sufficiently similar to the species and abiota extant in the previous location. Or consider the use of genetic engineering to increase the fitness of a species. For example, to protect American pica from an increasing warm climate, we might introduce foreign genes from more heat adapted populations or species into the endangered pica populations. These altered picas would be more or less native depending on the extent of their alteration.[5] The preference for native species would entail using genes from the most closely related populations, sup-species, or species.

This account also sheds light on how we might think about Pleistocene rewilding and de-extinction. There have been recent discussions about the existence of "ecological anachronisms," that is, extant species adapted to extinct species. For example, it is argued that the speed of antelope in the American West (over 60mph/95kmh!) evolved and was an adaptation to cheetah like predators who long ago went extinct in America. If many species today are still adapted to extinct species, then bringing those species back (say by cloning using ancient DNA) and locating them with many of their co-evolved species could count as return of natives or at least "somewhat native species." Attempts at back breeding to mimic extinct species might also produce somewhat native species, depending on how similar those species are to the extinct type and how similar the extant habitat is to what once existed.

While the degree phenomenon of nativeness means there are not sharp lines to be drawn and that borderline cases will have to be confronted, especially with the advent of climate change, at the extremes there are innumerable clear cases of native and non-natives on this account. Additionally, the preference for native species over non-natives remains useful even in cases where the distinction is not clear cut.

---

[5] See Palmer (2016) for a useful discussion.

22.3   The Prevalence of Non-Native Species

The assumption among environmentalists and others who prefer native species and harbor an antipathy towards non-natives is that non-natives represent disvalues of various sorts. To assess these value judgments, we need first to address some factual questions concerning non-natives. As it turns out, getting clear on the facts about non-natives is not straightforward for there is considerable scientific controversy concerning them.

How pervasive are non-natives? The numbers vary widely. The U.S. Fish and Wildlife service, citing a 2005 study by Pimentel et al., suggests there might be 50,000 non-native species in the U.S. Citing a 1999 study, it estimates that about 10% of them are invasive. Elizabeth Kolbert reports that "California alone acquires a new invasive species every sixty days" (Kolbert 2014, 211). The Global Invasive Species database lists 149 invasive species in the Netherlands (and 498 in the U.S.). A recent study of recorded "established" alien species comes up with about 17,000 worldwide and ominously concludes that:

> For all taxonomic groups, the increase in numbers of alien species does not show any sign of saturation and most taxa even show increases in the rate of first records over time. This highlights that past efforts to mitigate invasions have not been effective enough to keep up with increasing globalization. (Seebens et al. 2017, 1)

Climate change will undoubtedly increase the number of non-natives, perhaps exponentially. Given that we do not have a fixed idea on the total number of species on earth (estimates range from 3 million to 100 million to even a trillion–only 1.7 million have been described) the relative presence of non-native species compared to natives also cannot be determined. What we do know is that in some places they are very widespread. One study of central European cities found that 40% of plant species were alien (Pysek 1998). Another study suggests that "non-native plants and birds can make up 50% or more of species in some urban, insular, and old field environments" (Schlaepfer 2018).

While species invasion is an ancient and natural phenomenon essential for the flourishing of live on earth, recent human-caused introduction of non-natives is dramatically different, moving species far more rapidly and across greater distances and barriers:

> The human-induced rate not only of species extinction but also of species invasion has increased exponentially, in concert with the exponential growth of the human population over the last few hundred years. In addition, in more recent decades, global human travel and commerce have increased disproportionately relative to the increase in the sheer number of humans. Combined, these actors have produced burgeoning rates of nonindigenous species in every ecosystem that has been monitored…Although species invasions are natural, both the rate of their occurrence and the distances traversed by species now exceed by orders of magnitude those of only a few hundred years ago. (Lodge and Shrader-Frechette 2003)

In short, non-native species are pervasive in many ecosystems and human-introduction of non-natives continues at an unnatural and potentially alarming rate.

22.4 Judging Species by Their Origin

Should we be worried about this dramatic rise of the non-native? Many environmentalists are deeply alarmed. But an increasingly accepted idea is that the generalized negative attitude toward non-natives is prejudicial. An influential article in the prestigious science journal *Nature* proclaimed that we "ought not to judge species on their origin" (Davis et al. 2011). These critics suggest that the antipathy toward exotics involves the problematic (and even xenophobic) attitude of assuming that what hails from elsewhere is inherently bad. They think of the widespread efforts to control or remove non-natives as "persecution of the new just because it is new" and question "why our default attitude to novel biodiversity is antagonism or ambivalence" (Thomas 2013). One prominent invasion biologist suggests "the dominant paradigm in the field is still a 'when in doubt, kill them' sort of attitude" (Goode 2016).

Perhaps we should not be alarmed by non-natives if they behaved no differently than natives. But the evidence is that non-natives tend to be more ecologically disruptive than natives. While natives can irrupt and cause ecological upheaval, they do so less frequently than do non-natives. Non-native species are far more likely to have ecological and socio-economic impacts than do those native species that, for various reasons, undergo range expansions or increase in abundance to become 'weedy' (Simberloff et al. 2012). One study documents that non-native predators and herbivores had about 2 ½ times the impact on native prey than did native consumers (Paolucci et al. 2013), results that fit the common sense idea that native species will learn how to handle their native predators better than those with whom they have not interacted. Therefore:

> Ignoring biogeographic origins as a mediator of impact ignores the importance of evolutionary context in species interactions…The more 'alien' an established animal, plant or microbe is to its recipient community, the greater the likelihood it will be ecologically disruptive. (Richardson and Ricciardi 2013, 1463)

The defenders of non-natives can respond that "ecological disruption" is not necessarily a problem, or if the disruption is negative, how harmful these impacts are is open to question. They will argue that rather than assume a non-native will be problematic, that is, rather than approach non-natives as "guilty unless proven innocent," we should assess them individually in terms of their benefits and costs. Even though non-natives are more likely to cause disruption than natives, to assume that they will be disruptive (without specific evidence that they will be–other than their "alien" status), would only make sense if this likelihood was quite high. Given the "tens rule" it is not at all clear that it is.

Consider an analogy with human immigrants. Assume it true that a new immigrant to a country is more likely to be disruptive (e.g., culturally) than a citizen. It would only make sense to presume guilt (or to justify caution/skepticism) if this likelihood was extremely high or if the possible disruptive behavior was severely harmful. For example, even if immigrants present a greater risk of terrorism than do citizens, unless this risk was high or its possible consequences major (e.g., smuggling in nuclear bombs), it would not be acceptable to assume they will be problematic. So the ecological fact that non-natives are more likely to cause ecological disruption than natives does not, w/o further evidence, justify the "prejudicial" precautionary approach to aliens represented by the antipathy many environmentalists have toward them. Below I provide a rationale which I think does justify such a negative attitude, w/o specific evidence of potential for disruptive behavior.

22.5 Do Non-Natives Threaten Biodiversity?

A major reason behind the opposition to non-natives is the belief that they are a major threat to biodiversity. If it is true, as has been claimed by numerous published reports over the last 20 years or so, that non-native species are the second leading cause of species extinction world-wide, that would seem sufficient to justify the antipathy toward non-natives, especially given that humans are causing a mass extinction event on a scale not seen on the planet for 50 million years.

Whether, and to what extent, non-native species are a threat to biodiversity is a surprisingly complicated and debated topic. While repeated numerous times in scientific papers, environmental magazines, and news reports, the claim that non-native species are the second leading cause of extinction (behind habitat destruction) has been discredited and even ridiculed ("a canard," Davis and Chew 2017 call it). It is alleged to be based on shoddy science involving confirmation bias on the part of both committed conservation biologists and ideological environmentalists. Matthew Chew writes:

> While carefully recounting the origin, promotion, and deployment of the 'second greatest threat', I argue that its uncritical acceptance exemplifies confirmation bias in scientific advocacy: an overextended claim reflexively embraced by conservation practitioners and lay environmentalists because it apparently corroborated one particular, widely shared dismay about modern society's regrettable effects on nature. (Chew 2015, 1)

First asserted by E. O. Wilson in the 1990s and then rebutted by Mark Davis and 18 colleagues in their 2011 *Nature* paper, recent studies continue to document it. A 2016 study examined:

> The prevalence of alien species as a driver of recent extinctions in five major taxa (plants, amphibians, reptiles, birds and mammals)…Alien species are the second most common threat associated with species that have gone completely extinct from these taxa since AD 1500…Aliens are the most common threat associated with extinctions in three of the five taxa analyzed, and for vertebrate extinctions overall. (Bellard et al. 2016)

Mark Davis, who is a well-regarded invasion biologist, continues to dispute this claim by arguing that most of the extinctions documented have taken place on islands where species are far more vulnerable to non-natives than in other places. We cannot generalize these data to the extinction threats on land or in the seas, he suggests, because these biotas are far more resistant to competitive pressure.

A focus on species extinction alone, however, is a limited measure of the loss of biodiversity, for individual members of species can radically diminish in numbers, subspecies and populations can be extirpated, and genetic diversity within a species diminished, while the species itself continues to survive. For example, polar bears have not gone extinct as a species, but their numbers and diversity have seriously declined. This point, while important, must be treated with caution. For we cannot simply assume that the loss of individual members of a native species is an overall loss in biodiversity if they are being replaced with non-native individuals–unless, of course, we assume that the biodiversity provided by native individuals is more valuable than that provided by non-native ones. But this begs the question about the superiority of native species over non-native ones.

This point is made all the more poignant by the surprisingly diverse and multiple ways non-natives add to or promote biodiversity. For example, Britain has gained almost 2000 non-native species "without losing anything to the invaders" (Thomas 2013). The same paper reports that, because of hybridization (which can lead to new species), non-natives added to Britain have increased the global species count. Thomas goes on to suggest that warming temperatures resulting from climate change will increase regional diversity because warm-adapted species will invade more quickly than cold-adapted species move out.

Paradoxically (given the severe biodiversity loss attributed to non-natives arriving on the continent), Australia provides another example of how non-natives can add to biodiversity in certain dimensions. The continent lost all of its large (over 200 lbs.) megafauna by the end of the Pleistocene. Now it is home to eight introduced megafaunas, including the only wild population of dromedary camels in the world (Lundgren et al. 2018). In general, 1/3 of very large herbivorous species have wild populations outside their native ranges. While this study focuses on only a small group of species, namely giant herbivores, it is of interest that "the introduced herbivorous giant megafauna of the world have restored species richness across many continents to levels approaching the Pleistocene" (Lundgren et al. 2018, 865). Thus, if we only worry about endangered species in their native habitats, we will miss opportunities for preserving species more widely. Of course, the focus of preserving endangered species should be on preserving them in their native habitats. That they can survive in zoos or as introduced, non-native species in other habitats is of significantly diminished value.

Nevertheless, there are many examples of species endangered at home that are doing well as introduced species in non-native habitats. In some cases, endangered species are even considered invasive. The Monterey pine is endangered in California and Mexico but is treated as a pest in Australia, and Barbary sheep, which are endangered in Morocco, are allegedly overrunning the Canary Islands (Goode 2016).

We clearly should withhold a negative appraisal of non-native species that are also endangered if our objection to non-natives is based on concerns about biodiversity. As argued in an intriguing paper comparing human migrant ethics to the ethics of non-native species, just as refugees (who are endangered in their home countries) are accorded special status among immigrants, so endangered species should be treated as a special class of non-natives (Switzer and Angeli 2016). One scientist argues that the phenomenon of non-native but endangered species is widespread enough that if non-natives were included in biodiversity indices (which they are not now, but which he insists they should be), it would lower the extinction predictions for some species (Schlaepfer 2018).

Besides being endangered species in some cases, non-natives have been documented to provide numerous benefits for native species, including rare ones. In California, non-native eucalyptus trees provide habitat for Monarch butterflies. In Spain, non-native crayfish provide food for wetland birds, including some endangered ones (Goode 2016). A striking example is introduced, non-native donkeys in the Sonora desert of the Southwestern U.S. Where water is close to surface, they dig wells used by up to 30 species (including trees). The scientists documenting this phenomenon, argue that the non-native donkeys are increasing the functionality of these ecosystems and their resilience in the face of climate change (Lundgren et al. 2018).

Many point out the obvious fact that the introduction of a non-native species increases the local species count (by one!) (Sagoff 2000). If the number of non-natives species arriving outnumber the extinctions taking place, species richness will increase. In fact, because of the spread of non-natives, a general trend worldwide is that often local biodiversity (measured by a species count) is increasing even while overall global species numbers are in decline. "Empirical evidence points to ecological increases in the number of terrestrial species in most of the world's regions over recent decades and centuries, even though the total number of species on the planet is declining" (Thomas 2013). A study in the early 2000s of plants and birds on oceanic islands found that land birds species numbers remained constant (despite many extinctions) while plant species numbers doubled. Because those introduced plants and birds existed in other places, there was not global increase in species (Sax et al. 2002).

What are we to make of this tendency? At least in terms of biodiversity, increases in local biodiversity are valuable (unless we discount the biodiversity added by non-natives!). But if endemic species (those found only in one place) are being replaced with more cosmopolitan species, then the local gain in biodiversity is at the expense of overall global diversity. It seems clear that this is an overall loss in biodiversity. Consider an analogy with human cultural diversity. European humans arrive on an island with several distinct indigenous populations and promptly drive all but one extinct (perhaps they inadvertently infect the locals with diseases never before seen on the island). If enough different nationalities of Europeans arrive, the local diversity of cultures will have increased overall. But we should not count this as an overall increase in cultural diversity, even though there is now greater cultural diversity on this island.[6] Here is a description of the loss when non-natives replace local endemics:

> Native species have coevolved with one another and the physical environment, often resulting in intricate coadaptations. Loss of native species can erase unique evolutionary histories. Therefore, non-native species additions do not compensate for phylogenetic losses resulting from extinctions even if they increase overall local species diversity, because many non-native species erode diversity through local and global extinctions. Even if one is willing to offset the current losses of biodiversity with the promise of new biodiversity as non-native species

---

[6]This fictitious example may actually fairly well represent what happened in North America with the arrival of Europeans.

evolve and diverge, millions of years of biological adaptation and evolutionary history would
be lost. (Pauchard et al. 2018, 2)

While the situation is complicated, the overall threat to biodiversity due to non-
native species is real and, in my judgment, significant. Citing the World Wildlife
Fund's 2014 living planet index report, Davis and Chew (2017) claim, "More recent
assessments of biodiversity effects demote invasive species to a subsidiary role". That
study put "invasive species/genes" as the primary threat in 5.1% of the populations
studied, while climate change was primary in 7.1% of populations (WWF 2014, 20).
The 2018 WWF report lists invasive species/disease as again close to climate change
in the threat it poses to biodiversity (and for some taxa, a greater threat) (WWF
2018, 72). While true that compared to habitat degradation and direct exploitation,
non-native species play a "subsidiary role" in biodiversity loss, the negative effect
on biodiversity is close to the threat to biodiversity posed by climate change, and it
should not be ignored.

22.6 Homogenization

As with the question of whether non-native species reduce biodiversity, the related
question of whether or not they tend to homogenize the world's ecosystems is more
complex than one might first think. However, as with biodiversity loss, in the final
analysis, this concern has merit.

Biotic homogenization occurs when extant ecological assemblages lose their
distinctiveness and become more similar. Loss of biological distinctiveness occurs
at many levels, including species similarity, similarity of functional relationships,
loss of genetic distinctiveness, and similarity of evolutionary history. While there are
other causes of biotic homogenization (e.g., when flooding joins what had been two
once isolated bodies of water), the human transport of non-native species is a major
one.

A key factor in how biotic communities get their distinctiveness is their isolation
from one another. When this isolation is overcome, mixing begins, and one gets
homogenization. Humans are intentionally and unintentionally transporting species
around the globe, overcoming the natural isolation, barriers, and distances that have
helped created the spectacular diversity of earth's ecological assemblages. This is
a process driven by, and also similar to, the globalization of human economies and
cultures. In the human economy, quirky and distinctive "mom and pop" stores on
main street are driven out of business by the same big box stores, resulting in towns
and cities losing distinctiveness and become more similar. Take a major exit off a U.S.
highway interstate and you are likely to find the same two dozen retailers, whether
you are in the Pacific Northwest or 3000 miles away in the Southeast. Analogously,
the same weedy, generalist species, tolerant of diverse ecological conditions, appear
over and over again around the world, often replacing specialist, more sensitive
species. These species are ones with "broad diets and tolerances, rapid dispersal

and high reproduction" (McKinney and Lockwood 1999, 452). "Loser species" are being replace by "winners".[7] Just as McDonald's and Subways has spread all over the world, so have dandelions and rats. Invasion biologist Julian Olden (often one to defend non-native species) says: "From birds to plants to fish to mammals, there's strong evidence that things are becoming more similar" (Goode 2016). A number of studies document that biotic homogenization is occurring, especially in freshwater ecosystems (Petsch 2016). Some even suggest that humans "are creating a new Pangaea by bringing all the world's flora and fauna together" (Kolbert 2017).

Even when non-natives aren't weedy or invasive their presence in new habitats homogenizes by lessening floral and faunal distinctions among regions. Theoretically, non-natives introductions could decrease similarity as when two different fish species are introduced to lakes that otherwise have the same species, or when a non-native plant colonizes one habitat, but not another which is otherwise identical. But this is an anomaly, if successful non-natives tend to be generalists and widespread.

Consider the phenomenon of zoos (and perhaps also of gardens). Do they increase biodiversity or homogenize? As Holmes Rolston once pointed out, there are more species of animals in the Denver zoo than in all of Colorado. So, in one obvious sense, Denver, in virtue of its zoo, is much less similar to the rest of Colorado. But because of zoos, elephants, for example, are no longer to be found only in African and Asian grasslands and forests, but are now present in any major city on the planet! I think this is a helpful analogy for non-natives in general. While potentially increasing local diversity (until they wipe out native species), they homogenize the world by lessening the distinctiveness of different bioregions and assemblages.

A main rationale for rejecting the notion that non-natives simply homogenize is to emphasize the possibility of hybridization. Just as "the mixing and blending of cultural identities…lead to new forms of diversity" (Keulartz and Van der Weele 2009, 244) among humans, so too the mixing and blending of species resulting from non-native introduction can lead to new types of biodiversity. In an editorial arguing that the Anthropocene may well increase biodiversity, Thomas argues that "Hybridization is becoming particularly important as formerly separated species are brought into contact. The rates are astounding…Speciation by hybridization is likely to be a signature of the Anthropocene" (Thomas 2013).

But hybridization is a double-edged sword in terms of diversity. "Hybridization has been shown to be a major contemporary extinction force, especially when accompanied by habitat homogenization, causing species declines through introgression, genetic swamping and reproductive interference" (Richardson and Ricciardi 2013, 1463). Similarly, human hybridization can also create loss of diversity and homogenization. Think of the U.S. as a melting pot of peoples, homogenizing them into "Americans." Or what if those Jews who are committed to a distinct Jewish culture

---

[7]Some argue that we should start backing these "winners." "Conservationists must stop spending all their time backing loser species–the endangered and reclusive. They must start backing some winners" (Pearce 2015a, xvi). Of course, we already are, as we are the main cause of their movement to new habitats.

gave up their emphasis on Jews marrying other Jews and fully integrated with other religious cultures. Such "hybridization" would be a loss of a distinctive culture.

22.7 Naturalness Value and the Antipathy Toward Non-Natives

We have examined two related reasons for the antipathy toward non-natives: Their negative impact on biodiversity and their tendency to homogenize the world's ecological assemblages. There is another compelling reason for this antipathy: Non-native species seriously exacerbate the continued humanization of earthen nature. Non-native species, when they are human-introduced, as they are in the vast majority of cases, threaten and diminish a key environmental value, namely, naturalness. Respect for independent nature explains and justifies the antipathy toward non-native species. Human introduction of non-native species is a major way that humans are impacting the natural world. That many, perhaps most, human introductions of non-native species are unintentional does not lessen this point. Naturalness is compromised both by intentional and unintentional human actions.

There are a host of objections to the idea that protecting and restoring naturalness are important environmental obligations. Here I address a few and only in a cursory manner.[8] Some will argue that humans are natural and so whatever they do is natural, including spreading non-native species all around the globe. But by "natural" I mean the degree to which something is independent of human impact, and clearly spreading species around the globe is a human impact. Others might claim that there is no naturalness left to value or protect, that we are in the "Anthropocene," that anthropogenic climate change and global human pollution are so pervasive that we are (as McKibben claimed years ago) at "the end of nature" and so there is no more naturalness left to defend. But naturalness comes in degrees. Dimensions of nature are more or less impacted by human activities and so defenders of naturalness value urge us to value and protect the naturalness that remains. I have argued that we ought to "value naturalness in the Anthropocene; now more than ever" (Hettinger 2014), as rarity enhances preexisting value and makes it more precious. Those ecosystems which have not been overwhelmed by human-introduced non-natives have a special importance given their relative naturalness. Even ecosystem with significant non-native presence have remaining naturalness of significant value.[9] Naturalness can also return over time as humanization washes out of natural systems and humans can themselves speed along such recovery by active "rewilding," as when we remove trash or poisons from natural systems or tear down dams. Restoration of native ecosystems

---

[8] For a somewhat more vigorous defense of the value of naturalness and respect for independent nature, see Hettinger (2018).

[9] Contrast this idea with Keulartz and Van der Weele's (2009) suggestion that the more severely invaded an area is with non-natives, the less reason to worry about their presence.

by removal of human-introduced non-natives often increases the naturalness of those systems.

The connection between non-natives and being unnatural is contingent: what matters for naturalness value is not non-nativeness itself, but the human introduction of non-natives and the loss of naturalness that introduction instantiates. Because the vast majority of non-native species are human-introduced, this disvalue of non-natives, while not applying to non-natives per se, applies to the vast majority of them. Naturalness value therefore supports the "guilty until proven innocent idea" in a far stronger and secure manner than does the assumption that non-natives are going to cause ecological or other damage. As we have a seen, there is a relatively low probability that non-natives will cause ecological damage (such as species extinction). While the overall negative consequences non-natives have on biodiversity and increasing homogenization seem clear, whether a particular non-native has these effects is not so clear. The case is otherwise concerning the unnaturalness of non-native species. Because there is a very high probability that a non-native has been human-introduced, rather than being "open minded" about non-natives, naturalness supports a presumption against non-natives. So, the likely unnaturalness of non-natives counts against the increasingly widespread attitude that each introduction should be evaluated in its own right and in the particular area invaded (Seebens et al. 2017). It provides a compelling response to the many who think it "unclear why our default attitude toward ecological novelty is antagonism or ambivalence" (Thomas 2013). The antipathy toward non-natives results from and is justified by the overwhelming likelihood that they embody the ongoing humanization of nature. The preference for natives over non-natives when they are human introduced is a way of respecting independent nature. The default attitude toward the arrival of a non-native should be negative, until it is proven that it was not human introduced.

I now examine a series of important objections to the preference for natives over non-natives.

22.8 Is the Antipathy Toward Non-Natives Based on Misleading Popular Ecology?

The preference for natives over non-natives is a deeply ingrained value for many environmentalists and ordinary people as well. Critics argue that "many introduced populations are considered harmful, not because of their ecological effects per se, but because they challenge deep-seated ideologies about how nature should be" (Wallach et al. 2018, 1263). Ken Thompson, the author of *Where do camels belong? Why invasive species aren't all bad*, claims "It's almost a religious kind of belief, that things were put where they are by God and that's where they damn well ought to stay" (Goode 2016). I think the nativists can respond, "Guilty as charged": Respect for nature is a "deep-seated ideology" (though an eminently justified one) and that the opposition to humans taking over the world and "running nature" is a rejection of the idea that humans should play God with the earth.

It is often claimed that the antipathy toward non-natives is to be based on a set of scientifically dubious ideas which are nonetheless common in popular ecology. The nativists, it is claimed, believe that non-native introductions upset a balance of nature, that ecosystems tend toward an equilibrium that non-natives disrupt, and that change itself is a harm. Nativism based on valuing naturalness can accept that nature is dynamic, adaptive, and in flux, and that there is no one way ecosystems are supposed to be. Their objection is not to change as such, but only human-caused change. Nativism based on natural value favors removal of non-natives, not to prevent "biotic mixing," nor for the purpose of recreating beloved historical assemblages, but to remove or lessen human influence on ecosystems.

22.9 The Xenophobia Objection

A longstanding criticism of nativism (whether this be a preference for human or non-human natives) is that it is based on xenophobia, an irrational fear or dislike of the foreign and/or the unfamiliar, perhaps enhanced by feelings of superiority. It is a common (though not laudable) human tendency to identify with a group and to distinguishing group members with whom one feels comfortable from "outsiders" toward whom suspicion is aimed. It is also common to form opinions of others based on hasty generalizations and selectively-constructed stereotypes. When the foreign enters, the fear is of contamination and the desire is to cleanse the home and make it pure again. Critics of nativism believe that these attitudes underlie the antipathy toward non-native "alien" species and that there is a synergy between anti-human immigrant sentiments and the rejection of non-native species.

This criticism is perhaps made more salient in light of recent currents in world politics. A U.S. President argues for "America First," spews forth harsh anti-immigrant rhetoric ("murders, rapists, and drug smugglers are pouring into our country") and advocates policies to prevent and remove the "invaders" (the wall, family separation). These nationalist sentiments and anti-globalization attitudes (e.g., dissolution of the European Union) are not limited to the U.S., nor is the xenophobia which often underlies them. Critics charge that not only nativists in general, but also the field of invasion biologist itself uses biased ("alien species" "exotics") and militarist language ("invasion," "war against exotics") to fuel prejudicial and misleading attitudes and environmental policies toward non-natives.

Clearly there can be synergism between biological and cultural nativism/purism. As it is often pointed out, the Nazis had their own native plant movement purifying the biology of their country at the same time as they purged the human race of its supposed inferior elements. In South Africa, native-plant, gated communities cater to "suburbanites seeking to escape the increasingly mixed and threatening post-apartheid city…spaces that exclude problematic plants and people alike" (Ballard and Jones 2011, 1).

To avoid such a problematic xenophobia, we are urged to accept a cosmopolitan approach toward human immigration and culture and welcome foreigners into our societies. Similarly, we should accept non-native species into our ecosystems, celebrating their rich origins, their geographical and cultural histories, rather than persecuting them because they are not native (Kendle and Rose 2000).

The analogy between human immigration and non-native species introduction is both instructive and potentially misleading. Biological and cultural nativism can be mutually reinforcing, especially when the rhetoric used is similar or identical. This can be true whether or not xenophobic attitudes underlie the rhetoric. Nevertheless, it is important to point out that xenophobia need not underlie either type of nativism.

Consider the biological nativist's antipathy toward non-natives species. Those who prefer native plants in native habitats and eschew the entry of foreign species need not fear or dislike them, nor believe they are inferior. They might think quite highly of these species in their native habitats. For example, I don't want camels to roam the deserts of the southwestern U.S. But I don't fear or hate camels; they are neat animals! And again, I clearly admired that non-native mimosa tree when I planted it in my yard. As we have seen, opposition to non-native species can be based on concerns about harms to biodiversity, worries about homogenization of ecological assemblages, or opposition to the increasing humanization of the world.

It is harder to make the case that cultural nativism need not be based on xenophobia. People who don't want black people, Muslims, or Mexicans in their communities cannot plausibly say I like these folks fine in their own places, I just don't want them here. At least this is true if they are talking about individuals of these groups. However, if we take cultural integrity as a value, then there are non-xenophobic reasons for being concerned about "too much" immigration or "too many foreigners moving in." Mass immigration may threaten distinct and valuable ways of life. The desire to preserve unique cultures, as the desire to preserve unique biological communities, need not be xenophobic and can even be praiseworthy.

For example, the preservation of indigenous cultures is of significant value. Too many natives leaving the community, or too many non-natives entering, threatens this value, as does the ingression of western commercialism. When Southern parents send their children to Southern colleges, when Christian parents send their children to Christian schools, and when Jewish parents lobby their children to marry other Jews, these practices need not (though they obviously can) be based on xenophobic attitudes. Typically, they are based on desires to preserve important cultures or cultural attributes. I now live in a small community in the Rocky Mountain West with some wonderful small-town values: Cars stop for pedestrians, dogs are loved, there is strong community support of children and young women, and the community is extremely supportive of an active outdoor lifestyle. If enough people from large cities who are indifferent or antagonistic to such values moved into town it could destroy a cultural environment of great value. My opposition to the mass movement of such people into town is not xenophobic.

Such opposition to foreign entry is only legitimate when significant cultural values are in jeopardy. The anti-immigrant sentiment in the U.S. is not justified by a serious threat to the national's culture, especially because the U.S. has always prided itself a nation of immigrants. The debate about whether this nation of immigrants should promote assimilation (or homogenization!) for the sake of a unified culture or embrace a multi-cultural society is in some ways similar to debates about the importance of retaining unique biological communities in light of the concerns about homogenization.

These are complicated issues, but it should be clear from this discussion that biological nativists antipathy to non-natives cannot be summarily dismissed by accusations of xenophobia or contribution to xenophobia.

22.10 The Need for Non-Natives in the Anthropocene

A particularly provocative challenge to nativism is the idea that non-native species are increasingly essential to human and planetary flourishing. That they always have been becomes obvious if we focus on the human food supply. Pimentel et al. (2005) claims that "Introduced species, such as corn, wheat, rice, and other food crops, and cattle, poultry, and other livestock, now provide more than 98% of the U.S. food system." A more recent study suggested that worldwide an average of 70% of food crops were introduced from other regions (Khoury et al. 2016). But the critics are insisting on the importance of non-natives far beyond the agricultural domain so important to humans. It is alleged that wild nature needs them too, as indicated in the title of this recent book by Fred Pearce: *The new wild: Why invasive species will be nature's salvation*. Pearce writes:

> The more damage that humans do to nature— through climate change, pollution, and grabbing land for intensive agriculture and plantation forestry—the more important alien species and novel ecosystems will be to ensuring nature's survival. Aliens are rapidly changing from being part of the problem to part of the solution. (Pearce 2015a, 178)

Emma Marris makes a similar claim in her book *Rambunctious garden: Saving nature in a post-wild world*:

> As the planet warms and adapts to human domination, it is the exotic species of the world that are busy moving, evolving, and forming new ecological relationships. The despised invaders of today may well be the keystone species of the future's ecosystems, if we give them the space to adapt and don't rush in and tear them out. (Marris 2011, 109)

Even sober biologists think that:

> Non-native species might contribute to achieving conservation goals in the future because they may be more likely than native species to persist and provide ecosystem services in areas where climate and land use are changing rapidly and because they may evolve into new and endemic taxa. (Schlaepfer et al. 2011, 428)

We have seen that non-native species can and do contribute to biodiversity, although in ambiguous and ambivalent ways. They provide "ecosystem services" as when non-native sea grass provides a nursery and regulates water flow in places that have lost their native habitat formers. Other ecosystem services non-natives provide include food, pollination, water purification, seed dispersal, and the list goes on. Obviously, non-native species are biological creatures that can serve most all the biological functions that native species do. In a harsher and more unstable environment, they will more likely persist as they tend to be generalists rather than specialists and also better competitors. By refusing to accept non-natives species as legitimate parts of nature, the critics argue, people are blinded to these sorts of beneficial roles they can play (Wallach et al. 2018).

I believe there is room in the conservation agenda for both a focus on preserving and restoring native biodiversity, and occasional, though increasing, acceptance of non-natives in the specific cases when they provide sufficient benefits.

I agree with Marris that we must go beyond "black and white thinking on non-native species" (Marris 2014, 516). It is matter of weighing costs and benefits. It is important while highlighting the potential benefits of non-natives to remember how damaging non-native species can be. A few years ago, there was a report suggesting that "the annual combined economic cost of invasions worldwide exceeds that of natural disasters" and the authors argued that we should prepare for non-native species invasion in ways similar to how we prepare for (other?) natural disasters (Ricciardi et al. 2011). Perhaps this is hyperbole, but we should remember that the disasters perpetuated by non-native species are for the most part not "natural," but self-inflicted, as humans move these species around the globe. In cases where non-natives provide the sort of ecosystem services described above, we must weigh these biological benefits against the potential biological costs, both in terms of loss of biodiversity and homogenization. Even when this calculus turns out positive, we must compare these benefits to the loss of naturalness value that the human-introduced of non-natives involves. This is a classic case of conflict between two important conservation values, biodiversity and naturalness. I think despite the multi-faceted ways non-native species can be beneficial, non-natives bring with them sufficient probable disvalue to justify the generalized antipathy toward them and legitimizes the policy of treating them as guilty until proven innocent. The burden very much has to be on the defender of non-natives to prove they are not simply benign, but beneficial overall.

22.11   Non-Native Animals in Our Midst

In this last section I explore two related issues. First, how the conception of non-native species applies to animals, particularly "animals in our midst." Secondly, what sort of implications for our treatment of animals follows from the issues raised by the native/non-native species controversy?

As background, it should be noted (with outrage) that there has been, on average, a 60% decline in populations of vertebrate animals worldwide in last 40 years.[10]

---

[10]This does not mean a 60% decline in number of individual vertebrate animals on the planet because the sizes of the populations studied varies dramatically. For an explanation, see Brown (2018).

My focus will be on sentient animals (which I believe are roughly co-extensive with vertebrates) as it is with this (relatively small) subset of animals that especially powerful moral considerations arise and with them questions of treatment diverge in comparison to plants (and non-sentient animals).

How to conceptualize urban and rural animals, or wild animals, that wander or are thrust into human habitats is important in part because this phenomenon is increasing dramatically. Humans continue to encroach on animal habitats, and in response, they come into ours. For many, such as rats, coyotes, and song birds, human habitats are particularly compelling, providing food, shelter, and protection from predators. Such animals also raise interesting questions for the distinction between natives and non-natives.

The understanding of native species embraced herein is that native species are those that have significantly adapted/interacted with the local biota and abiota. Non-natives are those have not done so. Being native is a matter of degree and non-natives can naturalize over time. Applied to "animals in our midst," this suggest that new arrivals from habitats totally unlike the habitats they now inhabit should be conceived as non-native. So the "Golden-headed lion tamarins, squirrel sized monkeys," that "came out of the disappearing coastal forests of Brazil and found a new home in the suburbs of Rio de Janeiro" (Pearce 2015b) are not native to those suburbs. In contrast, squirrels who have lived in Rio for generations are native to the city as they have adapted/interacted with the flora and fauna (including humans) and abiota (houses, streets, soil) for quite some time. And the locals (including humans) have adapted/interacted with them. (For example, humans have developed and deployed squirrel-proof, bird feeders.) Further, squirrels brought into Rio de Janeiro from other cities would be more native than the monkeys, for they have adapted/interacted with humans and human abiotic environments more so than have the monkeys (although they might not have adapted to the particular species of flora and fauna in Rio). Or considers the house sparrow. Although European in origin, it has adapted to humans and human habitats world-wide and, on the conception developed here, should be considered a native species of these urban and suburban habitats. Coyotes in American suburbs and cities (one was seen stalking a fast-food restaurant in Chicago!) are an interesting case. Though once widespread in America, they are just recently returning to much of their former habitat and they and the locals (including humans) are just beginning the mutual adaptation/interaction that will revive their nativity.

Paul Knights has suggested a "cultural criterion" for being native to human communities, arguing that cultural relationships humans have with species helps ground their nativity (Knights 2008). The type of cultural relationships and associ-ations he has in mind include having common names for the species (Jack in the Pulpit), being used in play (buttercup under chin, bracelets made from daisies, kisses under mistletoe), being adopted in local cuisine (berries turned into jam, fish for dinner), being used for medicinal purposes, being used as a source for literary or artistic expression, and so on. Although Knights distinguishes these cultural rela-tionships from ecological relationships, for the most part these cultural relationships have an ecological character to them. Part of the criterion for nativeness I have proposed includes human adaptation/interaction to the species in these human habitats. Thus, these cultural relationships add to nativity, at least in so far as they involve an ecological component. If these cultural associations between species in human habitats and humans involves interaction/adaptation (which for the most part they do), then such associations do add to native characterization. When a species in a human environment is ignored by humans, this means there is less interaction and thus a diminished claim to being native.

Human treatment of animals in general is harsh, but peculiarly harsh toward animals perceived as non-native or animals considered to be invasive. Examples include, laser censors that spray poison on wildcats and viral diseases used to infect wild rabbits (Wallach et al. 2018). Sometimes species are labeled non-native as a way to help justify getting rid of them. English farmers have objected to beaver restoration on these grounds and coyotes are often tarred with the label non-native at the same time they are being persecuted. Mountain goats judged as not native to many U.S. National Parks are accused of "invading" the parks rather than simply beginning to colonize them. Do the 200 goats that live "in and adjacent to" Yellowstone National Park constitute an "invasion?" Yes, their presence is due to humans, their waste changes the soil chemistry in some sites, and they may compete with native bighorn sheep (and pass on disease to them). But is it really true that "This non-native species poses a threat to Yellowstone's alpine as well as bighorn sheep" (Yellowstone National Park 2019)? Perhaps. And perhaps they should be removed. But the non-native species invasion language inclines the discussion toward this outcome.

I have argued in favor of antipathy toward non-native species, including the "guilty until proven innocent" attitude. But sentient animals have a special value that other species lack. Because of this, I think the antipathy toward non-natives should be relaxed for sentient animal species and policies based on guilt unless proven innocent should be suspended. What justifies this differential treatment is that sentient animals–unlike insentient animals and plants–have feeling and desires. Their lives can go well or badly from their own perspective and they can suffer horribly. Because of these capacities, we should treat them as individuals and as having certain rights. Most clearly, they have the right not to have suffering inflicted on them, unless sufficiently strong justification for that suffering is presented, including the ruling out of less painful alternatives.[11] Perhaps sentient animals should even be given the presumption of freedom of movement.

Recall that what justifies the presumption of guilt against non-natives is that the vast majority of non-native species are human-introduced and such introduction decreases naturalness, something I have argued is an especially important value in our time of massive and ongoing humanization of earth. Concerns about loss of biodiversity and homogenization are also important, but cannot be simply assumed in the case of non-natives. Instead, they require investigation. But with sentient animals another definitive value is added to the mix. We should not shoot non-native mountain

---

[11]For example, if gene drives can be used to sterilize populations of invasive sentient animals threatening extinction of native species, they should be seriously considered as an alternative to traditional mechanisms (such as poison) that involve great suffering.

goats who stray into our national parks simply on grounds that it is exceeding likely they are human-introduced. I do not think the loss of naturalness value this (likely) represents is by itself sufficient to justify immediate dispatch. Their unnaturalness will count in favor of their removal, as will their possible effects on biodiversity and homogenization. But the burden of proof will have shifted because these species have a special value.[12] Note that with non-native plants or insentient animals (e.g., insects) the presumption of guilt remains and their expedited removal is permissible, for they lack the burden shifting value that comes with the rights of sentient animals.

22.12 Conclusion

This essay has defended biological nativism and its antipathy toward non-natives. Native species are those that have significantly adapted/interacted with local biota and abiota. Non-natives are increasingly pervasive, their presence indicative of and caused by massively increasing human global impact. While climate change puts pressure on the distinction, the categories of native and non-native remain relevant and important in environmental thought. There are serious concerns about and evidence supporting non-natives' negative impacts on biodiversity and their homogenization of ecosystems. However, there are numerous instances where non-natives contribute positively to biodiversity and such contributions are likely to increase. Nonetheless, the generalized antipathy toward non-natives is justified by respect for independent nature, as the vast majority of non-natives are introduced by humans and thus are part of the ever increasing and arrogant human domination of once natural dimensions of earth. Preference for natives need not be based on prejudicial dislike of the foreign or misconceived ideas about natural systems. Animals can become native to human habitats and the special value of sentient animals suggest that, for them alone, we should withhold our presumption in favor of expedited removal of non-native species. The preference for native species over non-natives continues to be a significant environmental value and one that is rationally defensible.

References

Ballard, R., and G. Jones. 2011. Natural neighbors: Indigenous landscapes and eco-estates in Durban, South Africa. *Annals of the Association of American Geographers* 101 (1): 131–148.

Bellard, C., P. Cassey, and T. Blackburn. 2016. Alien species as a driver of recent extinctions. *Biology Letters* 12 (2). https://royalsocietypublishing.org/doi/full/10.1098/rsbl.2015.0623.

Brown, E. 2018. Widely misinterpreted report still shows catastrophic animal decline. *National Geographic*, November 1.

---

[12]That the burden has shifted does not mean that non-native sentient animals should be allowed to wreak havoc on native species. Severely negative impacts on biodiversity (e.g., causing species extinction) clearly can justify killing and even causing suffering to sentient animals.

Chew, M. 2015. Ecologists, environmentalists, experts, and the invasion of the 'Second Greatest Threat'. In *International review of environmental history*, vol. 1, ed. J. Beattie, 7–40. Canberra: Australian National University Press.

Chew, M., and A. Hamilton. 2011. The rise and fall of biotic nativeness: A historical perspective. In *Fifty years of invasion ecology: The legacy of Charles Elton*, ed. D. Richardson, 36–47. Chichester, UK: Blackwell Publishing.

Crawford, A. 2018. Why we should rethink how we talk about 'alien' species. *Smithsonian Magazine*, January 9. https://www.smithsonianmag.com/science-nature/why-scientists-are-starting-ret hink-how-they-talk-about-alien-species-180967761. Accessed 23 March 2020.

Davis, M., and M. Chew. 2017. 'The denialists are coming!' Well, not exactly: A response to Russell and Blackburn. *Trends in Ecology & Evolution* 32 (4): 229–230.

Davis, M., M. Chew, R. Hobbes, A. Lugo, J. Ewel, G. Vermeij, et al. 2011. Don't judge species by their origin. *Nature* 474: 153–154.

Fox, D. 2007. Back to the no-analog future? *Science* 316 (5826): 823–825.

Goode, E. 2016. Invasive species aren't always unwanted. *New York Times*, February 29. https://www.nytimes.com/2016/03/01/science/invasive-species.html.

Hettinger, N. 2001. Exotic species, naturalization, and biological nativism. *Environmental Values* 10 (2): 193–224.

Hettinger, N. 2014. Valuing naturalness in the Anthropocene: Now more than ever. In *Keeping the wild: Against the domestication of earth*, ed. G. Wuerthner, E. Crist, and T. Butler, 174–179. Washington, DC: Island Press.

Hettinger, N. 2018. Naturalness, wild-animal suffering, and Palmer on laissez-faire. *Les Ateliers de l'éthique* 13 (1): 65–84. https://www.erudit.org/fr/revues/ateliers/2018-v13-n1-ateliers04192/1055118ar.pdf.

Justus, J. 2009. Exotic species. In *Encyclopedia of environmental ethics and philosophy*, ed. J.B. Callicott and R. Frodeman, 412–414. Macmillan Reference USA.

Kendle, A., and J. Rose. 2000. The aliens have landed! what are the justifications for 'native only' policies in landscape plantings?. *Landscape and Urban Planning* 47 (1): 19–31.

Keulartz, J., and C. van der Weele. 2009. Between nativism and cosmopolitanism: Framing and reframing in invasion biology. In *New visions of nature*, ed. M. Drenthen, J. Keulartz, and J. Proctor, 237–256. Dordrecht, The Netherlands: Springer.

Khoury, C., H. Achicanoy, A. Bjorkman, C. Navarro-Racines, L. Guarino, and X. Flores-Palacios. 2016. Origins of food crops connect countries worldwide. *Proceedings of the Royal Society B: Biological Sciences* 283: 20160792. http://dx.doi.org/10.1098/rspb.2016.0792.

Knights, P. 2008. Native species, human communities and cultural relationships. *Environmental Values* 17 (3): 353–373.

Kolbert, E. 2014. *The sixth extinction: An unnatural history*. New York: Henry Holt.

Kolbert, E. 2017. The fate of the earth. *The New Yorker*, October 12. https://www.newyorker.com/tech/annals-of-technology/the-fate-of-earth.

Lodge, D., and K. Shrader-Frechette. 2003. Nonindigenous species: Ecological explanation, environmental ethics, and public policy. *Conservation Biology* 17 (1): 31–37.

Lundgren, E.J., D. Ramp, W.J. Ripple, and A.D. Wallach. 2018. Introduced megafauna are rewilding the Anthropocene. *Ecography* 41: 857–866.

Marris, E. 2011. *Rambunctious garden: Saving nature in a post-wild world*. New York: Bloomsbury.

Marris, E. 2014. 'New conservation' is an expansion of approaches, not an ethical orientation. *Animal Conservation* 17 (6): 516–517.

McKinney, M., and J. Lockwood. 1999. Biotic homogenization: A few winners replacing many losers in the next mass extinction. *Trends in Ecology & Evolution* 4 (11): 450–453.

Palmer, C. 2016. Saving species but losing wildness: Should we genetically adapt wild animal species to help them respond to climate change? *Midwest Studies in Philosophy* 40 (1): 234–251.

Paolucci, M., H. MacIsaac, and A. Ricciardi. 2013. Origin matters: Alien consumers inflict greater damage on prey populations than do native consumers. *Diversity and Distributions* 19 (8): 988–995.

Pauchard, A, L.A. Meyerson, S. Bacher, T.M. Blackburn, G. Brundu, M.W. Cadotte et al. 2018. Biodiversity assessments: Origin matters. *PLoS Biol* 16 (11): e2006686. https://doi.org/10.1371/journal.pbio.2006686.

Pearce, F. 2015a. The new wild: Why invasive species will be nature's salvation. Boston: Beacon Press.

Pearce, F. 2015b. Invasive species will save us: The new way we must think about the environment now. *Salon*, April 11. https://www.salon.com/2015/04/11/invasive_species_will_save_us_the_new_way_we_must_think_about_the_environment_now/. Accessed 23 March 2020.

Petsch, D. 2016. Causes and consequences of biotic homogenization in freshwater ecosystems. *International Review of Hydrobiology* 101 (3–4): 113–122.

Pimentel, D., R. Zuniga, and D. Morrison. 2005. Update on the environmental and economic costs associated with alien-invasive species in the United States. *Ecological Economics* 52 (3): 273–288.

Pysek, P. 1998. Alien and native species in Central European urban floras: A quantitative comparison. *Journal of Biogeography* 25 (1): 155–163.

Ricciardi, A., M. Palmer, and N. Yan. 2011. Should biological invasions be managed as natural disasters? *BioScience* 61 (4): 312–317.

Richardson, D., and A. Ricciardi. 2013. Misleading criticisms of invasion science: A field guide. *Diversity and Distributions* 19: 1461–1467.

Russell, J., and T. Blackburn. 2017. The rise of invasive species denialism. *Trends in Ecology & Evolution* 32 (1): 3–6.

Sagoff, M. 2000. Why exotic species are not as bad as we fear. *Chronicle of Higher Education* 46 (42): B7.

Sax, D., S. Gaines, and J. Brown. 2002. Species invasions exceed extinctions on islands worldwide: A comparative study of plants and birds. *The American Naturalist* 160 (6). 766–783.

Schlaepfer, M. 2018. Do non-native species contribute to biodiversity? *PLoS Biol* 16 (4): e2005568. https://journals.plos.org/plosbiology/article?id=10.1371/journal.pbio.2005568.

Schlaepfer, M., D. Sax, and J. Olden. 2011. The potential conservation value of non-native species. *Conservation Biology* 25 (3): 428–437.

Seebens, H., T. Blackburn, E. Dyer, P. Genovesi, P. Hulme, and J. Jeschke. 2017. No saturation in the accumulation of alien species worldwide. *Nature Communications* 8: 14435. https://doi.org/10.1371/journal.pbio.2005568.

Simberloff, D., L. Souza, M. Nuñez, M. Barrios-Garcia, and A. Windy. 2012. The natives are restless, but not often and mostly when disturbed. *Ecology* 93 (3): 598–607.

Slezak, M. 2015. Cane toad has surprise effect on Australian ecosystem. *New Scientist*, March 19. https://www.newscientist.com/article/dn27199-cane-toad-has-surprise-effect-on-australian-ecosystem/.

Switzer, D., and N. Angeli. 2016. Human and non-human migration: Understanding species introduction and translocation through migration ethics. *Environmental Values* 25: 443–463.

Theodoropoulos, D.I. 2003. *Invasion biology: Critique of a pseudoscience*. Blythe, CA: Avvar Books.

Thomas, C. 2013. The Anthropocene could raise biological diversity. *Nature* 502 (7469): 7.

Thompson, K. 2014. *Where do camels belong? Why invasive species aren't all bad*. Vancouver, BC, Canada: Greystone Books.

Verburg, G. 2007. Policy memorandum on invasive exotic species. Letter by minister Verburg (LNV) describing the intentions for policy on invasive exotic species. https://www.government.nl/topics/environment/documents/parliamentary-documents/2009/10/15/policy-memorandum-on-invasive-exotic-species.

Wallach, A., A. Bekoff, C. Batavia, M. Nelson, and D. Ramp. 2018. Summoning compassion to address the challenges of conservation. *Conservation Biology* 32 (6): 1255–1265.

Woods, M., and P. Moriarty. 2001. Strangers in a strange land: The problem of exotic species. *Environmental Values* 10 (2): 163–191.

World Wildlife Fund (WWF). 2014. Living planet report. https://www.worldwildlife.org/pages/living-planet-report-2014.

World Wildlife Fund (WWF). 2018. Living planet report. https://www.worldwildlife.org/pages/liv
    ing-planet-report-2018.
Yellowstone National Park. 2019. Mountain Goat. https://www.nps.gov/yell/learn/nature/mountain-
    goat.htm. Accessed 23 March 2020.

Ned Hettinger is Professor Emeritus of Philosophy at the College of Charleston. His specialization is environmental philosophy, including environmental ethics, environmental aesthetics, and animal ethics. His papers have been published in *Philosophy and Public Affairs, Boston College Environmental Affairs Law Review, Environmental Ethics, Environmental Values*, and the *Journal of Aesthetic Education* among other places. Recent articles include "Naturalness, Wild-Animal Suffering, and Palmer on Laissez-faire" (2018) and "Evaluating Positive Aesthetics" (2017). His homepage is: http://hettingern.people.cofc.edu.

Chapter 23
Coexisting with Wolves in Cultural Landscapes: Fences as Communicative Devices

Martin Drenthen

Abstract This paper argues that many conflicts regarding the return of the wolf to the thoroughly humanized and densely populated cultural landscapes of Western Europe rest on the dualistic idea that culture and nature are two strictly separated realms of reality, and on the assumption that wild animals are primarily passive beings without proper agency. Once we acknowledge wolves as beings with agency with whom we share the landscape, we come to see that the challenge of coexistence with wild animals such as wolves is not primarily a matter of finding a compromise between human interests and the interests of wild animals. Rather, we have to learn and negotiate that the landscape is a space that is interpreted and inhabited by many different beings, with whom we are always already communicating, even if we are not always aware of it.

23.1 Wolves Recolonizing Europe

Throughout history, many animal species have sought the proximity of humans because of the opportunities this provided in terms of food and shelter. Domesticated animals have of course been members of the mixed community of humans and animals for thousands of years. The so-called liminal animals (Donaldson and Kymlicka 2011)—rats, foxes, seagulls and pigeons—have found ways to flourish in human spaces, making use of the various assets that cities provide. These non-domesticated species have also learned to live close to humans, in some cases they even prefer cities to the wild habitat in which they evolved.

However, in recent years, wild animals are also returning to our landscapes. Species such as Wolf, Golden Jackal, Lynx, and Wild Cat—animals associated with wild nature par excellence—have started to repopulate the thoroughly humanized and often densely populated cultural landscapes of Western Europe from which they

M. Drenthen (✉)

Institute for Science in Society, Radboud University, Nijmegen, The Netherlands

e-mail: martin.drenthen@ru.nl

© The Author(s) 2021

B. Bovenkerk and J. Keulartz (eds.), *Animals in Our Midst: The Challenges of Co-existing with Animals in the Anthropocene*, The International Library of Environmental, Agricultural and Food Ethics 33,

https://doi.org/10.1007/978-3-030-63523-7_23

went extinct only a few centuries ago. The spontaneous return of these wild animals sheds new light on the relation between humans and wild animals.

The return of the European Grey Wolf to Western Europe is surely the most controversial example of the 'return of the wild in Anthropocene' (Drenthen 2015). The wolf was eradicated in large parts of Western Europe in the eighteenth and nineteenth century. Only small populations remained in the East and the South (in the areas East of the former Iron Curtain, and remote parts of Italy, the Balkan and Spain). However, due to changes in land use, land abandonment in rural areas, increased legal protection, and the rise of more ecofriendly environmental attitudes towards nature (De Groot et al. 2011) wolves have been able to repopulate areas where they had disappeared earlier.

In the year 2000, Polish wolves repopulated the Lausitz region in Germany, south of Berlin. From there, the population gradually expanded to the North-West; now, less than twenty years later, they are recolonizing parts of Denmark, Belgium and the Netherlands.

In 2013, the first wolf visited the Netherlands. Each following year, the numbers of wolf sightings went up. In 2018, at least 10 wolves visited the Netherlands. In January 2019, the Dutch government officially declared that two female wolves had settled in the Veluwe, a large nature area in the center of the Netherlands. Only a few months later, in May, a camera trap revealed that one of these female wolves had a male companion, and in June it was announced that the first pups were born. This rapid colonization—from one individual in 2013 to an entire pack of wolves less than 6 years later—has caused a lot of debate about wolves in Dutch society.

23.2   Wolf Debates

Wherever wolves reappear, they give rise to social tensions. Most nature conservationists celebrate the return of the wolf as a success of European nature policies; they argue that predators such as the wolf are a welcome addition to the ecosystem and will have a positive effect on biodiversity. Many urbanites are fascinated by the idea that the wolf again lives in this densely populated part of the world. People in rural areas, on the other hand, are more worried that wolves will pose a danger to humans and livestock. However, many people appear to take a rather moderate and pragmatic approach towards the return of the wolf.

Surveys suggest that basic attitudes towards wolves among the Dutch population have not really changed much since their arrival. In 2012, just before the first confirmed wolf sighting in the Netherlands, a survey among Dutch citizens (Intomart 2012) found that about 45% of respondents would welcome the arrival of wolves, while 32% were against, and 23% took a neutral position. Ever since the actual arrival of the wolves in the Netherlands, these numbers seem not to have changed much. A survey in December 2018, found that 30% of respondents feel that wolves should not be restricted in their behavior, while an equally large group does not agree. However, 58% agree with the proposition that we should not chase away the wolf, but rather should be focusing on protecting ourselves and our livestock from the wolf (Kantar Public 2018).

In short, the majority seems to have a pragmatic attitude towards wolves, one that combines a willingness for coexistence with wolves with a cautious approach to possible risks associated with the presence of predators in our landscape. They are willing to give these animals a chance of finding a place in our landscape, and feel that we should seek to find a form of coexistence with wolves. Yet, most people also believe we should be on our guard for those individual wolves that cause disproportional damage to livestock and pose a threat to humans and their interests (Van Slobbe 2019).

However, in the debate about wolves, the extreme voices tend to be the loudest, and as a result, the debate easily becomes polarized. A small group of wolf lovers tends to romanticize the wolf. They refuse to acknowledge that the return of a large predator in a cultural landscape can be troublesome; any problem that might emerge will be solely to blame on humans (Drenthen 2016a). On the other side, some people are convinced that there is no room for wolves in a densely populated country like the Netherlands. They argue that the Dutch landscape is no longer suitable for these animals, because the Netherlands lacks the large-scale nature areas that most people associate with wolf habitat. Especially sheep farmers vehemently oppose the wolf; they see wolves as a threat to their livelihood.

23.3 Wolf Predation on Livestock

Young wolves leave their parents when they are two years old. They start to roam the landscape in search for new unoccupied habitat. Wolves can travel up to 80 km each night, and can migrate over many hundreds of kilometers from their birthplace. These wandering wolves try to keep a low profile while migrating through strange wolf territory in order to prevent conflicts with settled wolves. These wandering wolves are the biggest threat to domestic animals such as sheep. Wandering wolves tend to see sheep as an easy 'snack' along the way. As soon as wolves set up a new territory, they tend to shift their attention to wild prey such as deer and wild boar—probably because they prefer deer to wooly sheep, but also because they prefer to hunt away from human land. Moreover, young wolves learn how and what to hunt from their parents, and will normally learn that wild ungulates are the preferred food source. Yet, wandering wolves need to make a different trade-off and will not miss out unprotected sheep as an easy meal.

Because large carnivores have been absent for so long, sheep farmers in Western Europe abandoned the traditional methods of protecting livestock against predators that are still in use elsewhere. In those places where wolves never disappeared, people see wolves as belonging to the landscape and are used to dealing with them. They use so-called guard dogs or flock-protection dogs to protect the flocks against wolf attacks, and sometimes shepherds themselves stay with their sheep around the clock (White 2019).

In contrast, in those places where wolves went extinct, farmers abandoned the old practices that protected livestock against predators: sheep stay behind low fences designed to keep sheep in, but not to keep predators out. Moreover, without any natural predators, sheep are left to roam in relatively safe hillsides, without shepherds. Many sheep farmers only visit their livestock occasionally to check on their health. In the mountainous regions of the Alps, for instance, farmers stopped with the practice of full-time shepherding when large predators disappeared. Nowadays, in certain areas, the role of shepherds has become redundant (Hollely 2018).

The return of wolves to new territories forces shepherds to change their shepherding practices and start protecting their sheep again. Some proponents will argue that this should actually be a reason for shepherds to welcome wolves: wolves are good for shepherds, because without wolves, they would not have a job.

> Another positive effect of [the return of] the wolf is also that it is bringing back the shepherds. Shepherding has been part of Europe's cultural landscape for thousands of years. The increase in shepherds is now leading to the renovation of mountain huts, and increased job opportunities in rural communities. Switzerland has just opened up a new school in shepherding in recognition of the need for shepherds through Europe. (Hollely 2018)

Yet, it is also clear why sheep farmers in Western Europe dread the return of the wolf, since this change in keeping sheep, will make shepherding more time-consuming and less cost-effective, thus making their life more difficult.

23.4   The Cultural Conflict About Wolves

In order to help the acceptance of returning wolves in Europe—the wolf after all is a legally protected species—governments across Europe have come up with financial schemes that support farmers that suffer from livestock predation; they offer financial compensation for their losses and help with preventive measures such as electric fencing.

Financial compensation schemes certainly help the acceptance of wolves, but only up to a certain point. Most farmers do not just fear the financial costs of loss of sheep but also despise the killing of their animals by wild predators.

Wolfs do not just pose a financial risk; they also are a perceived threat to an entire way of life. Farmers in rural areas are usually well aware of the vulnerability of their livelihood vis-à-vis the threats posed by uncontrollable natural events: thunderstorms, floods, draughts—and predation. Part of the identity of being a farmer revolves around the idea that the land has to be 'worked' and cultivated, by controlling and managing wild nature if one is to make a living from the land.[1] Living a life that consists of facing up to the encroaching wild, farming communities typically value a sense of independence and autonomy. Farmers take pride to be able to grow their own food

---

[1]Research shows that this attitude also translates into an aesthetic landscape preference: rural populations typically appreciate more manicured and orderly landscapes compared to the type of landscape urbanites like (Van den Berg and Koole 2006).

and live a life of independence. For many of them, it is important to be autonomous and self-sufficient, and not dependent on the government, as so many urban dwellers are. The presence of a wolf undermines that feeling of being autonomous.

Moreover, the idea that the state will compensate for the financial losses due to wolf predation may itself be problematic to some. The idea of getting money from the state without having to work for it in itself does not fit well in this ideal life of self-sufficiency (Thorp 2014). Some wolf opponents complain that too much taxpayer's money is being spend on the losses due to wolves, even though they are the recipients of that money. Farmers opposing wolves do not want financial compensation; they want their livestock to be safe from predation.

It is a commonplace that human-wildlife conflicts are in fact most often human-human conflicts and the same applies here. It is not just the fear of livestock predation that fuels the opposition against the wolf, but also the unease about the overall societal movement that welcomes the wolves and argues for their protection. Those who oppose the wolf see its return as a threat to a way of life that exists in living a life of independence in the face of the encroaching wild, but they also feel that society fails to take their worries seriously.

There exists a deep sense of unease among wolf-sceptics with the way that urbanites tend to approach the issue of coexistence with wild animals, and with nuisance animals and large predators in particular. On Facebook accounts of anti-wolf groups such as No Wolves,[2] one can illustrate this sentiment: "City people are naïve, and fail to see that these animals are dangerous." "All these animal cuddlers in the city ignore the real danger." "These people don't know what misery the wolves will be bringing to us." "Eventually they will find out that we were right and they were wrong in trusting this animal, but by that time it will be too late." "Only when the first children will be killed by a wolf will they realize what they have inflicted upon us."

Opponents of the wolf say that the Dutch landscape is no longer suited for the wolf, and fear that they will be forced to change their way of life if large predators are a permanent presence in the landscape. Wolf sceptics worry that society fails to acknowledge the fact that some of these animals can be dangerous to human interests. More fundamentally, however, their worry is that society fails to consider their perspective. By focusing on wolves as vulnerable animals that need protection, society seems to ignore that living in rural areas means being vulnerable to forces of nature that the average city dweller does not have to face.

However, rather than bringing forward what may be justified feelings, many wolf opponents instead try to convince others by claiming that wolves do not belong in the landscape, that they cannot be controlled, and that their impact will be mostly negative, et cetera (Van Herzele et al. 2015). Whereas some conservationists see the extermination of the wolf in the nineteenth century as a wrong to be undone, some wolf opponents argue that there were good reasons to exterminate wolves in the past and that we should try to exterminate them again today. Others argue that the animal might have been in place in the past, but that the landscape has changed

---

[2]https://www.facebook.com/nowolvesnl/.

since then; that it may be that the wolf once lived in these regions, but that now it no longer belongs there. Still others claim that the wolves that are currently roaming the landscape are in fact not at all wild wolves. Rather, so the argumentation goes, these animals either have escaped from private zoos or have been released on purpose; they are probably hybrids and therefore should not be protected at all. As far as the wolf opponents are concerned, the wolf is not a native species taking back its rightful place in its original habitat, rather it is a dangerous intruder into human space.

In an attempt to circumvent the European protection schemes, some wolf opponents are claiming that wolves did not coming back spontaneously at all, but have been intentionally introduced by radical environmentalists. The narrative that the wolf resurgence is not at all a spontaneous occurrence, but rather the result of a secret introduction by environmentalists is very persistent, and can be witnessed in many places where wolves are returning after being absent for a while (Skogen et al. 2008).

As can already be gleaned from the above, the conflict about wolves appears to be linked with a much more general *cultural* conflict of worldviews. This cultural controversy prevents a pragmatic approach to wolf predation of livestock, because in that context, wolves become a symbol for the larger rapid changes in the landscape that people see as a threat to their way of life and their sense of identity. The wolf can easily become a symbol for the regional populist movements to signify a more general feeling of unease towards rapid disappearing of local identities and loss of control due to globalisation and the dominance of urban elites, along with other perceived threats like immigrants, European bureaucracy, and the environmental regulations in general. It may be only a matter of time before populist movements in the Netherlands discover the potential of the wolf case and decide to exploit the feeling of unease among rural voters against the urban elites for political purposes.

It seems that the legal protection schemes for wolves fuel the sense of unease wolf opponents, who feel that they cannot protect themselves and their livestock against a perceived threat. Capitalizing on this sense of disenfranchisement, some politicians with constituents in rural areas argue that the strict European legal ban on wolf hunting should be lifted, and the animals instead should be "managed", a euphemism for population control through lethal measures. In other words: a part of the rural population is aware of its vulnerability and feels that the danger is not taken seriously by those who see wolves as an asset to nature. Moreover, they do not believe that it will be possible to find ways to coexist with predators with little or no conflict.

23.5 The Stewardship Model as Underlying Cause of the Conflict

In order to counteract human animal abuse and the instrumental exploitation of animals by humans, both animal ethicists and conservation lawmakers have sought to articulate principles and values that should govern human behavior with respects to the intrinsic value of animals. In the case of wildlife, the dominant approach is to find ways in which wild animals can live their life in the wild, undisturbed by negative human impacts. However, in their focus on human wrongs, animal ethicists typically have difficulty dealing with those cases of conflict where wild animals pose a threat to the interests of humans or their livestock.

The European Habitat Directive demands that all member states have to ensure that endangered wild animals are in a "Favorable Conservation Status" (Epstein et al. 2015). It is illegal to hunt or disturb endangered animals.[3] This idea is relatively uncontested in case of those species that are relatively innocent and do not cause nuisance to humans. However, in the case of recolonizing predators such as the wolf, it is controversial, especially when these animals enter cultural landscapes inhabited by humans.

Much of our thinking about our relationship to wildlife revolves around the idea that animals are vulnerable and should be protected against intrusions and harms inflicted upon them by humans. Starting from the premise that wildlife typically lives in nature, nature protection often involves setting aside nature reserves, wildlife sanctuaries or national parks as habitat for animals. In this model, animals are mere passive recipients of human concern. The problem with this approach in the case of the returning wolf is that these animals surely appear to have agency and sometimes even pose a danger to our human interests. To those who fear its presence, the wolf is not an innocent victim, but a potential dangerous actor.

Therefore, it would be useful if we could develop an approach to wolves that recognizes the animal's agency, and the fact that there can be conflicts between animals and humans, and yet without framing nuisance animals as intruders or enemies.

23.6 Wolves as Sovereign Beings

In their 2011 book *Zoopolis*, Sue Donaldson and Will Kymlicka criticize the dominant framing of our relation to wild animals in terms of the 'stewardship model',[4] because that model fails to take seriously the agency of wild animals and sees them purely as passive objects of our care. Instead, their theory of animal sovereignty recognizes animals as having agency and acknowledges the independence and autonomy of wild animal communities.

---

[3]Most legislation also makes clear when circumstances are not normal, and permit lethal and nonlethal intervention, for instance when an individual animal behaves 'unnatural', and is showing aggressive behavior towards humans, or is specializing on predating protected livestock.

[4]Sue Donaldson and Will Kymlicka criticize this dominant "stewardship model", because even though in this model "human access and use might be strictly limited" this is "not as a recognition of animal sovereignty, but rather as an exercise of human management. This stewardship may be relatively interventionist or relatively hands off, but either way the relationship is conceptualized as one in which a human sovereign community has set aside a territory for a specific use, and to which the human community retains the right to unilaterally redefine boundaries and use" (Donaldson and Kymlicka 2011, 170).

Donaldson and Kymlicka distinguish three types of animals, each with a different relationship to humans. *Domesticated animals* are those animals that live in close proximity to humans and have intense relations of interdependence with human beings. According to Donaldson and Kymlicka, we should treat these animals as co-citizens, that is to say, as full members of a mixed community of humans and animals held together through all kinds of mutually beneficial relationships. Justice demands we grand them the appropriate rights and duties that belong to being part of such a community. The so-called *liminal animals* also live in the proximity of humans, but unlike domesticated animals, they are relative outsiders to our community. They have adapted to living amongst humans and even have become dependent on living in our proximity (for food or for housing), but they do not have intimate and reciprocal relationships with humans. Therefore, they do not possess all the rights and duties of full citizens, but they do deserve respectful treatment as co-inhabitants of the place we live.

Most relevant for us, though, is their view on *wild animals*: those animals who avoid humans and human settlements, and maintain a separate and independent existence (insofar as they are able to) in their own habitats or territory. Donaldson and Kymlicka (2011) suggest that we understand our relationship with wild animals in similar terms as our relationship with different human sovereign communities such as nation states. A fair relation between sovereign communities of humans and wild animals aims at a just allocation of harms and benefits between two communities. From this starting point, so they argue, we should recognize "that the flourishing of individual wild animals cannot be separated from the flourishing of communities, and…reframes the rights of wild animals in terms of fair interaction between communities" (p. 167).

This view on the relation between wild animals and humans can be useful to our discussion of wild wolf's comeback, because it not only acknowledges their agency, but also makes room to think about possible conflicts between animal and human interests. In highlighting animal communities as sovereign entities, it becomes possible to think of our relation with wild animals in a way that is more symmetrical. On the one hand, "the recognition of animal sovereignty limits our actions in terms of encroaching on wild animal territory, and imposes obligations on us to take reasonable precautions to limit our inadvertent harms to wild animals (e.g., by relocating shipping lanes, or building animal bypasses into road construction)." On the other hand, "it also limits our obligations in terms of positive assistance to wild animals."[5] Similarly, "it restricts the terms on which we can visit sovereign wild animal territory (or share overlapping territory), but at the same time it establishes terms for wild animals entering sovereign human societies." The sovereignty approach "obligates us to respect the basic rights of animals, but also protects us from violations in return" (ibid.).

Thinking of wild animals as sovereign entities implies that humans do not only have duties towards them, but also allows humans to pose limits on animals as long as these are part of a just arrangement that distributes costs and benefits fairly among

---

[5] See Clare Palmer's Chapter 21 in this volume.

different sovereign communities. Thus, the sovereignty theory seems much better equipped than the stewardship model to deal with the question of how to deal with returning wolves in cultural landscapes in a way that recognizes the worries of those who fear the arrival of wolves in cultural landscapes.

23.7 Parallel Sovereignties in a Shared Landscape

Even though resurging wolves do not actively seek out humans, it is inevitable that they run into contact with humans because they are traveling across huge distances through one of the most densely populated landscapes in Europe in search for territory (a pack needs 50–300 square kilometers). Moreover, because they are highly intelligent and flexible animals, they succeed surprisingly well in living in our humanized environments and in remaining unnoticed by most humans.[6] Since wolves are resurging in landscapes that were until recently exclusively human habitat, the questions regarding our relationship with wild wolves are of a different kind than those with wild animals that stay in their place within "nature".

Many problems in our relationship to nature in general and in our relation with predators in particular, derive from a dualistic way of thinking that assumes there is a clear boundary between "nature" and "culture".[7] Even the most outspoken opponents of wolves would be fine accepting that wolves should have a place "in the wilderness".

Problems appear when wolves show up in cultural landscapes. To many wolf opponents, the mere fact that wolves appear in a cultural landscape is a sign that there is something wrong with these animals. They argue that "a real wolf would never come this way" or that "a real wild wolf would never chose to walk on a sidewalk or use a human road or bridge." As soon as a wolf is comfortable navigating the human landscape, they see it as a sign that the animal must be a hybrid, an escapee from a zoo, or intentionally introduced by "rewilding activists". The simple fact that wild wolves appear in cultural landscapes is in itself already undermining the worldview that depends on a clear schematic separation of "nature" and "culture".

It is this dualistic frame of mind—in which wilderness and cultural landscapes are seen as two mutually exclusive domains of reality—that underlies many of the problems regarding wolves returning to cultural landscapes. For those who are against the wolf's return, the animal is an intruder, and the spontaneous resurgence of the wolf means a breach in the comfortable separation between wild lands and cultural landscapes (Drenthen 2015). However, recognizing wolves as real animals that live in the ecological and social context of our landscapes means that we have to acknowledge that sharing spaces with large carnivores will never be easy. We need to find

---

[6]In early 2019, a radio-collared wolf entered the Netherlands from Germany, walked all across the urbanized center of Netherlands to finally end up and settle in Belgium, all without being noticed by humans. See: https://www.wur.nl/nl/nieuws/Wolf-doorkruiste-Nederland-van-noord-naar-zuid.htm.

[7]See Cor van der Weele's Chapter 30 in this volume.

a *modus vivendi* that allows us to live together with wolves, and that will require some degree of management and control. Respecting wolves as sovereign entities, and more generally respecting nature's autonomy, also implies a willingness to live with wild creatures, not just when they are charismatic and cute, but also when they are a nuisance.

In reality, nature and culture are not different domains of reality, but rather two aspects of the same reality. "Culture" is what we have designed and control, "nature" is what we do not fully control or what constantly challenges our control. There is always nature in the cultural landscape (and that certainly applies to the wolf who will see every border as a challenge) and always culture in the so-called nature reserves (where we are not only occasionally present, but which we also manage and protect). Our landscape cannot simply be divided into two separate parts. Those cases where wild animals stay within a designated nature reserve are the exception. In principle, we always live in the same landscape as wild animals; more often than not, our territories will overlap.

> Different wild animal species occupy (and compete for) the same territory, and many species need to move large distances across territory occupied by other animals or by humans. Sovereignty, therefore, if it is to mean anything in practice, cannot be tied to a picture of neatly divided communities and territories. (Donaldson and Kymlicka 2011, 188)

This means that we need a perspective on the landscape that acknowledges that we inevitably share landscapes with multiple species. Recognizing justified claims of sovereign animal communities on the landscape can be complicated. Respect for the sovereignty of wild animals implies that we have to allow them to migrate through our territory. Conversely, we may use our own corridors to cross their territory or visit wild places, as long as we do so in such a way that damage to the sovereign animal community is kept to a minimum.[8] Sometimes, a strict territorial separation will be possible between wilderness areas with very restricted access for humans, and areas for humans and domestic animals that are restricted for wild animals. However, especially in the case of species with large territories, we need to acknowledge that our territories will inevitably overlap and thus abandon "an overly simplistic concept of territory-like the boundaries of a national park". Instead, to allow for "sustainable and cooperative parallel co-habitations" (ibid., 191) with animals with territories overlapping ours, we must think of the landscape as a "multidimensional" place.

---

[8]"Sovereignty is importantly tied to territory, since a community, especially most animal communities, cannot be ecologically viable, let alone autonomously self-regulating, without a land base to sustain it. But sovereignty need not be defined in terms of exclusive access or control over a particular territory, but rather in terms of the extent or nature of access and control necessary for a community to be autonomous and self-regulating." (ibid., 190).

23.8   Living in a Multidimensional Landscape

Sharing a landscape with wild animals means that that landscape has become a "multidimensional" place, where several landscape features will mean different things to different beings. Our territories may overlap in spatial terms, but we may nonetheless inhabit different worlds in semiotic terms. What for us is an office building or an apartment building may appear a rock face to a pigeon or a peregrine falcon. Similarly, wolves will interpret our cultural landscapes differently than we do.

The challenge is that these different semiotic worlds do not exist independently of each other. We are not the only ones to define the meaning and significance of features in the cultural landscape. These differences in interpretation may lead to conflicts if wild animals make decisions based on different interpretations that do not go well together with those of humans. Coexistence with wildlife is therefore not just about finding a compromise between human and animal interests. It is about figuring out *what it means* to inhabit a place that means different things for different inhabitants.

According to Susan Boonman-Berson (2018), co-habitation between humans and wild animals "requires that wildlife management be approached as an interactive and dynamic endeavor that focuses on the relation between humans, wild animals and landscape" (p. 64). She argues that such a relational perspective "implies an understanding of agency and subjectivity as emergent and as produced through learning in practice and through interactions between humans, wild animals, and the landscape" (p. 93). Both humans and animals interpret the landscape and change it—they 'read' and 'write' the landscape.[9] In order to understand how these readings and writings may interact with each other, she introduces the concept of 'multi-sensory writing and reading'. It this process, "signs (visual, olfactory, auditory, tactile), materialized in words, signals or things, are communicated between humans, between wild animals and between humans and wild animals through the writing and reading of these signs in the landscape" (pp. 70–71). The communication with wild animals does not take place directly, rather, it is itself based on material traces or signs. The shared reading and writing of the landscapes takes place on the basis of material aspects that can be accessed and interpreted by both humans and wild animals (p. 71).

Boonman-Berson points out that "communication with wild animals is apparent in most traditional hunter-gatherer practices where the success of a hunt may depend on the ability of a human hunter to 'think like' his or her quarry", which "requires years of experience in which animals are recognized as independent actors" (p. 71). Yet, she stresses the symmetry of the situation. Humans may read traces and signs left by animals, but animals do the same thing:

> Our notion of multi-sensory writing and reading recognizes that both humans and animals leave traces as well as trace, interpret and respond to these traces. Human writing in our case refers not only to the use of words, such as in policy and management plans that describe how to deal with human-wild animal interactions, but also to communication without words,

[9]In a similar vein, the ecosemiotic work of Morten Tønnessen (2011), clearly shows how the semiotic relationships between humans, wolves, and sheep have shifted and changed through time.

such as in placing fences to physically demarcate human from nonhuman spaces. Animal reading refers to the interpretation and enactment of these human writings as they become observable in the changing behavior and movements of the animals. The responses by animals are communicated through signals or things – animal writing –, such as footprints, left fur, faeces, and scent markings that indicate their presence in an area or demarcate their territories. (p. 72)

In order to find ways of dealing with the fact that we share a landscape with other beings, we have to understand how these other beings understand and navigate the landscape, and how they may understand and interact with the signs that we— consciously or unconsciously—leave behind in the landscape. Only if we can find ways of translating one species' interpretation into another, may we develop an understanding of the multidimensional landscape that can help avoid unnecessary conflicts.

23.9   Wildlife Management and the Biosemiotics of Borders and Fences

Living together in a layered multi-dimensional landscape with potentially dangerous predators such as the wolf, means that we will inevitably share the landscape and that we therefore have to come up with arrangements to avoid conflicts. Luckily, there are many alternative methods to lethal predator control that can solve the so-called "predator paradox": that in order to protect predators we need to protect humans against predator impacts (Shivik 2014).

One of those arrangements is the use of fences, in particular fences that keep predators from attacking livestock. Sheep are part of the mixed community of humans and domestic animals that is held together by mutually beneficial rules, with all parties having certain rights and duties towards each other. Humans have a duty to protect their sheep against predation and other harm. However, we also have a duty towards wild animals as sovereign beings to make sure they are able to live their lives and satisfy their essential needs. When wild animals predate on livestock, both duties come into conflict.

Yet there are ways in which we can navigate these conflicts and try to prevent them from playing out. In the case of potentially conflicting human land uses, we developed several ways that might inspire to find solutions in our relation to wild predators as well. For instance, in forests and other recreation areas, we developed ways to ensure that hikers do not run into conflict with mountain bikers and equestrians. Starting from an acknowledgment of the different needs and desires of these different groups of users, we devised a network of hiking trails, MTB trails and equestrian trails that avoid these groups from unnecessarily bumping into each other.[10] These arrangements only

---

[10]In addition to these spatial arrangements, one can also to make temporal arrangements, where different groups use the same location at different times of day. Humans and wild animals often use the same places at different times. Is it really such a big problem when wolves roam our streets

work as long as each user group accepts to stick to the designated trails. There is no absolute guarantee that all parties always abide to these rules; that would require stationing a park warden or police officer in every remote corner of the forest. In reality, these spatial arrangements usually also work without policing, because it is in the interest of each party to avoid unnecessary conflict and stick to the designated path. That only works, of course when the path design matches the needs of the particular user group, so that there is no need to break the implicit rules. If that is the case, we dare to trust each other, and are even willing to accept that occasionally an individual will violate that trust. The courage to trust each other is what enables different users to coexist and avoid unnecessary conflicts.

In a similar way, we could think of fences as means to prevent conflicts between predators and livestock. In the case of wolves, the problem is a lack of trust: many wolf opponents do not dare to rely on the possibility of coexistence without policing. They refuse to believe in a peaceful coexistence with wolves and claim that wild wolves will always attempt to climb over or dig under fences that wolf experts considered impenetrable to wolves.

It may be tempting to respond to this lack of trust by suggesting that fences can indeed impose a strict border between human and sheep on one side, and wolves on the other. In reality, however, once wolves are determined to cross a fence, most fences will not be able to stop them. The idea that it will always be possible to keep wolves out is based on a misunderstanding of what wolves are capable of. What is key, however, is that even if there is no absolute guarantee, in those landscapes where people and wolves have successfully learned to live together, most fences do seem to work nonetheless. Apparently, in these places human culture and wolf culture have somehow converged.

In those places where sheep farmers consistently protect their sheep with fences that are difficult enough for a wolf to cross, wolves typically have learned to change their behavior in response; they shift from preying on domestic livestock and focus instead on wild prey. Young wolves learn how and what to hunt from their parents, and will also tend to ignore livestock (provided that there is enough wild prey available). In time, a local wolf culture will develop in which young wolves are taught that it is much easier and safer for them to hunt wild prey and leave sheep be.

Evidence points out that this process is currently taking place in the new wolf territories in Germany. Recently, in the German state of Niedersachsen livestock predation has been going down while the number of wolves is still on the rise (Wolven in Nederland 2019). In Sweden, where wolves have been present for quite some time now, livestock predation is relatively low due to active measures such as state subsidies on prevention measures such as fencing. Yet, even though Swedish government subsidizes preventive measures and refuses to compensate farmers who did not protect livestock properly, some farmers even decide to not build fences because of

---

in those hours that we humans are asleep? It would mean, of course, that we would have to accept that there are times in which particular places are not exclusively ours, that there are certain times when we would be the ones out of place. Would it be possible that wolf and humans learn to deal with each other by negotiating mutual temporal divisions?

the low risk and the trouble it takes to build them. Wolves have stopped attacking sheep anyhow (Karlsson and Sjöström 2011).[11]

Yet, it is also clear that in practice fences will never be a fully impenetrable barrier that can uphold a strict separation between our cultural landscape and their wilderness. Instead, these fences are mere communicative devices that help parallel sovereignties navigate a multidimensional landscape and arrive at a common understanding.

23.10 Building Communities with Humans and Wolves

There is another dimension to fences that should be mentioned here. We have seen that one of the problems of the spatial arrangements is the lack of trust by some parties, both a lack of trust in the possibility of peaceful coexistence with wolves, and also a lack of trust in society to take seriously the worries of people in rural areas. In response to this problem, in some European countries, groups of volunteers have started to help farmers to build fences to protect their livestock.[12] Typically, these volunteers are people living in the city who want to help find a peaceful coexistence with the wolf. They know that there may be good reasons to like the idea of wolves returning, but also realize that the farmers will have to put up with the downsides of the presence of wolves in the landscape. These volunteer groups help farmers and sheep in order to help the wolf.

One of the sheep farmers helped by Wolf-fencing, Gijsbert Six, told a local newspaper how he views the wolf's return and the help from volunteers:

> I'm ambivalent about it. On the one hand, the wolf is a fascinating animal, on the other hand he kills sheep. And not in a gentle way, as I have seen for myself. This is quite a tough blow when it happens to you as a sheep farmer. Yet, for thousands of years we have eradicated everything that stood in the way. Should we also do that with wolves if their visit is not convenient for us? As long as it's not disruptive to society, I think we should learn to live with it. The fact that the volunteers of Wolf-Fencing helped me easily saves me a thousand euros in costs. *Plus, I feel like I'm not alone.*"[13]

As the embodiment of solidarity between farmers and non-farmers, the volunteering work of Wolf-fencing and similar groups contributes to sense of community. In the

---

[11]It should be noted, though, that experts warn that young wolves, like human adolescents, do like to experiment, and might be tempted to attack sheep anyhow if an easy opportunity presents itself. One should prevent them from developing a taste for sheep, and make sure that any attempt to attack a sheep will result in an unpleasant experience, such as an electric shock from an electric fence (Reinhardt et al. 2012).

[12]In Germany, a group called 'Wikiwolves' (http://www.wikiwolves.org/) started in 2015 in North-East Germany, and has grown ever since. Now they are active across the country (http://www.wik iwolves.org/). In 2018, a similar group called 'Wolf-fencing' (https://www.wolf-fencing.nl/) has begun work the Netherlands (building wolf-fences based on fences used in Sweden), one year later followed by Wolf Fencing Belgium (https://www.wolffencing.be/).

[13]https://www.wolf-fencing.nl/voor-wie. My emphasis, MD.

end, the biggest effect of this volunteering work may be that there is also a developing mutual understanding between the city dwellers and the farmers; both groups learn to understand the worries and concerns of the others in the process, friendships might even emerge.[14] The volunteers get to understand the fears and worries of livestock keepers, whereas the farmers may get a better understanding of the reasons for wanting to protect wildlife. Moreover, by working together on building fences, both groups get to know each other better, thus a sense of community can emerge that can counteract the feeling of being disenfranchised. These volunteering groups *and* the sheep farmers who decide to accept their assistance, together built a human culture of devotion to the very idea that living with predators is possible and a goal worthwhile pursuing. At the same time, by building fences, they also help develop a culture among wolves that is conducive to that coexistence.

Living with wolves of course also requires that we take into account the wolves' way of understanding the world, their needs and behaviour, and that we are willing to communicate with them. If we want to live with wolves, we must learn from and about wolves, and wolves must be able to learn from us how they can live with us. An analysis of the possibilities and limitations of the interspecies communication can help us understand how our actions not only have meanings for ourselves but also for the nonhumans we share the landscape with, and it can make us realize how we have changed the world of wolves.

In the end, however, the pragmatic approach to coexistence with wolves will only be possible if society recognizes that sharing the landscape with other beings is somehow meaningful, and therefore considers coexistence with other beings as a goal worthwhile pursuing.

23.11 The Meaning of Living with Wolves

An understanding of the landscape as a multidimensional space should become part of a broader world view that gives meaning to the fact that we co-inhabit the landscape. The pragmatic perspective on multidimensional landscapes has to be integrated with an overall *human* perspective on the meaning of coexistence with wolves, in which all the objective features are put into the interpretative narrative context and comes to *mean* something (Drenthen 2016b). We have to engage in a moral conversation about what the current situation means and what it requires. An understanding of the role of wolves in the landscape can make clear what the challenges are, and how they can be met. But what is also needed is a *reason why* we should be prepared to respond to those challenges in the first place. Why should we be willing to share our

---

[14]In a video clip on the wikiwolves website, a volunteer makes clear how the initial interest in wolf protection gradually changed into an interest in shepherding, and in helping shepherds with their problems. The aim is "to help the sheep farmers and let them know they are not alone." Euro LargeCarnivores (2019).

world with these sometimes annoying animals in the first place, rather than trying to finally get rid of them and have the world all to ourselves?

One reason might be that the world becomes more interesting and meaningful when there are other beings present. The world would be a lonely place if we would only encounter beings like ourselves.

A few years ago, I visited the Harz National Park in Germany, a wild remote area located along the former iron curtain in the center of Germany. On a hike in a forest, I met a park ranger and asked him if he knew whether wolves had already started to recolonize the area. "Not that I am aware of", he responded. "But while we are talking here, they might be watching us from behind the trees. In nature, you are never alone. In the forest you are not the only one watching, you are always also being watched. One pair of eyes is looking in, while a thousand pairs of eyes are looking out."

The ranger referred to an old German saying "Der Wald hat tausend Augen" ("The forest has a thousand eyes") that reminds hunters that the animals that they are hunting are aware of their presence. The saying articulates an awareness of the landscape as a multidimensional space, inhabited by a multiplicity of species that all have their own perspective on the world. The multidimensional landscape is a world full of meaning and wonder, because it is not just our world but also the world of other beings. In other words, the presence of other beings makes the world a larger, more interesting place.

Moreover, being confronted by the existence of other beings can make us aware of the finite nature of our own bodily existence in and perspective on the world; other beings open, as it were, a transcendent realm beyond our own daily mundane existence.

This idea is articulated more clearly by John Berger. Berger argues that the way we usually experience the world, is habitual, and is confirming our existing worldview. "What we habitually see confirms us" (Berger 2009, 9). However, every now and then we are presented with something that breaches this self-confirmation, and this sometimes happens when we encounter animals. Berger uses an interesting cinema metaphor to explain how animals inhabit a different world than we do:

> The speed of a cinema film is 25 frames per second. God knows how many frames per second flicker past our daily perception. But it is as if at the brief moments I'm talking about, suddenly and disconcertingly we see *between* two frames. We come upon a part of the visible which wasn't destined for us. Perhaps it was destined for night-birds, reindeer, ferrets, eels, whales... (Berger 2009, 10)

Because the world of animals is usually inaccessible to us, and yet is not entirely separated from us either, animal encounters can confront us with the fact that we cannot take our normal view for granted. The realization that "our customary visible order is not the only one: it co-exists with other orders" may be enriching: "Stories of fairies, sprites, ogres were a human attempt to come to terms with this co-existence" (ibid., 10) But the same experience can also be deeply confusing or unsettling. Seeing between the frames may be an uncanny experience.

Berger discusses the enigmatic work of Finnish photographer Pentti Sammallahti, in which dogs appear as beings that are "attuned both to the human order and to other visible orders" and as such they inhabit the "interstices," the world *in between* ours and theirs. As a result, in each picture

> the human order, still in sight, is nevertheless no longer central and is slipping away. The interstices are open. The result is unsettling: there is more solitude, more pain, more dereliction. At the same time, there is an expectancy which I have not experienced since childhood, since I talked to dogs, listened their secret and kept them to myself. (ibid., 10–11)

Berger suggests why it is that the presence of some animals seems to be so unsettling to some people and fascinating to others. Berger argues that even though we can appropriate animals by killing and eating them or by taming them, essentially they remain alien to us:

> But always its lack of common language, its silence, guarantees its distance, its distinctness, its exclusion, from and of man. Just because of this distinctness, however, an animal's life, never to be confused with a man's, can be seen to run parallel to his. Only in death do the two parallel lines converge and after death […]. With their parallel lives, animals offer man a companionship which is different from any offered by human exchange. Different because it is a companionship offered to the loneliness of man as a species. (ibid., 14–15)

Wolves may not be experts of interstices in the way dogs are, but they do have their own distinct perspective that is different from ours. Moreover, by performing their lives as sovereign beings independently of us, they are also putting into perspective the human order.

The fact that our landscape is being inhabited by wolves may be unsettling because it undermines the steady traditional human-centered view of the world in which all the meanings are a given, and our view of the world is being confirmed. Yet, to those who are open to it, the presence of these other creatures opens up a new perspective on the world and our place in it. This may explain why the responses to the resurgence of wolves are so extreme, and wide ranging: from deeply felt fear and hatred to wonder, excitement and awe.

According to environmental philosopher Glenn Deliège, this ambivalence has to do with a more fundamental characteristic of how we experience meaning altogether. Deliège (2016, 414) argues that we can only experience meaning when we are engaged with a reality that is external to ourselves:

> The reality towards which we are oriented in our quest for meaning and with which we hope to establish contact is only really external to our desires if it can negate those desires and resist full appropriation. It is only when our quest for meaning can be denied by the reality towards which we are oriented, that we know we are oriented towards a reality that is truly external to us. The presence of meaning is thus premised on the possibility of its counterpart: the denial of meaning.

Building on the work of Deliège, Mateusz Tokarski (2019, 121) argues that nature "as a domain independent from human control, an autonomous order of existence that follows its own purposes", can have "an importance in human life that transcends mere usefulness." But Tokarski also stresses the ambivalence of experiences of the unruliness of nature.

> There is [...] a clear ambivalence in the experience of such meaningfulness of nature: the possibility of meaning is dependent precisely on the possibility that we will not find meaning, that the meanings we seek in nature will not be confirmed. This ambivalence is not just a side effect we grudgingly accept. Rather, such ambivalence is intricately tied to how we come to experience meaning. (ibid.)

In his 'ethical guide to ecological discomforts', Tokarski concludes that the unruliness of nature appears to be a requisite for experiencing nature as meaningful altogether, and that for this reason, experiences of unruly nature are even more significant: "As such, ecological discomforts acquire a constitutive role in our interactions with the non-human world" (ibid., 121.).

The analysis of Berger, Deliège and Tokarski can explain both the feeling of unease with the presence of unruly animals such as the wolf in human landscapes, but also the feeling of wonder and excitement that the animals evokes. But even more so, it also confirms that living in a multidimensional landscape with unruly beings might be a complicated task, but also one that is deeply significant and worthwhile.

23.12   Conclusion

Living together with resurging wolves in a cultural landscape can only be achieved if we think of wolves not as merely vulnerable endangered species, but also recognize them as sovereign beings that belong to a sovereign community. Recognizing their sovereignty means accepting the legitimacy of their spatial claims, but that does not have to imply we give up our own legitimate claims. The challenge of coexistence however, is not a matter of finding a compromise between human interests and the interests of wild animals. Rather the challenge is to find a new understanding of the landscape as a multidimensional space inhabited by many parallel sovereignties. It means that we have to realize the fact that we are always already communicating with other beings, even if we are not always aware of it.

Our relation to these sovereign communities of wild animals will inevitably contain tensions, and a need to keep distance from one another, despite the fact we co-inhabit the landscape. Often we will be able to live next to each other in peace, sometimes our relationship will be more challenging, and in those instances, our respect for these sovereign beings will be the "respect" for a powerful opponent. However, if we learn to appreciate the game, we can also learn to wish our opponent all the best.

At the same time there is something to be won, our world can become bigger, more rewarding and more meaningful, knowing that we live in a landscape that is bigger than us, knowing that we are not the only ones using, knowing and understanding the land.

References

Berger, J. 2009. *Why look at animals?* London: Penguin Books.

Boonman-Berson, S. 2018. *Rethinking wildlife management: Living with wild animals.* Wageningen: Wageningen University.

De Groot, M., M. Drenthen, and W. de Groot. 2011. Public visions of the human/nature relationship and their implications for environmental ethics. *Environmental Ethics* 33 (1): 25–44.

Deliège, G. 2016. Contact! Contact! nature preservation as the preservation of meaning. *Environmental Values* 25 (4): 409–425.

Donaldson, S., and W. Kymlicka. 2011. *Zoopolis: A political theory of animal rights.* Oxford: Oxford University Press.

Drenthen, M. 2015. The return of the wild in the Anthropocene: Wolf resurgence in the Netherlands. *Ethics, Policy & Environment* 18 (3): 318–337.

Drenthen, M. 2016a. The wolf and the animal lover. In Animal ethics in the age of humans, ed. B. Bovenkerk and J. Keulartz, 189–202. Dordrecht: Springer.

Drenthen, M. 2016b. Understanding the return of the wolf: Ecosemiotics and landscape hermeneutics. In *Thinking about animals in the age of the Anthropocene*, ed. M. Tønnessen, K. Oma, and S. Rattasepp, 109–126. Washington, DC: Lexington Books.

Epstein, Y., J.V. López-Bao, and G. Chapron. 2015. A legal-ecological understanding of favorable conservation status for species in Europe. *Conservation Letters* 9 (2): 81–88.

Euro LargeCarnivores. 2019. October 23. Thomas Wagner—WikiWolves volunteers support shepherds [Video file]. Retrieved from https://youtu.be/ahkjkf3JCQc. Accessed 31 March 2020.

Hollely, R. 2018. Wolves are saving shepherds. *The European Wilderness Society*, September 21. https://wilderness-society.org/wolves-are-saving-shepherds. Accessed 23 March 2020.

Intomart. (2012). *Appreciatie-onderzoek naar de komst van de wolf*. Hilversum: Intomart.

Kantar Public. (2018). *De staat van het dier – publieksenquête. Rapportage in opdracht van de Raad van de Dieraangelegenheden.* https://www.rda.nl/publicaties/publicaties/2019/02/14.02. 2019-rapportage-publieksenquete-de-staatvan-het-dier-kantar-public/14.02.2019-rapportage-publieksenquete-de-staat-van-het-dier-kantar-public. Accessed 23 March 2020.

Karlsson, J., and M. Sjöström. 2011. Subsidized fencing of livestock as a means of increasing tolerance for wolves. *Ecology and Society* 16 (1): 16. http://www.ecologyandsociety.org/vol16/iss1/art16/.

Reinhardt, I., G. Rauer, G. Kluth, P. Kaczensky, F. Knauer, and U. Wotschikowsky. 2012. Livestock protection methods applicable for Germany—A country newly recolonized by wolves. *Hystrix, the Italian Journal of Mammalogy* 23 (1): 62–72.

Shivik, J. 2014. *The predator paradox. Ending the war with wolves, bears, cougars and coyotes.* Boston: Beacon Press.

Skogen, K., I. Mauz, and O. Krange. 2008. Cry wolf!: Narratives of wolf recovery in France and Norway. *Rural Sociology* 73: 105–133.

Thorp, T. 2014. Eating wolves. In *Old world and new world perspectives in environmental philosophy*, ed. M. Drenthen and J. Keulartz, 175–197. Dordrecht, The Netherlands: Springer.

Tokarksi, M. 2019. *Hermeneutics of human-animal relations in the wake of rewilding: The ethical guide to ecological discomforts.* Dordrecht, The Netherlands: Springer.

Tønnessen, Morten. 2011. Umwelt transition and Uexküllian phenomenology: An ecosemiotic analysis of Norwegian wolf management. Doctoral dissertation. University of Tartu, Department of Semiotics.

Van den Berg, A.E., and S.L. Koole. 2006. New wilderness in the Netherlands: An investigation of visual preferences for nature development landscapes. *Landscape and Urban Planning* 78 (4): 362–372.

Van Herzele, A., N. Aarts, and J. Casaer. 2015. Wildlife comeback in Flanders: Tracing the fault lines and dynamics of public debate. *European Journal of Wildlife Research* 61 (4): 539–555.

Van Slobbe, T. 2019. *Overijssel denkt… over de wolf.* Natuur en Milieu Overijssel.

White, P. (2019). Shepherd or no shepherd? https://www.wildtransylvania.com/2019/01/shepherd-or-no-shepherd.html. Accessed 23 March 2020.

Wolven in Nederland. 2019. Meer wolvenroedels, maar schade blijft achter in Neder-saksen. https://www.wolveninnederland.nl/nieuws/meer-wolvenroedels-maar-schade-blijft-ach ternedersaksen. Accessed 23 March 2020.

Martin Drenthen is Associate Professor of Environmental Philosophy at Radboud University in Nijmegen, the Netherlands. His research topics include the ethics of rewilding, ethics of place and environmental hermeneutics. He is the author of numerous articles and collections in the field of environmental philosophy. In 2003, he wrote *Grenzen aan wilheid* (in Dutch, Bordering wildness), in which he investigates, based on Nietzsche's work, the relationship between nature conservation and cultural critique and the paradoxical nature of the contemporary fascination with wilderness. He has edited several volumes: *New Visions of Nature. Complexity end Authenticity* (Springer 2009, with Jozef Keulartz and James Proctor), *Interpreting Nature. The Emerging Field of Environmental Hermeneutics* (Fordham University Press 2013, with Forrest Clingerman, Brian Treanor and David Utsler), *Environmental Aesthetics. Crossing Divides and Breaking Ground* (Fordham University Press 2014, with Jozef Keulartz) and *Old World and New World Perspectives in Environmental Philosophy* (Springer 2014, with Jozef Keulartz). In 2018, he published the book *Natuur in mensenland* (in Dutch: Nature in People's Land, 2018), in which he examines the ethical motives at play in controversies concerning rewilding and nature development in old cultural heritage landscapes. Lately he has been dealing with ethical questions surrounding the return of large mammals, especially the wolf, and the question of how we can coexist with wild animals in the Anthropocene. Together with others he wrote a book on this subject *De wolf is terug. Eng of enerverend?* (in Dutch: The wolf is back, scary or exiting?, 2015). In 2020, he published the book *Hek* (in Dutch: Fence, 2020) on the ethics of the border between farmland and nature area, and the question of coexistence with unruly wildlife.

Chapter 24
Consolations of Environmental Philosophy

Mateusz Tokarski

Abstract Due to successful protection and restoration efforts, humans and wild animals more and more often come to inhabit overlapping spaces. This is often experienced by humans as problematic, as animals may cause material damages to property and pose threats to humans and domesticated animals. These threats, as well as normative beliefs about belonging and culturally-based prejudices, often provoke distress or aggression towards animals. While philosophy has so far provided normative guidance as to what we should do in terms of developing proper relationships, the actual tools designed to facilitate the development of more peaceful cohabitation have been provided mostly by wildlife management and social sciences. In this contribution, I propose that environmental philosophy can provide conceptual tools easing the difficulties of cohabitation. One such tool is the practice of consolation. I begin by drawing a distinction between the contemporary and traditional forms of consolation. I further show that several common ethical arguments concerning cohabitation with wildlife can be seen as following the ancient concept of consolation. I close with some practical remarks regarding how environmental consolation could be practiced today in the context of difficult cohabitation with wildlife.

> 'Suppose that we are afraid, Socrates,' he said, 'and try to convince us. Or rather don't suppose that we are afraid. Probably even in us there is a little boy who has these childish terrors.'
>
> Plato, *Phaedo*

24.1 Introduction: The Difficult Coexistence

When it comes to human-animal relations, the Anthropocene is a time of paradoxes: while biodiversity and population numbers plummet worldwide, in many places

M. Tokarski (✉)
Independent researcher, Nijmegen, The Netherlands

© The Author(s) 2021
B. Bovenkerk and J. Keulartz (eds.), *Animals in Our Midst: The Challenges of Co-existing with Animals in the Anthropocene*, The International Library of Environmental, Agricultural and Food Ethics 33,
https://doi.org/10.1007/978-3-030-63523-7_24

445

encounters between humans and wild animals are becoming more numerous; while our understanding of animal agency deepens, technological and managerial manipulations are ever more present; while ethical sensibilities are changing taking into account animal interests, conflicts with wildlife are becoming increasingly common; while natural ecosystems are being destroyed, anthropogenic environments populated by humans are being actively colonized by non-human animals. Considering these tensions, the question 'How to share this planet with other living creatures?'— one of the central questions in environmental reflection—acquires new layers of complexity and poses new challenges.

One of these challenges, which is also among the central tensions in all the above-mentioned paradoxes, is that of respectfully accommodating animal agency in the context of direct coexistence. The difficulty here arises, first of all, from the fact that animal agency often proves undeniably troubling. This has to do with the material threats posed by non-humans, such as direct attacks, damages to property and infrastructure, or risk of disease (respectively: Linnell et al. 2002; Gordon 2009; Anthony et al. 2013). But no less important in determining attitudes to species living in proximity to humans are the less material issues. These include primarily emotional distress caused by fear of the animals (e.g. Flykt et al. 2013; Hiedanpää et al. 2016), which is often aggravated by culturally-determined prejudices (Lopez 1978), and symbolic issues linked to the experiences of transgressions and the unsettling of the established order of things (Knight 2000; Skogen et al. 2008; Cassidy and Mills 2012).[1]

In the past, in the western world, such conflicts and negative impacts would have been dealt with in a violent manner showing little recognition for animal well-being and much concern for human interests. Today, however, the situation seems to be changing. While the Anthropocene is marked by unprecedented human impacts across the planet, we are at the same time moving away from considerations of our Earth and its non-human inhabitants in terms of mere instrumentality. With respect to animals this means, among other things, experiencing other living beings as autonomous centers of life with their own sort of good. Consequently, while our experience of coexistence with wildlife often involves harm or distress, at the same time our changing moral sensitivity decries easy and violent solutions. As a result, we are confronted with the challenge of developing new forms of coexistence with animals—ones that would be respectful of their agency and at the same time would introduce ways enabling people to accommodate the destructive or distressing animal ways of living. Thus, in some sense, to recognize the agency and moral status of animals is not so much a solution to a problem (of past mistreatment of non-humans) but a beginning of a new one: how to reconcile moral concern with experiences of distress?

Several animal ethicists (for instance Michelfelder 2003; Acampora 2004; Donaldson and Kymlicka 2011) have recognized this issue, yet, following the major trends of environmentalist thinking, they largely focus on the threats that humans pose

---

[1] For an overview of research and a more extended discussion of these dimensions, including further references, see Tokarski (2019).

to animals. As such, the authors mostly limit themselves to arguing why we should respectfully accept our new neighbors. The moral arguments vary from strongly normative claims rooted in ethical theories, through instrumental considerations of benefits that animals bring, to pragmatic acknowledgments that we simply cannot get rid of wildlife spontaneously colonizing human-dominated spaces with the use of reasonable means. While such arguments lay directions for what we should strive for—a respectful cohabitation—and claim to determine what are the right things to do, they do not address the problematic psychological and experiential aspects of coexistence, and even less so the actual damages.

These latter two practical aspects are seen as primarily the tasks of environmental and wildlife management (which provides tools such as, e.g., fencing, culling, vaccinating, providing incentives and disincentives for humans, designing and re-designing infrastructure to take into account animals) (Woodroffe et al. 2005; Adams 2016). Common are also demands to change individual behaviors (animal-proofing houses, locking rubbish bins, keeping pets at home, driving carefully in wooded areas, etc.) and attempts to educate people for the new situation (how to act when encountering wild animals, replacing hearsay by factual knowledge).

The above practices are commonly based on collaborations between managers, who implement policies and use the tools, and social scientists, who study the human dimensions of coexistence in order to assess the best ways to apply these tools and measure their efficacy with respect to the changes in human behavior and attitudes (Baruch-Mordo et al. 2009). While not once questioning the adequacy and efficacy of such collaboration, in the following contribution I would like to focus on some practical means provided by philosophical-ethical discourses that could be of help in addressing the difficulties of coexistence with wildlife.

The kind of tools that I would like to focus on here are purely conceptual ones—strategies that target the ways in which humans make sense of themselves and the surrounding world, potentially altering those ways, and so consequently changing the perceptions of the world and particular events. As such, philosophical discourse would actually have a potential to address the negative psychological and experiential dimensions of animal impacts.

The specific philosophical discourse I would like to focus on in this contribution is the tradition of philosophical consolation. This is a type of philosophical discourse with origins in ancient Greece and is characterized by a clearly defined practical aim—that is of consoling those suffering. Given the distress experienced by many of those who find themselves coexisting with wildlife, and the role that perceptions of wildlife play in such distress, consolation seems strongly intuitively connected to the issue. Yet, despite a long disciplinary tradition, it has not been so far—to my knowledge—proposed explicitly as an approach by environmental or animal philosophers[2] in the context of wildlife impacts.[3] Therefore, I would like to make the first step on the way to possible applications of this tradition in the context of difficult coexistence with wildlife.

I will begin by presenting two concepts of consolation—first the dominant ideas and practices rooted in modern psychological scholarship and next the philosophical version of consolation as it was practiced by the ancient writers and as it still appears, albeit only rarely, in modern culture. I will then proceed to the discussion of the ethically-grounded responses to the instances of negative animal impacts. I will illustrate, firstly, how the philosophical-ethical discourses linked to the difficulties of cohabitation with wildlife can be in fact seen as consolatory in nature and, secondly, how we can only notice this potential if we refer to the ancient tradition of consolation rather than to its modern common-sense and academic analogues. I will conclude by providing some insights that arise from the confrontation of contemporary philosophical writings with the ancient practices, which, it is my hope, can be of some practical relevance for addressing the difficulties we face presently in attempting to develop new forms of human-animal coexistence.

24.2   The Dominant Concept of Consolation

While most commonly we associate consolation with the experience of grief after the loss of loved ones, psychologists studying consolation are quick to note that it is a much broader practice relevant in many different areas of life and addressing "distress caused by everything from daily hurts and hassles to major traumas" (Kunkel and Dennis 2003, 4). This extends all the way to an existential level, where "consolation is needed when a human being feels alienated from him or herself, from other people, from the world and from his or her ultimate source of meaning" (Tornøe et al. 2015, 8; see also: Norberg et al. 2001). Any such difficulties might provoke a need for consolation from others, and there is abundant evidence, both anecdotal and research-based, that consolation is efficacious in helping individuals and groups cope with challenging life situations (Kunkel and Dennis 2003, 4).

Consolation itself has been defined as "the type of communicative behavior having the intended function of alleviating, moderating, or salving the distressed emotional

---

[2]We might connect this to the general absence of this tradition from the discipline of philosophy throughout the modern period. And yet, despite the general falling out of favor of this tradition within academic philosophy, there are indications that today, as in the past, there is a significant demand for this sort of practice. One may mention the success of Alain de Botton's book The Consolations of Philosophy (2001) or the field of philosophical counseling (e.g. Schuster 1999).

[3]However, we must note that there is a significant number of texts on environmental losses which focus on melancholia and mourning as responses to the loss of species or degradation of ecosystems (e.g. Albrecht et al. 2007; Mortimer-Sandilands 2010; Lertzman 2015; Barnett 2019). While these belong to the same conceptual field as the argument developed here, the crucial difference is that while the aforementioned authors speak about losses of nature, I focus here on the losses to nature. While the comparison of these two approaches would be extremely interesting, it would require much more space than I have at my disposal here.

states of others" (Burleson 1984, 64). This, of course, is rather open, as it leaves much freedom with respect to the communicative means used and the ways of assessing the success of alleviating the distress. The latter might be connected to fairly super-ficial actions, as in the "modernist and medical concern to return the individual as rapidly as possible to efficient and autonomous functioning" (Walter 1996, 2; see also: Wambach 1985; Broadbent et al. 1990). When focused more on the individual rather than on the efficient performance of social roles, consolation may change the emotional states of the consoled: "The defining characteristic of solace is the sense of soothing. To be consoled is to be comforted. Solace is pleasure, enjoyment, or delight in the midst of sorrow's hopelessness and despair" (Klass 2014, 6). But it might also involve more fundamental, one could say existential, work, where it functions "as a form of healing that involves a changed perception of the world in suffering persons. This healing shift of perception enables suffering patients to set their suffering within a new pattern of meaning, in a new transcendent light" (Tornøe et al. 2015, 8; see also: Norberg et al. 2001).

The last point makes clear that consolation as a coping mechanism—in all its guises—is not focused on removing the external causes of distress but rather on attempts "to change what is perceived and how it is appraised" (Kunkel and Dennis 2003, 5).

Based on psychological research, several forms of emotional coping have been characterized that can achieve the above-summarized aims, and these include:

> (a) positive reappraisal (efforts to change, refocus, or reframe the meanings of an experience or event so that they are more positive, and less threatening); (b) distancing (efforts to detach oneself emotionally from the meanings of a stressful situation); (c) denial/suppression (choosing not to openly acknowledge stressors); and (d) escape/avoidance (trying not to think about what is troubling, and focusing instead on distractions). (Kunkel and Dennis 2003, 5–6)

Of course, not all of these coping mechanisms are deemed equally productive, and when it comes to actual practice of consolation the latter two are rather discouraged both for their lack of long-term efficacy and possibly problematic consequences of ignoring the distress (Kunkel and Dennis 2003, 8). These two aspects might be associated with the more informal, popular understanding of what consolation involves: the search for distraction in the midst of anxiety or the common enough assurances we hear that things 'will sort themselves out.'

The second aspect, the distancing, often in common parlance referred to as the 'letting go' of the departed or anxieties, has also been recently questioned, even though it forms the backbone of the traditional psychological means of coping with grief (Kunkel and Dennis 2003, 6). Instead, the first of the mentioned approaches is today deemed as the most adequate, and "proponents of the new paradigm of grief theory have re-emphasized the role of cognitive processes in emotional adjustment and recognized meaning reconstruction [...] as the central mechanisms in grieving" (Kunkel and Dennis 2003, 6). This takes the form of re-narrating the relationships and constructing new meanings, even as far as reshaping one's relationship to the world in such a way as to find a new place for the lost object of attachment (Neimeyer 2002, 302). This often involves "constructing a coherent narrative from a chaotic and troubling event" which makes the troubling situations "more accessible, more understandable, and less foreign" (Kunkel and Dennis 2003, 5).

While the above seems to be mostly an individual task of the person in distress, it can naturally be supported by others. Here we enter the intersubjective space of consolation which has received much attention and has been foregrounded in recent scholarship, at the same time being commonly underlined as important by those in need of consolation. Some have pointed out that:

> even etymologically consolation carries the sense [of] intersubjectivity. The word *comfort* is from the Latin *fortis*, strong or powerful, and the prefix *com*, that is from the Latin *cum*, that means *with*. To comfort means, then, to strengthen or find strength together. (Klass 2014, 6)

Much research on this dimension has been carried out in nursing studies, and here it has been pointed out that:

> spiritual and existential care interventions involve conveying empathy, active listening, being present with patients, helping patients to accept their thoughts and feelings around death and dying, showing respect and supporting patients' dignity. They also emphasize the importance of creating a compassionate and caring environment to bring hope, help patients to deal with the reality of death and to support their spiritual well-being in the terminal stage of life. (Tornøe et al. 2015, 2)

Thus, while the work of re-narrating relationships and searching for new meanings can only be carried out by the individual in distress, there seems to be a need for creating the environment supportive of such cognitive-narrative work, and this is precisely the responsibility of the consoler. One of the important elements raised in this context is for the consoler to open oneself to the pain of the other, and so to experience the pain together without passing judgments (Klass 2014, 8).

> To meet the other as another seems to indicate a vital aspect of what consolation is and what it is not. To meet the other as another is to create space to allow the person to be who he/she really is. This space includes the possibility to be able to suffer in one's own way. (Roxberg et al. 2008, 1085)

The last points, linked to the role of the consoler, underline that consolation is concerned in a significant way precisely with being able to suffer, with enabling one to suffer in one's own way and for one's own reasons, to acknowledge the rightness of this, and to open the space for the expression of grief and anxiety. Following this, I will not perhaps be wrong to state that within the present western culture there is a tendency to see distress over loss or troubling circumstances as something appropriate, even necessary. The strength of emotion is in some way seen as a corollary to the importance of the thing/person lost or the distress suffered.

However, it is precisely this basic assumption and prescription that stands in stark contrast to some of the most fundamental aspects of the philosophical tradition of consolation. I will presently move to the discussion of this alternative understanding of consolation, underlining the differences with the currently dominant ideas.

24.3   Philosophical Tradition of Consolation

We can safely assume that consolation as a simple act of comforting has always been a part of human relations.[4]

> Under rhetorical and philosophical influences a specialized literature began to develop in Greece, leading to the establishment of a tradition which persisted throughout Antiquity and continued into the Middle Ages... What unites the works in these various categories [of poetry, letters, philosophical treatises, funerary orations] is that they are all concerned, one way or another, with the treatment of grief, and that they draw to a large extent on a common stock of consolatory topics. (Scourfield 1993, 16)

Many of these stock tropes include ideas that we can easily recognize and even use ourselves. These include such advises as: be strong, consider how much good the other person has experienced, focus on the good memories, grief does not help in anything, etc. But beyond those, there is one element that forms the core of philosophical consolatory practice and which we may perhaps find rather surprising. This is to chastise our grief as inappropriate and to provide a new horizon of thought from which one can look at the world in such a way that the event which caused us anxiety no longer appears as an occasion for distress.

This approach was motivated by one of the key aims of ancient philosophy, which was to achieve a state of perfect calmness and satisfaction, a happiness that consisted primarily in the state of 'inner peace' (Hadot 1995). And "Since grief was considered an inconvenient and disruptive emotion, the rhetorical and philosophical methods, like the traditional approaches before, aimed to reduce excessive grief and contain the disruptive effects it had on family and society" (Baltussen 2013a, xiv–xv). As a consequence of this aim, grief, anxiety, and distress, as deeply unsettling emotions, were considered inappropriate in a strongly normative sense, even to the extent of being considered a fault (Boys-Stones 2013). Thus, philosophical consolation[5] became not so much a form of comforting but a way of eliciting a morally appropriate response to the trials and tribulations of life (Vickers 1993). If consolation is linked in our minds primarily with trying to cheer someone, to sympathize with them, acknowledge their pain, perhaps to distract them, the ancient and medieval consolation has much more to do with education, and even to some extent with chastising, admonishing, and reprimanding.

> The Greek word is παραμυθία, and παραμυθία means something much less like *comforting* and much more like *encouragement*. Specifically implied here is the idea that the addressee is being helped to take charge of themselves, to reassert control over an emotion that has run away with them... What they do is to challenge the griever to reconceptualise their grief as

---

[4]Indeed, there is currently an abundant literature on consolation even among animals, particularly primates (e.g. de Waal and Van Roosmalen 1979; Palagi et al. 2006; Fraser and Bugnyar 2010). Research on non-human consolation is so common that as a matter of fact it may be easier to find studies of animals than of human traditions.

[5]While this attitude and associated practices do not exhaust the content of actual extant consolatory texts, for the sake of brevity I will refer to this concern and the textual forms it gave rise to as philosophical consolations.

a different sort of problem, a problem of emotional susceptibility. (The solution—'getting a grip'—is, then, not just the way to overcome grief; it is the way to greater all-round emotional stability). (Boys-Stones 2013, 124)

While such control of emotions and passions is a paradigmatic Stoic strategy, it was present in virtually all the ancient schools (Hadot 1995, 86–87). The way this control of emotions is carried out is through rational work, that is through re-conceptualization of the way in which the sufferer thinks of his situation by reference to ethical or metaphysical theories. This is because the assumption is that grief is caused not so much by the situation in which one finds oneself but by a wrong concep-tualization of this situation, by a flaw of reasoning or inappropriate prejudices in light of which the situation is considered. This is well summarized by Epictetus: "People are not troubled by things, but by their judgments about things" (quoted in Hadot 1995, 193). Consequently, it is precisely the judgments that need to be changed, not the events in the world, and this is achieved through an attainment of a new perspective on things, a new theory of the world, which more adequately represents what we can expect and what we should be concerned with. In proposing this way of framing grief, ancient philosophers "broke new ground… in their effort to ratio-nalize the cause of grief and give meaning to it by exhorting the addressee to redefine or reconceptualize the event in order to enable him or her to move on" (Baltussen 2013a, xiv–xv).

While denying the adequacy of fear or sadness as a response to tragic events might strike us as insensitive—to say the least—the motif of reconceptualization might bear some resemblance to the cognitive-narrative work promoted in the contemporary research on grief. While superficially there might be some similarity, in fact there is a big difference. Presently, we are not dealing with a thorough reconstruction of the intellectual horizon so that the grief does not arise—indeed, such an attitude could be considered as callously insensitive—but with a reconstruction of the specific bond with the lost object/person so that our relationship to it is changed.

Here we meet with another issue. While the current reshaping of the perception of the situation relates to biographical re-narration and focuses on the subjective meaning of the event, object, or person, the ancient consolation was thoroughly a matter of a rational discourse which ultimately referred to theories about the nature of the world. As such, it related to universal claims regarding the universe and not to personal meanings and attachments. For Stoics, as well as for later Christian writers, this was connected, among other things, to the belief in a rational logos that rules the world, so that even the events experienced individually as tragic acquire a different meaning in the context of this higher rationality or plan. For Epicureans, the case was precisely the opposite—holding on to an atomistic view of the universe, they saw the world as a meaningless chaos with no pre-determined structures. While such a vision might seem to us like a source of anguish in itself, for Epicureans it extinguished many common fears and anxieties: of divine punishment, of afterlife, of not satisfying supposedly pre-existing social standards. At the same time, in such a chaotic world anything encountered in life that possessed a concrete form appeared as "a kind of miracle, a gratuitous, unexpected gift of nature, and existence... a wonderful celebration" (Hadot 1995, 209).

Another apparent similarity, shared between the classical and the modern idea of consolation, involves the element of participation. Today, this is connected to a revelation of vulnerability of both the consoler and the consoled and involves expression of individual feelings. In the philosophical consolation, while it could have involved expression of sympathy, the relationship between the two persons was rather like that between a teacher and a student. Their interaction would not be that of sharing experiences or feelings, as it is in modern practices, but rather would involve a rational discussion, and it very often involved "an exhortatory style which may not be palatable to modern sensitivity" (Baltussen 2013a, xvi).

All of these features (normativity, rationalization, and exhortation) are strikingly present in Boethius' *Consolation of Philosophy*, one of the most accomplished, if slightly unconventional, works of the genre. The treatise was written by Boethius in prison, where he was placed on false charges—an accusation which eventually resulted in his execution. In his treatise, Philosophy, embodied in the character of a woman, appears to a prisoner (whose situation is suspiciously similar to that of Boethius himself) and immediately expresses her surprise at his fallen spirit:

> 'Art thou that man,' she cries, 'who, erstwhile fed with the milk and reared upon the nour- ishment which is mine to give, had grown up to the full vigour of a manly spirit? And yet I had bestowed such armour on thee as would have proved an invincible defence, hadst thou not first cast it away. Dost thou know me? Why art thou silent? Is it shame or amazement that hath struck thee dumb? Would it were shame; but, as I see, a stupor hath seized upon thee.' (Boethius 2017, Prose II)

The defense of which she speaks is not that of any material means of protecting against enemies but rather a way of securing the spirit against all possible trials of fate through the appropriate understanding of the world. One of such shields was precisely the belief in the existence of a higher rationality that rules the world, giving meaning to even the most unjust actions. It is consequently not the imprisonment that Philosophy bemoans but the fact that the prisoner feels sorry for himself, which can only mean that he lacked the wisdom. Sometime later, rather than accusing his captors, she again points at the prisoner as the one who brought himself into misery:

> When I saw thee sorrowful, in tears, I straightway knew thee wretched and an exile. But how far distant that exile I should not know, had not thine own speech revealed it. Yet how far indeed from thy country hast thou, not been banished, but rather hast strayed; or, if thou wilt have it banishment, hast banished thyself! For no one else could ever lawfully have had this power over thee. (Boethius 2017, Prose V)

The true exile, in which the prisoner finds himself, is thus not from his family, home, possessions, or titles but from his wisdom—and this abandoning of wisdom, of the appropriate way of seeing the world, is what Philosophy bemoans:

> At bottom, she regards the prisoner's grief as the noxious exhaust of misguided belief about life, death, and value. Once our thoughts have the unity of a sound syllogism, she thinks, we can face the ravages of mortal existence with equanimity. (Campbell 2016, 449)

This might seem rather insensitive, perhaps even cruel, of her, to speak so to a man who had just been falsely accused, stripped of all his wealth, separated from family, denied honor, and made to spend days in a dungeon. (We are speaking here about the fictional prisoner, not Boethius himself, although it is difficult to stop oneself from establishing an identity.) And yet, this is precisely the message of Philosophy and the only true consolation she can grant—had his ideas about life and the universe been correct, he would never grieve over his situation nor face any anxiety. What she aims to do then, in the remaining part of the *Consolation*, following the process of rational discussion, is to remind Boethius about the appropriate way of looking at the situation in which he found himself.

We need not go into the details of the worldviews presented by the ancient consolations. Many of those will no longer be of much practical help to a modern seeker of consolation. The key observation at this junction concerns the structure of philosophical consolation: it is a process of rational reconstruction of the conceptual horizon of the sufferer that is motivated by ethical theories about the good life and grounded in metaphysical theories about the world. As a result of such reconstruction, one comes to see the world in a different way, so that the cause of suffering no longer appears as such. This is the philosophical process of consolation, and as such it might be distinguished from a psychological or a popular one.

24.4   Environmental Philosophy on Ecological Discomforts

At this point we may return to the question posed at the beginning of the chapter—how to live with wild animals taking into account their disruptive agency? Given my discussion of consolations above, we may ask now more specifically—can consolations be helpful in situations where we find ourselves distressed by the realities of coexistence?

Below, I will briefly present some of the most common, ethically-grounded strategies of engaging with cases of negative animal impacts, particularly in writings of environmental philosophers. The sources of these ideas are ethical texts meant to propose an appropriate way of engaging with wildlife. However, although they are not written with the intention of being consolations, I want to suggest that they can also be read and function as such. Consequently, they can perform a very important role in addressing human anxieties linked to coexistence. Indeed, perhaps implicitly they have already been performing this role.

Nevertheless, as will soon become apparent, they can only be considered as consolations if we make reference to the ancient philosophical strategies of consolation, strategies that, far from offering emotional support and catalyzing narrative work, engage in quite stern moral and rational argumentation. And just as the ancient consolations, rather than offering an acknowledgment of our suffering, they open a possibility of uprooting the very conceptual assumptions that are perceived as the sources of our distress.

These strategies can be divided into two broad groups. One of these combats anxieties through the promotion of a rational attitude predicated upon a moral theory, the other involves a perspectival shift away from the human point of view. Both of these strategies, if successfully embraced, may erase the very conditions upon which distress in the face of animal impacts is based.

With respect to the first group, the examples I will use come from moral arguments grounded in consequentialist-utilitarian and deontological theories.

Speaking about the former, even a cursory glance at the literature on human-animal conflicts reveals that many instances of animal impacts are discussed within the context of calculations of costs and benefits of coexistence with wildlife. The negative impacts (say, damaged crops or distress caused by a sense of threat) are tallied on the side of costs of cohabitation and are balanced against the potential benefits (for instance ecosystem services or the pleasure of watching wildlife). Thus, when one makes note of the damages that wildlife causes and the negative psychological effects this has, one is asked to take into account the fact that at the same time the animals are bringing with them a lot of benefits. This is only rational—if we already consider animal impacts in terms of their consequences, we should consider all possible consequences. Such are the requirements of consistency. What is more, following the principles of utilitarianism, we should consider the consequences of some event not just for ourselves, but for all those beings that can experience suffering or pleasure, and this, in more recent utilitarian accounts, should include also the costs and benefits for the animals involved.

The question of what to do when confronted with animal impacts is presented here as a matter of rational decision making based on an ethical theory that defines good and bad in terms of perceived overall utility. But once this stance is adopted in a consistent manner and internalized, it can also provide one with a way of freeing oneself from anxieties. This is, first of all, because the argument presents a framework which allows for tallying together all the individual values of particular experiences into a cumulative evaluation treating all the aspects and participants in principle equally (with such necessary corrections as intensity or quantity of effect). Given that such a cumulative evaluation of the coexistence with wildlife is commonly judged as good, from the perspective of such final evaluation it might be much easier to suffer some particular discomfort when one knows that it is merely an unavoidable element of an overall positive state of affairs.

A claim that bears some resemblance to this, is one often presented by the rights-oriented animal ethicists working in the context of theories of justice:

> At the moment, we are hypersensitive to any risk that liminal animals might pose to us getting sucked into airplane engines, causing car accidents, chewing insulated electrical wires. Or we wildly exaggerate threats, especially in the case of disease. Meanwhile, we ignore the countless risks we impose on liminals—cars, electrical transformers, tall structures and wires, window glass, backyard pools, pesticides, and many others. [...] it is unfair to have a zero-tolerance policy as regards animal risks to humans, while completely disregarding the risks we impose on them. (Donaldson and Kymlicka 2011, 244)

Here, justice is claimed to demand of us equality in the distribution of benefits as well as threats. Any specific damage, threat, or distress is thus presented not in the context of our individual experience but rather as an element in the fair balancing of rights, obligations, and harms. Once we internalize such an argument, our harm should appear to us in a different light—that is as an element of just distribution. While we might be personally to some extent disadvantaged, inconvenienced, or distressed, this personal experience acquires its full significance only in the light of the broader concern for distributive justice. And within the context of such framework, a loss or hurt we suffer is a positive step towards the ideal of justice (although it might also be an instance of injustice—as in the case of harms experienced by animals). While the negative impacts of animals might thus persist and inconvenience me, I should not be angered or frustrated by them because in fact they are an element of just distribution. At the very least, this should make one tolerant of events experienced as hurtful.

In both of the examples discussed above (utilitarian and deontological) we are asked to shape our attitude to negative animal impacts not following our emotional reactions or personal narratives but by rational deliberation following moral principles grounded in ethical theories. At the same time, such rational attitude carries a promise that when embraced it will provide us with a perspective from which what we experience as distressing will no longer appear so. The latter part is not an explicit element of philosophical argumentation as we presently encounter it in the writings within the field. Indeed, for the ethical theories presented it does not matter how we *feel* about the animal impacts. What matters is that we do the good/right thing. However, implicitly, these theories carry a potential for a philosophical consolation. This consolation functions not through the establishment of a personal narrative relationship to the situation but through the acceptance of a theory that claims to provide not so much a meaningful but rather a right or true picture of the situation.

The emotional disturbances that are primarily targeted in both, I would argue, are anger and frustration, which could then give rise to actions that seek retribution on animals. This reveals to us an important transition from the ancient practice. While traditionally philosophical consolations have been related primarily to the achievement of a right state of mind (that is ataraxia), and as such can be associated with concerns over personal character and excellence relating to virtue ethics, here we are focused on the right conduct towards others and the establishment of moral relations. Despite this important difference, the basic structure of grounding consolation in ethical (rather than psychological) considerations is maintained.

The above presented arguments demand a transition from the frame of reference of the individual experience to the consideration of rational, overarching principles with a claim to universal applicability. Such shift is even more pronounced, and more radical, in the second set of philosophical claims often presented in response to unease over animal impacts. Here, the idea is no longer connected to grounding our assessment of a situation in rational deliberation following ethical principles but rather in a change of perspective that can be very well termed metaphysical.

Holmes Rolston is one of the scholars who early on noticed and addressed the problem of ecological discomforts, developing a sort of ecological theodicy. The essence of Rolston's argument (Rolston 1983, 1992, 2015) is that while from an individual perspective certain features of life in nature might appear "evil" (suffering, death, predation, etc.), from a systemic perspective of a whole ecosystem (or even the whole biosphere) all these individual harms contribute to the development and flourishing of the system. More than that—from the perspective of that system the individual wrongs can be transvaluated in such a way that they become goods.

> Overall the myriad individual passages through life and death upgrade the system. Value has to be something more, something opposed to what any individual actor likes or selects, since even struggle and death which are never approved, are ingredients used instrumentally to produce still higher intrinsic values…This can seem in morally wild disregard for their individuality, treating each as a means to an end. But the whole system in turn generates more and higher individuality. Problem solving is a function of the system too as it recycles, pulls conflicts into harmony, and redeems life from an ever-pressing death. (Rolston 1983, 196–197)

It is important to understand in what way the reference to a system is introduced here. Quite often "in sorrow many people find consolation in the sense that they participate in something that transcends present space and time" (Klass 2014, 9). We could also interpret the framework proposed by Rolston as giving us a place within an ecological community that transcends individuality. This might be the interpretation embraced by many who come to see themselves as members of more-than-human communities. But the main thrust of Rolston's argument has to do with departing from one's individual perspective. One is not so much asked to see oneself as a member of a community but rather to take the perspective of such a community—a subtle but important difference—and only from 'such great heights' one can truly appreciate that everything, in the end, acquires a positive value. This is reminiscent of taking the "view from above," which was such a characteristic element of spiritual exercises of all the schools in ancient philosophy (Hadot 1995, 238–250) and is further connected to the belief in a higher rationality that stands as a guarantee of ultimate goodness of everything that happens in the world.[6]

To distinguish this from the shift involved in the previous two examples, we should note that here such transition is not based on ethical theory that determines the relationships between sentient creatures but on a different perspective of what constitutes the fundamental processes that organize the totality of life itself. The fundamental conceptual switch that is required here is not linked to ethical concepts (such as justice, good, or utility) but rather to metaphysical visions of what constitutes the privileged level of existence: wholes rather than individuals, processes and qualities rather than the separate entities they involve.

A similarly radical transformation of perspective can be found in Val Plumwood's writings motivated by her near-death encounter with a crocodile (Plumwood 2000, 2012). By trying to make sense of her experience, she arrived at an interpretation of the crocodile attack as a revelation of a form of justice much different from the one that is based on individual inviolability (like the one that is at work in Donaldson and Kymlicka's idea of justice) and that organizes exchanges between people.

> This is the universe represented in the food chain whose logic confounds our sense of justice because it presents a completely different sense of generosity. It is pervaded and organised

---

[6]It is perhaps not without significance that Rolston, among his titles, holds one in divinity and is an ordained minister of the Presbyterian Church.

by a generosity that takes a Heraclitean perspective, one in which our bodies flow with the food chain. They do not belong to us; rather they belong to all. A different kind of justice rules the food chain, one of sharing what has been provided by energy and matter and passing it on. (Plumwood 2012, 35)

The meaning of justice proposed here involves a completely different set of basic assumptions about the identity of individuals and the rules that govern the interactions between them. Here, there is no sense of individuality as a distinct identity separate from other creatures and processes. The living world is presented rather as a constant flux, and it is not without reason that Plumwood calls this a 'Heraclitean' world. In this framework individual loss no longer appears to us as a wrong or a harm. Rather, it becomes an appropriate way of existing, one which carries its own special sort of goodness. As such, when fully internalized, at least in theory, it should provide us with a way of assuaging the individual worries and anxieties, and that is because within the perspective based on unconditional generosity there is no stable sense of individuality which could experience the harm as a breach of its existence.[7]

On a smaller scale, though still following a similar principle, are the perspectival transitions involved in the appreciation of the intrinsic worth of other creatures, whether animals or plants, with their proper means of flourishing. For any entity we can discern its species-specific possibility of flourishing and evaluate the activities of the creature from that perspective rather than from the egocentric perspective of human self-interest (Taylor 1986). Such change allows for disinterested admiration, which in itself can become a sort of consolation. If we cease to look at the world from the perspectives of our own interests and desires and instead 'feel our way' into the flourishing of another creature (perhaps one that is inconveniencing us), the suffering, unease, or distress arising out of the confrontation with that creature should dissolve. This, I think, is the lifting of the curse of which John Muir spoke:

I cannot understand the nature of the curse, 'Thorns and thistles shall it bring forth to thee'. Is our world indeed the worse for this thistly curse? Are not all plants beautiful? Or in some way useful? Would not the world suffer by the banishment of a single weed? The curse must be within ourselves. (quoted in Norton 1994, 19)

To detach ourselves from our own judgments, cultural prejudices, and preferences is to find peace—to leave behind the curse of egotism, and, following further the biblical reference, to return to paradise. But we must remember it is not an earthly paradise where the lion will lay with the lamb but rather a state of mind, a peace with ourselves and the surrounding world.

The above-summarized ways of writing about ecological discomforts were not intended as consolations. Neither do they qualify as such if we look at the ways that consolation is commonly understood today. Indeed, in their focus on normativity,

---

[7]We must be careful here to note that Plumwood, while proposing this new framework of justice, claimed emphatically that it is necessary to retain also the human sense of justice based on individual inviolability. Unlike in Rolston, then, in her writings there is a continuing sense of tragic duality and consequently the consolation remains incomplete. Still, the perspective of justice as generosity, when viewed on its own, provides the sort of altered frame of reference in which individual loss no longer appears as a cause of distress.

impersonal principles, rationality, radical shift of perspective, and denial of appropriateness of grief and distress they may strike one as insensitive to the experiences of individual suffering. This is particularly so with Rolston, who takes the idea of systemic view of things to its logical consequence, something for which he has been in fact strongly criticized (Le Blanc 2001; Holland 2009; Plumwood 2012). However, if we look at the philosophical tradition of consolation, we can see striking affinities between the ancient and contemporary texts.

First of all, both ancient and environmental philosophers acknowledge that distress in face of difficult life experiences forms a problematic issue. For the ancients, this was connected to the disruptive potential of emotions, which drive one away from the ideal of serenity. Today, acknowledging troubling aspects of nature is often perceived as potentially undermining the motivation to protect nature (Ouderkirk 1999). While psychological practice strives towards establishing a relationship of sympathy and promotes emotional work, in ancient and environmental philosophy we find rational discourses in asymmetrical relationships. While presently we focus on assuaging of pain, the ancient philosophical consolation took diminishing of pain as secondary to the illumination of truth and concerned itself with bringing the individual back to the appropriate frame of mind—from which the diminishing of pain would follow naturally. Suffering was, therefore, not so much a 'disease' that needed treatment as a symptom of a deeper underlying problem—the falling away from truth, virtue, or appropriate conduct. I would venture proposing that this is what is also happening in environmental philosophy. The anxiety, fear, hatred, or disgust with animals are not in themselves the problem or are so only secondarily. These attitudes and emotions are rather symptoms of a fundamentally misguided worldview, most commonly characterized as anthropocentrism. Consequently, what is provided is a way to treat the actual problem, that is to propose an alternative metaphysical and ethical horizon. Discomfort, then, is a symptom of a deeper issue that needs to be treated; indeed, one that can only be treated not with sympathy but with fundamental normative-conceptual work. Such work takes the form of rational discourses and proposes a thorough transformation of the conceptual horizon from which one looks at the world.

The point I am making here is that as a consequence of this reshaping one can also expect the alleviation of grief, anxiety, or even the actual conflicts themselves—indeed, that is what seems implicitly to be promised in many ethical arguments:

> If we attempt to see animals apart from our anthropocentric projections and desires, might we be able to see these animals with some sort of clarity? If we understand the animals on their own terms, could we then minimize or end conflicts with problematic wildlife? (Nagy and Johnson 2013, 10)

As such, animal ethics and environmental philosophy are practical not only in providing directions for action, but also in that they provide conceptual tools for addressing certain problematic aspects of cohabitation with wildlife. This opens up an additional avenue for developing coexistence, one based not on material transformation of the environment or managerial regimes controlling the behavior of people and animals but on "spiritual" transformations effected by consolatory work.

Consequently, the analysis carried out here is not only an analytical framework, useful in analysis of philosophical positions, or another interpretative key that can be productively applied to nature writing. Indeed, the ancient examples show us precisely how this approach can be used in practice and how standard tropes (and we already have a number of those established in environmental thinking) can be creatively made use of to discuss individual instances of hurt, grief, or anxiety.

Unfortunately, some aspects of philosophical consolation discussed above might strike many modern-day readers as counter-intuitive. Moreover, the above-summarized moral arguments are not explicitly framed as consolatory. For these reasons, we might expect difficulties in the practical application of the discussed philosophical approach. Consequently, below I will discuss some lessons that we may derive from the ancient practice of philosophical consolation that may be of use in our present predicament.

24.5   The Scope of Consolation Is a Total Transformation

Ancient consolations often did not limit themselves to an issue immediately at hand. Because they touch the matter of despair and suffering, in many cases they eventually fall back upon one of the most fundamental questions of ancient—if not of modern—philosophy, that is what is happiness and how to attain it. Consequently, consolations usually became discourses on the good life—what constitutes it and how it can be achieved—and very often moved in the direction of basic metaphysical beliefs. The consequence of consolation is then envisioned as a deep spiritual transformation: "The ancient *consolatio* genre is a prime example of what Pierre Hadot calls 'spiritual exercises,' an engaged philosophy that seeks to form readers over an itinerary mapped by its arguments, rather than simply informing them" (Campbell 2016, 447).

The point here is to realize that philosophical consolation should not be treated merely instrumentally, as a sort of conceptual or symbolic analogue to more traditional means of wildlife management. We should also not expect that it simply aims to treat some separate instance of distress, while the rest of our life remains intact. Rather, this approach, far from being a mere tool, transforms the very situation in which it is applied as well as the person undergoing consolation. Ultimately it transforms us into people who think and act differently—as such it is existential rather than instrumental.

So how exactly does such a transformation take place?

24.6   Gentle and Strong Remedies

It is important to notice that ancient authors recognize two stages of consolation. Taking Boethius' *Consolation of Philosophy* as an example, we can distinguish 'gentler' and 'stronger' remedies, which "mirrors the ancient physician's approach to acclimate the patient to the medicines of increasing strength as one prepares oneself slowly and by degrees for moving out of darkness into bright light" (Phillips 2002). In the treatise, this means beginning from rhetorical arguments, poetry, and the kind of claims that are readily acceptable to the one suffering, as they still fit within her present worldview: "The 'gentler remedy' is not meant to cure but rather to strengthen the patient…so that he can take the stronger medicine later" (ibid.). In a similar vein, also Stoics underlined the importance of letting some time pass, and with it the worst of the grief, before undertaking serious philosophical consolatory work (Baltussen 2013b).

This does not mean we have to immediately break into song or engage in the kind of floral speeches we may encounter in, for instance, Cicero's *Tusculan Disputations*. Rather, and taking to heart the insights of modern consolation studies, the above drawn distinction would suggest beginning with the compassionate, narrative, and emotional grief-work that is currently seen as both most adequate and most efficacious. The first stage could be what we would commonly consider a consolation in that it takes on various forms of soothing, sympathizing, and acknowledging. The gentle means of compassion, such as encouraging expression and showing one's own vulnerability, might be the first step of assuaging grief and preparing the other to be ready to be consoled in a more radical manner, with the use of conceptual consolation that transforms the perspective on the situation. What is more, this initial step might help build trust and rapport in the context of issues that provoke strong partisan divisions. By encouraging expression it might also make the conceptual bases of anxiety more visible, which might be helpful in the later work of reformulating those through philosophical consolation.

While this might appear as contradictory—since it depends on confirming the attachment to the worldview that gives rise to distress—it might nevertheless be psychologically necessary. It is very difficult to enact a total transformation of a worldview at a moment of crisis where people seem to require stability the most and cling to the ideas that have so far organized their life. And since philosophical consolation strives for achieving actual results, it cannot afford to disregard the psychological structure of the experience it tries to address. Hence, to reach the pre-determined normatively defined goal, it often has to be pragmatic in the choice of means it employs. The writing cure of Cicero can be interpreted in this light as such first stage on the way to complete consolation, which would later include philosophical work that returns one to the appropriate way of perceiving the world. In general, it shows that philosophical consolation is always a balancing act and an art that requires sensitivity both to the psychological nature of the situation and the ethical goals of the practice.

24.7 The Individual and Private Is Universal and Public

Another important feature of consolation as a practice was that it most commonly took the form of a dialogue. The three great works of consolation: Plato's *Phaedo*,

Boethius' *Consolation of Philosophy*, and Cicero's *Tusculan Disputations*, all take the form of a dialogue. Consolations were also often part of epistolary exchanges—they were extended dialogues. Authors were often quite aware that such private letters would be made public; indeed, they often wrote with a larger audience in mind: "Pieces such as these [private letters by Seneca], though directed to a particular and personal situation, have the character rather of an essay than of a letter; here, as elsewhere, the distinction between letter and treatise becomes hard to define" (Scourfield 1993, 21).

This ambiguous status of consolatory texts connects the private and the public sphere of both grief and consolation and this has two important consequences. On the one hand, the texts needed to be made very individual, so as to address the specific suffering of the addressee. As such, the common tropes had to be adapted for the specific situation and with a particular person in mind. On the other hand, given they were meant for the greater public, they needed to provide consolation in a way that others could identify with. This is of course no mean feat.

One way to address this tension was to depend on stock arguments that developed together with the genre (Baltussen 2009, 71).

> These stock examples, far from being anemic, standardized commonplaces, served the user well in providing words at a time when many are at a loss for words. Rather than trivialize these commonplaces as "mere platitudes," we should acknowledge their power to express an individual's response to grief in a verbal form sanctioned by experience. (ibid., 91)

The skill of a consoler was visible precisely in the way that she managed to make the traditional tropes applicable to the situation at hand.

In a similar way we can endeavor to see whether there are such stock approaches in modern environmentalism. And indeed we can find those—above I have summarized several moral arguments that belong to the developing stock of tropes commonly used in situations of animal impacts by philosophers and broader publics alike. While they are 'stock arguments,' they can be at the same time transformed according to the need to fit a specific situation, and the skill of a consoler is visible precisely in the way that she can make the general, perhaps even universal, bear on the individual in a unique way. Perhaps this is why the writings of Val Plumwood on her experience with the crocodile are so powerful. She develops her philosophical ideas permanently referring to her own personal experience while at the same time managing to touch universal notes.

24.8   Conclusion—the Limits of Consolation

As any practice, also that of consolation has its limitations and it is important to be aware of those. There are several problematic features that philosophical consolation has struggled with from its very beginning.

While many philosophical works profess optimism as to the efficacy of this approach in expelling anguish, this has not been a common cultural presumption. For once, even in ancient Greece, there was a strong sentiment against consolation, which is visible in Greek tragedy that often underlines the impossibility of successful consolation (Chong-Gossard 2013). Private writings of philosophers, for instance Cicero's, show a discrepancy between the publicly professed belief in philosophy and personal experience. While he maintains that philosophical work has helped him show his grief less, he acknowledges that the actual pain remained unchanged (Baltussen 2013b, 74). Even some works of consolation, like Boethius' text, according to some scholars express doubts regarding the capacity of philosophy to actually provide the consolation it promises and can be read more as a satire (Relihan 1990, 2007).

The awareness of this limitation is important because it can also draw our attention to the limitations of the philosophical and rhetorical strategies used by environmentalists addressing the problematic issues surrounding coexistence with wildlife. It also opens the space for texts which, like ancient tragedies, directly address the refusal or the impossibility of thorough consolation and instead focus on the irreducibly tragic aspects of coexistence (Snyder 1990; Williams 1995; Steeves 1999; Jordan 2003; Tokarski 2019). We can heed this warning and include it more openly even in the philosophical, journalistic, and narrative texts which strive to address ecological discomforts.

Acknowledgment of the limited efficacy of consolation also helps us realize how immense work is asked of the grieving or distressed. Indeed, the consolations of the past were written most commonly as reminders of what one already believed but temporarily lost due to being overwhelmed by grief. They aimed to remind people already steeped in philosophical ways of perceiving the world how they should look at the tragic events that befell them. In some way, it was then preaching to the converted. In much of ancient consolation, one should have already been open to this way of thinking to accept it.

With this in mind, one can perhaps go as far as to note the presence of a paradox in the practice of consolation. Grief comes in a moment when our world is overturned by some tragedy, which might unsettle the beliefs we have held so far. But consolation in such instances might be nothing more than bringing back those beliefs to assuage our grief.[8] The very ideas and ideals that are put in question by a tragic event are brought up to set us more fundamentally in these beliefs. Not only a paradox, then, but even a circle. That this is not a vicious circle might be noted by observing that the ideals are now scrutinized in a different light, in a context of a different situation. It is significant in this context that most environmental philosophers do not bring new frameworks or principles into the question of negative animal impacts; they rather

---

[8]Naturally, this is only so for those who already hold given beliefs, in this case environmental ones. For those who do not, and who must be thoroughly transformed at the moment of tragedy, the work needed is even greater and does not involve this sort of circularity. But even for environmentalists there might be an element of transformation involved. This is well illustrated by the case of Val Plumwood and her near-death encounter with a crocodile. Plumwood notes that through this tragic even she came to realize how shallow her integration of the environmental beliefs actually was (Plumwood 2012). Consequently, tragic events and subsequent consolation might be the moment when the true internal transformation takes place, when one not only says the right things but also sees the world in a different way.

re-present their established theories by showing how they stand up to the challenge of tragic encounters.

Consequently, discomforting encounters might form something of a test even for professed environmentalists. Sometimes, as in the case of Val Plumwood, they may reveal the shallowness of one's own convictions and consequently lead to doubt. For others, they might lead to the collapse of ideals. But in a like manner, they might be a sort of trial-by-fire from which one emerges with a renewed conviction. They are what philosopher Paul Tillich calls "extreme situations" (1951)—it is only when a philosophical system can withstand such extreme situations that it is worth holding on to and following.[9] Belief and conversion at the time of crisis are perhaps a difficult task, but every system of belief must possess resources that could be employed at a time of a crisis—to provide guidance and consolation. Environmentalism and animal ethics have been heavy on guidance. Based on the above, we can see they also have much to offer in terms of consolation. This in turn can be of inestimable worth at a time such as the Anthropocene, when most of the beliefs our culture held most firmly are being overturned and challenged. If the Anthropocene requires radical rethinking of our relationship to the world, it will be not enough to employ sophisticated technologies and material tools provided by environmental management. We need something that will guide us through the existential crises that go hand in hand with fundamental conceptual transformations, and here the practice of philosophical consolation with its rich tradition can prove an important supporting element.

References

Acampora, R. 2004. Oikos and Domus: On constructive co-habitation with other creatures. *Philosophy & Geography* 7 (2): 219–235.

Adams, C.E. 2016. *Urban wildlife management*. Boca Raton: CRC Press.

Albrecht, G., G. Sartore, L. Connor, N. Higginbotham, S. Freeman, B. Kelly et al. 2007. Solastalgia: The distress caused by environmental change. *Australasian psychiatry: Bulletin of Royal Australian and New Zealand College of Psychiatrists* 15 (1): 95–98. https://doi.org/10.1080/10398560701701288.

Anthony, S.J., J.H. Epstein, K.A. Murray, I. Navarrete-Macias, C.M. Zambrana-Torrelio, A. Solovyov et al. 2013. A strategy to estimate unknown viral diversity in mammals. *mBio* 4 (5). https://doi.org/10.1128/mbio.00598-13.

Baltussen, H. 2009. Personal grief and public mourning in Plutarch's "Consolation to His Wife". *The American Journal of Philology* 130 (1): 67–98.

Baltussen, H. 2013a. Introduction. In *Greek and Roman consolations: Eight studies of a tradition and its afterlife*, ed. H. Baltussen, 13–25. Swansea: Classical Press of Wales.

Baltussen, H. 2013b. Cicero's *Consolatio ad se*: Character, purpose and impact of a curious treatise. In *Greek and Roman consolations: Eight studies of a tradition and its afterlife*, ed. H. Baltussen, 67–92. Swansea: Classical Press of Wales.

Barnett, J.T. 2019. Naming, mourning, and the work of earthly coexistence. *Environmental Communication* 13 (3): 287–299.

---

[9] In the context of grief work see also Klass 2014.

Baruch-Mordo, S., S.W. Breck, K.R. Wilson, and J. Broderick. 2009. A tool box half full: How social science can help solve human–wildlife conflict. *Human Dimensions of Wildlife* 14 (3): 219–223.

Boethius, Anicius Manlius Severinus. 2017. *The consolation of philosophy*, trans. H.R. James. Musaicum Books.

Boys-Stones, G.R. 2013. The *Consolatio ad Apollonium*: Therapy for the dead. In *Greek and Roman consolations: Eight studies of a tradition and its afterlife*, ed. H. Baltussen, 123–137. Swansea: Classical Press of Wales.

Broadbent, M., J. Sparks, and G. de Whalley. 1990. Bereavement groups. *Bereavement Care* 9 (2): 14–16. https://doi.org/10.1080/02682629008657243.

Burleson, B.R. 1984. Comforting communication. In *Communication by children and adults: Social cognitive and strategic processes*, ed. H.E. Sypher and J.L. Applegate, 63–104. Beverly Hills: Sage.

Campbell, A.L. 2016. Consolation in stitches. *The Journal of Religion* 96 (4): 439–466.

Cassidy, A., and B. Mills. 2012. Fox tots attack shock: Urban foxes, mass media and boundary breaching. *Environmental Communication: A Journal of Nature and Culture* 6 (4): 494–511.

Chong-Gossard, J. 2013. Mourning and consolation in Greek tragedy: The rejection of comfort. In *Greek and Roman consolations: Eight studies of a tradition and its afterlife*, ed. H. Baltussen, 37–66. Swansea: Classical Press of Wales.

De Botton, A. 2001. *The consolations of philosophy*. New York: Vintage International.

De Waal, F., and A. van Roosmalen. 1979. Reconciliation and consolation among chimpanzees. *A. Behavioral Ecology and Sociobiology* 5: 55–66. https://doi.org/10.1007/BF00302695.

Donaldson, S., and W. Kymlicka. 2011. *Zoopolis: A political theory of animal rights*. Oxford: Oxford University Press.

Flykt, A., M. Johansson, J. Karlsson, S. Lindeberg, and O.V. Lipp. 2013. Fear of wolves and bears: Physiological responses and negative associations in a Swedish sample. *Human Dimensions of Wildlife* 18 (6): 416–434.

Fraser, O.N., and T. Bugnyar. 2010. Do ravens show consolation? Responses to distressed others. *PLoS One* 5 (5): e10605. https://doi.org/10.1371/journal.pone.0010605.

Gordon, I.J. 2009. What is the future for wild, large herbivores in human-modified agricultural landscapes? *Wildlife Biology* 15 (1): 1–9.

Hadot, P. 1995. *Philosophy as a way of life: Spiritual exercises from Socrates to Foucault*. Oxford and Malden: Blackwell.

Hiedanpää, J., J. Pellikka, and S. Ojalammi. 2016. Meet the parents: Normative emotions in Finnish wolf politics. *TRACE Finnish Journal for Human-Animal Studies* 2 (1): 4–27.

Holland, A. 2009. Darwin and the meaning in life. *Environmental Values* 18 (4): 503–516.

Jordan, W.R.I.I.I. 2003. *The sunflower forest: Ecological restoration and the new communion with nature*. Berkeley: University of California.

Klass, D. 2014. Grief, consolation, and religions: A conceptual framework. *Omega—Journal of Death and Dying* 69 (1): 1–18. https://doi.org/10.2190/om.69.1.a.

Knight, J. (ed.). 2000. *Natural enemies: People-wildlife conflicts in anthropological perspective*. London: Routledge.

Kunkel, A.D., and M.R. Dennis. 2003. Grief consolation in eulogy rhetoric: An integrative framework. *Death Studies* 27: 1–38.

Le Blanc, J. 2001. A mystical response to disvalue in nature. *Philosophy Today* 45 (3): 254–265.

Lertzman, R. 2015. *Environmental melancholia: Psychoanalytic dimensions of engagement*. London and New York: Routledge.

Linnell, J.D.C., R. Andersen, Z. Andersone, L. Balciauskas, J.C. Blanco, L. Boitani, et al. 2002. *The fear of wolves: A review of wolf attacks on humans*. Trondheim: NINA.

Lopez, B.H. 1978. *Of wolves and men*. New York: Scribner Classics.

Michelfelder, D.P. 2003. Valuing wildlife populations in urban environments. *Journal of Social Philosophy* 34 (1): 79–90.

Mortimer-Sandilands, C. 2010. Melancholy natures, queer ecologies. In *Queer ecologies: Sex, nature, politics, desire*, ed. C. Mortimer-Sandilands and B. Erickson, 331–358. Bloomington and Indianapolis: Indiana University Press.

Nagy, K., and P.D. Johnson II. 2013. *Trash animals: How we live with nature's filthy, feral, invasive, and unwanted species*. Minneapolis, MN: University of Minnesota Press.

Neimeyer, R.A. 2002. Making sense of loss. In *Living with grief: Loss in later life*, ed. K.J. Doka, 295–311. Washington: Hospice Foundation of America.

Norberg, A., M. Bergsten, and B. Lundman. 2001. A model of consolation. *Nurs Ethics* 8 (6): 544–553.

Norton, B.G. 1994. *Toward unity among environmentalists*. Oxford: Oxford University Press.

Ouderkirk, W. 1999. Can nature be evil? Rolston, disvalue, and theodicy. *Environmental Ethics* 21 (2): 135–150.

Palagi, E., G. Cordoni, and S.B. Tarli. 2006. Possible roles of consolation in captive Chimpanzees (Pan troglodytes). *American Journal of Physical Anthropology* 129: 105–111. https://doi.org/10.1002/ajpa.20242.

Phillips, P.E. 2002. Lady philosophy's therapeutic method: The 'Gentler' and the 'Stronger' remedies in Boethius's "De Consolatione Philosophiae". *Medieval English Studies* 10 (2): 5–26.

Plumwood, V. 2000. Being prey. In *The ultimate journey: Inspiring stories of living and dying*, ed. J. O'Reilly, S. O'Reilly, and R. Sterling, 128–146. San Francisco: Travelers' Tales.

Plumwood, V. 2012. *Eye of the crocodile*. Canberra: Australian National University E Press. https://doi.org/10.22459/EC.11.2012.

Relihan, J.C. 1990. Old comedy, Menippean satire, and philosophy's tattered robes in Boethius' "consolation". *Illinois Classical Studies* 15 (1): 183–194.

Relihan, J.C. 2007. *The prisoner's philosophy life and death in Boethius's consolation*. Notre Dame: University of Notre Dame Press.

Rolston, H.I.I.I. 1983. Values gone wild. *Inquiry* 26 (2): 191–207.

Rolston, H.I.I.I. 1992. Disvalues in nature. *The Monist* 75 (2): 250–278.

Rolston, H.I.I.I. 2015. Rediscovering and rethinking Leopold's Green Fire. *Environmental Ethics* 37 (1): 45–55.

Roxberg, Å., K. Eriksson, A. Rehnsfeldt, and B. Fridlund. 2008. The meaning of consolation as experienced by nurses in a home-care setting. *Journal of Clinical Nursing* 17 (8): 1079–1087. https://doi.org/10.1111/j.1365-2702.2007.02127.x.

Schuster, S.C. 1999. *Philosophy practice: An alternative to counseling and psychotherapy*. Westport: Praeger Publishers.

Scourfield, J.H.D. 1993. *Consoling Heliodorus: A commentary on Jerome, Letter 60*. Oxford: Clarendon.

Skogen, K., I. Mauz, and O. Krange. 2008. Cry wolf!: Narratives of wolf recovery in France and Norway. *Rural Sociology* 73 (1): 105–133.

Snyder, G. 1990. *The practice of the wild*. New York: North Point Press.

Steeves, P.H. 1999. They say animal can smell fear. In *Animal others: On ethics, ontology and animal life*, ed. P.H. Steeves, 133–178. Albany: State University of New York Press.

Taylor, P. 1986. *Respect for nature: A theory of environmental ethics*. Princeton, NJ: Princeton University Press.

Tillich, P. 1951. *The courage to be*. New Haven: Yale University Press.

Tokarski, M. 2019. *Hermeneutics of human-animal relations in the wake of rewilding: The ethical guide to ecological discomforts*. Dordrecht, The Netherlands: Springer.

Tornøe, K.A., L.J. Danbolt, K. Kvigne, and V. Sørlie. 2015. The challenge of consolation: Nurses' experiences with spiritual and existential care for the dying—A phenomenological hermeneutical study. *BMC Nursing* 14 (1). https://doi.org/10.1186/s12912-015-0114-6.

Vickers, B. 1993. Shakespearian consolations. *The Proceedings of the British Academy* 82: 219–284.

Walter, T. 1996. A new model of grief: Bereavement and biography. *Mortality* 1 (1): 7–25.

Wambach, J.A. 1985. The grief process as a social construct. *Omega* 16 (3): 201–211.

Williams, B. 1995. Must a concern for the environment be centred on human beings? In *Making sense of humanity and other philosophical papers*, auth. B. Williams, 233–240. Cambridge: Cambridge University Press.

Woodroffe, R., S. Thirgood, and A. Rabinowitz. 2005. *People and wildlife, conflict or co-existence?* Cambridge: Cambridge University Press.

Mateusz Tokarski is an independent researcher, scientific and literary editor, and fiction writer. In his research he addresses environmental issues in an interdisciplinary fashion, bringing together philosophy, cultural studies, and semiotics. The main research themes he focuses on are human-animal conflicts, literary representations of nature, rewilding, and urban wildlife. He received his PhD from Radboud University, the Netherlands, on the dissertation titled Wild at Home. The Ethics of Living with Discomforting Wildlife. He obtained his MA from Aarhus University, Denmark, on the thesis titled Developing Motivation for Pro-Environmental Behaviour: Semiotic Leverage Points in Environmental Protection Programme. His homepage is mateusztokarski.eu.

Chapter 25
On Hunting: Lions and Humans as Hunters

Charles Foster

Abstract This is an interrogation of some commonly cited intuitions about killing animals, enjoying killing animals, and enjoying eating animals. It concludes that intuitions are the only possible philosophical guide through this territory. Accordingly if intuitions cannot be trusted, moral arguments about the killing of animals and related matters are likely to be fruitless.

25.1 Introduction

Lions hunt. Few would try to stop them doing so. But many try to stop humans hunting.

Why this difference? There are several possible reasons:

a. It might be said that lions have to hunt, because if they do not, they will starve. This is not true of most modern human hunters. I therefore exclude from this discussion those human hunters who would starve if they did not kill animals, simply noting as I exclude them that, by exempting such hunters from any blame, we are accepting that the loss of an animal life is justified if the loss is necessary in order to save a human life. At the start, then, we have a normative assumption, which is almost universally shared, that humans are more valuable than non-humans.

b. It might be said (and this is a point that relates closely and obviously to the first), that hunting is of the essence of the lion, whereas it is not of the essence of the hunter. One could not have a vegetarian lion, not only because a lion that ate only vegetation would die, but because killing is so quintessentially part of a lion's

C. Foster (✉)
Green Templeton College, Oxford, UK
e-mail: Charles.Foster@gtc.ox.ac.uk

University of Oxford, Oxford, UK

B. Bovenkerk and J. Keulartz (eds.), *Animals in Our Midst: The Challenges of Co-existing with Animals in the Anthropocene*, The International Library of Environmental, Agricultural and Food Ethics 33, https://doi.org/10.1007/978-3-030-63523-7_25

constitution that to make it vegetarian would be to un-wish it. A vegetarian lion would not be a dead lion; it would not be a lion at all.

This (the argument would go) is not true of humans. All of us know many human non-hunters, and they are not obviously non-human: it is not obvious that they are not thriving in the way that humans are intended to thrive.

The hunter's riposte would be simple contradiction. 'You are wrong', he (and it usually is a 'he') would say. 'Hunting and human identity are inextricably connected, in exactly the same way as hunting and lion-identity are connected. Excise the hunting from humans, and you will incurably damage human identity – and hence the ability of humans to thrive.'

Both the hunter's and the anti-hunter's claims are essentially empirical, but of course they are not the sort of empirical claims that can easily be empirically investigated. That is the justification for the form of my argument in this chapter, which is (unusually for a philosophy book) autobiographical and personally reflective.

c. It might be said (and this is a point that relates closely and obviously to the second) that the real moral offence in human hunting lies in two facts: (i) that the human *enjoys* the process; and (ii) that the enjoyment is the primary reason that the human hunts. Lions may enjoy the hunt, but that is not their main motivation. It is more usual to see this argument as a slogan than a carefully examined proposition. Rarely does the proponent of the argument mean that they are against human enjoyment per se. They would be perfectly content if the hunter gained pleasure equal to that of hunting from simply going for a walk in the countryside. The argument is instead that there is something illegitimate about the particular type of pleasure that results from hunting. Here, of course, there is again an empirical assumption about the type of pleasure that this is. And in the articulation of that assumption (on the rare occasions when articulation is demanded), slogan again often predominates. It is often argued, for instance, that it is obscene for humans to get pleasure from any activity that involves the (avoidable) death of a non-human animal. This argument can only consistently be maintained by those who also argue for vegetarianism (and possibly veganism too), for the eating of a steak is distinctly pleasurable to many, and yet involves the avoidable death of a non-human animal. But a steak-eating objector to hunting might nonetheless say that there is a particular form of obscenity involved in *active* participation in killing, or at least in proximity with the death itself. This argument, in its simple form, would seem to entail moral condemnation of slaughterhouse workers, but not fix with vicarious moral condemnation those who enjoy the fruits of the slaughter house. And that, surely, is problematic. But if we remember that the real objection is to enjoyment of the process of killing, that problem at least evaporates. Even if we enjoy steak, we would condemn a slaughterman who got a thrill when he slit the cow's throat. That kind of thrill, we would say, is not at all the same as the kind of pleasure that we get when we eat a steak. Since it is not the same kind of pleasure, it is not wrong on the grounds of commensurability of pleasures for us to enjoy a steak. (It may well be wrong on other grounds).

Note that the harm done to the cow by the thrill-seeking slaughterman is the same as that done by a reluctant slaughterman. Philosophers might say that the cow killed by the thrill-seeker has sustained a *wrong* in a way that the cow killed by the reluctant slaughterman has not—since the first cow has been a vehicle for an illegitimate pleasure. But this analysis—though traditional—takes us nowhere. The real objection to the thrill-seeking slaughterman, as to the thrill-seeking hunter, is one based squarely on another set of normative assumptions—this time about the kind of character that decent humans should have, and hence the kind of behaviour that they should demonstrate. Like most moral assertions, it rests on an intuition. In this case the intuition is that decent humans should not enjoy the process of killing another creature. It is hard to interrogate this intuition robustly without making some assumptions (which are religious or anthropological or both) about the sort of creatures that humans are and should be. And hence the debate between the hunters and the anti-hunters becomes shrill and intellectually uninteresting: there is simply a stand-off between those who insist that humans shouldn't enjoy the business of killing animals, and those who insist that there is nothing wrong with that enjoyment. It is an argument of the 'O yes it is', 'O no it isn't', type.

To break out of this tedious and sterile debate it is necessary to look harder at the underlying intuitions. I start by examining my own. I then move to a consideration of the work of José Ortega y Gasset. He took more seriously than any other modern philosopher the argument that the enjoyment of hunting might be a reason to commend rather than condemn hunting. He alone, then, addressed squarely the really problematic argument—my point (c).

I do not address here the argument that hunting, even if intrinsically morally offensive, is justifiable on other grounds—for instance that culling is necessary to ensure the health of an animal population or an ecosystem, or that it brings much needed funds into an economy. Such arguments will turn on the facts pertinent to the particular case being considered, on the philosophical worth that one attributes to the relevant utilitarian calculation, and on the theory of value that one uses in weighing the moral offence of the human hunter against the harms or wrongs that may result if the hunting is not permitted. The most interesting of these issues (the issue of the theory of value) is necessarily considered as I address the issue of enjoyment. Whether or not some variant of utilitarianism is the best way to approach these questions is a generic issue: hunting does not raise any novel difficulty in relation to the way the issue should be approached.[1]

---

[1] For overviews of the arguments regarding the ethics of sport hunting, see Dickson, B. 2009. The ethics of recreational hunting. Recreational hunting, conservation and rural livelihoods: *Science and Practice* 59–72; Gunn, A. S. 2001. Environmental ethics and trophy hunting. *Ethics and the Environment* 6 (1): 68–95; King, R.J. 1991. Environmental ethics and the case for hunting. *Environmental Ethics* 13 (1): 59–85; Varner, G. 2011. Environmental ethics, hunting, and the place of animals. In *The Oxford handbook of animal ethics*; Loftin, R.W. 1984. The morality of hunting. *Environmental Ethics* 6 (3): 241–250; Gibson, K. 2014. More than murder: Ethics and hunting in New Zealand. *Sociology of Sport Journal* 31 (4): 455–474; Causey, A. S. 1989. On the morality of hunting. *Environmental Ethics* 11 (4): 327–343.

25.2 Confession and Reflection

I have killed many animals. I started when I was very young. As a boy I shot rabbits and birds; I trapped, fished, and followed hunting hounds. I continued as an adult. I am too ashamed now to want to give many details, and do not need to confess for the sake of absolution. But I have ridden after foxhounds and staghounds, run after hare-hunting beagles, crawled through the African bush in search of plains game, shivered before dawn in trenches on the foreshore as I waited for the geese to come in from the sea, wandered around the hedgerows in the evening in hope of a pigeon or two, and (fanatically, year after year) stalked red deer in the highlands of Scotland. I have spent many an evening in pubs in the Lake District after a day's fell foxhunting, singing songs about epic hunts of old, and cheering when the hounds in the song caught up with their fox and 'broke him up' on the mountain side.

There are no doubt many levels of explanations for this behaviour, and probably my own view about why I engaged in it is the least likely of all views to be accurate. The more emphatic someone's assertion about their motive, the less likely it is to be credible. But it does seem to me that its root was a desire for an intimate connection with the natural world, and for the self-knowledge that is impossible without that connection.

This will sound perverted to many, and certainly there are perverted variants on the theme. If I said, for instance, that the death of a creature was the most tectonic, fundamental thing about its life apart from its birth, and thus to be an agent of its death was to be involved with the creature more intimately than any animal apart from its mother ever had been, I would rightly be characterised as a Nietzschean psychopath. But it wasn't that. It was a desire grounded in the basic Darwinian knowledge that these creatures were my close cousins, and I therefore needed to know them in order to know myself: to know what sort of creature I was, where I had come from, where I would be going (the same way as all those dead animals, in fact) and hence where my home really was. Only if I knew these things would real relationship be possible, and relationships, I knew even (or particularly) as a very young child, were the whole point of being alive. I hunted, therefore, to describe and to situate myself.

The suffering and death of sensate creatures might seem a high price to pay for this self-knowledge. To kill something to know oneself sounds monstrous: on an obscene par with Raskolnikov. I cannot pretend that it is not. All I can say is that that is how it is, and that the fact that that is how it is is a consequence of the interconnectedness of things.

No other way seemed possible. A man with binoculars isn't as *involved* with the deer he's watching as the same man watching the same deer through the sights of a loaded rifle. With the dreadful squeeze of the trigger finger comes the knowledge of shared destiny: one day the trigger will be squeezed on me. No evasion is then possible, either for the deer or for me: no physical or psychological evasion; no pretentious philosophising. The bullet tells it the way it is.

I tried less violent ways of getting close to the wild. I watched, collected, swam, crept, and slept out. I still do. My boyhood bedroom was full of skulls. Crudely stuffed birds, suspended on thread, hovered over my bed. I spent my pocket-money on glass eyes and formaldehyde. But none of it worked. My incurably reductionist, linguistically-tyrannised brain kept me from getting close to the sensory worlds occupied by non-human animals. I couldn't live in the same woods or rivers as they did. We could only really meet in the killing fields. We shared DNA and death, and not much else.

This all sounds dreadfully earnest: the stuff of morbid psychopathology. It was indeed serious. Though not (though I would say that, wouldn't I?) morbid. Partly the earnestness was because I knew that unless I learned about myself and about how to relate, I was done for. But there was a moral earnestness too, manifested in ceremony and fastidiousness. There was no Dionysiac revelling in my hunts: they were all sedate and Apolline. The death was always a source of real regret and remorse. I have always thought obscene those triumphalist photos of smiling hunters crouching fatly behind an animal, shot at no risk to them (cf. Kalof and Fritzgerald 2003). I can never quite choke down the thought that the animal lived a much more satisfactory life than those hunters ever could. I was not surprised to learn that many indigenous hunters pray before a hunt for the animal to be delivered up to them, and afterwards for protection from divine or ghostly anger. When, in Africa, I saw for the first time the Continental practice of putting a respectful sprig of vegetation on the dead animal, I gratefully recognised the sentiment. My childhood and adolescent hunting diaries are achingly meticulous. Every detail is laboriously recorded: weather; how I got there and back; which hounds made the running; even what was in my sandwiches. It seemed to me that I owed this care to the animal. It was no small thing to take its life: the least I could do was to document the death carefully.

I think that this ethos is unusual amongst modern western hunters. It is certainly the norm for most indigenous hunters, and I suspect that it was the norm for most of human history. If that suspicion is right it might be said to ground an ethical argument.

The argument would go something like this: Suppose that we have been behaviourally modern for around 40,000 years (which is more or less the consensus, at least in relation to humans in Europe). Assuming that a generation is 20 years, that is 2000 generations. Assume (wrongly, and over-generously to Neolithic people) that the Neolithic revolution (which involved settlement, planting, and the domestication of livestock) happened everywhere in the world simultaneously 10,000 years ago, and put an end to hunter-gatherer life styles then. There have, then, been 500 non-hunter-gatherer generations out of the 2000 behaviourally modern human generations. We spent our formative years as hunters. Our anatomy, physiology and psychology were all designed to hunt and to survive being hunted ourselves. We are both predator and prey. In our personal, modern, formative years we are hunters too: watch any child, uninhibited by tyrannous education. Our behaviour recapitulates our evolutionary history, just as Ernst Haeckel thought that embryos did. Since we are quintessentially hunters, we will not be properly ourselves unless we hunt. Any scheme of ethics has to deal with that inescapable fact. A scheme of ethics that presupposes that we are something other than what we are is pointless: ethics must deal with the facts as they are: with the world as it is; not with some pastiche. And therefore (goes the final step in the argument), no correct ethical code could unwish our predatory instincts. To do so would be to unwish ourselves: to cause the subject of our ethical deliberation to evaporate, making that deliberation pointless (cf. Cahoone 2009). It follows that the ethics of hunting should be directed towards the regulation of hunting, not its abolition.

I have some sympathy with this argument, but the sympathy does not (now) amount to agreement. To say that something is atavistic is to describe its origins: to give an account of origins says nothing necessarily about ethics: to explain is not necessarily to excuse. To a first degree of approximation, the whole business of ethics and law is about the reining in of tendencies, not their licensing. Unless it can be established (and it plainly cannot), that the hunting instinct is a type of automatism, then our biological history can only mitigate the moral offence of killing another sensate creature, rather than constitute a defence to that offence.

If the argument worked, it would amount to a blanket defence for hunters. Even those who hunted purely for enjoyment could avail themselves of it. But I do not think that it works, and I am back where we started: hunters who need to hunt need no defence, and hunters who hunt simply because they enjoy hunting need to look elsewhere for their justification.

I no longer hunt. My reasons for stopping were not (or not mainly) philosophical. Nor did I have an epiphany such as John Fowles had. Horrified by the suffering of a broken-winged bird that he had shot, he hung up his guns there and then. I would like to be able to claim such a conversion, but mine was less dramatic. I simply got tired of killing things. I began to think I had killed enough, and I understood 'enough' to mean that there was a diminishing return from the deaths: I was not learning sufficient new lessons from the deaths, or bolstering my knowledge of old lessons, to make the deaths morally justifiable. This presupposes, of course, that I had concluded that the deaths were morally significant. I had: and cannot remember a time when I did not believe this. This was not based on any conviction about animal personhood. I am now convinced that at least some animals should be regarded as persons, but it is not necessary to believe this in order to believe that to kill an animal requires moral justification. I have not been helped at any stage by the complex and tortured philosophical literature on animal personhood: none of it does any real work for me.

All my reflection on the reasons for giving up hunting was ex post facto, after I had given up. Perhaps this disqualifies me from commenting on the morality of hunting. Nonetheless, and mainly because I have been asked to, rather than because I feel any great psychological or intellectual imperative to do so, I will continue for a while to comment.

Today my only hunting is vicarious. I get other people (other hunters, farmers, and butchers) to do my killing for me. I try to eat the flesh only of animals that have been happy, and have died well, and I only eat it in the context of a big celebration. For a dead animal to be defensibly on the dinner table, there must be a great deal of consequential human pleasure. Meat is for high days and holidays, not for a slumped midweek dinner. That would be disrespectful: not consonant with the animal's dignity or mine.

It follows that I have no Kantian qualms about using an animal as a means to an end. My concern is to ensure that the end is proportionate to my valuation of the animal's life. Two things follow from this.

First: one justification of meat-eating (which I find convincing), and one justification of some types of hunting (which, when it applies, may sometimes be convincing), is that unless humans eat meat, agricultural animals (for example) will not exist. Since the lives of at least some of those animals have more pleasure in them than pain, their deaths are justified, since without those deaths (and the agricultural system which abets those deaths), there would be, net, less animal pleasure in the world.

Second: Hunting purely for fun isn't *necessarily* unethical. To establish that it is ethical the fun would (as in my dinner party example) have to be very intense fun, of a sort that itself increased the net amount of good/pleasure in the world (ruling out, for instance, sadistic pleasure in killing or hurting). There are two notable philosophers who have looked seriously at this possibility without becoming mired in the morass of animal personhood. They are Roger Scruton and Ortega y Gasset. Scruton's work is, to my eye, derivative from Ortega y Gasset, and so rushed and superficial a derivation that while it is eye-catching and useful for polemicists, it is unlikely to make many vegetarians put on their red coats, or strengthen anything other than the ardour of hunting's apologists (Scruton 1998). Ortega y Gasset is a different matter. He deals fearlessly and fundamentally with the suggestion that human pleasure in hunting might be a sufficiently potent justification. I come to him in a moment. But first a point must be made about herbivorous animals.

This is simply that, if the teleology can be forgiven, they exist for two purposes: to unlock the energy of the sun that is trapped by plants, and to transmit that energy to others. As to the first purpose, the energy is sequestered inside the cellulose walls of plant cells: herbivore digestive systems can break down those cell walls and release the energy. As to the second purpose, the digestive systems allow the energy to be transmitted to the herbivore itself, and then on, up the food chain, to anything that eats the herbivore. These two purposes have made herbivores what they are: have conferred their shape, their speed, their cunning, and all other aspects of their behaviour, their anatomy, and their physiology. Deer have been shaped by wolves (and wolves by deer). If deer were not edible, they would not be the way they are. They would not be deer at all. To unwish the teeth and the hunger of wolves is to unwish deer. It is not so inaccurate to say that deer exist to be killed. This biological fact (which turns out to be an ontological fact), is the hinterland from which Ortega y Gasset's *Meditations on Hunting* emerges.[2]

---

[2]Originally published as a Prologue to *Veinte Anos de Caza Mayor* (Edward, Count Yebes, Madrid, 1943) All citations here are from Ortega y Gasset 1972.

25.3 On Ortega Y Gasset's *Meditations on Hunting*

Ortega y Gasset refuses to embark on a systematic philosophical exploration of the subject of hunting. This refusal gives me confidence that my own refusal is intellectually reputable. Ortega y Gasset thinks that a rigorous exposition would be impossible because hunting, at bottom, is concerned with two issues that are definitively imponderable—death and Otherness: "...death is the least intelligible fact that man stumbles upon. In the morality of hunting, the enigma of death is multiplied by the enigma of the animal" (Ortega y Gasset 1972, 103). Add agency into the mix and the mystery becomes even more impenetrable: "...[D]eath is enigmatic enough when it comes of itself – through sickness, old age, and debilitation. But it is much more so when it does not come spontaneously, but instead is produced by another being" (ibid.).

With one important caveat, he successfully resists the temptation to philosophise, but he is clear about what needs to be defended—whether philosophically or in any other way: it is the ecstasy that comes from hunting.

> Dionysios is the hunting god: 'skilled cynegetic' Euripides calls him in the Bacchantes. 'Yes, yes' answers the chorus, 'the god is a hunter'. There is a universal vibration. Things that before were inert and flaccid have suddenly grown nerves, and they gesticulate, announce, foretell. There it is, there's the pack! Thick saliva, panting, chorus of jaws, and the arcs of tails excitedly whipping the countryside. (pp. 89–90)

This ecstasy is only possible if humans are they they are meant to be, and do what they are meant to do: they are meant to be fit, brave, resolute, and principled–even out in the wilderness where no one can see if they are adhering to the hunters' code of honour (p. 35).

It entails energetic commitment to an act, which Aristotle and all happiness theorists since have insisted is essential to human thriving. But the ecstasy does not *consist* of fitness, bravery, and so on. It consists instead in immersion in a hot numinous bath. Its effect is to put man in his place: to restore his relationship with the natural world: to truncate his hubris. It is, ironically, a cure for the presumption that can flow from Genesis 1's urge to dominate and subdue. The Dionysiac ecstasy is ecstasy in the literal sense: standing outside oneself–a process that diminishes self-obsession and increases a sense of connectedness with the rest of the world.

> Strictly speaking, the essence of sportive hunting is not raising the animal to the level of man, but something much more spiritual than that: a conscious and almost religious humbling of man which limits his superiority and lowers him towards the animal. (p. 111)

If hunting really produces this sort of ecologically realistic humility, there would be a compelling case for it. It would save many animals, many habitats, and many human souls. It would inhibit the drive to mastery: the hunter's natural place would not be astride a horse, but on his knees. If Ortega y Gasset's prescription works, it may be because (as my own experience sometimes suggested) there is in hunting a constant reminder of Mortality—and hence of one's own mortality. "Life is a terrible conflict", writes Ortega y Gasset, "a grandiose and atrocious confluence. Hunting submerges man deliberately in that formidable mystery and therefore contains something of religious rite and emotion in which homage is paid to what is divine, transcendent, in the laws of Nature" (p. 112).

It is easy to parody this. Indeed it is not easy, when one is attempting to argue rigorously, not to do so. It is also hard not to compare it unfavourably with the reality of many hunts: with the snobbery, the fat red faces, the big lunches, the screaming dismemberment, the broken-winged bird crouched in the bracken. Both the parody and the comparison should be resisted. There is a serious point being made here, about serious hunts by serious people. It is the red faces and the picnic basket that are the distortions (or so Ortega y Gasset would say).

Perhaps all this is not so far from my own speculation about humans as quintessential Upper Palaeolithic hunters, and hunting necessary as an expression of that nature, and an expression of that nature as necessary for personal integration and hence personal and corporate morality. I don't know, because Ortega y Gasset's formulation is by definition inaccessible to this (or any) sort of interrogation. It may be contemptibly weak as a result. It may also be indestructibly strong. How could we know?

There is one point at which Ortega y Gasset, neglecting his own injunction to himself not to philosophise, does move into territory where he is vulnerable to the ordinary dialectical weapons. In a passage reminiscent of C.S. Lewis's (1940) speculations about animal suffering, he asks: "…is it so certain that the beast is afraid? At least his fear is not at all like fear in man. In the animal fear is permanent; it is his way of life, his occupation. We are talking, then, about a professional fear, and when something becomes professional it is quite different." (pp. 90–91; cf. Scruton 2002).

One might have thought it tactically unwise for Ortega y Gasset to have exposed himself in this one place, since to be humiliatingly contradicted there might impeach the rest of his case, which otherwise would have remained subject to the absolute immunity to emphatic contradiction enjoyed by all metaphysics. But in fact it was shrewd, for his argument is surely both right in fact (it is, in essence, the same point that I have made above: herbivores are professional die-ers), and necessary to his frankly religious case—since in order to establish that his religion is good and not evil, and since nothing can be said about the goodness or evil of death, *something* has to be said about the magnitude of the animal suffering that has to be offset in the ethical calculus against the goods that result from the Dionysiac ecstasy.

Ortega y Gasset's non-argument is, it seems to me, the only argument that can be made for the ethics of sport-hunting. Yet the uncertainties inherent in it (indeed which are necessary to it) leave me queasily unconvinced. Ortega y Gasset himself seems to acknowledge that this will be (indeed probably should be) the case. "Every good hunter is uneasy in the depths of his conscience…He does not have the final and firm conviction that his conduct is correct. But neither…is he certain of the opposite…" (p. 102).

The precautionary principle does not help us out of this bind, since we do not know what is at stake. Perhaps by choosing not to hunt we are endangering ourselves and countless non-human animals. Perhaps by choosing to hunt we are killing persons who will revenge themselves eternally.

On the question of the ethics of *enjoying* hunting, then, we are no further forward. Argument tells us much less than our intuitions.

We shouldn't dismiss or denigrate our intuitions. They are ancient, and informed by a great multitude of sources. When assessing the value of an intuitive moral hunch, it is reassuring if the hunch is shared by others whose values and behaviour are generally commendable and have stood the test of time. Which takes me back to lions.

25.4 Another Look at Whether Lions Should Be Allowed to Hunt

If it is wrong for lions to hunt, even if they have been designed by Darwin to do so, and would die if they didn't, it seems probable that it is wrong for human hunters to hunt for mere fun.

There is an ancient and venerable line of authority that suggests that lions are not meant to hunt, that teeth are not meant to be sharp, and that co-operation and altruism, rather than competition, suffering, death, and waste, are the main fuel of the complexity-generating machine that is evolution. It is in the first of the two Hebrew creation stories in the book of Genesis. Genesis 1: 20–27 tells how all living things, including man, were created. In verse 18, man is given dominion. But very plainly this dominion does not include killing animals, for at this point *all* creatures are vegetarian. "And God said, 'Behold, I have given you every plant yielding seed which is upon the face of all the earth, and every tree with seed in its fruit; you shall have them for food. And to every beast of the earth, and to every bird of the air, and to everything that creeps on the earth, everything that has the breath of life, I have given every green plant for food.' And it was so" (Genesis 1: 29–30 [RSV]). In verse 31 God surveys everything that he has done, and concludes that 'it was very good'. What is 'very good' is a regime in which there is no predation, and where the only food is plants.[3]

But it all goes catastrophically wrong. Adam and Eve eat the forbidden fruit, and are expelled from the garden. The whole of the created order is warped by their disobedience. But this does not cause a fundamental change in the divine mind. It is not until Genesis 9 that Noah is told that he can eat flesh as well as plants, but this is by way of a rather grudging dispensation (Genesis 9:3). It is not the way things are meant to be. The old intention is remembered in solemn edicts about the value of life and the shedding of blood: "Only you shall not eat flesh with its life, that is, its blood. For your lifeblood I will surely require a reckoning; of every beast I will require it and of man; of every man's brother I will require the life of man." (Genesis 9: 4–5). Blood taboos are prominent in subsequent biblical edicts. They are

---

[3]The second of the two creation stories is silent on this issue: see Genesis 2: 4–25.

meant to be intrusive and inconvenient. Their purpose is to remind humans that the state of affairs in which blood is shed was not the original plan. No blood is to be eaten (Leviticus 17: 10–14); there are laws dealing with menstruation (Leviticus 15: 19–30) and peri-parturient bleeding (Leviticus 12: 1–8). Predation itself was never part of the plan, and the Hebrews are reminded of this by the prohibition on eating birds of prey (Leviticus 11: 13–19).

Far from teeth and claws being of the essence of lions, teeth and claws diminish them: de-lion them. How much more, if that's right, must a Parker Hale .308 de-humanise and diminish an omnivorous human.

I can't lift my own morality direct from these verses. They are too strange. Nor have they consciously informed my own thinking. But I am reassured by the concurrence of my intuition and the tradition. I note that the Noahide dispensation has never, in any of the subsequent Talmudic disputation, been thought to accommodate recreational hunting.

25.5 Hunting and the Anthropocene

Nobody denies that we are in a historically unprecedented and vertiginous age ('the Anthropocene'), characterised by anthropogenic ecocide, which (since the gods tend to punish hubris very ruthlessly and efficiently) may well lead to our own extinction. There is much discussion about when the Anthropocene started. The least popular suggestion is William Ruddiman's (2003): the Anthropocene started sometime in the Neolithic. The broad consensus is represented by the Anthropocene Working Group (AWG) of the Subcommission on Quaternary Stratigraphy of the International Commission on Stratigraphy, which voted for an as-yet-unspecified start date in the middle of the twentieth century (Subramanian 2019). Ruddiman is surely the nearest. Yet even his date is too late.

When *Homo sapiens* first arrived in South America and Australia (the dates are contested, but the date is almost certainly prior to Ruddiman's Anthropocene date) large animals were quickly decimated. The men were intoxicated by the sight and the taste of all that easy, lumbering flesh: the animals (unlike those of Africa and Eurasia who had known for millennia the sorts of creatures we are), were fatally naïve (Tudge 1999). If you were large and edible in late Pleistocene Argentina you would disagree with the AWG. The AWG's date seems to be a symptom of shifting baseline syndrome. The right baseline is not the already devastated world of the twentieth century, but the world of the Upper Palaeolithic, when you could have walked between Pacific islands using the backs of turtles as stepping stones.

The speed and size of those early American and Australian killings mark three massive and repercussive changes in human hunting history. First: The human populations concerned weren't large, and killings on that scale were unnecessary. Previous killings had been necessary. Second: because of the naivety of the animal populations, the killings were easy. Previous killings had been hard. And third (a guess): the killings (unlike the previous killings) weren't reverential, fearful, or sacramental.

There weren't prayers over the corpses, or oblations or sleepless nights. There were just too many dead animals for that sort of cult to be sustainable. Since cult sustains ethos and mindset, humans became casual killers. You can only kill casually something of which one is not a part, and so the casualness prised humans out of their place in the natural world. Now they stood outside it, arrogant and cruel. The Upper Palaeolithic hunters became petroleum executives, and the Anthropocene had begun.

It might be said that good, respectful, quaking, humble hunting might help to turn back the clock; to re-educate us; to re-forge proper relationships with the non-human world. And for some individuals it might. But it is too risky to advocate this as a strategy. Blood is heady stuff. It does unpredictable things to humans.

25.6   Conclusion

I agree with Ortega y Gasset: There can be no firm philosophical conclusion.

I disagree with y Gasset: sport hunting is not acceptable. I cannot demonstrate this conclusion: I can only give an account of the intuitions which tend towards it, and some of the arguments that, while not capable in themselves of making out the conclusion, might be said to buttress the intuitions.

I am uncomfortable about killing animals. I have sold my guns. This makes me feel better. It makes me feel more myself, not less. I can say little more than this.

References

Cahoone, L. 2009. Hunting as a moral good. *Environmental Values* 18 (1): 67–89.
Kalof, L., and A. Fitzgerald. 2003. Reading the trophy: Exploring the display of dead animals in hunting magazines. *Visual Studies* 18 (2): 112–122.
Lewis, C.S. 1940. *The problem of pain.* London: Centenary Press.
Ortega y Gasset, J. 1972. *Meditations on hunting,* trans H. B. Wescott. New York: Charles Scribner's Sons.
Ruddiman, W.F. 2003. The anthropogenic greenhouse era began thousands of years ago. *Climatic Change* 61 (3): 261–293.
Scruton, R. 1998. *On hunting.* London: Yellow Jersey Press.
Scruton, R. 2002. Ethics and welfare: The case of hunting. *Philosophy* 77 (4): 543–564.
Subramanian, M. 2019. Anthropocene now: Influential panel votes to recognise earth's new epoch. *Nature,* May 21. https://doi.org/10.1038/d41586-019-01641-5.
Tudge, C. 1999. *Neanderthals, bandits and farmers: How agriculture really began.* Connecticut, CT: Yale University Press.

Charles Foster is a Fellow of Green Templeton College, University of Oxford, and a Visiting Professor at the Faculty of Law, University of Oxford. He holds a PhD in Medical Law and Ethics from the University of Cambridge. He is the author of numerous books, including *Human Thriving and the Law, Identity and Personhood in the Law, Altruism, Welfare and the Law, Human Dignity in Bioethics and Law,* and *Being a Beast.*'

Chapter 26
Comment: Sharing Our World with Wild Animals

J. A. A. Swart

26.1 Wild Animals in the Anthropocene

The contributions of the authors in this section address some of the ethical, societal, philosophical, and ecological challenges of the Anthropocene with respect to wild animals. As explained in the introduction to this volume, animal habitats are becoming increasingly fragmented, polluted and disrupted by human activities such as transport, urbanization, agriculture, and overfishing. Climate warming is one of the most threatening aspects of the Anthropocene, as climate zones are moving towards the poles and up mountain slopes. Wild species that depend on the conditions in these zones must follow in order to survive. However, many cannot follow fast enough, or are hindered by agricultural lands, cities, industries, or roads. Some habitats may even disappear. For example, Clare Palmer describes in her contribution the sad case of the polar bear, a species that is threatened by melting ice sheets in the Arctic.

By contrast, some other species are doing relatively well, as they are able to adapt and exploit the opportunities of climate change, or to benefit from conservation and restoration efforts. For example, the wolf is recolonizing areas in Europe in which it has not been seen for hundreds of years. Martin Drenthen describes in his contribution the re-entrance of the wolf in the Netherlands. The beaver and the gray and common seal are also successful species in this country, as are storks and cormorants. A recent report lists nearly 40 mammal and bird species whose European populations are increasing, in particular in Northwestern Europe, due to successful nature conservation measures in recent decades (Deinet et al. 2013).

As well as reappearing native species, we are also seeing the establishment of new species in human landscapes. Some have come on their own, such as the Western

J. A. A. Swart (✉)
University of Groningen, Groningen, The Netherlands
e-mail: j.a.a.swart@rug.nl

© The Author(s) 2021
B. Bovenkerk and J. Keulartz (eds.), *Animals in Our Midst: The Challenges of Co-existing with Animals in the Anthropocene*, The International Library of Environmental, Agricultural and Food Ethics 33,
https://doi.org/10.1007/978-3-030-63523-7_26

Great Egret and the wildcat in the Netherlands, whereas others have been introduced deliberately or accidentally, such as the musk rat, the Egyptian goose, and more recently the raccoon dog. As argued by Ned Hettinger in his contribution, species introduced by humans have a much greater chance of threatening endemic species and disrupting existing ecosystems.

The appearance, or reappearance, of new animals in our landscapes, whether it is the result of successful conservation and restoration, the introduction by humans, or a consequence of climate warming, may meet resistance among the general public. This is described by Mateusz Tokarski and Martin Drenthen in their contributions. For example, Dutch sheep farmers fear the wolf, water managers are concerned about beavers disrupting watercourses, and garden owners get upset because wild boars plow their flower and vegetable beds. It is not just large mammals that cause unrest. Not so long ago, Dutch newspapers reported on field mice that were damaging farmer's gazing fields, and we have known of the problems caused by geese grazing the nutrient-rich grasslands of those same farmers for many years.

Not only vertebrate species concern us. The dramatic decline of some insect populations may disrupt ecosystems (Sanchez-Bayo and Wyckhuys 2019), while other insect species flourish and even turn into threatening invasive species, such as the oak processionary caterpillars that cause a severe skin irritation and asthma (Pieters 2019). Even more serious is the emergence of mosquito-borne infectious diseases such as Zika, malaria and dengue in moderate climate zones as a result of climate warming (Ryan et al. 2019).

The decline and emergence of populations of wild species have always played a role in human history, but this seems to be much more dramatic in the Anthropocene, a period that, according to most authors, began during the industrial revolution of the eighteenth century and accelerated in the middle of the twentieth century (Steffen et al. 2011). Charles Foster, however, argues in his contribution that the Anthropocene started as early as the Upper Paleocene, with the massive but unnecessary killing of large animals in Australia and America by our ancestors. His view implies that the Anthropocene, seen as the result of the human inclination to kill and take as much as possible from the earth's resources, is a *condition humaine*.

Charles Foster is right to stress the huge and early impact of modern man on biodiversity and its role in the irreversible loss of megafauna. However, we are currently crossing planetary limits within which humanity and countless other living creatures can safely exist (Steffen et al. 2018). According to some authors, we may have already gone beyond these limits (Rockström et al. 2009). These developments force us to answer a fundamental question: what kind of world do we want to live in, and how can we co-exist with other living creatures with whom we share the same earth?

26.2   Towards an Anthropocenic Animal Ethics

The contributions by the authors in this section may be regarded as attempts to address those questions, in particular by rethinking basic concepts relating to our relationships with wild animals. In this context, Clare Palmer demonstrates, using the case of the polar bear, that subjective welfare-oriented approaches fail to provide clear directions for coping with the challenge of climate warming with respect to wild animals. She therefore proposes conducting experiments to test intervention options such as supplementary food. These kinds of experiments are *wild experiments*, meaning that they are carried out in the real world to learn how to deal with actual challenges in the Anthropocene (see for example Lorrimer 2015).

Martin Drenthen's suggestion, which is to manage the human landscape and design artifacts such as fences so that they function as a means of communication to wolves, may also be considered a wild experiment. Of course, most species do not have the communicative skills of wolves, or they require quite different conditions for their subsistence, but his suggestion may be interpreted as a call to listen to what wild animals, as cohabitants of the world, are telling and asking us. These two cases also demonstrate the broad spectrum of anthropogenic effects in the Anthropocene, which range from threatening conditions for some wild species on the one hand, to favorable conditions for other species on the other.

The pleas for interventions to support threatened wild animals, and especially the appearance of wild animals in the human landscape, challenge the traditional vision of the nature-culture divide, between the human and the wild world, and between domesticated and wild animals. This calls into question the traditional view on animal ethics, which is that we should not interfere with the lives of wild animals, a view concisely worded by Tom Regan in 1983: "Let them be". Clare Palmer (2010) calls this "laissez faire intuition", which means that we do not have a duty to take care of animals in nature, unless we are responsible for their deteriorating circumstances. Similarly, it is argued that if we recognize the ethical value of wild animals and their populations in their natural habitats, we must provide these natural habitats (for example through nature protection measures) so that they can flourish as wild animals. This latter type of care is defined as non-specific care, to distinguish it from specific care, which relates to the individual needs of domestic animals that we keep in human society (Swart 2005; Keulartz and Swart 2012).

There are, however, a lot of wild animal species that do not fall into the categories of fully wild or domesticated animals, but instead somewhere in between (Klaver et al. 2002). We may think of hemerophiles (opportunistic human culture followers, e.g. many garden birds), feral animals (e.g. free roaming cats), and increasingly animals that appear in human landscapes as a consequence of the changing conditions in the Anthropocene. Such animals may be considered to be "semi-wild" (Swart 2005), as falling in the "contact zone" (Clare Palmer 2010), or as "liminal animals" (Donaldson and Kymlicka 2011). The presence of wild animals in the human landscape, whether due to climate change or other anthropogenic phenomena, is not expected to be temporary. We need animal ethics that recognize and explicate the moral standing of this group of animals.

In this context, Donaldson and Kymlicka (2011) have further elaborated the distinction between domestic, wild and liminal animals with the help of a political framework. Domestic animals are considered by them as fellow citizens in the human community with basically the same rights as humans. Wild animals, on the other hand, not being influenced by humans and living in the wild, are seen as members of sovereign communities that must be respected. People should in principle not be allowed to intervene in their communities, as if these communities were other nations. Liminal animals have a position in between. The authors compare them with denizens in human society such as refugees or immigrants who have certain basic rights but, for example, no voting rights in political matters. Similarly, liminal animals have a basic right not to be harmed or killed, but we do not have to feed them or provide housing as they are still wild animals. We should tolerate nuisance to a certain extent, but we may protect ourselves from serious damage, while still respecting their basic rights.

This approach is well-suited to the challenge of the Anthropocene, as it acknowledges the different relationships between humans and animals, ranging from domesticated animals in human society to wild animals living in their natural habitat, and it provides us with an underlying justification of different treatments of animals in different environments. It justifies the negative right of animals living in the wild not to be disturbed, but it also acknowledges positive rights, not only of domesticated animals kept by humans, but also of wild animals living in the human landscape. This is relevant because, as Donaldson and Kymlicka (2016) also indicate, countless wild species live in and are dependent on our rural and urban areas. This is especially true in the Netherlands and many other Western European countries with intensive agriculture and strong urbanization. For example, Dutch godwits breed in the spring on the Frisian meadows and spend the winter in West African rice fields, and barnacle geese overwinter in the humanized landscapes around the North Sea but breed around the Arctic Circle (Swart 2016).

26.3 A Heterogeneous, Coercive, Socioecological Network

However, we may wonder whether the concept of sovereign wild animal communities is an appropriate term to characterize animal communities in the wild. After all, it implies a human, sociological perspective of wild animals since it stems from a political theory. But, there are no sovereign rulers or institutions in the wild, nor shared objectives and values. As far as we can speak of a community, it is only "governed" by the behaviors of and interactions (e.g. predation, migration, reproduction) between members of multiple populations and species under particular biotic and abiotic circumstances such as climate, vegetation, soil composition, and so on. However, this is also the case for wild animal populations in man-dominated landscapes, where people and their artifacts belong to the biotic and abiotic conditions that animals have to deal with.

Wild animals, whether living in natural or human landscapes, often make opportunistic use of available and accessible natural, agricultural or urban resources, based on their species' characteristics. If biotic and abiotic conditions change, whether by natural or man-made causes, animals must develop new interactions with the new situation, for example through a change in food sources or by migration to another area, including human landscapes. Animals that are able to cope with such a heterogeneous and dynamic environment, including man-dominated landscapes, have a greater chance of survival.

Therefore, instead of considering wild animals as members of autonomous, sovereign communities and liminal animals as denizens in the human society and human landscapes, I prefer to consider both of them as nodes in a dynamic, heterogeneous, coercive, socioecological network of dependency relationships between abiotic and biotic factors which, in the Anthropocene, increasingly includes human society and its institutions (see for example Coeckelbergh 2012). Despite the existence of animal agency, i.e. ability of the animal to act in accordance with its species-specific drives, desires or will (see Meijer and Bovenkerk, this volume) and animal autonomy, i.e. the ability to put that agency into practice, wild animals, will nevertheless be strongly affected in their choices by the compelling circumstances of the biotic and abiotic environments, regardless of whether they live in natural or human landscapes (Swart 2005, 2016).

Recognizing the presence of wild animals in the human or humanized natural landscape, whether we call them wild, semi-wild, liminal, or contact-zone animals, and recognizing that they have certain negative and positive rights, raises the question of how we can live together and what this means. This is a difficult question to answer, because these animals, as explained above, form a highly heterogeneous group due to differences between species and their niches, and due to varying levels of adaptation to the human environment. The concepts of specific and non-specific care are not only very general, but also not aimed at this group of animals, except that they may apply to them to some extent. Neither does the approach of Donaldson and Kymlicka (2011) offer us clear suggestions, except that liminal animals' basic rights must be taken into account.

A more categorizing characterization of these animals may help. In this context, Swart and Keulartz (2011) distinguish two dimensions that relate to biological and sociological approaches of domestication. The first dimension is *adaptability*. This refers to the extent to which an animal has adapted or is able to adapt to humans or their environment. This is often evidenced by certain biological and behavioral characteristics, as is most visible in pets. The second one is *dependency*, a sociological dimension which is related to the extent to which an animal is dependent on the human system for its subsistence. Most pets are both strongly adapted to and dependent on humans, while wild animals living in undisturbed nature are not. However, wild animals that live in a human or humanized landscape may be characterized by these two dimensions to a certain extent. For example, some zoo animals can still be considered wild according to the first dimension, but are nevertheless completely

Adaptation to the
human system

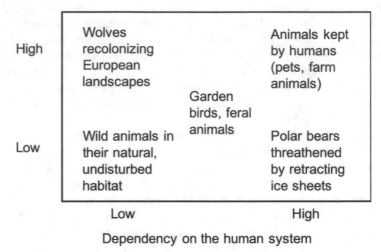

Fig. 26.1 A tentative scheme of adaptation to and dependency of animals on the human system

dependent on humans and domesticated in that sense. Many garden birds and feral
animals often take an intermediate position in both dimensions (see Fig. 26.1).

Applying these dimensions to the case described by Clare Palmer suggests that
the polar bear is in the lower right corner of Fig. 26.1. The population is seriously
affected by receding ice sheets in the Arctic, a situation to which polar bears cannot
adapt very well, so that they can become highly dependent on human interventions.
On the other hand, the recolonizing wolf, as described by Martin Drenthen, can be
placed in the upper left corner of Fig. 26.1 because these animals are well adapted to
the human landscapes of Europe while they still exist quite independently of humans.

The dimensions in Fig. 26.1 are not completely independent of each other, as
adaptation may lead to dependency and dependency may initiate a process of adaptation. Adaptation is anchored primarily in biology and is not easy to influence, or
only very slowly. On the other hand, dependency is a social dimension that we have
better control of. Most of the measures that we can take to influence the fate of
endangered animals or animals that live in the human landscape are therefore in this
dimension. For example, we should take measures to move animals from the lower
right-hand position of Fig. 26.1, which is a particularly undesirable situation, to the
left. On the other hand, if we want to prevent the further domestication of animals in
the upper left corner, we must take measures that keep them on the left and therefore
independent.

Ned Hettinger's concept of native and non-native animals seems to fit into this
gradual and two-dimensional perspective on wildness and domestication. He claims
that being native, which is a matter of degree, means that a species "has significantly
adapted to or interacted with the local biota and abiota (and vice versa). Non-natives

are species that have not significantly adapted to or interacted with the local inhabitants or abiota". His definition turns the concept of what a native species is from a historic into a contextual one, as whether or not a particular species can be considered native depends on the level of adaptation to the current set of biotic and abiotic conditions. Thus, from this perspective, the polar bear in the Arctic case has become less native as it is not well adapted to the new conditions, whereas the wolf in western Europe has become more native as it seems to be rather well adapted to that specific environment.

26.4   Non-Specific Care for Wild Animals in a Humanized World

In the Anthropocene, our expansive behavior means that human and animal worlds are increasingly merging and wild animals are becoming more and more dependent on human society. The contact zone is widening. Given this merging and our increasing dominance, we have strong obligations regarding the fate of wild animals, not only towards threatened animals in remote natural areas as the Arctic with its disappearing ice caps, but also towards wild animals that live in human landscapes.

We therefore need to adapt our basic attitudes to wild animals, as Mateusz Tokarski argues in his contribution. According to him, anxiety, fear, disgust or even hatred with regard to animals showing up in our humanized landscapes are actually "symptoms of a fundamentally misguided worldview, most commonly characterized as anthropocentrism". Making use of the stoic tradition, he argues that we need an environmental philosophy to develop an impersonal and rationalized worldview that may function as a source of consolation to such discomforts, and that we need to consider these discomforts as inevitable and acceptable in the context of our co-existence with wild animals. However, I doubt whether such a universal and impersonal worldview is practically feasible in our current world and I also wonder whether a stoic way of thinking can lead to disinterest or even apathy with regard to the dramatic fate of wild animals in the Anthropocene. Nevertheless, I agree that we need to adjust our basic attitudes to wild animals, an adjustment that may enrich our existence through a better understanding of the interconnectedness and interdependencies of all forms of life, including ourselves. Charles Foster's desire for an intimate connection with the natural world to acquire self-knowledge fits into this perspective, although in my opinion it contradicts his defense of hunting.

I believe that we do not have exclusive rights to the earth and that we should recognize and respect wild animals as fellow-earthlings. Because of our impact, this implies an empathic, nonspecific care perspective towards wild animals. I have previously described non-specific care as interventions focusing on the natural environment of animals so that they can live their natural lives. Unfortunately, this is no longer enough. In the Anthropocene, such non-specific care should also focus on the conditions for wild animals in the human world so that they can live there according to their capabilities as much as possible. As the authors in this section show, this means giving space to the wild animals that appear in the human landscape, and taking measures to restore natural areas and adjust our landscapes and infrastructures to make their lives here possible. It requires wild experiments but also a reconsideration of our ethics, philosophies, culture, institutions, and politics.

References

Coeckelbergh, M. 2012. *Growing moral relations: Critique of moral status ascription.* New York: Palgrave Macmillan.

Deinet, S. et al. 2013. *Wildlife comeback in Europe: The recovery of selected mammal and bird species.* BirdLife International and the European Bird Census Council. London: ZSL.

Donaldson, S., and W. Kymlicka. 2011. *Zoopolis: A political theory of animal rights.* Oxford: Oxford University Press.

Donaldson, S., and W. Kymlicka. 2016. Comment: Between wild and domesticated: Rethinking categories and boundaries in response to animal agency. In *Animal ethics in the age of humans: Blurring boundaries in human-animal relationships*, ed. B. Bovenkerk and J. Keulartz, 225–239. Dordrecht, The Netherlands: Springer.

Keulartz, J., and J.A.A. Swart. 2012. Animal flourishing and capabilities in an era of global change. In *Ethical adaptation to climate change: Human virtues of the future*, ed. A. Thompson and J. Bendik-Keymer, 123–144. Cambridge, MA: MIT Press.

Klaver, I., J. Keulartz, H. van den Belt, and B. Gremmen. 2002. Born to be wild. A pluralistic ethics concerning introduced large herbivores in the Netherlands. *Environmental Ethics* 24 (1): 3–23.

Lorrimer, J. 2015. *Wildlife in the Anthropocene: Conservation after nature.* Minneapolis, MN: University of Minnesota press.

Palmer, C. 2010. *Animal ethics in context.* New York: Columbia University Press.

Pieters J. 2019. Poisonous caterpillars putting a damper on outdoor fun. *New York Times*, June 28. https://nltimes.nl/2019/06/28/poisonous-caterpillars-putting-damper-outdoor-fun.

Regan, T. 1983. *The case for animal rights.* Berkeley and Los Angeles: University of California Press.

Rockström, J., W. Steffen, K.J. Noone, A. Persson, F.S. Chapin III, E.F. Lambin, et al. 2009. A safe operating space for humanity. *Nature* 461: 472–475.

Ryan, S.J., C.J. Carlson, E.A. Mordecai, and L.R. Johnson. 2019. Global expansion and redistribution of Aedes-borne virus transmission risk with climate change. *PLoS Neglected Tropical Diseases.* https://doi.org/10.1371/journal.pntd.0007213.

Sánchez-Bayoa, F., and K.A.G. Wyckhuys. 2019. Worldwide decline of the entomofauna: A review of its drivers. *Biological Conservation* 232: 8–27.

Steffen, W., A. Persson, L. Deutsch, J. Zalasiewicz, M. Williams, K. Richardson, et al. 2011. The Anthropocene: From global change to planetary stewardship. *Ambio* 40: 739–761.

Steffen, W., J. Rockström, K. Richardson, T.M. Lenton, C. Folke, D. Liverman, et al. 2018. Trajectories of the earth system in the Anthropocene. *Proceedings of the National Academy of Science* 115 (33): 8252–8259.

Swart, J.A.A. 2005. Care for the wild. Dealing with a pluralistic practice. *Environmental Values* 14 (2): 251–263.

Swart, J.A.A. 2016. Care for the wild in the Anthropocene. In *Animal ethics in the age of humans: Blurring boundaries in human-animal relationships*, ed. B. Bovenkerk and J. Keulartz, 173–188. Dordrecht, The Netherlands: Springer.

Swart, J.A.A., and J. Keulartz. 2011. The wild animal in our society. A contextual approach to the intrinsic value of animals. *Acta Biotheoretica* 59: 185–200.

Dr J.A.A. (Sjaak) Swart (1951) studied agriculture, biology, and philosophy. He received a PhD in medical biology and has worked for more than 25 years at the Science and Society Group (SSG) of the Faculty of Science and Engineering of the University of Groningen, in the Netherlands. Currently, he is associated to the Energy and Sustainability Research Institute Groningen (ESRIG) of this faculty. His research projects focus on societal and ethical aspects of biotechnology, ecological restoration, nature conservation, indigenous knowledge, breeding, and animal experimentation.

Part V
Animal Artefacts

Chapter 27
De-extinction and Gene Drives: The Engineering of Anthropocene Organisms

Christopher J. Preston

Abstract Advances in gene reading, gene synthesis, and genome editing are making possible a number of radical new practices for transforming animal futures in the Anthropocene. De-extinction may make it possible to bring back lost species. Gene drives may enable the sending of desirable traits through wild populations of organisms. The hype accompanying these promises can make each of them look ethically irresistible. This chapter investigates the 'speculative ethics' that has arisen around these technologies, asking questions about both their viability and the approach to animals they contain. Reductive and non-relational thinking is identified as one potential problem with the thinking behind these techniques. The neglect of non-human agency is identified as another. After indicating some of the problems these two ways of conceptualizing an animal and its genome can create, a brief suggestion is made about how to better conceptualize animal futures in the Anthropocene.

27.1 Synthetic Animal Futures

From the perspective of evolutionary biology, we live in radical times. Since the conclusion of the Human Genome Project in 2003, a formidable toolbox has been assembled for reworking the products of Darwinian evolution. The powerful trio of gene sequencing, gene synthesis, and gene editing together appear to make possible the genomic reconstruction of any biological lifeform. Genotypes are becoming moldable in ways that were inconceivable just a few years ago. From *Mycoplasma mycoides* to *Homo sapiens*, what you find at the genetic level no longer has to be what you keep.

Research time for this chapter was supported by Critical Scientists Switzerland

C. J. Preston (✉)
University of Montana, Missoula, MT, USA
e-mail: christopher.preston@mso.umt.edu

© The Author(s) 2021
B. Bovenkerk and J. Keulartz (eds.), *Animals in Our Midst: The Challenges of Co-existing with Animals in the Anthropocene*, The International Library of Environmental, Agricultural and Food Ethics 33,
https://doi.org/10.1007/978-3-030-63523-7_27

Today genomicists don't just sequence genomes, they build them. The turn of the century project known as "Human Genome Read" has given way to "Human Genome Write." The genomes of entirely new organisms are being built from scratch in labs. The genomes of existing ones are being significantly tweaked for traits deemed necessary by human designers. Reductions in costs have outpaced Moore's law (more than halving every two years). The possibilities, say the advocates, are endless. Never before has such a responsibility for the shape of the biological world fallen so completely into one species' hands.

These are head-spinning times for bio-ethicists and particularly challenging times for environmental ethicists. Some of the bedrock positions in the field are being thrown into question. The idea that there is moral significance to longstanding evolutionary processes (Rolston III 1988; Leopold 1949) is having to confront the fact that the independence of these processes from human design is becoming increasingly rare. It is not just that we are in the Anthropocene, the epoch in which unplanned human impacts have undeniably become global. Even more than entering an Anthropocene, the planet is running at some speed towards a biological "synthetic age" (Preston 2018). Thanks to these new tools in biotechnology, unthinkable prospects like the de-extinction of lost species and the re-engineering of wild animal populations are moving towards the realm of possibility. Before long, the ecologies within which our species dwells may be populated by a host of synthetic organisms and carefully re-engineered life forms.

To dismiss these deep technological changes solely on the basis that they replace natural organisms with synthetic ones (Lee 1999) is clearly unacceptable. At a conceptual level, the complexity and ambiguity of a term like 'natural organism' makes such a blanket condemnation philosophically dubious (Siipi 2008; Lie 2016). At a practical level, a position resting on hard-to-defend metaphysical commitments like 'naturalness' looks like a poor basis for policy (Kaebnick 2009). Rhetorically, there is a risk of appearing like a troll or a luddite.

Furthermore, even if there once was an intuitive preference for the natural over synthetic organisms this preference may no longer be determinative. One of the most striking things about these developments in biotechnology from an ethical point of view is how the arguments in favor of deploying them to solve difficult problems in conservation and public health look incredibly compelling. De-extinction, for example, offers the tantalizing promise of returning a number of charismatic and ecologically important animals to the planet's roster of species. Gene drives dangle the carrot of a vast reduction in human suffering in the face of diseases like malaria, dengue fever, and zika. Gene drives are also claimed to offer the possibility of eradicating invasive mammals from island ecosystems without poisons, traps, and the accompanying suffering to animals these methods cause. The moral arguments in favor of the new techniques can look unimpeachable. From a consequentialist standpoint, who would question such desirable outcomes?

What even a cursory inspection reveals, however, is that the promises driving some of these research trajectories rest on what Alfred Nordmann has called in a different technological domain "a speculative ethics" (Nordmann 2007). A speculative ethics is one in which a hypothetical prospect is floated in front of decision-makers and their publics in such a way that "an imagined future overwhelms the present" (Nordmann 2007, 32). Speculative thinking expresses confidence that *if* a certain futuristic research trajectory is followed, *then* a highly desirable result will follow. "What looks like an improbable, merely possible future in the first half of the [if-then] sentence, appears in the second half as something inevitable" (ibid.).

In the case of animals, a speculative biotechnological ethics appears to be creating a misleading mandate for humans to reconstruct the life forms that surround them. In the following pages, I show how the discourse around both de-extinction and gene drives is creating a harmful speculative ethics of animal biotechnology. I put part of the blame for this harmful direction on two types of faulty thinking. One is the methodological problem of non-relational and reductive thinking. The other is the ethical problem of the neglect of non-human agency. In identifying this second fault, I float the possibility of expanding the idea of agency in organisms to include the agency of the genome. I conclude by pointing briefly to a couple of the costs of this inappropriately speculative ethics in two arenas in which its applications are most often touted.

27.2 Speculations About de-Extinction

De-extinction demands the careful application of all three of the powerful tools gilding today's biotechnology toolbox. In order to de-extinct a lost species, a researcher would need an accurate account of the genome of the absent organism. The sequencing techniques perfected during the Human Genome Project, pioneered by Craig Venter's 'shotgun approach,' have pushed the door open to this possibility (Venter et al. 1998).

In many cases, sequencing requires analysis of actual tissue from an extinct animal. Sometimes this is surprisingly easy to find. Wooly mammoth tusks, bones, hair, and even whole flanks of their frozen flesh have been dug up in various locations across northern Europe and the Arctic. Numerous passenger pigeon specimens exist in U.S. natural history museums. Stellar sea cow bones are regularly being freed from sea ice to wash up on northern shores. Thirteen different thylacine fetuses (known as 'joeys') sit preserved in bottles of ethanol in zoos and museums in Hobart, Melbourne, Sydney, and Prague.[1] As a result of these existing DNA fragments, wooly mammoth, passenger pigeon, and thylacine genomes have all already been sequenced.

Once an accurate read of the genome is available, this read can be compared to the genome of a closely related species. A wooly mammoth, for example, can be compared to an Asian elephant, a Pyrenean ibex to a domestic goat, a passenger pigeon to a band-tailed pigeon. In many of these cases, the gene sequences will have an overlap in the high ninety percent range. An Asian elephant, for example,

---

[1]In future the task of gathering threatened or extinct animal DNA may be much easier. A number of "Frozen Zoos" have been set up to deliberately preserve cellular tissues so DNA is available for future research needs.

is already 99.96% wooly mammoth (Revive & Restore 2018). This considerable overlap, however, does not make things altogether easy. In mammoths, even this small difference amounts to approximately 1.4 million different nucleotides (Shapiro 2017). But when the main genetic differences between the living and the extinct species have been identified, gene synthesis and gene editing come into play.

Gene synthesis techniques allow genes and genome sequences to be engineered in a laboratory out of their constituent chemicals (a nucleobase, phosphate, and sugar). Although the history of gene synthesis dates back to 1955 and the work of Sir Alexander Todd, it is only recently that computerized sequencing tools and plunging costs have turned a technique that earned its creators Nobel Prizes into a something you can practice in your garage. Synthetically made nucleotides have already been stitched together to form a whole bacterial genome with more than a million base pairs (Gibson et al. 2010) and the construction of the much larger genome of the first eukaryote is in process (Richardson et al. 2017). With typical animal genes possessing anything from a few hundred to a couple of million bases, the individual genes responsible for the defining characteristics of an extinct animal are within the grasp of today's gene synthesis experts.

With the critical genes synthesized, the third technique required for de-extinction is a reliable gene editing mechanism that can be used to transform the genome of the living species into something that resembles the genome of the extinct species. Since 2012, the CRISPR Cas9 editing system has been available to cut living genomes at precisely targeted places. The editing system uses RNA guide molecules to identify the spot and a 'CRISPR package' can remove the targeted genes from the genome and insert new, extinct-animal-genes in their place. A Harvard lab run by George Church has already used these techniques to insert several of the genes that create characteristic wooly mammoth traits—hairiness, cold resistance, and copious quantities of subcutaneous fat—into an Asian elephant genome (Lewis 2015).[2]

At a future time, improved versions of these techniques might allow de-extinct animal embryos to be created by combining gene synthesis and gene editing with somatic cell nuclear transfer, stem cell embryogenesis, or germ line transmission. Resulting embryos could be implanted into surrogate mothers or placed into artificial wombs for the gestation period. The resulting organisms would be close approximations of extinct animals. These could be hybridized or further edited to become more accurate reproductions. Using Asian elephant eggs, some synthesized genetic material, an accurate gene editing system, and a surrogate mother, these techniques might soon allow a newborn wooly mammoth to roam the earth for the first time in four thousand years (Shapiro 2015).

---

[2] At this stage, none of the editing has resulted in living animals. Rather, the researchers are using Asian elephant tissue cultures sustained in a petri dish to experiment on. These modified cultures have been induced to grow into various tissue types allowing the study of the effects of the gene editing on cell characteristics.

27.3 Speculations About Gene Drives

Gene drives use the same sequencing, synthesis, and editing techniques to shape wild populations of fast-breeding sexually-reproducing animals. A significant practical barrier to altering wild populations has always been the inability of a scientist to ever reach all the members of the population she wanted to change. An evolutionary barrier has been the fact that traits chosen by scientists are unlikely to be naturally selected for in wild animals. Evolution, as co-discoverer of DNA Francis Crick pointed out, is much smarter than we are. Even if a deliberately selected trait did, against all odds, turn out to be neutral for an organism, the laws of Mendelian inheritance would ensure the trait is only passed on fifty percent of the time. With these odds it is almost impossible for the inserted gene to become prevalent in a population.

Gene drives work by imitating selfish mechanisms occasionally found in nature which provide genes a better than average chance of being inherited (Burt 2003). A CRISPR Cas9 package can be placed into an organism's germ cells so that a trait selected by the scientist is edited into both chromosomes. When the chromosomes split during meiosis, each chromosome will contain a copy of the selected trait. In this way, after fertilization with the germ cell of the other parent, the zygote is guaranteed to possess the gene in at least one of its chromosomes. Since the editing takes place in the germline the whole editing package is carried into the next generation to perform the same trick again (Esvelt et al. 2014). Recent tests in contained populations of mosquitoes have demonstrated the ability to spread a trait through a population in as few as seven to eleven generations (Kyrou et al. 2018). Such an effective upending of Mendelian rules could theoretically allow for the propagation of any chosen genetic construct through a wild population of organisms even though most of the population never set foot anywhere near the lab.

There are a number of reasons why it might be desirable to drive such a change. One of them is to knock down the numbers of a population of a species using what has been called a "suppression drive." A suppression drive could progressively alter the sex ratio of multiple generations of mosquito until the population crashed (Kyrou et al. 2018). Alternatively, a suppression drive could be used to make one sex of a troublesome population sterile (Zhuo et al. 2018). These technique have the potential to reduce—or even eliminate—particular species of mosquitoes that serve as disease vectors (e.g. *Aedes aegypti, Aedes gambiae*). There is also speculation that similar suppression drives will prove capable of extinguishing populations of invasive rodents from island ecosystems. An organization called Predator Free 2050 together with the New Zealand Department of Conservation are supporting the testing and field assessment of gene drives to eliminate a range of introduced predators from the country within three decades (New Zealand Department of Conservation 2019).

Different forms of gene drive could be inserted with the intention not of suppressing a population but of simply sending a particular non-lethal trait through it. These so-called 'replacement' drives would remake the existing genome sequence of the targeted organism into something that suits a particular human purpose (National Academies of Sciences 2016, 16). Replacement drives could potentially change the genotypes of wild animals to prevent them from being disease vectors. They could make an organism susceptible to a particular chemical. More speculatively, they could change an organism's taste for a particular prey or make it smarter at evading predators. Because both suppression and replacement drives would be self-propagating, they would allow the human reach to extend further out from the lab and into the wild than it has ever extended before.

27.4   The Problem with Promising Big

In recent years, numerous problems have emerged with the idea of using CRISPR Cas9 to edit genomes, problems that will plague gene drives, de-extinction, and numerous other gene editing ambitions for public health, conservation, and agriculture. The CRISPR Cas9 gene editing system has been found to sometimes create unpredicted changes both at the cut site and at other parts of the genome far away from where the editing is taking place. Studies have found CRISPR Cas9 causing "large deletions and more complex genomic rearrangements at the targeted sites" as well as completely unintended "lesions distal to the cut site" (Kosicki et al. 2018) (see also Shin et al. 2017; Mou et al. 2017). "We have been lulled into the view that editing is small and local and controllable," says a co-author of one of these studies, "but the reality of DNA repair in a cell is much more complex" (Rusk 2018, 569).[3]

Further problems with CRISPR have been appearing with alarming regularity. The *Wall Street Journal* reports that gene edits carried out in China to promote muscle growth in pigs resulted in the appearance of extra vertebrae, while similar edits on rabbits created disproportionately large tongues. Goats edited to produce longer wool grew too large during fetal development to undergo natural childbirth (WSJ 2018). When gene drives using the CRISPR Cas9 were inserted into mice by a different research group to see if they could spread a certain coat color, researchers found to their surprise that the drives only worked on female mice. They also found that the DNA break caused by CRISPR was not always repaired correctly, and it affected genomes at a much lower percentage than predicted (Grunwald et al. 2018). Overall, the predictability of the effects of using CRISPR Cas9 appears to be much lower today than researchers originally thought it would be. According to development geneticist Paul Thomas, enthusiasts for CRISPR Cas9 should find all of these results both "sobering" and "a reality check" (Callaway 2018).

To add to the list of complications, CRISPR has also been found to work best when a tumor suppressing protein named p53 is absent (Ihry et al. 2018; Haapaniemi et al. 2018). Having CRISPR work only when p53 is absent is highly inconvenient. The absence of this protein could lead to increased occurrence of cancers in organisms whose genomes are successfully edited, creating a potential problem for the goal of

---

[3] According to one researcher from Imperial College London, these problems are not as widespread in mosquitoes (personal communication, June 2019).

de-extinction (and for a gene drive if it is designed to replace but not eliminate a population).

Even if the technology worked as hoped, the idea that you can remove or replace one gene and then predictably change a phenotype in a replacement drive can also be overstated. Individual genes often work only against a background of other genetic elements, which may be comprised of hundreds of other genes. In many cases, the absence of this background will prevent the change from having the desired effect (Lehner 2013). Many traits are also 'polygenic,' requiring a number of genes to be actualized before the trait is displayed.

Other complications abound. 'Pleiotropic effects' occur when single gene creates more than one unrelated effect. Researchers using CRISPR Cas9 to make an edit to change the pigmentation of a butterfly's wing were surprised to find that the edit changed both the color *and* the structure of the wing's scales (Matsuoka and Monteiro 2018). Even if a researcher could reach a degree of confidence about the effect of an edit, uncertainty would still be present about the future stability of the edited genome (Shapiro 2017). While it may be possible to swap out a gene under laboratory conditions, it is not known what the consequences of this tinkering may be over the full lifetime of an organism, especially one living not in the controlled conditions of the lab but under the vicissitudes of a wild environment.

In the case of gene drives, the uncertainties are not only about the phenotype produced after using CRISPR Cas9. Above the genetic and organismic level, a great deal remains unclear about how effectively a gene drive would work in the wild. Organisms exist in complex ecological and species relationships. Whether the drive would spread as predicted, whether a natural resistance would evolve, whether hybridization between target and non-target populations might occur, whether horizontal transfer of altered genes to other species might ensue, and whether gene drives could be limited to the environment in which they were designed to operate are all unanswered questions (Biotecknologiradet 2017).

Population dynamics are almost certain to impede how a gene drive works. A study done by the US National Academies of Science expressed caution, concluding that there are "considerable gaps in knowledge regarding the implications of gene drives for an organism's fitness, gene flow in and among populations, and the dispersal of individuals, and how factors such as mating behavior, population sub-structure, and generation time might influence a gene drive's effectiveness" (National Academies of Sciences 2016, 42). Research about how to effect the required molecular changes is where most of the attention has been placed thus far. Far less research has been done on its operation at the population and ecosystem level.

Even before these surprises about CRISPR Cas9 had started to appear, those looking closely at de-extinction were already clued into some of the problems attending the whole idea of recreating lost species. From the very start, a certain amount of doubt had circulated about whether de-extinction was even the right label to use to motivate the research program. In April 2014, a de-extinction task force was created by the International Union for the Conservation of Nature (IUCN) under its Species Survival Commission. The task force examined both the promises and the risks created by thinking about de-extinction for conservation purposes.

As soon as the task force began its work, it recognized that any 'extinct' species brought back to life through biotechnology would be only an approximation of the species that humans had previously wiped out. The technical difficulties of recreating an exact replica of an extinct species' genome, the effects of missing mitochondrial DNA on fetal development, the lack of the original microbiome of the extinct species, the absence of appropriate ecosystemic conditions, and the lack of an extant animal culture from which a de-extincted individual could learn appropriate behavior meant the organisms recreated would at best be only approximations of the missing animal and not the missing animal itself. As a result, the report decided to characterize the whole endeavor not as an attempt to bring back lost species but as an attempt to create populations of organisms *resembling* extinct species for the purpose of serving important ecological functions. De-extinction then became "a conservation translocation issue that seeks to re-establish populations of *proxy species* in suitable areas of habitat to achieve *ecosystem conservation benefits*" (IUCN/SSC 2016, 5) (emphasis added). To reflect this retreat from the idea of de-extinction, the report was titled 'Guiding Principles on Creating Proxies of Extinct Species for Conservation Benefit.' The report urged readers to be clear on its starting assumption that "none of the current [technological] pathways will result in a faithful replica of any extinct species, due to genetic, epigenetic, behavioural, physiological, and other differences" (IUCN/SSC 2016, 5).

What this quick survey of the problems attending the use of CRISPR Cas9 and the difficulties of recreating extinct organisms reveal is that both gene drives and de-extinction remain highly speculative prospects. DNA appears to be neither manipulable enough nor determinative enough to be confident either de-extinction or gene drives can deliver on their promises. Genomes can certainly be edited using CRISPR Cas9 with remarkable accuracy. The resulting organism, however, together with its effect on the population in which it resides, will have a lot about it that we do not know. The lesson, perhaps, is that animals may not be reducible to chemistry of their genomes.

27.5   Reductionism and Thinking Relationally

One way to characterize doubts and uncertainties about de-extinction and gene drives revealed above would be to suggest that not enough research has been done and the genetic mechanisms at work are not yet fully understood. The genome is a complicated machine. Perhaps we simply don't yet fully understand it. More work must be needed to iron out the remaining uncertainties.

A slightly different way to characterize the doubts is to suggest the whole picture about the relationship between genes and their expression in an organism is faulty. As the discussion above illustrates, a key part of the problem appears to be the reductive approach to genomes that gene editing and gene synthesis assume. The properties of genetic parts cannot be so easily isolated from the larger contexts in which they operate. Context will determine the appropriateness of connecting a certain genomic rearrangement to a predicted behavior. Reducing a genome to its constituent parts may be a useful exercise when trying to draw correlations between individual genes and diseases. Reading a genome can indeed make it possible to offer statistical predictions about the likelihood of certain diseases appearing in individuals.[4] But it looks like reductionism and non-relational thinking about genomes can be seriously misleading when used as the grounds on which to try de-extincting whole organisms or introducing selected traits across wild populations.

The recent surprises that have emerged around CRISPR gene editing make it clear that the larger genetic, organismic, and ecosystemic contexts in which a single gene operates all affect the phenotype displayed. The broader genetic context can create unexpected deletions attending CRISPR gene edits, both at the targeted editing site and locations on the genome distal to it. The genetic context might mean tumor growth if p53 is absent. It might involve additional unknown effects on the phenotype if the gene is pleiotropic. It might also mean the impotence of the gene drive if resistance to the drive evolves within the genome.

The whole organism context is also relevant. This context might lead to the gene edit not having the desired effect if the typical microbiome, and other epigenetic factors, are not present (Morris 2012). The sex of the organism might impact the effectiveness of the gene drive edit. The stage of development of the organism might also change the likelihood of a gene drive working.

The ecosystemic context can also not be ignored. This context, external to the organism itself, might mean the gene drive not spreading as anticipated if the surrounding population dynamics change. The gene drive might be deployed in an ecosystem that allows it to move into non-target populations of the target species. It might also affect non-target sub-species if the organism with the gene drive hybridizes. Genomic, organismic, and ecosystemic contexts all influence the consequences of a CRISPR gene edit.

The many layers of context that influence how information encoded in a gene gets expressed in a fully-fledged organism make genome editing into a tricky proposition at the best of times. To make matters worse for gene drives and de-extinction, many of these essential contexts are fluid and dynamic. The minute alterations to the organism's genome constantly taking place through mutations as the generations proceed, the organism's constantly shifting microbiome, and the roving populations of surrounding symbionts and competitors are all dynamic factors influencing both genome expression and the genome's prospects in the future. In the case of gene drives, each of these shifting contexts will influence the way the drive spreads—or fails to spread—across the wild world. A non-relational, reductive approach that considers the different chemical structures contained within an organism's genome as interchangeable parts to be engineered at will seems ill-suited for thinking about whole organisms operating in larger ecosystemic contexts.

The multiple contexts influencing the organism point to the importance of thinking about the ontology of a gene in more relational terms. With a relational approach to

---

[4]Several thousand "single gene" diseases have been identified.

ontology, an entity becomes what it is through the relationships in which it partic-ipates. In relational worldviews, organisms, processes, or individual genes operate through a multitude of networks, interconnections, and feedback loops. Thinking about them in isolation from their relationships leads to a flawed understanding not only of what they are, but also of how they work.

In a relational approach to genomes, efforts to isolate genetic information and transpose it into completely different genomic, cellular, organismal, and ecolog-ical environments will create unpredictable effects. Genes express themselves only through highly specific interactions between various elements found within the many different layers of a given system. Altering the relational context creates different results. This is not simply because the gene now undergoes a different set of interac-tions. There is a deeper point to be made. When the relational network is changed, so is the gene itself. The gene is constituted by its relations as much as by its chemistry. A subscriber to a relational approach to genes would chuckle at the idea that the chemical components of a living organisms' genome can be edited and exchanged at will with predictable results.

27.6   Genomes and Non-human Agencies

If a reductive, non-relational approach to genes is one part of the problem, another part is the neglect of genomic, organismic, and ecosystemic agency. At a very basic level, gene drives and de-extinction are each an attempt to impose human agency onto the world through deliberate changes to an organism's genome. Since genomes are chemical structures made of up nitrogenous bases, phosphates, and sugars, it appears reasonable to suppose it might be possible to excise selected portions of their molecular structure, rebuild others, and swap out undesirable sequences for desirable ones.

The idea of isolating properties of the material world in order to understand how they function before imposing a change upon them is, of course, a central and important part of established scientific and technological practice. In numerous contexts, the approach works. However, in order to ensure that enough predictability and control follow from any imposition of human will, a key assumption must be made. While agency is ascribed to the scientist who is instigating the intervention, it must be assumed the material she is manipulating is in relevant respects passive (or at least passive enough not to thwart the engineer's plans). Once the intervention has occurred, the behavior of the manipulated material must follow linear and known laws.[5] If predictability is to be possible, any independent agency in the manipulated

---

[5]Engineering does not always require the produced object to exist in stasis. While engineers who create autonomous vehicles do not know everything about the future states and positions of the engineered object, they rely on the belief that the machine they have intentionally designed or modified will behave as they anticipated.

material must fall below a threshold that would allow a future to unfold contrary to the engineer's designs. If it didn't, it would not be engineering but a game of chance.

The idea of using CRISPR Cas9 to engineer biological forms appears to fall short in this regard. There are several different forms of agency that a genetic construct inserted into a genome for gene drives or de-extinction has to encounter. Bacterial agency in the microbiome, the agency of members of the same species in its population, and agency in any symbionts, competitors, or hybrids the organism encounters are all possible disruptors. These dynamic factors influence genome expression as well as the genome's future prospects. Each of these shifting contexts will influence how the gene drive spreads—or fails to spread—across the wild world. For potentially de-extincted animals such as the northern white rhino, the agency within the social and cultural structures leading to the organisms' extinction in the first place remains highly relevant.

In addition to the agency surrounding the altered genome, it is also possible to talk about agency within the genome itself. Although typically thought of as a property of individual organisms and larger collectives, genomes themselves are frequently thought of in agential terms. Dawkins calls them "selfish," Thaler calls them "intelligent," Wills calls them "wise," and Rolston calls them "smart" (Thaler 1994; Wills 1989; Rolston III 2012; Dawkins 2006). While some of these uses are deliberately metaphorical, the problems attending CRISPR Cas9 raise a new set of questions concerning how to think about genetic agency.

It is clear there is spontaneous behavior in the genome that is not the result of linear causes. Before, after, and during both meiosis and mitosis, a range of mutations can occur. The mutations can involve insertions or deletions of genetic material, inversion or translocation of a segment of DNA within a chromosome, duplications of a DNA sequence, changes in individual nucleobases, and other types of unscripted event. In the long run, of course, it is a good thing that these mutations occur. Without them, evolution would never take place. Random changes in the genome are what create the selective advantages (as well as disadvantages) which shape all ecologies.

Mutation can be viewed as a form of spontaneous agency in a genome. Some mutations are caused by external factors or mutagens such as chemicals or radiation. These ones are traceable to external causal factors and are non-agential as far as the genome itself is concerned. Others simply appear during cell division when the chromosomes do not replicate perfectly or when the inherently instability of the bases causes, for example, a cytosine base to become uracil. These latter sorts of changes are entirely spontaneous and unpredictable. They are not traceable to a proximate cause but emerge out of the genome itself. If agency is defined broadly in terms of the capacity for spontaneous and unpredictable action, then mutations will count as a form of agency and this agency clearly belongs to the genome itself.

Even though the notion of genomic agency may be counter-intuitive, the idea of agency existing below the level of the complete organism should not come as a complete surprise. Somewhere between the chemistry of the nucleobases and the complex whole that is the living organism, agency in the more familiar sense emerges. Properties emerging at higher levels can have precursors at lower ones. A variety of forms of agency may be present at levels beneath the whole organism. If random mutation is a minimal form of agency, one should perhaps regard the unpredictable insertions and deletions following from CRISPR Cas9 gene edits as forms of genomic agency. It is certainly clear the genome will not behave as the entirely passive recipient of an inserted gene. The genome responds spontaneously and, in some cases, unpredictably to the new material.

The neglect of agency thus appears alongside non-relational thinking as a second form of faulty philosophy often accompanying the discussion of gene editing for de-extinction and for gene drives. The kind of passivity required for precise engineering of organisms in order that they perform designated roles in an environment is simply not there in biology. To think this way incorrectly creates the impression that the research scientist is the only active agent in the gene editing process, while the chemical structures on which she operates are entirely passive. Yet it is clear that the scientist is not the only element of the system whose spontaneous agency influences the outcome. These two forms of faulty thinking are part of the reason why the discussion of gene drives and de-extinction lends itself to a misleading speculative ethics.

27.7   Speculative Biotechnology and Anthropocene Organisms

When Alfred Nordmann warned of the dangers of a speculative ethics with powerful emerging technologies, he was concerned about unrealistic projections of what a technology can accomplish having unfortunate consequences. One of the most obvious of these is that a speculative ethics can funnel both funding and enthusiasm towards a research direction that may not in the end deserve it. Over-eager researchers employ a speculative ethics "to invent a mandate for action" (Nordmann 2007, 33). So desirable (and inevitable) does a consequence of a particular technology appear that it becomes morally problematic, and perhaps even inhumane, to even consider not pursuing it. Arguably the current speculations about ending malaria with gene drives or eradicating invasive mammals from island ecosystems fall into this camp. The promise of moral restitution or of huge benefits for conservation by reintroducing de-extincted species (or their proxies) comes not far behind.

Nordmann adds to his argument that speculative ethics can also be damaging because of the opportunity costs it engenders. An over-eager pursuit of speculative silver bullets in the face of complicated environmental and public health challenges creates real dangers. These dangers arise when speculative ethics "squanders the scarce and valuable resource of ethical concern…and deflects consideration from the transformative technologies of the present" (Nordmann 2007, 31).

In cases of public health, there may be less risky interventions with considerable power to curtail the spread of diseases than a speculative gene drive. Without diminishing in any way the substantial challenges still posed by diseases spread through insect vectors, traditional and emerging approaches to control can have promising results.[6] On the island of Zanzibar, symptomatic malaria cases reported at health clinics decreased 94% from 2003–2015 (Björkman et al. 2019). India has reduced its malaria deaths by two thirds since 2000 with a 24% drop in cases between 2016 and 2017 (World Health Organization 2018). The Indian government maintains its goal of complete malaria eradication by 2030.

While these numbers in no way suggest that malaria is not still a significant and tragic public health issue, they might provide a reason to pause before the premature embrace of a highly speculative technology such as a gene drive. In situations where the traditional methods involve dramatic improvements in public health facilities, education, and funding, there might also be a broad range of co-benefits to pursuing the technologically less flashy solution over the more risky, speculative one.

Similarly in cases of conservation, devoting resources to the proximate causes of a species becoming endangered or extinct may be more beneficial in the end than devoting resources to a technologically speculative one. Even if there is no reason to believe it is a zero sum game, strategies such as the protection of good habitat, the provision of economic security for local populations, and improved attention to the international forces contributing to the original extinction threat may provide more lasting benefits than spending considerable amounts of money attempting to recreate an extinct animal through a speculative technology. Without the needed social, economic, and cultural changes, the de-extincted animal is unlikely to face any better a fate than its predecessors. Furthermore, implementing these traditional strategies will usually provide benefits for other organisms who share the habitat of the endangered or extinct one. In each of these cases, the speculative ethics may be creating the distraction Nordmann warns about.

* * *

Synthetic biology—the discipline within which gene reading, gene synthesis, and gene editing operate—is often characterized as "an engineer's approach to biology" (Breithaupt 2006). De-extinction and gene drives look very much like the transference of the techniques of mechanical engineering from physics over into to biology. If developed successfully, their proponents argue, they will not only solve some major public health and conservation problems, they might also remove some of the mystery that surrounds the concept of life by reducing it to its constituent parts. For some, these developments are both inevitable and desirable. Yuval Harari describes gene editing as playing a key role in the next revolution in human history, one in which "intelligent design becomes the basic principle of life" (Harari 2015, timeline).

The breakthroughs that have taken place in gene reading, synthesis, and editing in the last decade and a half certainly represent a remarkable scientific achievement. A great deal of good is likely to come from their careful application in appropriate domains. The moral benefit of solving some of the problems they seek to address would undoubtedly be high. What recent findings about gene editing reveal,

---

[6]Climate change may intensify these challenges.

however, is that approaching the genome with an engineer's mentality has a number of shortcomings.

When it comes to building or amending whole organisms, treating life as modular at the genetic level may in the end prove to be highly misguided. Organisms have always existed in highly complex genomic and ecological contexts that present significant barriers to the engineering approach. Plants and animals are bound up in multiple layers of constitutive relationships. Their lives are also entangled with multiple forms of agency that lend them their shape. At a time when new research paradigms such as multi-species studies are starting to highlight the significance of these relationships and agencies (Haraway 2016; Tsing 2015; Van Dooren 2016), attempts to engineer biology with gene synthesis and gene editing risk neglecting these provocative new insights. A speculative ethics of biotechnology consistently risks steering the conversation about de-extinction and gene drives away from contemporary understandings about the embeddedness of all animals in multiple human and non-human contexts. This non-relational thinking has certainly pushed the conversation about these new techniques far ahead of where it currently deserves to be.

The presence of lively, spontaneous animals with whom our species shares this planet has provided mystery, challenge, and inspiration throughout the two and a half million year history of our kind. In an increasingly crowded Anthropocene world, the presence of independent animals will provide a vital counterpoint to a harmful 'species narcissism' as ever more of the earth's surface becomes stamped by human designs (Mark 2015). Some forms of biotechnology appear to express this narcissism by being blind to the relationships and agencies of the animal others with whom all of our lives are entwined.

The linking of biology with engineering is symptomatic of this attitude. This linking may not simply be a rhetorical mistake. As National Medal of Science and Crafoord Prize winner Carl Woese suggested, "a society that permits biology to become an engineering discipline, that allows science to slip into the role of changing the living world without trying to understand it, is a danger to itself" (Woese 2004, 173).

References

Biotecknologiradet. 2017. *Statement on gene drives*. The Norwegian Biotechnology Advisory Board. Oslo. http://www.bioteknologiradet.no/filarkiv/2017/02/Statement-on-gene-drives.pdf.
Björkman, A., D. Shakely, A.S. Ali, U. Morris, H. Mkali, A.K. Abbas, et al. 2019. From high to low malaria transmission in Zanzibar—Challenges and opportunities to achieve elimination. *BMC Medicine* 17 (1): 14. https://doi.org/10.1186/s12916-018-1243-z.
Breithaupt, H. 2006. The engineer's approach to biology. *EMBO Reports* 7 (1): 21–23.
Burt, A. 2003. Site-specific selfish genes as tools for the control and genetic engineering of natural populations. *Proceedings of the Royal Society of London. Series B: Biological Sciences* 270 (1518): 921–928. https://doi.org/10.1098/rspb.2002.2319.
Callaway, E. 2018. Controversial CRISPR "gene drives" tested in mammals for the first time. *Nature*, July 10. https://www.scientificamerican.com/article/controversial-crispr-gene-drives-tested-inm ammals-for-the-first-time/.

Dawkins, R. 2006. *The selfish gene*, 30th Anniversary Edition. New York: Oxford University Press.

Esvelt, K.M., A.L. Smidler, F. Catteruccia, and G.M. Church. 2014. Concerning RNA-guided gene drives for the alteration of wild populations. *eLife* 3: e03401. https://doi.org/10.7554/eLife.03401.

Gibson, D.G., J.I. Glass, C. Lartigue, V.N. Noskov, R.-Y. Chuang, M.A. Algire, et al. 2010. Creation of a bacterial cell controlled by a chemically synthesized genome. *Science* 329 (5987): 52–56. https://doi.org/10.1126/science.1190719.

Grunwald, H.A., V.M. Gantz, G. Poplawski, X.S. Xu, E. Bier, and K.L. Cooper. 2018. Super-Mendelian inheritance mediated by CRISPR/Cas9 in the female mouse germline. *Nature* 566 (7742): 105–109. https://doi.org/10.1101/362558.

Haapaniemi, E., S. Botla, J. Persson, B. Schmierer, and J. Taipale. 2018. CRISPR-Cas9 genome editing induces a P53-mediated DNA damage response. *Nature Medicine* 24 (7): 927–930. https://doi.org/10.1038/s41591-018-0049-z.

Harari, Y.N. 2015. *Sapiens: A brief history of humankind*, 1st ed. New York: HarperCollins.

Haraway, D.J. 2016. *Staying with the trouble: Making kin in the Chthulucene*. Durham, NC: Duke University Press.

Ihry, R.J., K.A. Worringer, M.R. Salick, E. Frias, D. Ho, K. Theriault, et al. 2018. P53 inhibits CRISPR-Cas9 engineering in human pluripotent stem cells. *Nature Medicine* 24 (7): 939–946. https://doi.org/10.1038/s41591-018-0050-6.

IUCN/SSC. 2016. Guiding principles on de-Extinction for conservation benefit (Version 1.0). Gland, Switzerland.

Kaebnick, G.E. 2009. Should moral objections to synthetic biology affect public policy? *Nature Biotechnology* 27 (12): 1106–1108. https://doi.org/10.1038/nbt1209-1106.

Kosicki, M., K. Tomberg, and A. Bradley. 2018. Repair of double-strand breaks induced by CRISPR-Cas9 leads to large deletions and complex rearrangements. *Nature Biotechnology* 36: 765–771. https://doi.org/10.1038/nbt.4192.

Kyrou, K., A.M. Hammond, R. Galizi, N. Kranjc, A. Burt, A.K. Beaghton, et al. 2018. A CRISPR-Cas9 gene drive targeting doublesex causes complete population suppression in caged *Anopheles gambiae* Mosquitoes. *Nature Biotechnology* 36 (11): 1062–1066. https://doi.org/10.1038/nbt.4245.

Lee, K. 1999. *The natural and the artefactual: The implications of deep science and deep technology for environmental philosophy*. Lanham, MD: Lexington Books.

Lehner, B. 2013. Genotype to phenotype: Lessons from model organisms for human genetics. *Nature Reviews Genetics* 14 (3): 168–178. https://doi.org/10.1038/nrg3404.

Leopold, A. 1949. *A sand county almanac*. New York: Oxford University Press.

Lewis, T. 2015. Wooly Mammoth genes inserted into Elephant cells. *LiveScience*, March 26. https://www.livescience.com/50275-bringing-back-woolly-mammoth-dna.html.

Lie, S.A.N. 2016. *Philosophy of nature: Rethinking naturalness*. Abingdon, UK: Taylor and Francis. https://www.routledge.com/Philosophy-of-Nature-Rethinking-naturalness-1st-Edition/Lie/p/book/9781138792883.

Mark, J. 2015. *Satellites in the high country: Searching for the wild in the age of man*. Washington, DC: Island Press.

Matsuoka, Y., and A. Monteiro. 2018. Melanin pathway genes regulate color and morphology of butterfly wing scales. *Cell Reports* 24 (1): 56–65. https://doi.org/10.1016/j.celrep.2018.05.092.

Morris, K.V. (ed.). 2012. *Non-coding RNAs and epigenetic regulation of gene expression: Drivers of natural selection*. Poole, UK: Caister Academic Press.

Mou, H., J.L. Smith, L. Peng, H. Yin, J. Moore, X. Zhang, et al. 2017. CRISPR/Cas9-mediated genome editing induces exon skipping by alternative splicing or exon deletion. *Genome Biology* 18 (1): 108. https://doi.org/10.1186/s13059-017-1237-8.

National Academies of Sciences, Engineering, and Medicine. 2016. *Gene drives on the horizon*. Washington, DC: National Academies Press. https://doi.org/10.17226/23405.

New Zealand Department of Conservation. 2019. Strategic Priorities. https://www.doc.govt.nz/about-us/statutory-and-advisory-bodies/nz-conservation-authority/strategic-priorities/.

Nordmann, A. 2007. If and then: A critique of speculative nanoethics. *NanoEthics* 1 (1): 31–46. https://doi.org/10.1007/s11569-007-0007-6.

Preston, C.J. 2018. *The synthetic age: Outdesigning evolution, resurrecting species, and reengineering our world*. https://mitpress.mit.edu/books/synthetic-age.

Revive & Restore. 2018. Wooly Mammoth project: Progress to date. https://reviverestore.org/projects/woolly-mammoth/progress-to-date/.

Richardson, S.M., L.A. Mitchell, G. Stracquadanio, K. Yang, J.S. Dymond, J.E. DiCarlo, et al. 2017. Design of a synthetic yeast genome. *Science* 355 (6329): 1040–1044. https://doi.org/10.1126/science.aaf4557.

Rolston III, H. 1988. *Environmentale ethics: Duties to and values in the natural world*. Philadelphia, MA: Temple University Press.

Rolston III, H. 2012. *A new environmental ethics: The next millenium for life on earth*, 1st ed. New York: Routledge.

Rusk, N. 2018. Surprising CRISPR roadblocks. *Nature Methods* 15 (8): 569. https://doi.org/10.1038/s41592-018-0097-9.

Shapiro, B. 2017. Pathways to de-extinction: How close can we get to resurrection of an extinct species? *Functional Ecology* 31 (5): 996–1002. https://doi.org/10.1111/1365-2435.12705.

Shapiro, B. 2015. *How to clone a Mammoth: The science of de-extinction*, 1st ed. Princeton, NJ: Princeton University Press.

Shin, H.Y., C. Wang, H.K. Lee, K.H. Yoo, X. Zeng, T. Kuhns, et al. 2017. CRISPR/Cas9 targeting events cause complex deletions and insertions at 17 sites in the mouse genome. *Nature Communications* 8: 15464. https://doi.org/10.1038/ncomms15464.

Siipi, H. 2008. Dimensions of naturalness. *Ethics and the Environment* 13 (1): 71–103.

Thaler, D.S. 1994. The evolution of genetic intelligence. *Science* 264 (5156): 224–225. https://doi.org/10.1126/science.8146652.

Tsing, A.L. 2015. *The mushroom at the end of the world: On the possibility of life in capitalist ruins*. Princeton, NJ: Princeton University Press.

Van Dooren, T. 2016. *Flight ways: Life and loss at the edge of extinction*. New York: Columbia University Press.

Venter, J.C., M. Adams, G.G. Sutton, A.R. Kerlavage, H.O. Smith, M. Hunkapiller, et al. 1998. Shotgun sequencing of the human genome. *Science* 280 (5369): 1540–1542. https://doi.org/10.1126/science.280.5369.1540.

Wills, C. 1989. *The wisdom of the genes: New pathways in evolution*. New York: Basic Books.

Woese, C.R. 2004. A new biology for a new century. *Microbiology and Molecular Biology Reviews* 68 (2): 173–186. https://doi.org/10.1128/MMBR.68.2.173-186.2004.

World Health Organization. n.d. World malaria report 2018. Accessed March 7, 2019. www.who.int/malaria.

Zhuo, J.-C., Q.-L. Hu, H.-H. Zhang, M.-Q. Zhang, S.B. Jo, and C.-X. Zhang. 2018. Identification and functional analysis of the doublesex gene in the sexual development of a hemimetabolous insect, the Brown Planthopper. *Insect Biochemistry and Molecular Biology* 102: 31–42. https://doi.org/10.1016/j.ibmb.2018.09.007.

Chapter 28
Does Justice Require De-extinction of the Heath Hen?

Jennifer Welchman

Abstract It is often argued that we "owe it" to species driven to extinction "to bring them back." Can justice really require us to make restitution for anthropogenic extinctions? Can it require de-extinction? And if so, can justice require us to attempt the North American Heath Hen's de-extinction? I will first review the types of de-extinction technologies currently available. I will then discuss the criteria used to determine when restitution is owed for injuries as well as the special challenges arising when (i) victims are wild animals and (ii) are extinct. After arguing that restitution may be due for some extinctions and that de-extinction would sometimes be an appropriate means, I apply these arguments to the case of the Heath Hen.

28.1 Introduction

> By all means, bring the Heath Hen back and undo the horrible mistake of letting this animal go extinct in the first place. We caused its extinction; we are responsible for bringing it back. (Dan 2014)

Human beings are responsible for many species' extinctions since our own first appeared on the planet. However, it was not until the late eighteenth century that scientists realized that species could become globally extinct (Barrow 2009). It was another century before the possibility of anthropogenic extinctions was recognized (Cowles 2012). After the disappearance of the Dodo and the Great Auk, events in which humans were clearly implicated, government regulations to control over-hunting and fishing of game birds and animals were expanded in many European countries and extended to their colonies abroad but did little to diminish the impact of habitat disruption and over-hunting on many species. By the end of the nineteenth century, societies for the protection of wildlife, especially birds, began to

J. Welchman (✉)
University of Alberta, Edmonton, AB, Canada
e-mail: welchman@ualberta.ca

form in Europe and North America. Great Britain's Society for the Protection of Birds (later the Royal Society for Protection of Birds) began in 1892, followed by the Vogelbescherming Nederland (1899), the Audubon Society (1905), and La Ligue pour la protection des oiseaux (1912) (Boardman 2006).

In 1908, the United States began one of its first serious attempts to prevent an extinction with the creation of a sanctuary for the Heath Hen on the island of Martha's Vineyard (Massachusetts). The Heath Hen, a ground-dwelling bird, related to Prairie Chickens, had been extirpated from the mainland in the 1870s. Predictions for the remnant population on Martha's Vineyard were so dire that a sanctuary was created to reduce the pressure of hunting and habitat disturbance. The sanctuary reduced these pressures but only within its 612 acres. In the 1920s, the Heath Hen population went into a decline from which it never recovered. The last Heath Hen, Booming Ben, disappeared in 1932. Henry Hough, editor of a local newspaper, blamed human chauvinism for the bird's demise.

> The gospel of conservation, it is said, has won the day. We know this is not true…Is nothing to follow the extinction of this bird except one more lesson in conservation for school books and a sentimental mourning? (Hough 1933; see Barrow 2009)

The Revive & Restore Organization has been investigating an alternative: de-extinction, the genetic engineering of an approximation of the Heath Hen for eventual release onto islands around Martha's Vineyard (Revive and Restore).

Revive and Restore's Heath Hen Project is one of several "de-extinction" ventures whose goal is the replication of extinct species. Supporters argue that they offer many potential benefits. First, conservationists could have new tools for slowing or reversing biodiversity losses. Second, de-extinct Ivory-Billed Woodpeckers, Moa, and Tasmanian Tigers would stimulate eco-tourism whose profits could bankroll protections for endangered species. Third, improvements in gene-editing techniques could have spillover benefits for human health, boosting the development of gene therapies. Supporters have also argued that de-extinction is worth pursuing for another kind of reason, as a means to right historical wrongs. As Dr. Jeffrey Johnson, a Heath Hen Project advisor put it, "These species have either gone extinct, locally extinct, or on the verge of extinction primarily due to the actions of man, so I feel obligated that we should try to do whatever we can to prevent further extinction, and now with this technology maybe even bring back a species" (Brown 2015b).

This last argument is different in kind from the other three. The first three appeal to our desires for the future outcomes de-extinction is said to offer. Their persuasiveness is dependent upon our desiring those outcomes (and believing that de-extinction can achieve them). By contrast, the last argument is "backward-looking". It invites conscientious reflection on our past actions from a moral point of view. If some anthropogenic extinctions morally wronged their victims or third parties, justice might demand restitution for them. And if de-extinction was an effective means of doing so in some cases, justice might require us to embrace de-extinction, independent of any desires we might have for the future.

Backward-looking appeals to restitutive justice are commonly made by champions of de-extinction projects. "If we can retrieve the animals or retrieve at least the appearance of the quagga," says a director of a South African effort to replicate this extinct zebra, "then we can say we've righted a wrong" (Page and Hancock 2016).[1] The Australian Lazarus Project, whose founders hope to clone the extinct Gastric Brooding Frog, has been characterized as fulfilling "an ethical responsibility...to undo the harm that we have done in contributing to its extinction" (Smith 2017). Reversal of the anthropogenic extinctions of Tasmanian Tiger (or thylacine) and Passenger Pigeon have likewise been justified in moral terms: "Having hunted the thylacine to extinction, we owe it to the species to bring it back" and "we caused the extinction of the species...Now we have a moral obligation to bring [the Passenger Pigeon] back" (Pickrell 2018; Bethge 2013).[2] Increasingly, these appeals are drawing attention from environmental philosophers.[3] One of de-extinction's most persistent philosophical critics, Ben Minteer, has described it as "the most powerful" argument in favor of de-extinction offered by its proponents (Minteer 2014).

Minteer's claim is surprising when considered in light of recent debates in contemporary intergenerational ethics. Arguments for restitution to victims of historical injustices are always highly contentious when all the victims are dead (even when the victims are human beings). It is not clear how we can make restitution to the dead. Sometimes the victims of historical injustices have descendants. In these cases, claims for restitution, though politically controversial, are at least taken seriously. But while some extinct species have living relatives, few have any direct descendants. Is it plausible that we could have duties of justice to make restitution for anthropogenic extinctions? Would they require de-extinction? More specifically, might we have such a duty to the Heath Hen and, if so, would it require us to attempt the Heath Hen's de-extinction? These are the questions I will try to answer in this chapter. First, I will briefly review the three families of de-extinction techniques currently available. Next I will discuss the criteria used to determine when and to whom restitution is owed for injuries done to others, and the special challenges that arise when (i) victims are members of wild species and (ii) there are no living survivors. After arguing that restitution may be due for some kinds of anthropogenic extinctions and when de-extinction technologies might be the best means to employ, I will consider whether the Heath Hen's extinction is such a case.

The focus of this chapter throughout will be on the backward looking reparations argument for de-extinction as a duty of justice. So I shall not discuss forward-looking rationales, such as the potential to preserve biodiversity, stimulate eco-tourism, or improve human health. Nor shall I discuss the usual objections to them; that a de-extinct species might threaten those which have inherited its former territory or that de-extinction projects may siphon public funding away from traditional conservation

---

[1] For information about the Quagga Project, visit: https://quaggaproject.org/.

[2] Interestingly, Ben Novak, the Passenger Pigeon Project scientist quoted in Bethge (2013), has been more hesitant about the application of this reasoning to other species. See Mitchell (2018).

[3] Most discussions ultimately conclude reparations arguments for de-extinction face insurmountable challenges of the sort to be discussed below, but not all. Pessimistic assessments include Rohwer and Marris (2018), Sandler (2014), Palmer (2012), Cohen (2014), Diehm (2015), Campbell and Whittle (2017), and Minteer (2014, 2015, 2019). A more optimistic assessment is offered in Jebari (2016).

efforts. Focusing on the Heath Hen project will help avoid distraction by these kinds of concerns. No candidate for de-extinction has less potential to become an ecological pest. The birds produced by the Heath Hen project would be non-migratory, ground-dwelling birds initially restricted to barrier islands off the coast of Martha's Vineyard, islands free of any related species with which they might compete. Reintroduction in the species' historical range on the island of Martha's Vineyard would follow only after sufficient habitat was prepared. Currently, the project is privately funded. Were any public funds spent later to restore habitat for their reintroduction on Martha's Vineyard, traditional conservation goals would not suffer.[4]

28.2   De-Extinction Techniques

Modern medical technology is sometimes credited with restoring the 'clinically dead' back to life. What is actually reversed is a dying process, not death. Modern medicine can sometimes bring the dying back from death's door but cannot restore those who have already passed through it. Press reports sometimes describe de-extinction techniques as if they literally restored extinct animal species to life. As we will see, they too may sometimes help bring dying species back from extinction's door (no mean feat in itself) but cannot restore those which have already passed through it.[5] "De-extinction" is an umbrella term which covers three families of breeding techniques developed to replicate the phenotype (form and appearance) and/or genotype of an extinct animal population. These include (i) back-breeding, (ii) cloning, and (iii) genome editing (Shapiro 2016).

Back-breeding is possible when members of a living species can trace their ancestry in part to an extinct species. Back-breeding projects try to replicate the phenotype of lost species by cross-breeding their hybrid descendants. The South African Quagga Project is one example. Its goal is to replicate the Quagga, an extinct subspecies of Plains Zebra, by cross-breeding hybrid Plains Zebra who retain traits from their distant Quagga ancestors. Back breeding has also been proposed to replicate the Pinta Island Tortoise, known internationally for its last survivor, "Lonesome George". Hybrid tortoises have recently been discovered on Floreana Island with Pinta Island Tortoise genes. Cross-breeding over several generations could produce offspring that closely replicate their Pinta Island ancestors (Beeler 2015).

Cloning projects try to replicate the genotype of a lost species by implanting the nucleus of a preserved cell from an extinct species into the egg cell of a related species (a process known as somatic cell nuclear transfer or SCNT). If the reconfigured egg cell begins dividing and develops into an embryo, the embryo can then implanted

---

[4]To prevent further distraction, I will also assume that the welfare interests of all the birds involved in the research are adequately met.

[5]I will use the term 'species' interchangeably with 'population' through out this paper as collective nouns for a group of closely genetically related animals that can interbreed. I shall not assume 'species' as wholes have interests or welfare distinct from that of their members.

in the uterus of a related species for gestation. Cloning has been tried on a few occasions, with only partial success. An embryo was developed from preserved cells of the extinct Pyrenean Ibex but the clone died from congenital abnormalities shortly after birth (Kupferschmidt 2014). The Lazarus Project created living embryos from preserved tissue of extinct Gastric-Brooding Frogs, but none developed into adult frogs (Smith 2017).

Genome editing is a third technique that might be used in cases where there are no living hybrid descendants to be back-bred, no specimens sufficiently well preserved to make cloning possible, and/or, as in birds, where the target species' eggs are not amenable to SCNT. Scientists would edit the genome of a close living relative in order to replicate sequences distinctive of the extinct species they hoped to replicate. The edited cells would later be used to produce viable embryos. This is the method being researched by Revive & Restore for the de-extinction of the Heath Hen.

Each method would replicate the extinct species' phenotype and/or genotype to some extent, but in no case would the offspring possess all and only the traits of the extinct species. Back-breeding hybrid Pinta Island tortoises cannot fully erase the genetic signature of their Floreana ancestors.[6] The complexity of gene editing suggests that offspring would at best approximate the genomes of the target species. Heath Hens produced by editing the genome of Prairie Chickens would possess the traits of a Heath Hen–Prairie Chicken hybrid rather than all and only those of historical Heath Hens. Even cloned animals will have hybrid traits, as they would inherit maternal mitochondria from the eggs used in the SCNT process.

Critics of de-extinction projects have argued that for this reason, de-extinct animals are unworthy for release into natural environments. De-extinct animals are variously described as hybrids, inauthentic, and worse: "engineered dopplegangers", "franken species and eco-zombies" and "technological artefacts, not members of any natural species" (Minteer 2015; Shultz 2016; Campbell 2017). We should not allow ourselves to be influenced by this sort of language. First, it is rooted in outdated ideas about species that have underwritten racist ideologies for centuries. The assumption that "miscegenation", the mixing of different species or races of beings, including human beings, results in unnatural or inferior "mongrel" offspring is a survival of pre-Darwinian essentialist ideas about the nature of species—which sadly persisted in Darwin's own work, where hybrids are described as outcomes of "illegitimate unions" and "unnatural crossing" (Darwin 1896). On this view, any so-called "human being" who has inherited Neanderthal genes is actually an impure hybrid who ought not be allow to procreate with purer, more authentic, human beings.

Second, it relies on equally questionable assumptions about the value of the "natural" versus the "artificial". If de-extinct Heath Hens are technological artefacts, what then are the thousands of children born every year from in vitro fertilized donor eggs? Pseudo-human franken-children who should be kept apart from their more authentic, naturally-conceived counterparts? Of course we should be concerned about

---

[6]Humans have not had Neanderthal reproductive partners in over 30,000 years, yet from 1 to 4% of the genetic material of their hybrid human descendants remains Neanderthal (Sánchez-Quinto and Lalueza-Fox 2015).

the potential consequences of releasing replicated Heath Hens into natural areas. From an ecological perspective, it matters enormously whether they can behave as good ecological citizens. However, the purity of their lineage and the manner of their conception have no comparable significance. Thus for current purposes, I will assume that at least some replicas of extinct species would be good environmental citizens and that a de-extinct Heath Hen could be one of them.

28.3   Can Restitutive Justice Be Extended to Wild Animals?

Common sense conceptions of justice require that those responsible for causing avoidable and unjustifiable harm to others' welfare offer restitution. If the injury took the form of misappropriation of material goods, the person responsible should *restore* those goods to the individuals entitled to them. If, on the other hand, direct restoration is impossible, then the person responsible should offer *reparations* to compensate the victim by other means. Material reparations are not due unless four criteria are met: the perpetrator was a competent moral agent, the injury could have been avoided, the victim was an individual capable of suffering, and making material reparations is possible. Wild deer do not owe farmers reparations for the field crops they damage because wild deer are not moral agents. If the farmers' employees could not have prevented the deer from entering those fields, they do not owe their employers reparations. The damaged crops are owed nothing because they are incapable of suffering. And finally, because 'ought' implies 'can', material reparations are not owed if they cannot be made.

Most now agree that if an animal is capable of suffering, we have a moral duty to avoid causing it to suffer. Many go further, arguing that animals have rights not to be made to suffer or be confined by human beings. However, most of these discussions focus on domesticated animals. Sentient wild animals possess the same claims to moral standing as their domestic counterparts. But some doubt that we can extend these duties to wild animals without contradiction. Many wild animals suffer from illness and predation, conditions that could be mitigated if we forced sick animals to accept medical treatment or confined and/or exterminated predators. In other words, as Martha Nussbaum points out, when it comes to wild animals, acting for the good of one species often entails acting against the good of another. She asks:

> Should humans police the animal world, protecting vulnerable animals from predators? … The death of a gazelle after painful torture is just as bad for the gazelle when torture is inflicted by a tiger as when it is done by a human being. That does not mean that death by tiger is as blameworthy; obviously it is not. But it does suggest that we have similar reasons to prevent it, if we can do so without doing greater harms…The problem is that the needs of the predatory animal must also be considered, and we do not have the option of giving the tiger in the wild a nice ball on a string to play with. (Nussbaum 2006, 379)

Clare Palmer argues this is just one of "a number of difficulties raised by the importing of claims about justice from human/human to human/animal relations".[7] Our concepts of justice arose in response to our need to manage human social relationships. Many domestic animals have roles in those relationships, and to the extent they do, she argues, their roles entitle them to consideration. This is not the case with wild animals. Palmer (2010, 89) writes:

> Fully wild animals…do not have such relations with humans; so duties to assist them are not generated on these grounds. There is no analogy to current or historical unfairness or injustice (or, indeed, fairness or justice) about the states in which wild animals find themselves. Inasmuch as they live without human contact, they are outside the realm of justice altogether. (Palmer 2010, 87)

Since wild animals fall outside the scope of our concepts of justice, Palmer suggests that human beings do not owe wild animals reparations either for harms inflicted by other wildlife or by human interventions in their environments. In her view, talk of human beings owing reparations to wild animals is confused or metaphoric. She argues that at most we can have "reparation-like *special* obligations" to act benevolently towards those animals, chiefly domesticated, whom we have made vulnerable through our interventions in their lives (Palmer 2010, 96).

However, as Palmer notes, "justice" is used to refer to a host of related principles: economic, distributive, retributive, restorative, social and political. There would be enormous difficulties in trying to extend all of these many principles of justice to animals generally, let alone wild animals specifically. The basic common-sense principle of restitutive justice seems another matter. Meeting the criteria for claims of restitutive justice on sentient animals' behalf does not seem inherently problematic – at least not in cases where their injuries were avoidable, no overriding moral reasons compelled the acts responsible, and there was some practical means of making restitution to the animals concerned.

Palmer might reply that this is an exaggeration. Meeting all four criteria when the "victim" is a wild animal is not as simple as I have made it appear. In most cases, we face huge epistemic challenges in determining whether and to what extent our avoidable interactions with wild animals harm them. Most wild creatures' vulnerability to suffering is consequence of life in the wild, not their interactions with human beings. Premature death from illness, injuries, and predation is a norm in the wild. As Mark Sagoff famously remarked:

> The principle of natural selection is not obviously a humanitarian principle; the predator-prey relation does not depend on moral empathy. Nature ruthlessly limits animal populations by doing violence to virtually every individual before it reaches maturity. (Sagoff 1984, 299)

Some of the animals pursued by hunters and anglers may go on to lead long, reasonably healthy, and enjoyable lives if not captured and killed. Others would not. They would die soon anyway from any of a myriad of causes, many of them painful. In shooting a deer, a hunter may regrettably shorten a life worth continuing or,

---

[7]See also a related argument in Anderson (2004).

alternately, mercifully end its suffering from disease, injuries, or parasites. The well-intentioned angler who throws a trout back into a stream has no real way of knowing if she has spared its life or thrown it into the mouth of a predator.

There are two ways one might respond to the issues Nussbaum and Palmer raise. One would be to argue that we cannot have the same sorts of moral duties towards wild animals that we have towards domesticated animals, because we know too little of the circumstances of any particular wild creature to determine whether our interventions in its affairs would be helpful or harmful overall. 'Ought' implies 'can'. If we can never be sure that any given act will actually injure a wild animal's life prospects overall, we cannot be obliged to avoid harming them nor owe them restitution if we do. The other would be to argue that the epistemic challenges wild animals present do not change the kinds of duties we have to them, *vis a vis* domesticated animals. It simply changes the methods we are warranted in using to evaluate harms we cause. Our familiarity with and control over the lives of domesticated and captive animals usually ensures we can estimate how any particular action is likely to affect them. Our lack of knowledge about or control over fully wild animals routinely ensures we cannot. But this do not mean there are no measures available to us to evaluate our conduct towards wild animals. Population measures can provide a proxy by which to make rough estimations of whether particular practices are harmful to wildlife.

Different measures of harm can be used to determine what justice requires when we injure others. These measures fall into two broad categories: comparative and non-comparative. Comparative measures determine how much worse off an individual is after an injurious event, either by comparing that individual's condition prior to and following the event (a historical comparison) or by comparing her condition after the event with how it would have been had the event not occurred (a counter-factual comparison).[8] Provided one has sufficient information about an individual's circumstances to make comparisons of either type, one can arrive at fine-grained analyses of the degree of harm a party has suffered from another's intentional or negligent actions. When the victims are individuals or members of groups about whose prior histories and particular circumstances we know little, historical and counter-factual comparisons become impractical. In these cases, we turn to non-comparative measures. We assume that certain states are intrinsically bad states for individuals (e.g., physical debilitation, mental suffering, loss of autonomy) and if the intentional or negligent act has imposed such states on those affected, they are said to be harmed. (Shiffrin 2012; Harman 2009; Hanser 2008) Because non-comparative measures rely on pre-determined lists of intrinsically bad states and do not measure ancillary harms victims may undergo, their results are less fine-grained than comparative measures. To do victims full justice in assessing claims for restitution, comparative measures are preferable. But when we lack the information required to employ them, non-comparative measures become an alternate if cruder means of proceeding.

We rarely know enough about the lives of wild species to measure the effects of our acts upon individual animals comparatively, even those which have received

---

[8]For background, see Perry (2003). For criticism and defense of comparative accounts of harm, see, respectively, Norcross (2005) and Klocksiem (2012).

scientific study. Often our only basis for determining the effects of human activity upon them are rough estimates of population trends.[9] But this kind of information does provide a non-comparative means by which to gage statistically whether our interventions are imposing intrinsic evils on a species' members. Inability to obtain sufficient food, water, and shelter to grow to a reproductively viable age is surely an intrinsically bad state for any sentient being, human or nonhuman. Whenever it is evident that human practices are preventing a population from sustaining its numbers, we can be sure we are harming them in a non-comparative sense by denying its members, on average, access to the basic goods necessary to sustain their numbers. If those practices are avoidable, not required by overriding moral considerations, and mitigatable, I would argue all four criteria for a duty of restitutive justice to sentient species of wild animals can be met.

Using the effects of human activity on population numbers means that we will not owe reparations for every act that kills or displaces an individual animal, provided it is a member of a species whose individual welfare we cannot accurately assess. It will be the effects of our practices on that species' population levels that determine whether reparations can reasonably be claimed on that species' behalf. So it would seem that if a restricted season for hunting or angling did not diminish populations of the target species, restricted hunting and angling would not warrant reparations. Similarly, if human development of a natural area does not diminish populations of any wildlife displaced, then the displacement would not warrant restitution or reparations. However, if hunting, fishing, and development practices evidently do deny a population the means to sustain itself, those practices are harmful and may warrant reparations.

So we need not wait until a species becomes extinct to demand restitutive justice on its behalf. Extinction is not itself a kind of harm. Extinction is an indicator that a wild population has been harmed; i.e., suffered intrinsically bad states through denial of the basic natural goods by which it sustained itself. Conservation advocates have prima facie grounds for demanding restorative justice on a species' behalf as soon as (i) the population's numbers are shrinking and (ii) its losses are traceable to culpable human activity.

Restitutive justice claims become more urgent as a species' numbers shrink to levels below statistical norms for survival, especially when its genetic diversity has shrunk along with its numbers.[10] Such a species is at risk of entering an extinction vortex; a state in which "an insidious mutual reinforcement can occur among biotic and abiotic processes such as environmental stochasticity, demographic stochasticity, inbreeding, and behavioural failures, driving population size downward to extinction" (Fagan and Holmes, 2006). If this harm is traceable to avoidable human behavior, to do justice to the victims, action should be taken to restore the basic natural goods to

---

[9]For an extended argument to this effect, see Delon and Purves (2018). See also Johanssen's 2019 discussion. I am prepared to grant that there are *some* species about which we have sufficient information to be sure that particular kinds of human interventions are directly harmful, but these are exceptions not the rule when it comes to the totality of sentient wild species on this planet.

[10]For an overview of the history and current methods of developing minimum viability population measures, including the "50/500" rule, see Stephens (2016).

levels sufficient to allow the survivors to recover and sustain their typical numbers. In rare cases, justice might even require us to use de-extinction technologies to rescue a species already entering an extinction vortex.

One basic good essential for sexually reproducing species to sustain their numbers are viable reproductive partners. Currently neither of the two surviving female Northern White Rhinos has access to a viable breeding partner because no males exist. As this state is a direct result of culpable human action, some believe that frozen tissue from Northern White Rhinos should be used to clone replacement breeding stock sufficient for the population to recover to sustainable levels.[11] A similar argument has been made regarding Black-footed Ferrets. As all surviving Black-footed Ferrets are descended from just seven survivors, it is feared that their low numbers and diminished genetic diversity put them at imminent risk of inbreeding depression and reproductive failure. In other words, they may be on the brink of an "extinction vortex", even while their needs for other basic natural goods are being met. This might be forestalled by providing Black-Footed Ferrets with the genetically diverse reproductive partners they currently lack, if it becomes possible to produce offspring by cloning the tissue of preserved ferrets. Another candidate might be the Ivory-Billed Woodpecker. Unconfirmed observations have persuaded some that the species may persist but in numbers so low, extinction must be imminent. Were it possible to edit the genome of a related species of woodpecker in order to provide surviving Ivory-Billed Woodpeckers with viable partners, their extinction might be forestalled. Had these technologies been available in the late 1920s, we might have used them to provide reproductive partners for the last three Heath Hens, all male. As we were responsible for their plight, we would have owed it to them to make the attempt. Might we owe it them now?

28.4   Special Challenges Posed by Historical Injustices

Prior to the Heath Hen's extinction, our duties of restitutive justice might have been fulfilled by restoring the basic goods denied them by human development on Martha's Vineyard, such as food, shelter from predation, reproductive partners, and the like. As simple restoration of these goods is no longer possible, the question of compensatory reparations arises. Whether reparations could be due in this case is complicated by the fact that the injustice done is 'historical', i.e., the injustice occurred so long ago none of the parties directly involved remain alive. As with other cases of historical injustice, it may not be clear that all four criteria for material reparations can be met. Who can be obliged to offer reparations and to whom would they be due? If we cannot

---

[11]"They are at the brink of extinction only due to human activity," says Jan Stejskal, director of communication and international projects at the Dvůr Králové Zoo in the Czech Republic, where Sudan lived from 1975 until 2009. "If we have the techniques or methods to assist them to survive, I think it is our responsibility to utilize them" (quoted in Potenza 2018).

provide adequate solutions to these challenges, attempts to provide reparations for anthropogenic extinctions would be purely symbolic.[12]

Consider the loosely analogous case of the injustices perpetrated by the European colonizers of another island, Newfoundland, north of Martha's Vineyard, off Canada's Atlantic coast. European settlement of Newfoundland in the eighteenth and nineteenth centuries harmed the indigenous Beothuk community by displacing them from fishing grounds on which they depended for their survival, competing with them for forest resources, and introducing diseases to which the Beothuk had no resistance. Shanawdithit, the last known survivor of Beothuk island community died in 1829. Some would argue that any attempt to make reparations to the Beothuk community now would be merely symbolic. The Beothuk community is gone as are those responsible for the injuries that caused the community's dissolution. The dead no longer exist. To extract reparations from the colonists and transfer them to the victims, would involve "backward causation". In other words, we would have to be able to go back in time and change the past. As this is impossible, no agent, individual or collective, owes material reparations to the Beothuk. By the same logic, one might argue, no one is now obliged to make material reparations for the wrongs done the Heath Hen by the European colonizers of Martha's Vineyard.

The developing literature on intergenerational ethics suggests that these are not insurmountable challenges. Consider the claim that no one is now responsible for the wrongs done the Beothuk. This might seem correct regarding infractions committed by deceased individuals acting alone. It is not so obviously correct when those responsible acted on behalf of, at the behest of, and/or with the consent of transgenerational collective agents, such as corporations, societies, and political states. These kinds of collectives are not 'natural' persons and so do not naturally possess either moral agency or rights. Nevertheless many are granted artificial agency and rights, by legislation or regulation, to facilitate their provision of valuable services and to ensure accountability for injuries they cause. The agency granted is moreover often transgenerational, a long standing legal and social practice which we have good reason to support (Thompson 2009). Were there no institutions whose agency did not transcend generational change, the force of treaties, bequests, and corporate and other contracts would not survive the lives of their signatories.

This means that if a historical injustice is traceable to members of transgenerational institutions who were acting in accordance with that institution's policies or directives and that transgenerational institution still exists, then there will be no difficulty in determining who owes reparations in that case. Applied to the case of historical injustice done the Beothuk, the responsible party would be Canada. Even though Canada did not exist in its current political form until 1867 or include Newfoundland until 1949, the nation was created by and from transgenerational institutions responsible for colonizing Newfoundland. As well as inheriting various rights,

---

[12]This does not entail that symbolic reparations could not or should not be made. The symbolic benefits of governmental apologies for past injustices, even those leaving no survivors, can have enormous social benefits. However, symbolic gestures are not meaningful for animals, so I shall not discuss them further here.

possessions, and territories of its predecessors, it also inherited certain obligations. As Tracy Isaacs (2014) argues:

Canada is an enduring and identifiable collective entity, causally connected to the history of wrongdoing against First Nations. For these reasons, Canada and its [non-indigenous] citizens share in blame for past wrongs.

If we apply the same reasoning to the Heath Hen, then the State of Massachusetts and the local governments of Martha's Vineyard, are the transgenerational agents responsible both for the harms leading to the Heath Hen's demise and any reparations owed.

A second challenge in cases of historical injustice is determining who is entitled to claim reparations after the original victims are dead. In some cases, the original victims were themselves members of transgenerational collectives that still exist. In these cases, the recipients to whom reparations should be paid are easily identified. For example the indigenous residents of Martha's Vineyard, the Wampanoag Tribe of Gay Head (Aquinnah), is such a collective. In 1987, the Wampanoag won a settlement from the United States government for historical injustices done their ancestors, including wrongful appropriation of territory by earlier generations of non-indigenous settlers. As the Heath Hen played a role in the traditional diet of the Wampanoag, arguably the United States government's failure to preserve the Heath Hen was a further wrong for which the Wampanoag should receive compensation. Unfortunately, we cannot use this approach to establish a duty of reparations to the Heath Hen themselves (as opposed to a duty to the Wampanoag community *regarding* the Heath Hen.) Populations of wild animals are not organized collectives or institutions able to exercise transgenerational agency.

Living descendants of the victims of historical injustices (including descendants of indirect victims who suffered collateral damages) are another potential source of legitimate claimants to reparations. For example, the United Nations 2005 resolution, "Basic Principles and Guidelines on the Right to a Remedy and Reparation for Victims of Gross Violations of International Human Rights Law and Serious Violations of International Humanitarian Law", states that "where appropriate, and in accordance with domestic law, the term 'victim' also includes the immediate family or dependants of the direct victim" (United Nations 2005). In practice, legal enforcement is piecemeal and highly variable, however the principle that descendants have legitimate moral claims for reparations for wrongs done their predecessors is widely accepted. There are at least two ways one might apply this principle to the case of the Beothuk. First one could follow up reports from the coastal Mi'kmaq that many Beothuk fled Newfoundland before the final dissolution of their community. It may yet be possible to identify descendants of these refugees who would have a moral claim to reparations. Second, one could follow up on the effects of the dissolution of the Beothuk community on its Mi'kmaq neighbors. Losing access to a former trading partner very likely had negative consequences for neighboring Mi'kmaq communities, undermining their welfare and social and economic independence. If so, then in virtue of this collateral damage, descendants of these Mi'kmaq families would have a moral claim to material reparations.

Clearly we cannot use the first of these two strategies to generate reparations claims on behalf of extinct species, such as the Heath Hen, which have no descendants. Nevertheless, the second could potentially warrant moral claims for some non-human species affected by the anthropogenic extinctions of others. Imagine the human practices responsible for the Heath Hen's demise inadvertently harmed other sentient species in ways that continue to depress their descendants' abilities to maintain their populations. Alternately, imagine the Heath Hen provided ecosystem services important to a second species whose numbers have been depressed ever since the Heath Hen disappeared. In each scenario, there would be descendants of individuals who had suffered collateral damages, directly or indirectly, from the culpable harm done the Heath Hen.

Some critics of reparations arguments for historical injustices have worried that these strategies are undermined by the infamous "non-identity problem". Derek Parfit (1987) has shown that when we use comparative conceptions of harm, trying to establish that individuals were harmed by events responsible for those individuals being born may be difficult or impossible. On the comparative measures of harm we employ for most human interactions, a person is harmed if she is made worse off than she was—or worse off than she would have been—had the harmful event not occurred. If we apply a comparative measures of harm to the situations of surviving descendants of Beothuk or to descendants of the Mi'kmaq affected by the dissolution of the Beothuk community, it might be impossible to establish that any were made worse off comparatively speaking. These individuals would probably never have been born had European settlers not driven the last Beothuk from Newfoundland. Their ancestors' lives would have been sufficiently different to ensure that most would have had children with different reproductive partners in each of the intervening generations. By this time, very probably, none of their descendants would ever have existed but for the wrongs done the Beothuk. So unless their lives are so bad as to be not worth living at all, it would seem that they cannot be said to be worse off comparatively speaking.[13] Critics of reparations claims for anthropogenic extinctions believe the same is true for descendants of non-human species impacted by anthropogenic extinctions. If their identities were fixed by the events leading to those extinctions, we could not now claim those individuals were worse off comparatively speaking, unless their lives were so bad as not to be worth living at all (Palmer 2010, 2012; Rohwer and Marris 2018).

The non-identity problem does pose a challenge to some kinds of reparations claims involving human beings, specifically those which rely on comparative measures of harm. However, this problem rarely arises with historical injustices to wild animals. As noted above, with most wild animal populations, the only measures available to us are non-comparative. On non-comparative measures, all we need to determine is whether there is statistical evidence that the members of an animal population have been subjected to intrinsically harmful states. Population numbers are

---

[13] We are, of course, not restricted to using comparative measures even with human beings. When paradoxical conclusions like this result from employing comparative measures, others should be adopted.

used for this purpose. If human practices are causing a population of wild animals to shrink, we have evidence that its members, by and large, are being denied access to vital basic goods. And if this harm could have been avoided, and was not compelled by overriding moral considerations, we also have grounds for claiming reparations on those animals' behalf.

28.5 De-Extinction, Reparations, and the Heath Hen

The efforts made to save the Heath Hen were piecemeal and inadequate. Hunting was never banned outside the grasslands refuge created in 1908. Ill-advised fire suppression reduced the quality of the habitat the refuge provided. Feral cats were allowed to roam freely, preying on fledglings. Chickens and introduced game birds competed for resources.

Other native ground dwelling birds, such as the Ruffed Grouse, have also diminished in numbers. Bobwhite Quail numbers have been maintained but only by restocking with captive reared birds. As the human practices that caused the Heath Hen's demise are evidently also harming descendants of these other species, then we appear to have grounds for recognizing a duty to make reparations to them; controlling feral cats for example, eliminating hunting perhaps, and improving fire management. At the same time, we have reason to dismiss this rationale as irrelevant. Our obligations to reform our practices are over-determined. If existing wildlife populations are struggling because of harmful human practices, we already have sufficient reason for concluding we should reform them; the fact that they are harmful now. We would have exactly the same reason, even if our practices had not caused the Heath Hen's extinction. These reasons suggest we have obligations to restore habitat, control feral cats, improve fire suppression techniques and reduce human development in areas critical to these species' survival, rather than recreating the Heath Hen.

De-extinction could be morally obligatory if it turned out that Heath Hens were a foundational species for the grassland ecosystems on which the Heath Hen and other species depended. Jeffrey Yule has argued that in cases where no other species is able to fill the ecological gap left following an extinction, we may have an obligation to provide the ecological communities impacted with "a sort of ecological artificial limb or prosthesis… to replace the amputated original" (Yule 2002). When we think of de-extinction as a means to provide damaged ecological communities and co-dependent species with a prosthesis, "the quality of the replacement in duplicating the missing original would likely be a key consideration" (ibid.). That said, authenticity matters less than functionality. So it would not matter that recreated Heath Hens did not perfectly replicate their extinct "ancestors", provided they behaved in substantially similar ways. As he notes, "ecosystems, like people, are arguably far better off with prosthetics, particularly high quality prosthetics, than without them" (ibid.). Currently there is no other related species that can thrive in the conditions that suited the Heath Hen. This suggests that if replacing the Heath Hen's ecological function is essential to right wrongs done to other species on Martha's Vineyard, we might owe it to them (albeit not to Heath Hens themselves) to genetically recreate the Heath Hen. As it happens, there is no evidence that the Heath Hen played a foundational role in the grassland ecology of Martha's Vineyard.

That would seem to spell an end to the possibility of justifying its replication as an act of reparation. On closer examination, there is another party to whom reparations might be owed: the human residents of Martha's Vineyard. It is a standing grief to many that they will never enjoy the sight or sound of Heath Hens strutting and booming on their breeding grounds as earlier generations did. It is likewise a standing grief to others that they will never see a living Tasmanian Tiger, Passenger Pigeon, or Gastric Brooding Frog. Commenting on the Heath Hen Project, a local naturalist, Tom Chase confessed, "The truth is that my reasons for supporting this are completely emotional and value based" (Brown 2015b). Chase mourned the steady decline in avian biodiversity since his childhood:

> Kestrels were so abundant on Martha's Vineyard one flew in my house once and perched at the end of my bed," he said. "It was a time, like my father's time, of freedom and abundance. So when I see a project like this," Chase remarked, "as a conservationist I'm tired of fighting for things and always losing. I want to get on the proactive side and not the reactive side. (Brown 2015b)

Chase is just one of hundreds of thousands of people around the world who mourn the aesthetic, cultural, emotional, and material losses that avoidable anthropogenic extinctions have caused them. That means that he and the others like him can claim to be *worse-off* than they would have been had their predecessors passed on the same natural legacy of biodiversity they themselves had received from earlier generations. When extinctions are as recent as the Heath Hen's, the non-identity problem is not an issue. There is little likelihood that all those mourning the heath Hen's extinction would not have been born had its extinction *not* occurred.

There may be species whose cultural importance is so great as to warrant reparations claims against any transgenerational agencies responsible for their (avoidable) extinction. And in some of these cases, replicating the lost species might be the most effective means of righting the wrongs done. There is little reason to think the Heath Hen played such a role in the lives of residents of Martha's Vineyard. The people most likely to have formed such a relationship with the bird are those who co-existed with it the longest, i.e., the Wampanoag. Had the Heath Hen played a particularly significant role in their culture, Massachusetts might owe it to them to support the Heath Hen project. As the Wampanoag have never made such claims, I take it that the Heath Hen's loss was not so culturally injurious as to warrant de-extinction in reparation.

In a letter to the *Vineyard Gazette*, which Minteer discusses, Stephen Kellert provides a different moral defense for the Heath Hen Project:

> Restoring the heath hen offers us the chance for a moral reawakening. It provides us with an affirmative opportunity to restore our connection to the earth and contribute to the healing and beauty of the land. It offers Martha's Vineyard the extraordinary opportunity to lead as an inspiring example to America of how by living in right relation to nature we may flourish and achieve an ineffable and deepening connection to the larger community of life. (2014)

Minteer is dismissive because he thinks Kellert is recommending the Heath Hen Project as a techno-fix, i.e., a technological 'bandage' that mitigates symptoms of a social problem but does not address its cause. Thus he fears that Kellert's recommendation may reinforce rather than correct the vices responsible for anthropogenic extinctions and environmental degradation. Minteer writes:

> In fomenting the fantasy that we can erase the environmental abuses of the past by pursuing high-tech species revival technologies, promoters of de-extinction are inadvertently undermining the responsibility to learn the lessons of our environmental history. (2015, 15)

Minteer's point is well-taken but I think he is missing Kellert's. Kellert does not present Heath Hen replication as a techno-fix that would make atoning for and reforming long-standing vicious attitudes towards nature unnecessary. What Kellert sees as an affirmative opportunity for a moral reawakening is what would happen afterwards: learning to live *with* Heath Hens. To live with Heath Hens, non-indigenous islanders would have to develop new virtues of care for this species and its environment. Virtuous dispositions cannot be created by mental fiat. It takes time and practice to cultivate new virtues and to express them effectively. Means of identifying one's blind spots and positive reinforcement for correcting them are both important for success. The project of learning to live with Heath Hens might be an excellent way of going about this. Gaps in community caring would be visible to all. Backsliding could not be hidden. Success would likewise be visible and a point of pride for the community, reinforcing their commitment to self-reform. Learning to live *with* Heath Hens could indeed be morally, culturally, and indeed spiritually transformative.

Some fear that such projects will prove morally hazardous by inducing public apathy towards rising rates of species extinctions from the mistaken belief that extinctions are always reversible. There is no empirical evidence to support this. By contrast, there is a great deal of evidence to support worries about the effects on public attitudes of feeling powerless in the face of large-scale environmental problems, which include apathy, disengagement, and denial (Moser 2008; Moser and Dilling 2012). Anthropogenic extinction is this kind of problem. What can I do to prevent mass extinction? Which alterations in my character or behavior would make a difference? How could I know if I was succeeding? If the residents of Martha's Vineyard were to commit themselves to learning to live with Heath Hens, they would have concrete answers to these questions. They would have to learn to treat these birds as fellow residents of the Martha's Vineyard, entitled, as human residents are, to respect for their ways of life, fair shares in the natural goods the island affords, and community support in times of need. Were the communities of Martha's Vineyard to succeed, the Heath Hen would be a source of pride that could sustain them in other conservation endeavors. Perhaps then, Hough would feel that the gospel of conservation was finally winning the day on Martha's Vineyard.

Kellert might be right. Perhaps the best way to find out is to give it try. If the Heath Hen project has the benefits he suggests, de-extinction could be well worth the effort. But as we are now in the terrain of forward-looking reasons for replicating species, I will leave such questions to others.

28.6   Conclusion

I have argued that that we can have basic duties of restitutive justice to wild animals and that in some cases, these duties might require us to make reparations for harming them. The harms for which we can be morally liable are those we could predict before (and so avoid) or after the fact (and so make reparations). These include the imposition of states intrinsically bad for any sentient creature, caused by the denial of basic natural goods essential for life, as revealed by statistical measures of population trends. If a population is not succeeding in maintaining its numbers over time, and human practices are responsible, we have a prima facie case that our practices are causing that population harm. If it turns out that our practices were avoidable and not excusable in light of other more pressing moral claims, then there is a prima facie case for judging the harm caused as morally culpable and warranting material reparations.

As we cannot make material reparations to the dead, we do not owe reparations directly to extinct animals. We can make reparations to descendants of human and wild animal populations indirectly harmed by a culpable anthropogenic extinctions, so we may sometimes have an obligation to do so. In two kinds of cases, our obligations might require employing de-extinction techniques to generate replicas of extinct species if adequately developed techniques are available: cases in which recreating a species is the only way to make reparations for harm to another species of wildlife and cases in which recreating a species is the only way to make reparations for undermining a human culture. In both sorts of cases, duties to recreate extinct species are owed to the living members of other species, *not* the extinct species itself.

In only one case in which we could owe it directly to a species itself to 'bring it back'. When we are responsible for driving a species into an extinction vortex for lack of viable reproductive partners, we could in principle owe it to that species to 'bring it back'—albeit from extinction's door rather than extinction proper. In this rather small subset of cases, the reparations argument for using de-extinction techniques has real teeth.

Regrettably, the Heath Hen's extinction falls into none of these categories. No species on Martha's Vineyard will become extinct if we do not breed a new variety of Heath Hen. No human culture will fail. Genetic engineering cannot now rescue the Heath Hen from extinction. This does not mean that there are no good forward-looking consequentialist arguments for breeding a new variety of Heath Hen. There may be many. What it does mean is that we are under absolutely no moral obligation to try. So until good forward-looking reasons present themselves, *requiesce in pace Tympanuchus Cupido Cupido*.

References

Anderson, E. 2004. Animal rights and the values of nonhuman life. In *Animal rights: Current debates and new directions*, ed. C.R. Sunstein and M.C. Nussbaum, 277–298. New York: Oxford University Press.

Barrow Jr., M.V. 2009. *Nature's ghosts: Confronting extinction from the age of Jefferson to the age of ecology*. Chicago: University of Chicago Press.

Beeler, C. 2015. How an extinct species is being revived on the Galapagos Islands. *Public Radio International* (online). https://www.pri.org/stories/2015-12-15/how-extinct-spe cies-being-revived-galapagos-islands. Accessed July 4, 2019.

Bethge, P. 2013. Raising passenger pigeons from the dead. *Spiegel Online*. Available at https://www.spiegel.de/international/zeitgeist/scientists-aim-to-bring-the-passenger-pig eon-back-from-extinction-a-893744.html. Accessed July 4, 2019.

Boardman, R. 2006. *The International politics of bird conservation*. Cheltenham, UK: Edward Elgar.

Brown, S. 2014. Never say never; Heath Hen may get its boom back. *Vineyard Gazette*, April 3. https://vineyardgazette.com/news/2014/04/03/never-say-never-heath-hen-may-get-its-boom-back?k=vg533e0e76735bb&r=1. Accessed July 4, 2019.

Brown, S. 2015a. Heath Hen project advances quickly. *Vineyard Gazette*, March 5. https://vineya rdgazette.com/news/2015/03/05/heath-hen-project-advances-quickly. Accessed July 5, 2019.

Brown, S. 2015b. Heath Hen as gateway bird for de-extinction inches closer to reality. *Vineyard Gazette*, August 20. https://vineyardgazette.com/news/2015/08/20/heath-hen-gateway-bird-de-extinction-inches-closer-reality. Accessed July 4, 2019.

Campbell, D. 2017. On the authenticity of de-extinct organisms, and the Genesis argument. *Animal Studies Journal* 6(1): 61–79. http://ro.uow.edu.au/asj/vol6/iss1/5. Accessed July 5, 2019.

Campbell, D.I., and P.M. Whittle. 2017. *Resurrecting extinct species*. Cham, Switzerland: Palgrave Macmillan.

Cohen, S. 2014. The ethics of de-extinction. *NanoEthics* 8 (2): 165–178.

Cowles, H. 2012. A Victorian extinction: Alfred Newton and the evolution of animal protection. *The British Journal for the History of Science* 46 (4): 695–714.

Dan. 2014. Comment on Brown's Never say never; Heath Hen may get its boom back. *Vineyard Gazette*. https://vineyardgazette.com/news/2014/04/03/never-say-never-heath-hen-may-get-its-boom-back?k=vg533e0e76735bb&r=1. Accessed July 5, 2019.

Darwin, C. 1896. *The origin of species by means of natural selection or the preservation of favored races in the struggle for life*, vol. 2. New York: D. Appleton & Co. https://oll.libertyfund.org/tit les/darwin-the-origin-of-species-vol-2. Accessed December 4, 2019.

Delon, N., and D. Purves. 2018. Wild animal suffering is intractable. *Journal of Agricultural and Environmental Ethics* 31 (2): 239–260.

Diehm, C. 2015. Should extinction be forever? Restitution, restoration, and reviving extinct species. *Environmental Ethics* 37 (2): 131–143.

Fagan, W.F., and E.E. Holmes. 2006. Quantifying the extinction vortex. *Ecology Letters* 9: 51–60.

Hanser, M. 2008. The metaphysics of harm. *Philosophy and Phenomenological Research* 77 (2): 421–450.

Harman, E. 2009. Harm as causing harm. In *Harming future persons: Ethics, genetics and the nonidentity problem*, ed. M.A. Roberts and D.T. Wasserman, 137–154. Dordrecht: Springer.

Hough, H.B. 1933. *The Heath Hen's journey to extinction, 1792–1933*. Tisbury, MA: Dukes County Historical Society.

Isaacs, T. 2014. Collective responsibility and collective obligation. *Midwest Studies in Philosophy* 38 (1): 40–57.

Jebari, K. 2016. Should extinction be forever? *Philosophy of Technology* 29: 211–222.

Johanssen, K. 2019. To assist or not to assist? Assessing the potential moral costs of humanitarian intervention in Nature. *Environmental Values* 29(10): 29–45.

Kellert, S. 2014. Inspirational Project. *Vineyard Gazette*, July 31. https://vineyardgazette.com/news/2014/07/31/inspirational-project. Accessed July 4, 2019.

Klocksiem, J. 2012. A defense of the counterfactual comparative account of harm. *American Philosophical Quarterly* 49 (4): 285–300.

Kupferschmidt, K. 2014. Can cloning revive Spain's extinct mountain goat? *Science* 344(6180): 137–138. https://science.sciencemag.org/content/344/6180/137.full. Accessed July 5, 2019.

List, C., and P. Pettit. 2011. *Group agency: The possibility, design, and status of corporate agents*. Oxford: Oxford University Press.

Mitchell, N. 2018. If extinct animals could be brought back from the dead, should we do it? *ABC News* [Australian Broadcasting Company online] https://www.abc.net.au/news/science/2018-12-16/de-extinction-species-thylacine-woolly-mammoth-passenger-pigeon/10616670. Accessed July 5, 2019.

Minteer, B.A. 2014. Is it right to reverse extinction? *Nature* 509 (7500): 261. https://doi.org/10.1038/509261a.

Minteer, B.A. 2015. The perils of de-extinction. *Minding Nature* 8(1): 11–17.

Minteer, B.A. 2019. Promethean dreams. In *The fall of the wild: Extinction, de-extinction, and the ethics of conservation*, author B.A. Minteer, 97–118. New York: Columbia University Press.

Moser, S.C. 2008. More bad news: The risk of neglecting emotional responses to climate change information. In *Creating a climate for change: Communicating climate change and facilitating social change*, ed. S.C. Moser and L. Dilling, 64–80. Cambridge: Cambridge University Press.

Moser, S., and L. Dilling. 2012. Communicating climate change: Closing the science-action gap. In *The Oxford handbook of climate change and society*, ed. R. Norgaard, D. Schlosberg, and J. Dryzek, 161–174. Oxford: Oxford University Press.

Norcross, A. 2005. Harming in context. *Philosophical Studies* 123: 149–173.

Nussbaum, M.C. 2006. *Frontiers of justice: Disability, nationality, species membership*. Cambridge, MA: Harvard University Press.

Page, T., and C. Hancock. 2016. Zebra cousin went extinct 100 years ago: Now, it's back. *CNN online*. https://www.cnn.com/2016/01/25/africa/quagga-project-zebra-conservation-extinct-south-africa/index.html. Accessed July 5, 2019.

Palmer, C. 2012. Can—And should—We make reparation to Nature?'. In *The environment: Philosophy, science, ethics*, ed. W.P. Kabasenche, M. O'Rourke, and M. Slater, 201–222. Cambridge, MA: MIT Press.

Palmer, C. 2010. *Animal ethics in context*. New York: Columbia University Press.

Parfit, D. 1987. *Reasons and persons*. Oxford: Clarendon Press.

Perry, S. R. 2003. Harm, history, and counterfactuals. *Faculty Scholarship*. Paper 1108. http://scholarship.law.upenn.edu/faculty_scholarship/1108.

Pickrell, J. 2018. Return of the living thylacine. *Cosmos Magazine*, July 4. https://cosmosmagazine.com/biology/return-of-the-living-thylacine. Accessed July 5, 2019.

Potenza, A. 2018. Inside the high-tech plot to save the northern white rhino from extinction. *The Verge*, April 6. https://www.theverge.com/2018/4/6/17175936/northern-white-rhino-de-extinction-stem-cells-sudan. Accessed July 5, 2019.

Revive & Restore. n.d. Heath Hen project. https://reviverestore.org/projects/heath-hen-project/. Accessed July 6, 2019.

Rohwer, Y., and E. Marris. 2018. An analysis of potential ethical justifications for mammoth de-extinction and a call for empirical research. *Ethics, Policy & Environment* 21 (1): 127–142.

Sagoff, M. 1984. Animal liberation and environmental ethics: Bad marriage, quick divorce. *Osgoode Hall Law Journal* 22(2): 297–307. http://digitalcommons.osgoode.yorku.ca/ohlj/vol22/iss2/5. Accessed July 5, 2019.

Sánchez-Quinto, F and C. Lalueza-Fox. 2015. Almost 20 years of Neanderthal palaeogenetics: Adaptation, admixture, diversity, demography and extinction. *Philosophical Transactions of the Royal Society B* 370: 20130374. https://doi.org/10.1098/rstb.2013.0374. Accessed January 22, 2021.

Sandler, R. 2014. The ethics of reviving long extinct species. *Conservation Biology* 28 (2): 354–360.

Shiffrin, S.V. 2012. Harm and its moral significance. *Legal Theory* 18: 357–398.

Shultz, D. 2016. Should we bring extinct species back from the dead? *Science* (online). https://www.sciencemag.org/news/2016/09/should-we-bring-extinct-species-back-dead. Accessed July 5, 2019.

Shapiro, B. 2016. Pathways to de-extinction: How close can we get to resurrection of an extinct species? *Functional Ecology* 31 (5): 996–1002.

Smith, D. 2017. De-extinction research to feature on ABC TV. *University of New South Wales Newsroom website.* https://newsroom.unsw.edu.au/news/science-tech/de-extinction-research-feature-abc-tv. Accessed July 5, 2019.

Stephens, P. 2016. Population viability analysis. In *Oxford bibliographies in ecology*, ed. David Gibson. New York: Oxford University Press. https://doi.org/10.1093/obo/9780199830060-0142.

Thompson, J. 2009. *Intergenerational Justice.* New York: Routledge.

Tuck, J.A. 2006. Beothuk. In *The Canadian encyclopedia: Historica Canada.* Last Edited June 20, 2019. https://www.thecanadianencyclopedia.ca/en/article/beothuk. Accessed July 5, 2019.

United Nations. 2005. Basic principles and guidelines on the right to a remedy and reparation for victims of gross violations of international human rights law and serious violations of international humanitarian law, U.N. Doc. E/CN. 4/2005/L. 48. https://www.ohchr.org/EN/ProfessionalInterest/Pages/RemedyAndReparation.aspx.

Yule, J.V. 2002. Cloning the extinct: Restoration as ecological prostheses. *Common Ground* 1 (2): 6–9.

Jennifer Welchman (Phd Johns Hopkins University) is a professor of philosophy at the University of Alberta, Canada. Her main areas of research are in ethics, the history of ethics, and environmental ethics. Her chief philosophical indulgence is aesthetics (including environmental aesthetics). Current research is focused upon virtue ethics, the moral philosophy of David Hume, and the ethical implications of new biotechnologies for conservation. She is the author of Dewey's Ethical Thought and editor of The Practice of Virtue: Classic and Contemporary Readings in Virtue Ethics. She is currently co-director of the Canadian Society for Environmental Philosophy.

Chapter 29
The Welfarist Account of Disenhancement as Applied to Nonhuman Animals

Adam Shriver

Abstract I criticize the current usage of the terms "enhancement" and "disenhancement" in the debate over the genetic modification of animals and propose an alternative definition of these terms based on how modifications affect animals' welfare in particular contexts. The critique largely follows a similar criticism of the use of the term "enhancements" in the human bioethics literature. I first describe how the term "disenhancement" has been used in debates thus far, and argue that the present lack of a shared definition is problematic. I then consider some potential definitions of "disenhancement" that can be adapted from the human bioethics literature and argue that most of these uses are flawed for the purposes of using the term in current ethical debates. Finally, I elaborate on the welfarist conception of disenhancement and consider some potential objections, using examples from the literature to illustrate key points.

29.1 Introduction

Consider the following scenarios from the ethics literature about genetically modifying animals:

Football Birds: Using gene-editing, chicken DNA is altered so radically that it results in headless (football-shaped) "birds" that are merely fed nutrients through tubes and produce edible eggs. The organisms completely lack anything resembling a brain and are completely insentient (Comstock 2000, 152).

Live Fast, Die Young: In order to avoid the act of killing livestock, certain animals' genes are altered such that they die painlessly shortly after reaching adulthood (McMahan 2008).

A. Shriver (✉)
Uehiro Centre for Practical Ethics, Oxford, UK
e-mail: adam.shriver@philosophy.ox.ac.uk

© The Author(s) 2021
B. Bovenkerk and J. Keulartz (eds.), *Animals in Our Midst: The Challenges of Co-existing with Animals in the Anthropocene*, The International Library of Environmental, Agricultural and Food Ethics 33,
https://doi.org/10.1007/978-3-030-63523-7_29

Polled Cattle: Cows are genetically modified to lack horns. This eliminates the practice of painful dehorning surgery. This could be accomplished via selective breeding over many decades at great expense, but using genetic technology dramatically speeds up the process (McConnachie et al. 2019).

Dino-Chickens: Scientists insert DNA into modern chicken eggs to reproduce their ancient evolutionary ancestors, which happen to resemble dinosaurs because they possess snouts rather than beaks. The lack of beaks results in decreased welfare problems from pecking or debeaking (Shriver and McConnachie 2018).

Painless: Pigs are modified to not feel the unpleasantness of pains by eliminating a particular neurotransmitter in part of their brains. They are still able to reflexively respond to pains (Shriver 2009).

Blind Chickens: Chickens are modified so that they are blind, which results in decreased welfare problems from pecking or debeaking (Sandøe et al. 1999; Ali and Cheng 1985).

Each of these scenarios raises slightly different ethical issues. But all of the animals involved would, in contemporary debates, be described by many authors as "disenhanced" animals. Use of the term "disenhancements" isn't intended to settle debates definitively about whether particular practices are right or wrong. Rather, disenhancements was introduced into the discussion of animal ethics in order to connect these debates to those in human bioethics about using technologies (genetic, pharmaceutical, bionic) to "enhance" humans by making them smarter, stronger, more loyal, etc. But just as in the human bioethics debate one might claim that enhancement can be impermissible or permissible depending upon circumstances, in the debate about animal ethics, one might believe that a particular modification is a disenhancement but nevertheless hold that it is permissible, or perhaps even obligatory in some circumstances while impermissible in others. Nevertheless, the words we use are important, and the term disenhancements implies that we are taking something away that would otherwise be present. For this reason, applying the term to describe changes to animals carries a strong connotation that such practices ought to be avoided, all else being equal, at least for those who think that animals' interests ought to be taken into consideration.

In what follows, I criticize the current usage of the terms "enhancement" and "disenhancement" in the debate over the genetic modification of animals and propose an alternative definition of the terms based on how modifications affect animals' welfare in particular contexts. My critique will largely follow a similar criticism of the use of the term "enhancements" in the human bioethics literature. The strategy will be as follows: I first describe how the term disenhancement has been used in debates thus far, and argue that the present lack of a shared definition is problematic. I then consider some potential definitions of "disenhancement" that can be adapted from the human bioethics literature and argue that most of these uses are flawed for the purposes of using the term in ethical discussions. Finally, I elaborate on the welfarist conception of disenhancement and consider some potential objections, using the scenarios above to illustrate key points.

29.2 The "Opposite of Enhancement"

The term disenhancement has become ubiquitous in discussions of gene editing animals over the past decade, largely popularized by a 2008 article by Paul Thompson entitled "The opposite of human enhancement" and a series of responses published in the journal *Nanoethics*. The term has become a useful shorthand for what seems to be an intuitively graspable concept relating to modifying an animal by "taking something away." Early examples included the idea of breeding blind chickens to reduce welfare problems that result from cannibalism in confined quarters, or creating the completely insentient "football birds" described above. However, there has not been much of an attempt in the literature to rigorously define the term "disenhancement" as it applies to animals. Perhaps this is due to authors being largely content to rely on what they perceive to be a shared folk understanding of enhancement, or perhaps the assumption is that the term "enhancement" has been sufficiently defined in the human literature such that disenhancement can simply be characterized as the opposite of enhancement. However, I will argue that both of these assumptions are flawed.

Regarding reliance on folk conceptions for key ethical terms, we might surmise that the assumption is that there is sufficient agreement on the reference of the term "enhancement" among the population or some particular fixed group in the population such that the term can be used to refer to uncontroversial cases in the ethics literature. But proceeding without a definition of a key term is generally speaking contrary to contemporary analytic philosophical practices, and for good reason. For one thing, it provides no methodology for deciding on controversial cases. Perhaps technological modifications that simultaneously result in loss of a function, decreased fitness, and decreased welfare can easily be considered disenhancements, but what about mixed cases where different dimensions are varied in opposite directions (e.g. increased fitness but decreased welfare)? Relying on folk intuitions about uncontroversial cases leaves us without a good sorting mechanism for more difficult cases.

Moreover, we should not merely assume widespread agreement even on the cases philosophers believe are uncontroversial. The experimental philosophy movement, despite its flaws, has shown fairly conclusively that philosophers often assume that "the folk" agree on certain concepts or intuitions without sufficient evidence. Philosophers' assumptions about "the folk conception of X" are often mistaken. Moreover, even if the assumptions are correct for a certain population, there might be cultural, economic, or gender differences in application. As such, it is risky to base arguments or claims on unverified assumptions about the folk conception of enhancement in the absence of empirical research.

And finally, it often turns out that the way the general population uses certain terms is inconsistent and even incoherent and therefore is not well-suited for use in philosophical arguments. The general population might for example believe both that pains are "mental events" *and* that pains are located in body parts, despite the fact that these two beliefs cannot be easily reconciled. Similarly, folk conceptions of something like disenhancement might sometimes link it to evolutionary fitness, sometimes to what is "species typical," and sometimes to welfare, shifting their reference point in response to different salient features. But clearly these different conceptions come apart: a given change might result in any combination of increases or decreases along those dimensions, so a definition that lumps them all together will not be able to handle cases that diverge along these dimensions.

The upshot of these problems is that (1) we should not assume, in the absence of empirical evidence, that philosophers can accurately capture the folk definition of disenhancement in a single definition and (2) that it's extremely unlikely that the folk definition, assumed to represent an approximate agreement across the entire population, will be able to play the role it needs to in a proper philosophical analysis. To say something meaningful about enhancement in the context of ethical debate, we need a term that reliably and meaningfully captures a set of shared properties.

This leaves us with the definitions of enhancement provided in the human bioethics literature. However, the situation in the bioethics literature is anything but settled. Savulescu et al. (2011, 3) write, "Although there is much debate about the ethical implications of new technologies, only a few authors have attempted to provide an explicit definition of enhancement." Moreover, they have pointed out a number of flaws in the definitions on offer, which led them to propose a welfarist conception of enhancement.

In making their argument, Savulescu et al. (2011) usefully divide definitions of enhancements into two categories which they describe as functional enhancements and human enhancements. Functional enhancements refer to enhancements of particular capacities, capacities such as strength, intelligence, or memory. Thus, for example, taking certain medications such as Adderall might be a functional enhancer of attention, and anabolic steroids might be functional enhancers of strength or endurance. But some changes might enhance certain functional capacities while detracting from more holistic measures of the flourishing of the organism. We can think of cases where increases in particular capacities actually leads to negative consequences for the human or animal (Earp et al. 2014). This is why we need a second category, human enhancement, to capture changes related to the overall performance of the organism. Examples of this type of enhancement could be the person's health or well-being. Savulescu et al. (2011) use "human enhancement" to describe these changes, but I think differentiating "domain-specific enhancements" from "holistic enhancements" more accurately captures the relevant difference and does so in a way where the terminology can be easily extended to nonhuman animals.

One seemingly straightforward way of connecting domain-specific enhancements to holistic enhancements would be to define the latter as follows: A holistic enhancement is just any instance where an organism has one or more domain-specific enhancements. Similarly, we could say that a holistic disenhancement is just any instance where an organism has one or more domain-specific disenhancements. The problem, however, is that we can imagine cases where the same changes are enhancements in relation to one domain but disenhancements in regards to a different domain. Consider, for example, a change that resulted in greater strength but less fine motor control. Since at the holistic level we presumably don't want it to be the case that particular changes are both enhancements and disenhancements simultaneously, the move from domain-specific to holistic enhancements won't serve the purposes ethicists need in having a clear definition of disenhancement.

Among the holistic definitions of enhancement, there are several possible conceptions that can be sussed from the literature. The first is the "not medicine" or "more than treatment" approach. This approach, coming from a specific history in the bioethics literature that was preoccupied with questions about over-prescription of psychiatric medication, defines enhancement as improvements to human form or functioning that go beyond what is necessary to "sustain or restore good health." Thus, on this usage, taking Adderall to counteract attention deficit hyperactivity disorder is treatment; taking it to study for a test is enhancement. Similarly, using transcranial direct-current stimulation to treat depression would count as a treatment, but tDCS to provide relaxation would be enhancement.

One of the challenges of the "more than treatment" approach is the requirement of coming up with additional definitions for arguably equally contentious ideas such as "good health" and "disease" (Zohny 2014). And this challenge is exacerbated when we try to use the definition of enhancement to create a definition of disenhancement. In the case of defining an enhancement according to performance along one particular domain, it's easy to get "results in a *decrease* in performance along domain X" as a definition of disenhancement if your definition of enhancement is "results in an *increase* in performance along one particular domain." But if your definition of enhancement is "increases performance beyond what is necessary to sustain or restore good health," then additional questions arise. Should disenhancement be regarded as anything that impairs health back down to average good health, or anything that drops an organism below good health, both, or something else entirely?

One way of answering this complication would be to utilize the definition provided by Sabin and Daniels (1994), which defines enhancement roughly as follows:

> Normal species-functioning definition of enhancement: Any change in the biology or psychology of an organism which increases species typical normal functioning above some statistically defined level.

Correspondingly, we can define the opposite of enhancement as:

> Normal species-functioning definition of disenhancement: Any change in the biology or psychology of an organism which decreases species typical normal functioning below some statistically defined level.

Of course, deciding exactly where we set these statistically defined levels raises its own challenges, but hopefully the idea at least is clear enough for present purposes.

Finally, I will consider two other potential definitions of holistic enhancement. These types of measures define enhancement and disenhancement in reference to a single property, but a property that applies to the organism as a whole rather than to a specific capacity. The property Savulescu et al. (2011) propose in regards to humans, and that I will be advocating for in regards to nonhuman animals, is well-being or welfare. They propose the following definition of enhancement:

> Welfarist definition of enhancement: Any change in the biology or psychology of an [animal] which increases the chance of leading a good life in the relevant set of circumstances.

And thus we can also propose:

> Welfarist definition of disenhancement: Any change in the biology or psychology of an animal which decreases the chance of leading a good life in the relevant set of circumstances.

However, there are other possible candidates which become especially salient when we think about nonhuman animals. In particular, though we generally don't think of evolutionary fitness as a measure for the flourishing of modern humans, this is a term that is more easily applied to nonhuman animals, particularly for animals living in the wild, outside of direct human influence. This would look something like this:

> Evolutionary fitness definition of disenhancement: Any change in the biology or psychology of an organism which decreases the organism's evolutionary fitness.

Thus, for example, taking the "Blind Chicken" example from above, it seems relatively straightforward to see how one might intuitively consider this change a disenhancement insofar as it deprives the chickens of a capacity that was important for their survival in the past.

29.3   Normal Species Functioning and Fitness Are Irrelevant for Animals Under Human Supervision

Thus far, I've argued that we can't defer to presumed folk intuitions in place of a definition of disenhancement and also cannot extend the domain-specific definition of enhancement into a makeshift holistic version. That leaves us with the following possible contenders for a definition of holistic disenhancement: the normal species-functioning definition, the welfarist definition, and an evolutionary fitness definition. In this section, I argue that both the normal species functioning and fitness definitions are irrelevant for key areas of discussion in the animal ethics literature.

First, consider evolutionary fitness. There is of course considerable debate as to whether evolutionary fitness has any ultimate intrinsic normative value with, I believe, most ethicists opposing the idea. However, in circumstances where animals are under direct human supervision, the notion of fitness seems especially irrelevant. In cases like the keeping of companion animals, or livestock, or animals put to use for labour, animals' health and opportunities for reproduction are almost entirely controlled by humans. Many traits that would lead to greater ability to survive or reproduce in wilder environments are irrelevant for animals under the direct supervision (and control) of humans.

Consider a trait like aggression in animals such as chickens or pigs. In more naturalistic environments, such a trait obviously would play a role in fighting over resources or protecting territory. However, in many modern confined feeding operations these traits aren't' helpful: at least in well-designed systems, the animals are just as likely (or unlikely, as it were) to survive and have their genes passed on whether or not they are aggressive. So it would be bizarre to label hyper-aggressive animals "enhanced" and passive animals "disenhanced" in a confined feeding operation based on the fact that these traits might benefit the fitness of their wild counterparts in particular circumstances.

We might, alternatively, think that the traits that enhance fitness are now traits that make humans more likely to try to pass on the genes of particular animals. In other words, for livestock, we could describe traits such as passivity, decreased aggression towards humans, cuteness, fast growth, etc., as "fitness enhancements" since people may intentionally select for these traits in animals for their own benefit. But this seems like a distortion of the term; the "Football Chickens," for example, would count not only as "not disenhanced" but would in fact be "enhanced animals", since humans would likely be all-too-happy to keep these genes in circulation.

Similarly, "normal species functioning" seems like a challenging idea to apply to the lives of contemporary livestock. First, there's a question of how "normalcy" is determined; are we using an average based on what the lives of most current animals look like, in which case it would be heavily skewed towards the lives of animals already in confinement for most species used as livestock, or are we using a version of "normality" linked to what some earlier wild version of the animals would have looked like? Using the former seems bizarre: imagine trying to define "normal species functioning" of humans based on averages derived entirely from prison populations. But the lives of "normal" evolutionary predecessors again seem far removed from the modern context. Traits that were presumably helpful in flourishing outside of captivity, such as the ability to detect predators (increased vigilance), fight off rivals (aggression), and reproduce (high sex drive) can all be plausibly imagined to be detrimental for the animals in extreme confinement. Does it make any sense to call an animal modified to have less anxiety in an environment where it will never encounter a predator or non-human threat a "disenhanced" animal?

Unlike these criteria, the welfarist conception of enhancement and disenhancement is relevant in both relatively natural environments and in manmade and highly controlled environments. We can talk plausibly about changes that increase or decrease the welfare of wild animals and of animals in highly artificial environments. And in both cases we have reason to think that these changes matter morally. I'll consider some possible counterexamples below, but I hope it is clear that welfare continues to matter in artificial environments in a way that the other holistic criteria do not.

29.4   Elaborating the Welfarist Conception of Disenhancement and Responding to Objections

As noted above, the welfarist definition of disenhancement is as follows: Any change in the biology or psychology of an animal which decreases the chance of leading a good life in the relevant set of circumstances. Before discussing the virtues and vices of this definition in more detail, it's worth highlighting a particularly crucial component of the definition. Namely, in order to be successful, welfare must be defined *in relation to a particular set of circumstances.*

To see why, consider the Polled Cattle example in two different environments. In both cases, assume that the horns possible role in temperature regulation does not sufficiently influence welfare (we can assume both environments have consistently mild weather). However, in one environment, there exists a particular type of parasite that is completely absent in the other environment. The horns both of the animals themselves and of conspecifics are remarkably effective at limiting the pervasiveness of this parasite, which in turn prevents unpleasant itching sensation and possible infections from wounds. In such cases, would we consider using genetic technology to create a polled variant of cattle to be an example of disenhancement? It seems clear that taking away horns can count as a disenhancement in the parasite environment, since the change would result in decreased welfare, but not in the parasite-free environment, since the change would there be welfare-neutral. It is a virtue of the welfarist conception that it can capture this divergence. As such, the welfarist definition of disenhancement is necessarily tied to particular sorts of circumstances.

Returning to the examples from the beginning can help illustrate some of the features of how the welfarist conception works. The Football Birds, contrary to current usage, would not count as "disenhanced" animals on the welfarist conception. But they also would not be "enhanced." Since they have no welfare at all, these terms do not apply to them. For almost all of the other cases, the answer as to whether or not they are enhanced or disenhanced is "it depends." Is the lack of an ability to feel pain an enhancement or disenhancement? It depends on whether the animal is in circumstances where avoiding certain normally pain-inducing features can help prevent further states of negative welfare. If they simply don't feel pain in some situations but don't suffer any further negative consequences as a result, this would count as an enhancement on the welfarist view.

Even the Live Fast Die Young example wouldn't necessarily be considered a disenhancement on this account. If the animals were living lives that were, on balance, full of positive well-being, then causing them to die early would be a disenhancement. However, if their lives were unhappy or even welfare neutral, then dying young would not be a "disenhancement."

I'm sure there is disagreement about whether it truly makes sense to call a pain-free or short-lived animal "enhanced." But hopefully it is reasonably close enough to common usage, or at least easy enough to fit to modern usage with some argumentation and clarification. Nevertheless, there are some additionally fairly counterintuitive implications of the welfarist view that need to be acknowledged, and these come out in some of the potential objections to welfarist views. One objection is that welfarist definitions make enhancement and disenhancement too ubiquitous, as they seem to apply to an extremely wide class of changes. Taking antibiotics to treat a disease? That meets the welfarist definition of enhancement. Breaking a leg in a fight? Disenhancement. Even, say, exercising to improve physical fitness could be regarded as a form of enhancement. The terms would no longer pick out only a very small and specific type of proposed changes to animals but would instead refer to a wide class of changes.

But this is a problem only if the original usage was picking out changes that are genuinely qualitatively different from those in the expansive definition and which thereby need to be treated separately. As Zohny 2014 has detailed, many different ideas have been proposed to distinguish enhancements from "natural changes," such as suggesting that the latter require more work or concentration, but none of the suggestions put forward thus far ultimately hold up against scrutiny. In any event, it remains for proponents of other uses of enhancement to suggest why other types of changes should be regarded as qualitatively different from changes that are brought about through drugs, neurointerventions, or genetic modification, and this would require a more thorough attempt at providing a clear definition.

Another potential criticism considered in the human bioethics literature is that using a welfarist conception of "disenhancement" prematurely settles the moral debate. By linking enhancement with improvement to well-being and disenhancement to decreased well-being, the terminology is such that all enhancements are regarded as permissible and all disenchantments as impermissible.

While it certainly seems true that using the welfarist definition would establish, for many, a presumption in favour of enhancement and against disenhancement, the new terminology in no way should be seen as settling the question of whether a given disenhancement is permissible or not. There might be some enhancements that cause harm to others or to the natural environment, that reflect badly on our character, or that make us complicit in regards to an unjust practice. In all such cases, changes might be deemed "enhancements" but nevertheless considered morally wrong. Similarly, there may be cases where a particular "disenhancement" is dramatically outweighed by other benefits that result, such as a case where a particular change is very good for the environment but results in mild decrease in welfare. The term disenhancement has a negative connotation; however, unlike alternative definitions including current usage, the welfarist definition of disenhancement has connotations that actually track something of (dis)value in a way that provides a useful, though not definitive, assessment.

One final criticism from debates in the human bioethics literature I will consider is that this definition of disenhancement is too dependent upon contested notions of well-being. The philosophy literature has numerous competing accounts of well-being, generally grouped into hedonistic, desire fulfilment, and objective list theories. How can we apply a welfarist definition to enhancement if we haven't reached agreement among philosophers as to which theory is preferable? As Zohny (2014) says, the differences between these views can be exaggerated...often certain changes count as obvious increases or losses to welfare on any of the types of theories. But, perhaps even more clearly in the case of animals than with humans (where other considerations such as autonomy, rights, or justice are often centred in moral debates), challenges with evaluating welfare in animals will inevitably need to be addressed in a thorough moral debate. Assessing welfare will always be a crucial part of evaluating the morality of policies and practices that involve animals; as such, avoiding questions about characterizing well-being in regards to enhancement simply pushes those questions to a different stage of the debate. In other words, deciding upon the best conception of welfare is certainly challenging, but the difficultly doesn't allow us to avoid the questions when aiming for the best actions.

29.5   Agency and Disenhancement

My discussion thus far has been largely consequentialist, focusing on particular states of the altered animals as the only relevant possible criteria for definitions of enhancement and disenhancement. However, considering an agency-centred approach raises additional questions.[1] First, one might suggest that particular technological changes can impair or improve the agency of the altered animals and that these changes are relevant to whether the changes count as enhancements or disenhancements. For example: creating "football birds" deprives organisms of any ability to make decisions or to choose to interact with the environment in particular ways, and we may think that in losing those things the chickens have lost something of value.

However, even if one believes this is an important consideration, on many accounts this notion of agency can simply be incorporated into the notion of welfare at stake. One can claim that it is *good for* an organism to have agency, and that decreasing agency thereby lowers welfare and hence can count as a disenhancement, all things being equal. A crude way of making this argument might simply state that having agency *feels good*. Other views hold that agency can contribute to well-being independently of how it makes one feel. But the important thing, for my purposes, is that the value of agency can still be included in a welfarist conception of disenhancement. It is true that some may hold that depriving animals of agency is bad even if doing so is not bad *for the animals,* but these views can be classified as instances of impermissible enhancements without jeopardizing the utility of a welfarist definition of the term.

A very different sort of agent-centred concern has to do not with the agency of the altered animal, but of the individual or group doing the altering. On this type of account, what matters are the intentions behind the changes; if one *intends* to benefit an organism in a particular way, then we call it an enhancement. On the other hand, if the intentions behind a particular change, like the Dino-Chicken, are simply to save expenses and increase the efficiency of an operation, it might seem strange to call it an "cnhancement" simply because it happened to improve welfare by accident. Surely we shouldn't reward exploitative practices that have no concern for animals by calling them "enhancements" simply because they had unintended benefits!

Though such an approach has a certain intuitive appeal, there are simply too many complications with relying on presumed motivations to determine whether changes count as enhancements or disenhancements. First, people are notoriously bad at determining the motivations of others, and political allegiances would likely play a large role in determining whether or not the stated intentions of a particular change are believed. But even putting aside the possibility of dishonest statements of

---

[1]The following discussion is based on helpful suggestions from Christopher Preston.

intentions, people are also notoriously bad at identifying even our own motivations for particular actions and have a strong tendency to create self-serving narratives. As such, relying on the intent of technological changes rather than the actual effect would be epistemologically infeasible and could also lend itself to distortion through self-serving motivations. And finally, though we might imagine an intention-focused criteria allowing for cases of enhancement by those trying to improve the lives of animals, it seems extremely unlikely that anything would count as a disenhancement if it were required that the change was made with the intention of depriving an organism of some beneficial capacity. Any negative changes are most likely to be side-effects of attempts to increase profit, or benefit human health, etc. and so relying on the primary intention to determine the definition would mean that almost no proposed changes would count as disenhancements.

29.6   Conclusion: Why a Welfarist Account?

As technology advances, humans are increasingly proposing modifications to nonhuman animals. This is perhaps most prevalent in the realm of agriculture, but proposals have also been made to use genetic technology to stop disease-carrying species from reproducing or to alter the balance of specific ecosystems. And if past behaviour is any guide, humans unfortunately are likely to also attempt to use technology to change companion animals or to design new roles for altered animals in military engagements. Ethical debates about such uses of animals are at an early stage, and have not yet caught up to the technological possibilities. But in order to have a robust debate, it is important that key terms are clearly defined.

The term disenhancement has been increasingly used to describe potential modifications of nonhuman animals across a range of contexts. However, the term has not been clearly defined and the corresponding human bioethics literature has also struggled to come up with a widely accepted definition of enhancement. Given that we need a clear definition in order to properly frame and discuss philosophical debates about human interactions with animals, I have attempted to outline a case for using a welfarist definition of enhancement. This choice of terminology, of course, does not settle any ethical debates, but it does imply that some cases such as creating animals incapable of suffering might be better regarded as enhancement rather than disenhancement.

References

Ali, A., and K.M. Cheng. 1985. Early egg production in genetically blind (rc/rc) chick-ens in comparison with sighted (rc +/rc) controls. *Poultry Science Reviews* 64: 789–794.
Comstock, G. 2000. *Vexing nature? On the ethical case against agricultural biotechnology.* Norwell, MA: Kluwer Academic.

Earp, B.D., A. Sandberg, G. Kahane, and J. Savulescu. 2014. When is diminishment a form of enhancement? Rethinking the enhancement debate in biomedical ethics. *Frontiers in Systems Neuroscience* 8 (12): 1–8.

McConnachie, E., M.J. Hötzel, J.A. Robbins, A. Shriver, D.M. Weary, and M.A. von Keyserlingk. 2019. Public attitudes towards genetically modified polled cattle. *PLoS ONE* 14 (5): e0216542.

McMahan, J. 2008. Eating animals the nice way. *Daedalus* 137 (1): 66–76.

Sabin, J.E., and N. Daniels. 1994. Determining "medical necessity" in mental health practice. *Hastings Center Report* 24 (6): 5–13.

Sandøe, P.B., L. Nielsen, L.G. Christensen, and P. Sørensen. 1999. Staying good while playing God—The ethics of breeding farm animals. *Animal Welfare* 8 (4): 313–328.

Savulescu, J., A. Sandberg, and G. Kahane. 2011. Well-being and enhancement. In *Enhancing human capacities*, ed. J. Savulescu, R. ter Meulen, and G. Kahane, 3–18. Oxford: Wiley-Blackwell.

Shriver, A. 2009. Knocking out pain in livestock: Can technology succeed where morality has stalled? *Neuroethics* 2 (3): 115–124.

Shriver, A., and E. McConnachie. 2018. Genetically modifying livestock for improved welfare: A path forward. *Journal of Agricultural and Environmental Ethics* 31 (2): 161–180.

Thompson, P.B. 2008. The opposite of human enhancement: Nanotechnology and the blind chicken problem. *Nanoethics* 2 (3): 305–316.

Zohny, H. 2014. A defence of the welfarist account of enhancement. *Performance Enhancement & Health* 3 (3–4): 123–129.

Adam Shriver is a Research Fellow at the Oxford Uehiro Centre for Practical Ethics. Adam's research examines the intersection of ethics and cognitive science and he has written multiple articles about human well-being and animal welfare. In particular, Adam's research has examined the significance of the dissociation between the affective and sensory components of pain for philosophical theories of ethics and well-being. He also has research examining the ethics of using genetic modifications in livestock. Previously, Adam organized a workshop on neuroethics and animals, and is a co-editor on the book Neuroethics and Nonhuman Animals.

Chapter 30
How to Save Cultured Meat from Ecomodernism? Selective Attention and the Art of Dealing with Ambivalence

Cor van der Weele

Abstract As a highly technological innovation, cultured meat is the subject of techno-optimistic as well as techno-sceptical evaluations. The chapter discusses this opposition and connects it with arguments about seeing the world in the right way. Both sides not only call upon us to see the world in a very particular light, but also point to mechanisms of selective attention in order to explain how others can be so biased. I will argue that attention mechanisms are indeed relevant for dealing with the Anthropocene, but that dualism has paralysing effects. In a dualistic framework, cultured meat is associated with ecomodernist optimism, bold technological control over nature and alienation from animals. But interested citizens and farmers in focus groups rather envisioned the future of cultured meat through small scale production on farms combined with intensive relations with animals. Such scenarios, involving elements from both sides of the dualistic gap, depend on constructive ways of dealing with dualisms and ambivalence.

30.1 Intro: Wizards and Prophets

We need to eat less meat. Although the urgency is growing, the call is not new. One starting point is Ruth Harrison, who published *Animal machines* (1964) in protest against the then upcoming factory farming. A few years later, in *Diet for a small planet* (1971), Frances Moore Lappé stressed how inefficient meat is for feeding the world. The book is full of recipes illustrating how the combination of pulses and grains is a wholesome alternative for meat. Yet beans and lentils did not take over the world. On the contrary: global meat consumption kept rising, global pulse consumption kept falling, and plant-based meat-imitations did not take off either; consumers found them too different from meat (Hoek 2010). This is why the idea of cultured meat—an idea that had been lingering in the sidewings ever since tissue culturing technology came

C. van der Weele (✉)
Wageningen University, Wageningen, The Netherlands
e-mail: cor.vanderweele@wur.nl

© The Author(s) 2021
B. Bovenkerk and J. Keulartz (eds.), *Animals in Our Midst: The Challenges of Co-existing with Animals in the Anthropocene*, The International Library of Environmental, Agricultural and Food Ethics 33,
https://doi.org/10.1007/978-3-030-63523-7_30

up in the beginning of the twentieth century—could suddenly look so promising when it was rediscovered around the turn of the century. It was going to be made of animal cells, it would ('really') be meat but without its downsides. This might finally help.

As a highly technological option, cultured meat was an immediate subject for discussions on the pros and cons of technofixes.[1] Will this technology open up new directions of socio-technical change (Driessen and Korthals 2012), or is it rather a narrowly framed pseudo-solution that distracts from the need for fundamental shifts in our lifestyles? While quite a few animal protection organizations, after some hesitation, pragmatically embraced it as a more hopeful strategy for improving the fate of animals than raising public awareness,[2] others, including many who campaigned for more natural food, were skeptical. Simon Fairlie was one of them. He not only criticized the technological character of cultured meat but also how this fitted in with vegan perspectives bound to further alienate us from nature. In his book *Meat, a benign extravagance* (2010), Fairlie starts out by rejecting the view that veganism is the best ethical response to the problems of meat. Environmentally speaking, eating small amounts of meat is actually more sustainable than eating no meat at all, he argued, because some of our resources (some types of grassland, some forms of waste), are best used by raising animals (in animal-friendly ways).[3] So the vegan solution is simplistic, but apart from that it will lead us to artificial food and estrangement from nature. While the organic sector is campaigning for "slow food, real meat and fresh local produce", he wrote, vegans and vegetarians are pointing in "the opposite direction", namely of *factory food, and cultured meat lies at the end of that road (Fairlie 2010, 228). In his view, this technological enthusiasm additionally includes attempts to genetically engineer factory-farmed livestock without the capacity to suffer. In fact, he suspects, we might be seeing the first signs of "a convergence of interests between factory farming, veganism and genetic engineering" which will lead us to the "brave new world of transhumanists". Against the dominant trends of urbanization, Fairlie argues for ruralization, for small scale human settlements and

---

[1] Including skepticism about unrealistic expectations raised by technologists. However, cultured meat was atypical in this respect. The enthusiasm was more a matter of ethical pull than of technology push. At a time when no researchers were asking for money (there had been a NASA project focusing on goldfish, but it was terminated), the first lobby organization for cultured meat, New Harvest, was founded by the American student Jason Matheny. He had been so shocked by the industrial chicken breeding he had witnessed that he decided something must be done and that cultured meat—he read about the NASA project—looked like a promising idea. When visiting the Netherlands, he met with protein researchers, with the Minister of agriculture and with businessman and long term cultured meat promotor Willem van Eelen. Cultured meat research in the Netherlands began with the funding of three PhD projects. See https://www.new-harvest.org/about.

[2] The American organization PETA, for example, discussed whether or not an organization that included many vegans should support something that made use of animal cells. The outcome was that PETA embraced cultured meat as the most promising quick road to a better future for animals and announced a $1 million prize to the first laboratory to create commercially viable test-tube chicken.

[3] This view has been corroborated by more recent research at Wageningen University. See e.g. Van Kernebeek et al. (2016), Van Zanten et al. (2018).

permaculture, with people living closer to nature and closer to the sources of their food.

Fairlie's way of picturing a diametrical opposition between rural sustainable farming and urban estrangement from nature has been affirmed and strengthened by later discussions, especially since the appearance of the *Ecomodernist Manifesto* (Asafu-Adjaye et al. 2015). This manifesto stated that many human activities need to be intensified with the help of technology, in order to "decouple" human well-being from environmental destruction. Ecomodernism comes with a plea for urbanization, under the assumption that this is more sustainable: "cities both drive and symbolize the decoupling of humanity from nature." When Chris Smaje responded on the website of *Dark Mountain*, a movement with a somewhat older manifesto of its own to which I will return, he concludes that the ecomodernist decoupling of human well-being from natural impacts also implies a physical decoupling of people from nature. In his summary: "to us the city and the minimum amount of farmland necessary to support it, to the rest of creation the wilderness" (Smaje 2015). He could not disagree more.

Does this debate have a deep or essential core? Charles Mann's recent analysis certainly suggests so. Under the title *The Wizard and the Prophet* (2018), he focuses on the opposition between "techno-optimist" Norman Borlaug, a hero of eco-modernists who believed in human ingenuity for solving all problems, and "green luddite" or "catastrophic environmentalist" William Vogt, who believed in living more sustainable lives and eating lower down the food chain. Mann describes how Borlaug's path leads to urbanization and labor-extensive industrialized farming, while the prophets strive for small scale labor-intensive agriculture that takes care of the earth and the soil. In Mann's analysis, they appear as fundamentally and unchangeably different views of how to deal with the future of our planet. So perhaps it need not surprise that they now also serve as opposing perspectives on climate change and the Anthropocene: technology versus lifestyle as the way forward.

This dualism clearly comes with an emphasis on opposed strategies in agricultural thinking and research, while detracting from examples that might undermine the divide, such as the use of GPS and drones in organic arable farming, or the use of milking robots in nature-oriented animal farming. Below, I will explore how cultured meat could strategically be developed in a way that undermines rather than reinforces the seemingly cast-in-stone dichotomy between techno-optimists and green luddites. But let me first focus on one additional aspect of the dualism, namely how both sides call upon us to pay attention to the world in very specific ways. While wizards/ecomodernists encourage us to see hopeful developments all around and a bright future on the horizon, their prophet-opponents, with Dark Mountaineers prominently among them, want us to open our eyes to the things that go wrong and that will lead to inevitable collapse. Both sides regard their own way of seeing the world as an accomplishment that takes effort and courage, and both point to mechanisms of selective attention that explain why others have such a biased perspective.

Arguments about selective attention and perception are prominent in both camps. Directing the spotlight on this difference will first seem to enlarge the gap, but it will then also afford leads for dealing with it differently, namely by taking a closer look at attention mechanisms and how to deal with them in more constructive ways.

30.2 Selective Attention

Ecomodernists propose that if human wellbeing is decoupled from environmental impacts, technology—with a focus on energy technology—will keep human consumption going: "meaningful climate mitigation is fundamentally a technological challenge" (Asafu-Adjaye et al. 2015). We can live in the cities, decrease the burden on environments and regreen the earth by such processes as urbanization, agricultural intensification, genetic engineering, nuclear power and desalination; as the examples show, they derive hope from the prospect of bold frontier technologies. They explicitly embrace an optimistic view towards the future, which received a boost by what is now called the "new optimist" movement, with authors such as Pinker, Norberg, Ridley and Roser (Nisbet 2018). New optimists, including the ecomodernists, have been inspired by Hans Rosling, whose 2006-TED talk *The best stats you've ever seen* has been designated as the birth of the new optimism (Burkeman 2017). Rosling devoted the last part of his life to making facts about global improvement more accessible through a form of visual statistics called *Gapminder*. The book that appeared in 2018, after his death, is titled *Factfulness; ten reasons why you are wrong about the world and why things are better than you think*. It explains that we tend to severely underestimate for example how many children in the world are vaccinated or how many years of school education girls on average receive, globally. The book discusses ten 'instincts' that explain why our view of the world tends to be far too pessimistic. Several of them are closely related to the character of the news, with its strong focus on what goes wrong. This selective focus can be explained by seeing how we have evolved as beings who need to be on the lookout for things that threaten us. Things that are going well or that slowly become better are not newsworthy, while in reality many things go well or are slowly but steadily getting better. Slow change is hard to see at any specific moment. Yet time makes it visible: our world is completely different form that of our grandparents.

The world also faces real threats, Rosling acknowledges. Climate change is one and we need to be deeply concerned. But it won't help to focus on worst case scenarios; we need to quickly move on from fear and endless talking and use that energy to solve the problem. "So, what's the solution? Well it's easy. Anyone emitting lots of greenhouse gas must stop doing that as soon as possible." (Rosling 2018, 231). He adds that although the planet's common resources can only be governed by a globally respected authority (the United Nations), it can be done: we already did it with ozone depletion and with lead in gasoline (ibid., 239).

His overall advice is not to count on journalists for a good view of the world, since they are caught in the attention-grabbing drama business. Rather than burdening them with unrealistic expectations, let us realize that the news fits in with our evolved habit to automatically look for threats, and that it is simply not very useful for understanding the world. Instead, we'd better learn to attend to the facts. This is also what optimists inspired by Hans Rosling hope to teach us: evolution has not built us to pay attention to what goes well or is slowly improving, but we can train ourselves to embrace a fact-based worldview. When it comes to climate change, ecomodernists and new optimists add that the development of helpful technology is not inevitable; it will require effort, as the ecomodernist manifesto states, from both states and the private sector.

While ecomodernists tend to see optimism as a moral duty, its opposite is now also propagated as the most responsible way forward; let me turn to this very different view of the world.

Climate change can be described as a super wicked problem (Levin et al. 2012), since time is running out, many parties still have an interest in the status quo and there is no central authority for effective interference. Now that climate change is increasingly experienced as a reality, emotions such as fear, grief and despair are on the rise. Apart from being psychological phenomena, pessimism and depression have also become starting points of new forms of activism, for example by environmental activists and journalists Paul Kingsnorth and Dougald Hine, who lost belief in effective change and together founded the *Dark Mountain* project. The name is derived from a line by Robinson Jeffers (1887–1962), a Californian poet who withdrew from civilization in order to seek harmony with nature.

The *Dark Mountain Manifesto* (Kingsnorth and Hine 2009) starts by explaining that all civilizations must go under, and that this is hard to see or predict from the inside, since "the pattern of ordinary life, in which so much stays the same from one day to the next, disguises the fragility of its fabric". Civilizations are held together by belief. But once such belief starts to crumble, collapse may become unstoppable.

The dark mountaineers have given up belief in our civilization. They see what they say most of us are "unable to see"—since we have all been trained not to pay attention to basic things about our civilization: "its fundamental destructive features": its myths of progress, of human centrality and of our separation from nature. The problem is not one of outright and explicit denial. Yes there are some remaining climate skeptics, but focusing on them distracts from a far larger and more important form of denial, which is about the emotional inability to really connect to what we know about ecological unravelling and to take it seriously. "Ecological and economic collapse unfold before us and, if we acknowledge them at all, we act as if this were a temporary problem, a technical glitch."

Yet these uncanny signs are not technical glitches, but inherent to our civilization. In order to find responses—for when the world as we know it no longer exists—the writers think that science and technology are not going to be helpful and neither are politics, ideology or activism. What we need instead is art. They are especially interested in new storytelling in the genre of *Uncivilized writing*. Such writing attempts to shift our worldview and find a position outside the bubble of civilization, a position at which we can carefully pay attention to the nonhuman world and re-engage with it, grounded in a sense of place and time.

Dark mountaineers try to teach us that because we cling to the false safety of our belief in day-to-day existence in a civilization that is built on destructive attitudes to nature, we are heading for collapse. The courage we need is to really look the painful facts in the face, accept grief and despair, and from there on search for new beginnings.

In the next section I will go into somewhat more detail into the mechanisms of selective attention sketched by both sides of the dualism. While such mechanisms are real, they are just a small subset of potentially relevant mechanisms. Dualistic framing is itself an attention mechanism, and in the paragraph after the next I will focus on its power and some of its dangers. My overall suggestion is that we need more awareness of attention mechanisms in general and of their pitfalls, in order to find ways of dealing with them more imaginatively.

30.3   How Daily Life Blinds Us in Different Ways

My experience is what I agree to attend to, William James famously said. Though such 'agreement' will often not be conscious, the implications that attention is necessarily selective and that it determines how we experience the world seem right. Both ecomodernists and dark mountaineers argue that the seemingly self-evident nature of daily life blinds us to what is really going on. Ecomodernists think that because we have evolved to selectively attend to what goes wrong, we fail to see how good our daily life really is. Dark Mountaineers think that because of our emotional rootedness in our seemingly safe routines we fail to see their fragility as well as their destructive externalities. Daily life is apparently able to blind us in different ways: it makes us neglect slow changes in the good direction, it also makes us resist unwelcome facts.

Both types of bias have also been noted by philosophers and social scientists studying patterns of attention, though typically in less absolutist ways. Analyses of the selective focus of the news and the media tend to note similar things as Rosling: the news, depending as it is on triggering our attention, is a biased way of learning about what happens in the world (e.g. Wijnberg 2018; De Botton 2014). News tends to focus on incidents, while it is generally silent on what goes well or what is slowly changing, because that is not newsworthy. What the optimistic account seems to miss, meanwhile, is that slow change can also be for the worse.

The type of neglect observed by Dark Mountain too is the object of study and reflection, and increasingly so in the context of climate change. In the book *Living in denial; climate change, emotions and everyday life*, Kari Norgaard (2011) called attention to mechanisms of everyday denial. She sets out noting that the idea that we would respond more adequately to climate change 'if only we knew' (the idea that knowing the facts will make us act rightly) cannot explain the finding that people's interest may decline as more information becomes available; people have been found to stop paying attention to climate change when they realized there is no easy solution (Norgaard 2011, 2). Not paying attention is different from climate skepticism in that it is not a denial of climate change as such but a way to protect daily life, in a situation in which we feel powerless or confused, by avoiding the issue. The book documents the social organization of such protection mechanisms in Norway. Information is not the key variable; emotions of helplessness and guilt are key, and they estrange us from the realities of our lives, argues Norgaard. Getting or remaining in touch with such disturbing realities is a real struggle.

Similar analyses abound. It has amply been documented how cognitive dissonance leads to coping mechanisms in which people adapt their beliefs, for example about the cognitive abilities of the animals they eat (Bastian et al. 2012). I have been writing about a related way of evasive coping in the context of meat consumption, 'strategic ignorance': not wanting to know too much about meat and the way it is produced in order to avoid awkward choices. Strategic ignorance has long been confused with indifference, because on the level of behavior it looks the same. However, strategic ignorance is in fact based on ambivalence rather than indifference (Onwezen and Van der Weele 2016); it is a mechanism of coping with the psychological unrest and indecisiveness of great tensions. In a psychoanalytically inspired study of climate apathy, Renée Lertzman (2015) likewise challenges the view that apathy and denial typically result from a lack of concern; instead of a lack, she observes a surplus of concern or affect. In her interviews in the vicinity of an industry that was both good for employment and disastrous for the environment, she encountered much ambivalence (both love and hate) towards this industry. Suppression of the hate-aspect led to unresolved mourning and apathy. These findings evidently have much in common with what Dark Mountaineers describe.

Such studies confirm that phenomena of selective attention that ecomodernist wizards and dark prophets talk about are real and troubling. We tend to ignore slow and unspectacular progress because of our habitual dependence on the news. We tend to ignore things that go wrong when they are too emotionally uncomfortable. And daily life gives rise to more mechanisms of selective attention, partly intertwined with the ones just discussed. Not only the news, but habits and routines too blind us to the conditions of normal life; we tend to only pause and reflect, often reluctantly, when things go wrong, disappear or otherwise change. From Plato onward, waking up in wonder to our self-evidences has been seen as the beginning of philosophy. But what do we see when we wake up and reflect—may we perhaps be unhealthily attracted to dualism?

30.4   Dualisms as Paralyzing Attention Tools

Apart from being a world-making mechanism, attention has also become a scarce and valuable commodity in our age of social acceleration and exploding digital information. Business models in the attention economy, centring on harvesting our attention and then selling it, are not only extremely profitable, they have also increasingly been criticized as turning customers at least partly into products. In his book *The attention merchants*, Tim Wu (2016) traces this business model back to the nineteenth century, when newspapers started to make money by selling their readers' attention to advertisers. He also notes that since the model completely depends on gaining and holding attention, it strongly encourages extreme content, as this is likely to engage 'automatic' attention.

Even though 'dichotomous thinking' is known as a personality disorder, presenting the availabilities as two options that look like polar opposites is in fact an eternally tempting model of thought, which creates order in a simple way. Think of black and white, all or nothing, hate or love, nature versus nurture, male and female, technological solutions versus lifestyle solutions or optimism versus pessimism—dualistic and polarizing ways of thinking are forceful tools of attracting and selectively framing attention and creating meaning and order.

Dualism has also been discovered as an attention mechanism by tech companies. Roger McNamee (2019) has documented how the battle for attention put tech companies in Silicon valley on that track; extreme views have been discovered and strategically used as attention capturing devices, as they stir emotion and keep people engaged. They are part of what Tim Harris (2019) calls the 'extractive attention economy'. Worrisome levels of polarization result from this race to the bottom.

Because selective attention creates our realities, attention tools deserve philosophical as well as psychological scrutiny. This has become more urgent in the era of the attention economy, with tech companies putting much effort in manipulating our attention. 'When tech knows you better than you know yourself' (Thompson 2018), the challenge is to understand and face what makes us so vulnerable. The Anthropocene and its debates likewise illustrate the urgency of becoming acquainted with attention mechanisms, including the role played by emotions and dualistic distinctions. How we are being misled by the news, how people are prone to resist painful subjects by avoiding them if they can, how emerging patterns of attention are socially organized, or how dualisms encourage us to think that we should choose between tech solutions and lifestyle changes or between pessimism and optimism.

This takes us back to the dualism that towers over environmental and Anthropocene debates as described by Charles Mann. How can we choose between the poles of such dualisms? Mann himself illustrates how hard this is, saying that he oscillates between the stances: "On Monday, Wednesday and Friday I think Vogt was correct, On Tuesday, Thursday and Saturday, I go for Borlaug. And on Sunday, I don't know." (Mann 2018, 13). Mann's wavering ambivalence seems to lead to a kind of paralysis, quite comparable to the proverbial indecisiveness of Buridan's ass—the donkey that dies from hunger and thirst between a pile of hay and a pail of water, because it is equally hungry and thirsty and cannot choose between the two. Ambivalence notoriously undermines our ability to choose and it perhaps need not surprise that it has received a bad press in both philosophy and psychology. Psychologically, ambivalence is often extremely uncomfortable. In philosophy, Harry Frankfurt (1988) has influentially argued that ambivalence stands in the way of being a wholehearted and free person. This perhaps helps to explain why Mann thinks that in the end we will have to answer the question who is right, wizard or prophet; "our children will have to answer it" (Mann 2018, 9).

Yet ambivalence is a normal and ubiquitous aspect of life, especially in times of change. From this perspective, we might need better options to deal with it than making a forced choice, being stuck in paralysis or trying to avoid the subject altogether (Rorty 2009; Razinsky 2017; Van der Weele and Driessen 2019). One important starting point for more constructive attitudes towards ambivalence is that the need to make an either-or choice between precisely two (opposed) options may actually be a very rare phenomenon. Take optimism and pessimism, for example: we can be partly or moderately optimistic or pessimistic, alternately pessimistic and optimistic, or perhaps live with being ambivalently optimistic and pessimistic at the same time. In the final paragraph, I will now return to meat and cultured meat. Both abundantly give rise to ambivalence, and I will argue that constructive skills for dealing with ambivalence can help to avoid dichotomous stalemates and find more constructive solutions.

30.5   Cultured Meat and the Pig in the Backyard

In dichotomic terms, cultured meat seems to clearly belong to the wizard or ecomodernism pole, as has indeed been argued repeatedly. Fairlie's book, in which cultured meat is portrayed as an urban technological strategy that will estrange us from nature, already did so *avant la lettre*, and similar analyses abound. Wyatt Galusky (2014, 945) writes that cultured meat as a technological solution will lead to the disappearance and invisibility of the animal, and that "rather than confront the ethical questions of engaging animals and humans and ecologies in the context of meat, we turn those questions into engineering ones." Valan Anthos (2018) explicitly associates cultured meat with ecomodernism; it does not focus on lifestyle change but on its alternative: fulfilling the demand for meat, with the aim of technologically decoupling meat from its harmful consequences." They do not like this alternative. Modern industrial farming already exemplifies an approach in which our relations to the nonhuman world are viewed in terms of control, efficiency and usefulness, according to both Galusky and Anthos. Cultured meat obeys to the same logic, exerting even deeper control over nonhuman nature, even fostering the unhealthy illusion of total control over a nonhuman world. Our relations with animals, which should be "on the forefront of our confrontations with meat" (Anthos 2018, 46), meanwhile fade into the background: cultured meat sidesteps crucial underlying questions about our relations to animals, because there no longer is any relation with animals to consider. The assumption, clearly, is that cultured meat will make human relations with animals disappear. And this, Fairlie could have added, is precisely the wrongheaded aim of many vegans.

Most people probably prefer better relations with animals to the absence of animals, and will find this prospect scary. But is it indeed the only or the most plausible prospect? This suggestion depends on picturing cultured meat in the framework of the dualism of technological control-cum-alienation versus lifestyle changes; we have either technology to replace animals or moral reflection on our relations with animals, not both. Already in 2008, Hopkins and Dacey responded to such dichotomous thinking in a paper on the ethics of cultured meat, noticing not only that cultured meat is informed by moral considerations, but also that technology and morality do not represent separate but interactive roads to change. They speculated that cultured meat may change our relations with animals so that "people in the future find eating meat from living animals unbearably barbaric" (Hopkins and Dacey 2008, 589).

The question also rises whether this dichotomic way of thinking produces the only or most plausible scenario for cultured meat. It is certainly not the only one, as a very different scenario has been emerging within my own research, at least partly in response to fears of alienation. In that scenario, our relations with animals do not disappear, on the contrary: cultured meat finally enables us to develop loving relations with the animals we eat. Through focus groups, Clemens Driessen and I have been exploring responses to cultured meat and its relations with meat (Van der Weele and Driessen 2013, 2019). We always found much ambivalence, both about cultured meat and about meat. The remark that cultured meat is unnatural, for example, always made someone else wonder how natural our ordinary meat actually is—from animals kept in confinement, and/or containing added water, preservatives, antibiotics etc. Both factory farmed meat and cultured meat were in fact associated with unnaturalness, technology, and alienation from our food. But concerning cultured meat, such downsides vanished completely with a scenario that spontaneously emerged from one discussion and that we termed 'the pig in the backyard'. Participants who had started out being quite hesitant about the idea of cultured meat at some point started to envision its production through a local and small scale industry: cells from free ranging pigs, in backyards or urban farms, would be taken through biopsies every now and then and cultured into meat in neighborhood factories. The idea immediately warmed the participants to cultured meat. The scenario integrated ideas of local and urban food production, good relations with (farm) animals, and a neighborhood scale combination of production and consumption. The idea of cultured meat being unnatural, alienating or too technological had vanished completely; in fact, this scenario seemed almost too good to be true (Van der Weele and Driessen 2013).

In later groups, this scenario was welcomed as very sympathetic ("in this way you can experience the animal as a living being and love it") but also often as more or less implausible, unrealistic, and going against rules and regulations (Van der Weele and Driessen 2019). Perhaps, some participants wondered, doing it on small farms instead of backyards would already be somewhat more realistic.

In a follow up project, this idea of small scale cultured meat production on farms has taken the form of the question whether cultured meat might perhaps be an opportunity, instead of a mere threat, for farmers. In the first part of this project, we held focus groups with farmers. The project is still ongoing, but some results already emerged beforehand, for example about the interest of farmers for this subject. Many farmers were skeptical, so that it was not easy to find enough participants for the focus groups. But there were also a few farmers who were eager not just to participate but to be involved in cultured meat production as soon as possible. Apart from a chance of contributing to circular agriculture, these farmers saw cultured meat as an opportunity to foster new and better connections with consumers, with animals, and with society in general. The combination that attracts them is keeping free-range animals, producing local kinds of cultured meat and invite consumers to come and visit. A few of them did not wait for the outcomes of the project to already take further initiatives by seeking connections with start-up companies and find out more about technological possibilities.

It remains to be seen what comes of famer-scale cultured meat production. The important point for this chapter is that the activities in this direction do not conform to the either-or schema that dominates so many discussions. Envision a farm with relatively small (e.g. 2000 L) cultured meat bioreactors. It might be a farm that keeps animals of special breeds in free-range conditions. While the animals live their good lives, biopsies are taken from them now and then, to make cultured meat that is then branded as a specialty from this farm and/or from these special breed animals. Other activities on the farm might include elements from a wide range of other options, varying from very technological (milking robots, scanning drones) to very traditional (a small shop, care activities) or new and experimental (a food forest, recycling dung, pixel farming).[4]

Such activities are neither purely ecomodernist nor the opposite, and possible attempts to 'disambiguate' (Chiles 2013) and reinterpret them in terms of pure positions seem to be beside the point. Rather, the pure positions themselves have to give way to activities which—just as in the pig in the backyard scenario—combine very traditional elements with new technological options. Such a 'tinkering' approach starts from the tension and ambivalences between different values, wishes, and available options, and looks for ways that sidestep or go beyond paralyzing oppositions, thus undermining the dualistic framing. This more generally illustrates that doing and trying need not conform to the dividing lines set out in societal and academic debates, and that progress can be a matter of making new combinations rather than drastic choices.

One obstacle for such small scale initiatives is that they may go at least partly against dominant economic rationality. Yet the dominance of dichotomous thinking may be an at least as powerful obstacle, as it takes attention away from working with the tensions in more constructive ways. While debates between ecomodernists and Dark Mountaineers rage on, some farmers are constructing quite different options, and their efforts deserve more conceptual as well as practical attention. As I have been suggesting in the last paragraph, one attention skill that we urgently need for avoiding dualistic stalemates, is the art of dealing with dualistic ambivalence through imaginative tinkering instead of choosing.

References

Anthos, V. 2018. Meat reimagined: The ethics of cultured meat. Graduate school, University of Montana. https://scholarworks.umt.edu/etd/11203/.

---

[4]Pixel farming involves working with very small plots in order to avoid the use for pesticides associated with monocultures. See e.g. https://www.futurefarming.com/Smart-farmers/Articles/2020/2/Pixel-farming-plots-of-10-by-10-centimeters-532286E/.

Asafu-Adjaye, J., L. Blomqvist, S. Brand, B. Brook, R. Defries, E. Ellis et al. 2015. An ecomodernist manifesto. http://www.ecomodernism.org/manifesto-english.

Bastian, B., S. Loughnan, S. Haslam, and H.R.M. Radke. 2012. Don't mind meat? The denial of mind to animals used for human consumption. *Personality and Social Psychology Bulletin* 38 (2): 247–256. https://doi.org/10.1177/0146167211424291.

Burkeman, O. 2017. Is the world really better than ever? *The Guardian*, July 28. https://www.theguardian.com/news/2017/jul/28/is-the-world-really-better-than-ever-the-new-optimists.

Chiles, R. 2013. Intertwined ambiguities: Meat, in vitro meat, and the ideological construction of the marketplace. *Journal of Consumer Behaviour* 12 (6): 472–482. https://doi.org/10.1002/cb.1447.

De Botton, A. 2014. *The news, a user's manual*. London: Penguin.

Driessen, C., and M. Korthals. 2012. Pig towers and in vitro meat: Disclosing moral worlds by design. *Social Studies of Science* 42 (6): 797–820.

Fairlie, S. 2010. *Meat, a benign extravagance*. White River Junction, VT: Chelsea Green.

Frankfurt, H. 1988. *The importance of what we care about*. Cambridge, NY: Cambridge University Press.

Galusky, W. 2014. Technology as responsibility: Failure, food animals and lab-grown meat. *Journal of Agricultural and Environmental Ethics* 27 (6): 931–948.

Harris, T. 2019. A new agenda for tech. *Center for Humane Technology*, April 23. https://humanetech.com/newagenda/.

Harrison, R. 1964. *Animal machines: The new factory farming industry*. London: Vincent Stuart.

Hoek, A. 2010. Will novel protein foods beat meat? Consumer acceptance of meat substitutes, a multidisciplinary research approach. PhD thesis, Wageningen University.

Hopkins, P.D., and A. Dacey. 2008. Vegetarian meat: could technology save animals and satisfy meat eaters? *Journal of Agricultural and Environmental Ethics* 21 (6): 579–596.

Kingsnorth, P., and D. Hine. 2009. The dark mountain manifesto. https://dark-mountain.net/about/manifesto/.

Lappé, F.M. 1971. *Diet for a small planet*. New York: Ballantine Books.

Lertzman, R. 2015. *Environmental melancholia: Psychoanalytic dimensions of engagement*. London: Routledge.

Levin, K., B. Cashore, S. Bernstein, and G. Auld. 2012. Overcoming the tragedy of super wicked problems: constraining our future selves to ameliorate global climate change. *Policy Science* 45: 123–152.

Mann, C. 2018. *The wizard and the prophet*. Basingstoke: MacMillan.

McNamee, R. 2019. *Zucked: Waking up to the Facebook catastrophe*. New York: HarperCollins.

Nisbet, M. 2018. The ecomodernists: A new way of thinking about climate change and human Progress. *Skeptical Enquirer* 42(6): 20–24. https://skepticalinquirer.org/2018/11/the-ecomodernists-a-new-way-of-thinking-about-climate-change-and-human-prog/.

Norgaard, K.M. 2011. *Living in denial; climate change, emotions and everyday life*. Cambridge, MA: MIT Press.

Onwezen, M.C., and C.N. van der Weele. 2016. When indifference is ambivalence: Strategic ignorance about meat consumption. *Food Quality and Preference* 52: 96–105. https://doi.org/10.1016/j.foodqual.2016.04.001.

Razinsky, H. 2017. *Ambivalence: A philosophical exploration*. London: Rowman & Littlefield.

Rorty, A. 2009. A plea for ambivalence. In *The Oxford handbook of philosophy of emotion*, ed. P. Goldie, 425–444. Oxford: Oxford University Press.

Rosling, H. 2018. *Factfulness*. London: Hodder & Stoughton.

Smaje, C. 2015. Dark thoughts on ecomodernism. *The Dark Mountain Project*, August 12. https://dark-mountain.net/dark-thoughts-on-ecomodernism-2/.

Thompson, N. 2018. When tech knows you better than you know yourself. *Wired*, October 4. https://www.wired.com/story/artificial-intelligence-yuval-noah-harari-tristan-harris/.

Van der Weele, C., and C. Driessen. 2013. Emerging profiles for cultured meat; Ethics through and as design. *Animals* 3 (3): 647–662.

Van der Weele, C., and C. Driessen. 2019. How normal meat becomes stranger as cultured meat becomes more normal. *Frontiers in Sustainable Food Systems* 3 (69). https://doi.org/10.3389/fsufs.2019.00069.

Van Kernebeek, H.R.J., et al. 2016. Saving land to feed a growing population: Consequences for consumption of crop and livestock products. *International Journal of Life Cycle Assessment* 21 (5): 677–687.

Van Zanten, H., M. Herrero, O. van Hal, E. Röös, A. Muller, T. Garnett, et al. 2018. Defining a land boundary for sustainable livestock consumption. *Global Change Biology* 24 (9): 4185–4194.

Wijnberg, R. 2018. Unbreaking news: the problem with real news and what we can do about it. *De Correspondent*, September 12. https://medium.com/de-correspondent/the-problem-with-real-news-and-what-we-can-do-about-it-f29aca95c2ea.

Wu, T. 2016. *The attention merchants*. London: Atlantic Books.

Cor van der Weele is special professor of humanistic philosophy at Wageningen University. She has a background in biology as well as philosophy. Her PhD thesis, in the philosophy of biology, about selective scientific attention in understanding animal embryological development, was titled *Images of development* and published in 1999 (Suny). Since then she has remained interested in the entanglement of conceptual and empirical issues and in boundary areas of philosophy and biology - as well as psychology and sociology. Her interest in cultured meat started in 2007. It has led, partly in collaboration with Clemens Driessen, to a series of publications on that subject, in the context of change processes concerning meat. In the course of these studies, ambivalence emerged as an important phenomenon for understanding such processes. As the chapter in this book illustrates, she is also still interested in selective attention.

Chapter 31
Comment: Evolution 2.0—Rewriting the Biosphere

Henk van den Belt

CRISPR-Cas9 was only discovered in 2012, but this new genome editing tool immediately raised high expectations about the virtually unlimited range of applications it brought into view as it turned previously remote possibilities into realistic options. Apart from applications in human medicine and industrial biotechnology, its most obvious uses are in the genetic improvement of crops and domestic animals, where it can replace the older, more cumbersome and less versatile techniques of genetic modification (aka recombinant-DNA technology). But CRISPR also aims at applications beyond conventional agriculture. It makes the prospect of reconstructing extinct species ('de-extinction') as a new approach to conservation seem more feasible and also lies at the basis of the ingenious 'gene drive' technique, which enables us to quickly spread desired traits through wild populations. As co-discoverer Jennifer Doudna declares with no false modesty: "CRISPR gives us the power to radically and irreversibly alter the biosphere that we inhabit by providing a way to rewrite the very molecules of life any way we wish" (Doudna and Sternberg 2017, 119).

Environmental philosopher Christopher Preston considers de-extinction and gene drives, alongside nanotechnology, synthetic biology, geo-engineering and the creation of so-called novel ecosystems, as pre-eminent technologies of the Anthropocene or Synthetic Age, as he prefers to call the new era (Preston 2018). What these technologies have in common is that they all reach very deeply into the 'metabolism' between man and nature: processes that are basic to the functioning of terrestrial systems are increasingly replaced by processes that are directly controlled by humans. Evolution, for instance, is no longer left to the blind process of random mutation and natural selection but becomes an object of deliberate 'evolutionary engineering' through synthetic biology, de-extinction efforts and gene drives; the earthly climate

H. van den Belt (✉)
University of Western Australia, Crawley, WA, Australia

© The Author(s) 2021
B. Bovenkerk and J. Keulartz (eds.), *Animals in Our Midst: The Challenges of Co-existing with Animals in the Anthropocene*, The International Library of Environmental, Agricultural and Food Ethics 33,
https://doi.org/10.1007/978-3-030-63523-7_31

is no longer exclusively determined by the amount of solar irradiation but becomes an object of direct manipulation through geo-engineering.

Standing at the threshold of these unprecedented possibilities, many would counsel caution and urge to pause for reflection before rushing headlong into the brave new world of Synthetic Age technologies. However, there is also a very influential group of technology optimists known as ecomodernists who eagerly embrace the 'good Anthropocene' and cannot wait to go full speed ahead with the new technologies, as these in their view hold out the promise of continued increase in human wellbeing while simultaneously addressing environmental problems like climate change and biodiversity loss. Given its prominence in contemporary debates, it is not surprising that all chapters in this section engage with and confront ecomodernism in one way or another, even if only implicitly. This engagement will also provide the red thread in my commentary to connect the various themes discussed in the chapters under consideration.

31.1   Gene Editing, Gene Drives and De-extinction

In his contribution to this volume Preston has chosen a different approach than in *The Synthetic Age*. In his book he attempted to define the epochal significance of the new technologies that might enable us to take over some of Nature's most basic operations, accepting the sweeping claims made by the protagonists of those technologies and their ecomodernist supporters more or less at face value. Now he takes a different tack. Following Alfred Nordmann's criticism of 'speculative ethics', he focuses on the credibility and tenability of the claims made with regard to the potential performance of de-extinction and gene drives, concluding that these claims are overblown and that "the discourse around gene drives and de-extinction is creating a harmful speculative ethics". Such a discourse might lead to a premature fixation on highly speculative technologies, thereby obscuring alternative, less glamorous but ultimately more promising approaches from view.

Preston explains that the success of de-extinction and gene drives in actual practice crucially depends on a key assumption of reductionist molecular science, to wit, that it is possible to obtain *predictable* effects on the level of the behaviour of organisms within their natural environments by intervening on the level of their genomes. He adduces many findings from biological research which indicate that this assumption of control and predictability may not actually hold. Although CRISPR has been advertised as an extremely accurate and precise gene editing tool, in practice researchers have meanwhile been confronted with many unpleasant surprises like off-target effects, unforeseen genomic deletions and rearrangements, and other unpredicted changes in the genome. It is also well-known that there is no one-to-one correspondence between genotype en phenotype; a single genetic alteration may have more than one phenotypic effect. Preston mentions a Chinese attempt to use gene editing for promoting muscle growth in pigs, which also resulted in the unforeseen appearance of extra vertebrae. Furthermore, the stability of the altered genome may also be problematic. Neither should we forget that constructed life forms, despite their man-made origin, become subject to the Darwinian pressures of mutation and selection after environmental release. There is also no guarantee that a gene drive can be confined to the target population, as hybridization with related species remains a realistic possibility. For all these and other reasons, Preston holds that gene drives and de-extinction (and also gene editing in general) are far more questionable than they are usually held to be. He gleefully cites the sobering conclusion of the IUCN 'de-extinction taskforce' of 2014 that modern biotechnology cannot actually bring back a faithful replica of a lost species but at best only a 'proxy species' or approximation.

The main problem behind the enthusiasm for gene drives and de-extinction as conservation tools, Preston suggests in his chapter, is the delusion of predictability and control to which their protagonists and supporters fall so easily prey, and the pretence that *agency* is exclusively located on the side of humans and that the rest of living nature constitutes no more than *passive* matter. This emphasis on epistemological (and also ontological) criticism is slightly at odds with the thrust of Preston's earlier analysis in his book. There he was mainly concerned about the potential loss of the 'otherness' of Nature (or her genuine 'wildness') as a consequence of our unremitting attempts to impose our own designs on her workings. In the end, however, there is no real contradiction, as wildness not only refers to the autonomy of animals and landscapes beyond human endeavours, but also connotes an essential lack of predictability:

> In its fickleness, its unpredictability, and its capacity continually to exceed our expectations, wildness will ensure that remaking the earth will always remain a game of high chance. When we insert ourselves so deeply into the workings of a planet, we are unlikely to be able to predict all of the consequences of our actions. There are serious risks to letting ourselves be seduced by the sublime beauties of technology. (Preston 2018, 178)

31.2   Resurrecting the Heath Hen

In her contribution Jennifer Welchman offers a detailed scrutiny of the so-called reparations argument in favour of de-extinction by extensively considering the case of the heath hen, a ground-dwelling bird that went extinct in 1932. The heath hen has been adopted by the Revive and Restore organization as a candidate for a de-extinction project on Martha's Vineyard, an island before the coast of Massachusetts. Revive and Restore was founded in 2012 by the self-declared ecomodernists Stewart Brand and Ryan Phelan as a nonprofit organization for the "genetic rescue" of endangered and extinct species. The name of the organization suggests that it is actually possible to "revive" an extinct species. As it explains on its website, "the trick [with de-extinction] will be to transfer the genes that define the extinct species into the genome of the related species, effectively converting it into a living version of the extinct creature" (Revive and Restore, n.d.). The fact that a "revived" species can at best be only a proxy or approximation but not a faithful replica of the extinct animal is somewhat downplayed here. Various rationales are invoked to justify de-extinction projects, among which we can also recognize the reparations argument: "to undo harm that humans have caused" (Brand 2014). However, this argument is just briefly mentioned but not further elaborated and expanded upon.

Welchman rightly notes that the reparations argument, unlike many of the other, mostly utilitarian or consequentialist reasons invoked for de-extinction, is not forward-looking but backward-looking. In fact there is something odd about the fact that de-extinction has been embraced as a worthy 'conservation' goal by ecomodernists, because the aim of bringing back extinct species from the dead looks itself extremely backward-looking. It appears to run counter to the usual ecomodernists' appeal to the Anthropocene as a new dispensation that precludes any possibility of returning to the ecological past. Yet, as Ronald Sandler duly observes, "de-extinction aims to recapture something lost and […] is *highly* nostalgic" (Sandler 2017). Thus we find Stewart Brand feeling thrilled by the prospect of recreating the majestic spectacle of "clouds of passenger pigeons once again darkening the sun" (Brand 2014)—a spectacle famously evoked earlier by conservationist Aldo Leopold to mourn the loss of this characteristic American bird (Leopold 1947). But we may wonder whether the thrill is really about the possible return of a scene of natural sublimity or rather about the technological marvels of human ingenuity. The new 'resurrectionists' seem intent on stealing the thunder from the old conservation movement and turning the unceasing succession of sad news stories on the loss of one species after another into some kind of good news show. Indeed, Brand sees huge strategic advantage in a more positive approach to conservation: "The conservation story could shift from negative to positive, from constant whining and guilt-tripping to high fives and new excitement" (Brand 2014).

More traditional conservationists like Paul Ehrlich are understandably worried that an emphasis on de-extinction will create a moral hazard leading to diminished public support for conventional efforts to prevent extinction: "The problem is that if people begin to take a 'Jurassic Park' future seriously, they will do even less to stem the building sixth great mass extinction event" (Ehrlich 2014). This moral hazard would be another illustration of Alfred Nordmann's speculative ethics that was discussed by Preston. Environmental philosopher Ben Minteer maintains that 'resurrectionists' refuse to accept natural limits and that their projects reflect "a new kind of Promethean spirit that attempts to leverage our boundless cleverness and powerful tools for conservation" (Minteer 2014).

In her chapter Welchman touches only indirectly on these 'ideological' background disputes. Her focus is on the reparations argument and her main aim is to constructively elaborate the rather rudimentary version of this argument as it is found in the literature into a more full-fledged and defensible form, so as to provide a normative standard by which to judge the particular case of the de-extinction project around the heath hen. The argument that 'we'—or in practice some collective agency like the United States or the State of Massachusetts—have a moral duty of (restitutive) justice to undo the harm of anthropogenic extinction by creating a replica of the extinct species, though often invoked, is much more difficult and problematic than might appear at first sight. Even if it is granted (against Clare Palmer) that humans can have duties of justice vis-à-vis wild animals, the reparations argument is still confronted with many challenges. A key question is *who* could be owed reparations. In comparable cases of historical injustice inflicted on certain human populations (e.g. colonialism and slavery), it is in principle possible to right such wrongs by compensating the descendants of the dead victims. But extinct animals are all dead and have no living descendants. However, reparations might also be owed to the few remaining members of a severely endangered species, e.g. when the breeding population size is no longer viable. Thus in the case of the northern white rhino and of the black-footed ferret, it would be appropriate (or perhaps even morally obligatory) to create additional mating partners by using cloning techniques (which are also used in de-extinction programs). Strictly speaking, of course, this is not an instance of de-extinction. As Welchman writes, "when we are responsible for driving a species into an extinction vortex for lack of viable reproductve partners, we could in principle owe it to that species to 'bring it back' – albeit from extinction's door rather than extinction proper." She holds that the reparations argument most suitably applies to such rare and somewhat atypical cases.

Interestingly, Welchman also contemplates the possibility that the duty to undo the harm of extinction might be owed to humans who mourn the loss of extinct animals: "It is a standing grief to many that they will never enjoy the sight or sound of Heath Hens strutting and booming on their breeding grounds as earlier generations did." Welchman cites the views of a local naturalist from Martha's Vineyard, Tom Chase, who mourns the progressive loss of a rich bird life since his early childhood and is enthusiastic about the prospect of genetically replicating the heath hen: "[A]s a conservatonist, I'm tired of fighting for things and always losing. I want to get on the proactive side and not on the reactive side" (quoted in Welchman). Another local naturalist, Stephen Kellert, sees in the heath hen de-extinction project an extraordinary opportunity for the communities on Martha's Vineyard to restore their connection to nature and to deepen their relation with the larger community of life.

Such views have been severely criticized by Ben Minteer. He condemns de-extinction as a technological fix that cannot "atone" for the harm that has been done by driving the heath hen to extinction. It is, in his judgment, a Promethean celebration of human technological prowess. By pursuing high-tech species revival technologies, the protagonists and supporters of de-extinction undermine our responsibility for preventing mass extinction by promoting the dangerous illusion that any harms we might have inflicted and are still inflicting can ultimately be fixed again.

Welchman holds that Minteer's criticism is unduly dismissive and also based on a misinterpretation. The replication of the heath hen as a technical achievement would not as such "atone" for past abuses. Rather, it is seen as providing an opportunity for the communities on Martha's Vineyard to reconnect with nature; the focus is on what would happen *after* successful replication, when the islanders must learn to live *with* (replicas of) heath hens. To succeed, they "would have to develop new virtues of care for this species and its environment".

Welchman is also sceptical of the moral hazard argument. She claims that there is no empirical evidence for the expectation that the public would become indifferent to the continuing loss of biodiversity because of the erroneous belief that extinctions can always be remedied. By contrast, she points out that there is strong evidence to suggest that the public's feeling of powerlessness in the face of large-scale environmental problems (such as anthropogenic extinction) can easily lead to apathy, disengagement and denial. Thus she cautiously and partly subscribes to the ecomodernists' stress on the need for some positive message. For her, however, the focus is not primarily on the technical achievement of replication. De-extinction of the heath hen would enable the residents of Martha's Vineyard to find out what they can do themselves in concrete ways to mitigate, if ever so slightly, the colossal problem of anthropogenic extinction. If they succeed in learning to live with heath hens and to grant these fellow creatures the chance to flourish on their island, this example might stimulate other de-extinction and conservation efforts. Perhaps the best thing the islanders can do is simply give it a try. But here Welchman notices that she has moved beyond the reparations argument and entered the domain of forward-looking reasons for replicating extinct species.

31.3   Cultured Meat

Cor van der Weele's chapter is about cultured meat (or 'clean meat', as it is sometimes labelled). It is proposed as a technological solution for the huge environmental and animal welfare problems created by our current mode of meat production through the raising and slaughtering of livestock. If successfully rolled out on a global scale, it would fundamentally alter the human-animal relationship. Van der Weele points out that the impetus to the development of cultured meat is more a matter of 'ethical pull' than of 'technology push'. However, it would also neatly fit the tenets of ecomodernism: "Cultured meat is a clear example of *decoupling* [as favoured by ecomoderists] since it attempts to employ technology to produce the same vast amounts of meat with significantly less environmental damage" (Anthos 2018, 20). According to some early estimates, the potential for reduced resource use is enormous, allegedly allowing 7–45% lower energy use, 78–96% lower greenhouse gas emissions, 99% lower land use, and 82–96% lower water use (Tuomisto and Texeira de Mattos 2011). Another key ethical argument for cultured meat is that it would put an end to the massive suffering of livestock on 'factory farms'.

The current scale of 'industrialized' farming can itself be seen as indicative of the Anthropocene epoch: "Livestock now constitute 60% of the mammalian biomass and humans another 36%. Only 4% remains for the more than 5000 species of wild mammals" (Baillie and Zhang 2018). In terms of avian biomass, poultry currently makes up 70% of all birds on earth, leaving only 30% for wild birds (Carrington 2018). These incredible percentages show the inordinate size of human claims on the biosphere. It is important to point out that the present size of the human footprint has been made possible only through a whole series of scientific and technological advances in the areas of animal and plant breeding, nutrition science, microbiology, antibiotics, etcetera (Boyd 2001). One important aspect is the immensely increased use of nirtogen fertilizer to grow feed crops, enabled by the fixation of nitrogen through the Haber-Bosch process, which led to a huge rechanneling of the nitrogen cycle on planet Earth—itself a major indicator of the Anthropocene (Elser 2011).

It is clear that animal farming in its current size and form, the end result of a long process of technological advance and modernization, is simply unsustainable. Ecomodernists, however, hold that we need more technology to solve the problems created by earlier technology. Thus Nordhaus and Shellenberger declare: "The solution to the unintended consequences of modernity is, and has always been, more modernity – just as the solution to the unintended consequences of our technologies has always been more technology" (Nordhaus and Shellenberger 2012). Cultured meat seems to ideally fit the ecomodernists' bill. No wonder, then, that commercial projections optimistically forecast that by 2040 some 35% of the demand for 'meat' will be covered by cultured meat (and 25% by plant-based vegan replacements) (Carrington 2019).

One may wonder whether the technology of 'culturing' meat can be considered a Synthetic Age technology in Christopher Preston's sense insofar as it entails a fundamental change in the 'metabolism' between humans and nonhuman nature. Compared to more conventional animal husbandry, it surely involves a radically new step by taking control down to the cellular level, virtually amounting to a "second domestication" (Shapiro 2017). However, it can also be seen as a further continuation of the ongoing industrialization of animal farming, culminating finally in a Hegelian sublation and negation of the animal itself. While in current systems of factory farming, animals are increasingly reduced to living protein machines in order to maximize the production of edible meat, 'culturing' meat would take this development to its ultimate conclusion of "just a protein machine, without the animal" (Galusky 2014, 932). For Simon Fairly, an advocate of sustainable (and modest) meat consumption, cultured meat also lies at the end of a road to factory food and human alineation from nature, away from the trend in the organic sector towards "slow food, real meat and fresh local produce" (quoted by Van der Weele).

Critics like Anthos and Galusky question whether the expected benefits of cultured meat will ultimately be realized and also point out that this new technology requires an extreme level of human control over the biological processes of muscle tissue growth—a level of control that might well be illusionary (compare Preston's critique of gene editing and gene drives). Somewhat paradoxically, however, their biggest fear seems to be that this technology might nonetheless ultimately prove feasible. For this would mean that we could 'solve' our problematic relationship with animals and the natural environment by a simple 'technofix', without changing our lifestyle. In that case, Galusky holds, "the ethical questions surrounding eating meat are not so much engaged as eliminated" (Galusky 2014, 937). Anthos similarly remarks that "[o]ut of concern for the animals, the relationship to the farm animal disappears, reduced to a more abstract idea that sentient beings are no longer sufferng for this meat" (Anthos 2018, 38). The disappearance of our relationship with farm animals would in his view amount to a further stage in the alienation of humans from nonhuman nature.

In her contribution, Cor van der Weele aims to save cultured meat from ecomodernism (as the title of her chapter indicates), but also from its detractors. She very much deplores the dualistic way of thinking that appears to hold sway whenever we reflect on the future of agriculture and humanity's relation to nature, as manifested, for instance, in the polarization between 'wizards' and 'prophets' described by Charles Mann. This characteristic opposition of views, pitting optimistic faith in technological progress against an emphasis on lifestyle changes, is also found in the polarization between the ecomodernists, on the one hand, and the subscribers to the Dark Mountain Manifesto, on the other. Van der Weele points out that both parties have their own accounts of mechanisms of selective attention, which are held to explain why the other party is seemingly blind to the obvious truth. These two accounts are almost each other's mirror image and they are used in a rather self-serving way to bolster belief in one's own view of the world (illustrating the biblical saying that you look at the speck in your brother's eye but fail to notice the beam in your own eye). Van der Weele holds that selective attention is indeed an extremely important phenomenon and that its underlying mechanisms are worthy of in-depth study. However, this study should be done in a more impartial and less asymmetric way. She also emphasizes that a dualistic way of thinking is itself a very influential mechanism of selective attention, which may have paralysing effects.

To criticize the dualism between technology and lifestyle, Van der Weele argues against the ecomodernist framing of cultured meat as the only plausible scenario of the future. In her research with focus groups, an interesting alternative scenario has been suggested by one of those groups, dubbed "the pig in the backyard". In this scenario, humans would still have relations with farm animals: "Participants who had started out being quite hesitant about the idea of cultured meat at some point started to envision its production through a local and small scale industry: cells from free ranging pigs, in backyards or urban farms, would be taken through biopsies every now and then and cultured into meat in neighbourhood factories. The idea immediately warmed the participants to cultured meat." In an ongoing follow-up project, Van der Weele further elaborates this scenario with farmers who see opportunities to combine small-scale farming with free-ranging animals, better connections with consumers and society, and production of local kinds of cultured meat.

Van der Weele is far from claiming that these alternative scenarios necessarily point the way to the future. They are rather inspiring images of a desirable future that in the end may still flounder on the dominant constraints of economic efficiency. But meanwhile they can break the hold that the ecomodernist framing of cultured meat has on our minds.

31.4 Enhancement, Disenhancement and Animal Welfare

Adam Shriver does not explicitly engage with ecomodernism, but the topic of his chapter is very relevant given the increasing levels of technological control humans exercise over ever more animal lives, especially through new technologies of gene editing. With much analytical finesse, Shriver attempts to develop a rigorous account of animal (dis)enhancement. He argues that popular intuitions about what constitues animal disenhancement—e.g. the widespread view that it involves taking something away from the animal that would otherwise be present—are unreliable and misleading. Such notions are invoked, for instance, in ethical debates about blind chickens and polled cattle and about more futuristic examples like 'dino-chickens' and 'football birds'. Shriver holds that we need a holistic rather than domain-specific conception of (dis)enhancement and that at first sight there are three possible candidates for such a conception, to wit, the evolutionary fitness, the normal species functioning and the welfarist account. He dismisses the former two accounts rather unceremoniously as irrelevant for dealing with animals under direct human supervision, thus leaving the welfarist definition of (dis)enhancement as the only tenable option.

It is not hard to see why evolutionary fitness is deemed irrelevant for domesticated animals held in controlled environments. While a trait like aggression may once have been conducive to the survival of their wild ancestors, it no longer is for animals living under conditions determined by humans. It would therefore make no sense, Shriver claims, to call an unusually aggressive animal 'enhanced' or an unusually passive animal 'disenhanced'. Similar considerations pertain to the normal species functioning definition, which obviously begs the question to which population the standard of normalcy applies. Shriver's reasoning thus looks unobjectionable. However, by rejecting the two alternative accounts of animal (dis)enhancement, he thereby also declares the evolutionary past of kept animals to be radically irrelevant for judging their present situation. Or so it seems. Is this apparent break with previous evolutionary history just another manifestation of the advent of the Anthropocene?

Take the well-known example of the chickens that have been genetically modified to make them blind, so as to mitigate problems resulting from pecking and debeaking. One might perhaps concede that under confined conditions such an intervention would increase chicken welfare, but still hold that it shows little respect for the (evolved) 'nature of the beast' and not hesitate to use the term 'disenhancement' in this connection. But, of course, it is precisely this usage that is proscribed by Shriver's reasoning, which seems to justify any further technological erasure of the genetic heritage passed down to our domesticated animals.

Shriver's welfarist account of animal (dis)enhancement is not without its own problems, some of which he duly mentions. His definition of disenhancement is as follows: "Any change in the biology or psychology of an animal which decreases the chance of leading a good life in the relevant set of circumstances". The reference to the relevant set of circumstances constitutes, in his view, an essential component of the definition. However, it also illustrates a crucial weakness of Shriver's welfarist approach, beyond the problems he himself signals. In searching for a proper account of (dis)enhancement that is relevant for the situation of domesticated animals, he apparently accepts the actual conditions in which animals are held simply as given, or as "the relevant set of circumstances". In this way, it would seem, he robs the concept of animal welfare of its normative bite, as the inserted clause ensures that it can no longer function as an independent basis from which to criticize actually existing husbandry conditions.

It is significant that in 5 of the 6 concrete cases of more or less radically changed animals discussed by Shriver, the answer whether the change under consideration amounts to an enhancement or a disenhancement is: "It depends". Indeed, the answers vary, depending on what particular circumstances are assumed. But this shows that his conception lacks the normative power to judge the circumstances themseves.

Only in the extreme case of the 'football birds' is there a definite answer. This case would neither count as an example of enhancement nor of disenhancement, simply because these hypothetical 'birds' would by definition be completely insentient and thus have no welfare at all. (One might consider them as a logically transitional stage toward cultured meat, although the actual development looks bound to skip this transitional stage.) But even this clear answer is later qualified. Taking up Preston's suggestion that the creation of 'football birds' might involve a loss of agency on the part of the animal and therefore still constitute a case of disenhancement, Shriver appears ultimately willing to accept this conclusion by incorporating agency into his concept of welfare. By considering such remote futuristic cases as serious possibilities, however, we are indulging in speculative ethics.

References

Anthos, V. 2018. *Meat reimagined: the ethics of cultured meat*. Graduate Student Thesis University of Montana.

Baillie, J., and Y.-P. Zhang. 2018. Space for nature. *Science* 361 (6407): 1051.

Boyd, W. 2001. Making meat: Science, technology, and American poultry production. *Technology and Culture* 42 (2): 633–664.

Brand, S. 2014. The case for de-extinction: why we should bring back the woolly mammoth. *Yale Environment 360*, Januari 13. https://e360.yale.edu/features/the_case_for_de-extinction_why_ we_should_bring_back_the_woolly_mammoth.

Carrington, D. 2018. Humans just 0.01% of all life but have destroyed over 80% of wild animals. *The Guardian*, May 21.

Carrington D. 2019. Most 'meat' in 2040 will not come from dead animals, says report. *The Guardian*, June 12.

Doudna, J., and S. Sternberg. 2017. *A crack in creation: the new power to control evolution*. London: Vintage.

Ehrlich, P.R. 2014. The case against de-extinction: it's a fascinating but dumb idea. *Yale Environment 360*, January 13. https://e360.yale.edu/features/the case_against_de-extinction_its_a_fasc inating_but_dumb_idea.

Elser, J.J. 2011. A world awash with nitrogen. *Science* 334: 1504–1505.

Galusky, W. 2014. Technology as rtesponsibility: Failure, food animals, and lab-grown meat. *Journal of Agricultural and Environmental Ethics* 27: 931–948.

Leopold, A. 1947. On a monument to the Pigeon. In *A Sand County Almanac*, 108–112. https:// www.birdwatchingdaily.com/news/conservation/monument-pigeon-aldo-leopold/.

Minteer, B. 2014. Is it right to reverse extinction? *Nature* 509 (7500): 261.

Nordhaus, T., and M. Shellenberger. 2012. Evolve: The case for modernization as the road for salvation. *The Breakthrough Journal*, March 26.

Preston, C.J. 2018. *The synthetic age: Outdesigning evolution, resurrecting species, and reengineering our world*. Cambridge, MA: MIT Press.

Revive & Restore. n.d. What "genetic rescue" means. https://reviverestore.org/what-we-do/genetic-rescue/.

Sandler, R. 2017. De-extinction and conservation ethics in the Anthropocene. *Hastings Center Report* 47 (S22): S43–S47.

Shapiro, P. 2017. Lab-grown meat is on the way. *Scientific American*, December 19.

Tuomisto, H.L., and J. Texeira de Mattos. 2011. Environmental impact of cultured meat production. *Environmental Science and Technology* 45 (14): 6117–6123.

Henk van den Belt was assistant professor at Wageningen University, The Netherlands, until his retirement in May 2019. He has done research in the broad area of the history, sociology and philosophy of science, in particular with regard to the modern life sciences. His more recent interest is in the literary and scientific imagination of possible futures. He is now an Honorary Research Fellow in the School of Humanities at the University of Western Australia in Perth.

Correction to: Animals in Our Midst: The Challenges of Co-existing with Animals in the Anthropocene

Bernice Bovenkerk and Jozef Keulartz

Correction to:
B. Bovenkerk and J. Keulartz (eds.),
Animals in Our Midst: The Challenges of Co-existing
with Animals in the Anthropocene, **The International Library**
of Environmental, Agricultural and Food Ethics 33,
https://doi.org/10.1007/978-3-030-63523-7

The original version of this chapter was inadvertently published with incorrect author name in Chapters 1 and 3 reference citations and list. The author's name is corrected from "O'Neill, J.S. and M.H. Hastings" to "David A. Leavens". The book has been updated with the changes.

The updated version of the book can be found at
https://doi.org/10.1007/978-3-030-63523-7

© The Author(s) 2021
B. Bovenkerk and J. Keulartz (eds.), *Animals in Our Midst: The Challenges of Co-existing with Animals in the Anthropocene*, The International Library of Environmental, Agricultural and Food Ethics 33, https://doi.org/10.1007/978-3-030-63523-7_32

Index

© The Editor(s) (if applicable) and The Author(s) 2021
B. Bovenkerk and J. Keulartz (eds.), *Animals in Our Midst: The Challenges
of Co-existing with Animals in the Anthropocene*, The International Library
of Environmental, Agricultural and Food Ethics 33,
https://doi.org/10.1007/978-3-030-63523-7

Printed in the United States
by Baker & Taylor Publisher Services